2nd International Conference on Agricultural Engineering for Sustainable Agricultural Production (AESAP 2017)

IOP Conference Series: Earth and Environmental Science
Volume 147

Bogor, Indonesia
23 - 25 October 2017

Editors:

Leopold Oscar Nelwan
Usman Ahmad
Slamet Widodo
I. Wayan Astika
Anisur Rahman
Shinichiro Kuroki
Wan Nor Zanariah Binti Zainol

ISBN: 978-1-5108-6279-1
ISSN: 1755-1307

Printed from e-media with permission by:

Curran Associates, Inc.
57 Morehouse Lane
Red Hook, NY 12571

CURRAN ASSOCIATES INC.
proceedings.com

Some format issues inherent in the e-media version may also appear in this print version.

Printed by Curran Associates, Inc. (2018)

For permission requests, please contact the Institute of Physics
at the address below.

Institute of Physics
Dirac House, Temple Back
Bristol BS1 6BE UK

Phone: 44 1 17 929 7481
Fax: 44 1 17 920 0979

techtracking@iop.org

Additional copies of this publication are available from:

Curran Associates, Inc.
57 Morehouse Lane
Red Hook, NY 12571 USA
Phone: 845-758-0400
Fax: 845-758-2633
Email: curran@proceedings.com
Web: www.proceedings.com

Table of contents

Volume 147

The 2nd International Conference on Agricultural Engineering for Sustainable Agricultural Production (AESAP 2017)

23–25 October 2017, Bogor, Indonesia

Accepted papers received: 16 April 2018
Published online: 14 May 2018

Instrumentation and Monitoring System

IOP Conf. Series: Earth and Environmental Science **147** (2018) 011001 doi:10.1088/1755-1315/147/1/011001

Preface

Proceedings of AESAP 2017 contains papers presented in technical session of the 2nd International Conference on Agricultural Engineering for Sustainable Agriculture Production, held at Bogor Agricultural University (IPB) in 23-25 October 2017.

This proceedings consists of 48 papers presented in the technical session of the conference. The papers cover a broad area in Agricultural Engineering. The papers discuss the topics in postharvest and food engineering, energy and environment, land and water resources engineering, system and management in agriculture, instrumentation and monitoring system, and agricultural machinery.

We would like to thank all authors for their efforts in preparing their papers. A great appreciation is also given to the reviewers for their assistance in reviewing the manuscripts. Special thanks to secretariat members for their assistance in formatting the layout of the proceedings.

Dr Desrial
Chairman of AESAP 2017

IOP Conf. Series: Earth and Environmental Science **147** (2018) 011001 doi:10.1088/1755-1315/147/1/011001

Editors

Dr. Leopold Oscar Nelwan (COORDINATOR)
Department of Mechanical and Biosystem Engineering
Faculty of Agricultural Engineering and Technology
Bogor Agricultural University, Indonesia
lonelwan@yahoo.com

Dr. Usman Ahmad
Department of Mechanical and Biosystem Engineering
Faculty of Agricultural Engineering and Technology
Bogor Agricultural University, Indonesia

Dr. Slamet Widodo
Department of Mechanical and Biosystem Engineering
Faculty of Agricultural Engineering and Technology
Bogor Agricultural University, Indonesia

Dr. I Wayan Astika
Department of Mechanical and Biosystem Engineering
Faculty of Agricultural Engineering and Technology
Bogor Agricultural University, Indonesia

Dr. Anisur Rahman
Postdoctoral Research Associate
Department of Biosystems Machinery Engineering
Chungnam National University, South Korea
anis@cnu.ac.kr

Dr. Shinichiro Kuroki
Assistant Professor
Department of Agricultural Engineering and Socio-Economics
Kobe University, Japan
skuroki@dragon.kobe-u.ac.jp

Dr. Wan Nor Zanariah Binti Zainol @ Abdullah
Senior Lecturer
Faculty of Agriculture and Food Sciences
Universitas Putra Malaysia, Malaysia
wnzz@upm.edu.my

IOP Conf. Series: Earth and Environmental Science **147** (2018) 011001 doi:10.1088/1755-1315/147/1/011001

AESAP 2017
Agricultural Engineering for Sustainable Agriculture Production

Hosted By:

INSTITUT PERTANIAN BOGOR

DEPARTMENT OF MECHANICAL AND BIOSYSTEM ENGINEERING
FACULTY OF AGRICULTURAL ENGINEERING AND TECHOLOGY
BOGOR AGRICULTURAL UNIVERSITY

PERTETA
BOGOR CHAPTER

Sponsored By:

YARI-IPB
INDONESIA

Sugar Group
Companies

Peer review statement

All papers published in this volume of *IOP Conference Series: Earth and Environmental Science* have been peer reviewed through processes administered by the proceedings Editors. Reviews were conducted by expert referees to the professional and scientific standards expected of a proceedings journal published by IOP Publishing.

Ultrasonic Technique for Predicting Grittiness of Salted Duck Egg

S Erawan[1], I W Budiastra[1] and I D M Subrata[1]

[1] Department of Mechanical and Biosystem Engineering, Bogor Agricultural University, Indonesia

E-mail: semedhi@gmail.com

Abstract. Grittiness of egg yolk is a major factor in consumer acceptance of salted duck egg product. Commonly, the grittiness level is determined by the destructive method. Salted egg industries need a grading system that can judge the grittiness accurately and nondestructively. The purpose of this study was to develop a method for determining grittiness of salted duck eggs nondestructively based on ultrasonic method. This study used 100 samples of salted duck eggs with 7,10,14 and 21 days of salting age. Velocity and attenuation were measured by an ultrasonic system at frequency 50 kHz, followed by physicochemical properties measurement (hardness of egg yolks and salt content), and organoleptic test. Ultrasonic wave velocity in salted duck eggs ranged from 620.6 m/s to 1334.6 m/s, while the coefficient of attenuation value ranged from –0.76 dB/m to -0.51 dB/m. Yolk hardness was 2.68 N at 7 days to 5.54 N at 21 days of salting age. Salt content was 1.81 % at 7 days to 5.71 % at 21 days of salting age. Highest scores of organoleptic tests on salted duck eggs were 4.23 and 4.18 for 10 and 14 days of salting age, respectively. Discriminant function using ultrasonic velocity variables in minor and major diameter could predict grittiness with 95 % accuracy.

1. Introduction

Production of salted duck eggs from 2011 to 2015 increased from 256.2 thousand tons to 282.6 thousand tons [1]. Duck eggs contributed 15.56 % of the national egg production in 2015. Duck eggs are generally preserved and processed into salted eggs and very popular in Indonesia and Asia. Chi and Tseng [2] reported that generally salted eggs are made from duck eggs because of protein and fat content in egg yolks 19.08 % and 36.53 % respectively higher than chicken eggs 16.32 % and 32.25 %.

Conventionally, egg is coated with a red soil paste mixed with salt, or it is immersed in the salt solution. During salting, the yolk gradually becomes solidified and hardened and could form a gritty texture [2], [3]. The grittiness quality of salted duck egg is obtained from the yolk. The grittiness is a major factor affecting consumer acceptance of salted egg product [2]. The grittiness level of the salted duck egg is determined by destructive method. The salted duck egg industries need a grading system that can judge the grittiness accurately and nondestructively.

The purpose of this study was to develop a method for determining grittiness of salted duck egg nondestructively based on ultrasonic method. More specific purpose of this study are: a) to determine characteristics of ultrasonic wave transmissions and physicochemical characteristics of salted duck egg at four of salting age, b) to determine the relationships between ultrasonic properties and the physicochemical properties of salted duck eggs, and c) to predict grittiness of salted duck egg based on ultrasonic properties.

2. Materials and methods

Salted duck egg used in this study was obtained from Sidoarjo district, East Java, Indonesia. One hundred samples of cooked salted egg consisted of 7,10,14 and 21 days of salting age were measured by an ultrasonic system at frequency 50 kHz [4], this frequency was recommended as an appropriate one for characterization of agricultural materials [5], followed by physicochemical properties measurement (hardness of egg yolks and salt content), and organoleptic test. Ultrasonic properties (velocity and attenuation) in minor and major diameter were measured (figure.1) and calculated using ultrasonic equations [6].

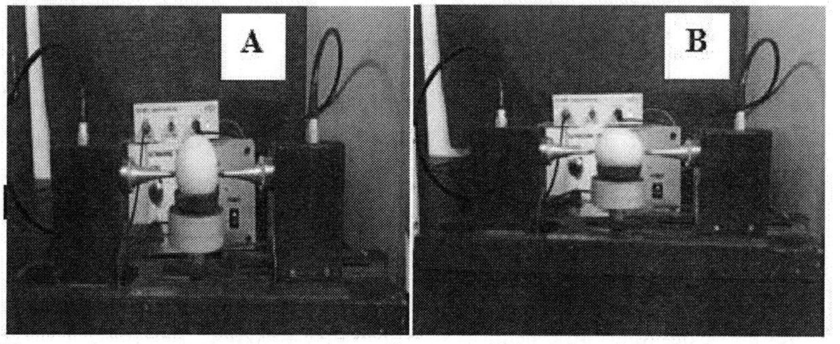

Figure 1. Ultrasonic measurement at minor diameter (A) and major diameter (B) of salted egg

Yolk hardness was measured using CR 300 DX-L Model Rheometer, with experimental set up as follows; mode of 20, maximum load of 20 mm/minutes and probe diameter of 5 mm. Salt content was determined by rapid analysis of modified Mohr method [7]. An organoleptic test was conducted with 22 trained panelists. The organoleptic test parameters were general appearance, salting taste of egg white and gritty texture of egg yolk with the scoring scale consists of: (1) very dislike; (2) dislike; (3) neutral; (4) like; (5) very like [8]. Relationships between ultrasonic and physicochemical properties were analyzed using linear regression. Discriminant function of ultrasonic properties of the salted egg was developed to classify the grittiness of salted duck egg.

3. Results and discussion

3.1. Ultrasonic and Physicochemical Characteristics of Salted Duck Egg

Ultrasonic properties of salted duck egg are shown in table 1. The velocity of ultrasound passing through the salted duck egg increased from 7 days to 21 days of salting age. Attenuation coefficient did not change significantly. Physicochemical properties of the salted duck egg are showed in table 2. The hardness of egg yolk was 2.68 N at 7 days to 5.54 N at 21 days of salting age. Salt content was 1.81 % at 7 days to 5.71 % at 21 days of salting age.

Table 1. Ultrasonic properties of salted duck egg at four of salting age

Ultrasonic Properties	7 Day	10 Day	14 Day	21 Day
Velocity in minor diameter (m/s)	720.30^{d}	863.42^{c}	995.40^{b}	1334.60^{a}
Velocity in major diameter (m/s)	620.60^{d}	730.90^{c}	914.80^{b}	1057.30^{a}
Attenuation in minor diameter (dB/m)	-0.71^{a}	-0.75^{a}	-0.76^{a}	-0.71^{a}
Attenuation in major diameter (dB/m)	-0.51^{a}	-0.65^{c}	-0.59^{b}	-0.63^{bc}

Notes: Different superscripts in the same row indicate significant differences (p<0.05)

Table 2. Physicochemical properties of salted duck egg at four of salting age

Physicochemical Properties	7 Day	10 Day	14 Day	21 Day
Hardness of egg yolk (N)	2.68[c]	3.05[bc]	3.46[b]	5.54[a]
Hardness of egg white (N)	3.98[a]	3.87[a]	3.08[b]	0.29[c]
Hardness of egg shell (N)	38.46[a]	39.28[a]	40.02[a]	39.67[a]
Salt content (%)	1.81[b]	3.31[ab]	4.97[ab]	5.71[a]
Moisture content (%)	66.94[a]	67.17[a]	65.22[a]	59.10[a]
Lipid (%)	12.98[a]	12.97[a]	13.83[a]	17.49[a]
Protein (%)	13.02[a]	12.76[a]	10.90[a]	10.91[a]

Notes: Different superscripts in the same row indicate significant differences ($p<0.05$)

3.2. Organoleptic test result of salted duck egg

The organoleptic test result of salted duck egg at four of salting age is shown in table 3. Highest scores of grittiness level of organoleptic tests on salted duck eggs were 4.23 and 4.18 for 10 and 14 days of salting age, respectively. Since the organoleptic score of 10 and 14 days is not significantly different, these salting age will be classified into one group. Beside grittiness level, the level of consumer preference in salted duck eggs depends on the salty taste of egg whites and appearance of an interesting orange color of egg yolk (table 3).

3.3. Relationship between ultrasonic and physicochemical characteristic

The relationship between ultrasonic and physicochemical properties is shown in table 4. The velocity of ultrasound passing through the salted duck egg had significant linear correlations with physicochemical properties of salted duck egg. But, the attenuation coefficient had no significant linear correlation with physicochemical properties of salted duck egg. During salting, the egg yolk gradually becomes solidified and hardened [1]. Kaewanee *et al.* [3] reported that the microstructure of salted egg yolk by SEM (Scanning Electron Micrograph) indicated that polyhedral granules were more closely localized than those in fresh egg yolk. Ultrasonic wave more easily passes through a solid medium than non-solid medium [9]. Changes of physical properties in egg material (changes in hardness of egg yolk) have a direct impact on the velocity of ultrasound propagating through the egg material.

Table 3. Organoleptic test result of salted duck egg

Scoring Scale	7 Day	10 Day	14 Day	21 Day
General appearance	3.14[b]	4.18[a]	3.82[ab]	2.32[c]
Salty taste of egg whites	3.18[b]	4.09[a]	3.55[ab]	1.50[c]
Gritty texture of egg yolk	2.05[c]	4.23[a]	4.18[a]	3.14[b]

Notes: Different superscripts in the same row indicate significant differences ($p<0.05$)

Table 4. Relationship (r) between ultrasonic and physicochemical characteristic of salted egg

Variables	Hardness of egg yolk (n=100)	Salt Content (n=12)
Velocity in minor diameter	0.7756	0.8171
Velocity in major diameter	0.6021	0.8868
Attenuation in minor diameter	0.0360	0.0162
Attenuation in major diameter	-0.2920	-0.3266

Notes: n= number of data analyzed

3.4. Grittiness prediction of salted duck egg based on ultrasonic characteristic

The various discriminant functions of ultrasonic properties were established to classify grittiness of salted egg into 3 groups, that is D_3 (not gritty) D_4 (gritty) and D_5 (very gritty) (table 5). The best of classification of the grittiness of salted duck egg was obtained using the discriminant function of velocity variables in minor and major diameter with an accuracy of 95 %.

Table 5. Classification results of salted duck egg using discriminant function of ultrasonic variables

Variables	Formed Function (D_i), n=80	Validation (n=20)
Velocity in minor diameter (X_1) and attenuation in minor diameter (Y_1)	D_3=-177.6761+0.2068X_1 -281.8322Y_1 -0.0001X_1^2+0.0330X_1Y_1-181.4139Y_1^2 D_4=-42.9160+0.0451X_1 -25.2319Y_1 +X_1^2-0043X_1Y_1-21.5219Y_1^2 D_5=-144.7076+0.1287X_1 -221.0496Y_1 -0.0001X_1^2-0.0294X_1Y_1-165.5990Y_1^2	90 %
Velocity in minor diameter (X_1) and velocity in major diameter (X_2)	D_3=-79.3011+0.2188X_1 -0.0318X_2 -0.0002X_1^2+0.0002X_1X_2-0.0001X_2^2 D_4=-58.7536+0.0417X_1 +0.0383X_2 +X_1^2+X_1X_2+X_2^2 D_5=-79.5193+0.1412X_1 +0.0102X_2 -0.0001X_1^2+X_1X_2+X_2^2	95 %
Velocity in major diameter (X_2) and attenuation in major diameter (Y_2)	D_3= -26.1298+0.0379X_2 -40.5928Y_2 +X_2^2 -0.0293X_2Y_2 -56.8370Y_2^2 D_4= -263.4847+0.0856X_2 -6757862Y_2 +X_2^2 +0.0566X_2Y_2 -483.2326Y_2^2 D_5= -124.0024+0.0808X_2 -283.0569Y_2 +X_2^2 +0.0519X_2Y_2 -192.7464Y_2^2	70%
Velocity and attenuation in minor diameter (X_1 ,Y_1) and velocity and attenuation in major diameter (X_2, Y_2)	D_3= -202.3230+ 0.2957X_1 -0.0785 X_2-289.3941Y_1 -47.3170Y_2 -0.0002X_1^2 +0.0002X_1X_2 +0.0701 X_1Y_1+ 0.0388 X_1Y_2-0.0001X_2^2 -0.0287X_2 Y_1 – 0.0444 X_2Y_2 -187.8915Y_1^2 + 20.5335Y_1Y_2 -59.3858Y_2^2 D_4= -366.0625+ 0.0826X_1 +0.0959 X_2+57.2340Y_1 -867.1685Y_2 +X_1^2 +X_1X_2 +0.0079 X_1Y_1+ 0.0588X_1Y_2+X_2^2 -0.0183X_2 Y_1 +0.0865X_2Y_2 – 27.2632Y_1^2 + 104.6285Y_1Y_2 -607.5533Y_2^2 D_5= -391.2293+ 0.3568X_1 +0.0251X_2-108.6289Y_1 -548.1938Y_2 -0.0001X_1^2 +X_1X_2 -0.0642 X_1Y_1+ 0.2501 X_1Y_2+X_2^2 -0.0346X_2 Y_1+0.0364 X_2Y_2 – 84.8597Y_1^2 + 129,5048Y_1Y_2 -307.2995Y_2^2	90 %

Notes: D_i=Discriminant scoring of salted duck eggs (where: i=3; not gritty, i=4; gritty and i=5; very gritty) and n was number of data analyzed

4. Conclusions

Ultrasonic wave velocity in salted duck eggs ranged from 620.6 m/s to 1334.6 m/s, while the coefficient of attenuation value ranged from – 0.76 dB/m to -0.51 dB/m. Yolk hardness was 2.68 N at 7 days to 5.54 N at 21 days of salting age. Salt content was 1.81 % at 7 days to 5.71 % at 21 days of salting age. Highest scores of organoleptic tests on salted duck eggs were 4.23 and 4.18 for 10 and 14 days of salting age, respectively. The velocity of ultrasound passing through the salted duck egg had a significant correlation with yolk hardness and salt content, while attenuation coefficient had no significant correlation with physicochemical characteristics. Discriminant function using velocity variables in minor and major diameter could predict grittiness with 95 % accuracy.

5. References

[1] Director General of Livestock and Animal Health Services 2015 *Statistic of livestocks and Animal* (in Indonesian). Available at: http:// ditjennak.pertanian.go.id.

[2] Chi S P and Tseng K H 1998 Physicochemical properties of salted pickled yolk from duck and chicken eggs *Journal Food Science* **63**(1): 27–30.

[3] Kaewmanee T, Benjakul S and Visessanguan W 2009 Changes in chemical composition, physical properties and microstructure of duck egg as influenced by salting *Food Chemistry* **112**(3): 560–569

[4] Juansah J, Budiastra I W and Suroso 2005 Development of ultrasonic measurement system for manggis (Gracinia Mangostana L) (in Indonesian) *Jurnal Keteknikan Pertanian* **20**(2):167-178

[5] Mizrach A, Galili N and Rosenhouse G 1989 Determination of fruit and vegetable properties by ultrasonic excitation *Trans. ASAE* **32**: 2053–58.

[6] Krautkramer J, Krautkramer H 1990 *Ultrasonic testing of materials* (Heidelberg: Springer-Verlag)

[7] Faridah N, Kusnandar F, Herawati D, Kusmaningrum HD,Wulandari N and Indrasti D 2010 Laboratory Guide for Food Analysis (in Indonesian). (Bogor: Graduate School of Bogor Agricultural University)

[8] Setyaningsi D, Apriantono A and Puspitasari M 2010 *Sensoris analysis for food and agro industry* (Bogor: IPB Press)

[9] Gooberman G L 1968 *Ultrasonic Theory and Application* (London: The English Universities Press Ltd)

Determination of the Internal Damage of Purple Sweet Potatoes Non-Destructively Using Ultrasonic Wave Characteristics

Sutrisno[1] and Fauzi R M[1]

[1] Department of Mechanical and Biosystem Engineering, Bogor Agricultural University, Indonesia

E-mail: kensutrisno@yahoo.com

Abstract. This study aimed to analyze the ultrasonic wave characterization of Ayamurasaki purple sweet potatoes to predict the internal damage caused by weevil attack non-destructively. Sixty of health and 45 of infected purple sweet potatoes from farmers in Ciampea, Bogor were used for the experiment. Sample divided into 2 groups, the first group of 2/3 part was used in developing the model, and the second group of 1/3 part was used for validation. Ultrasonic wave characterization of purple sweet potatoes that harvested was tested using ultrasonic wave measurement apparatus with 50 KHz frequency. The result shows that the average wave velocity of infected purple sweet potatoes was 247.32 ms-1 and Mo value was 19.45, while the wave velocity of healthy purple sweet potatoes was 235.15 ms-1 and Mo value was 18.72. The average attenuation coefficient of infected purple sweet potatoes was 40.06 Np/m, while in health purple sweet potatoes was 42.69 Np/m. The result of discriminant analysis obtained a linear discriminant model that can distinctly separate between health and infected purple sweet potatoes. Accuracy of this grouping was high enough, around 85.71% so that the discriminant function can be used for grouping the health and infected purple sweet potatoes

1. Introduction

Purple sweet potato has a high economic value, is one of the main sources of prebiotic oligosaccharides and anthocyanin, and has great potential as a food industry in the country. One of the main obstacles of purple sweet potato cultivation is the pest attack (*Clay fomicarius* Fabricus), or in Indonesia well known as "lanas". Lanas attack the purple sweet potatoes either during cultivation, transportation, or during storage. The lanas pest, known as the "boleng", attacks the root epidermis, the stem and the outer surface of the tubers by making a hole that produces a toxic terpene and causing a bitter taste that cannot be consumed [1].

Sorting is an important step that done in postharvest handling of purple sweet potato to cut off the lanas pest attack. One of the obstacles to stop pest attack is the low application of technology by farmers during the sorting process, by mean that sorting is done manually, i.e., by looking at the outer surface of the tubers or by cutting both ends of the bulbs. Manual sorting is not appropriately applied to agricultural products, especially in purple sweet potatoes to detect pest attacks, as it is difficult to observe attacks on tuber meat from the surface so that the eggs or larvae of pests are carried along and cause further attacks during storage. Sorting destructively by cutting off both ends of the tubers also creates considerable losses in tuber meat and causes a reduction in income of farmers.

It is necessary to estimate the damage of purple sweet potatoes to effective pest attack without damaging the material. One method that can be applied is the use of ultrasonic waves, i.e., by observing

IOP Conf. Series: Earth and Environmental Science **147** (2018) 012002 doi:10.1088/1755-1315/147/1/012002

the acoustic properties of waves propagated in purple sweet potatoes, such as wave velocity, coefficient of attenuation, and zero moment power. Non-destructive test analysis using ultrasonic waves can be applied in the determination of the quality of fruits [2], such as detecting the ripeness level of sweet starfruit [3], knowing the physicochemical characteristics of dragon fruits [4], detecting the damage of 'Arumanis' mango due to the fruit fly [5], and predicting pest attack on "Cilembu" sweet potato [6]. Furthermore, [7] states that ultrasonic waves have penetrating power that exceeds the NIR wave and the cost of investments is cheaper than X-ray waves. The purpose of this research was to study the characteristics of ultrasonic wave transmission to estimate the damage of purple sweet potato caused by the 'lanas' pest, non-destructively.

2. Materials and methods

2.1. Materials and tools

The material used in this study was purple sweet potato varieties of Ayamurasaki obtained from farmers in Ciampea Bogor with harvest age ± 3.5 months after planting. After harvest, sweet potatoes were cleaned (without using water) to omit the soil and dirt attached to the skin of the tubers. The purple sweet potato samples were selected manually as many as 105 tubers, which were 60 healthy purple and 45 purple tubers pest attack. This 45 pest attacked tubers were investigated visually and categorized from light to heavy attack according to [8]. A total of 2/3 parts of the sample was used in developing a model for estimating the damage of purple tubers as a result of pest attack, while 1/3 part was used as validation of the model that has been formulated.

The main equipment used for this research were ultrasonic wave measuring instrument that consisting of personal computer, digital oscilloscope, ultrasonic transmitter, transducer (transmitter and receiver) [9]. Frequency of transducer used was 50 kHz, 1:16 of sweep setting, time base of 400 μs/div, sample data for each measurement were 2048 data, with a sampling rate of 512 kHz. Other equipment used was a rheometer (Sun Rheometer type CR 300 DX-L) with a 5 mm diameter of cylinder probe to determine the hardness of purple sweet potato, the caliper for measuring the diameter of tubers, digital scales, and knife for destructive testing.

2.2. Measurement of Ultrasonic Wave characteristics and Physical Properties of Tubers

The measurement of ultrasonic wave acoustic properties was performed to obtain the data of wave velocity, coefficient of attenuation, and zero moment power, and done at the bottom, center, and end of tubers.

Measurement of hardness of tubers was done after the measurement of wave acoustic properties and carried out at the base, middle, and end of the tubers around the points of the measurement of the acoustic properties of the wave. Measurements were done using a rheometer that was set in the mode: 1, R/H (hold): 9.99 kg, P/T (press): 10 mm/m with a maximum pressure of 10 kg, and using 5 mm diameter cylinder probes.

Destructive testing and measurement were done to see the pest attack on tuber flesh, i.e., by peeling and cutting tuber flesh to observe the presence of lanas pest attack in the form of black spots, larvae pest larvae, and hole.

2.3. Data analysis

2.3.1. Ultrasonic Wave Velocity.

The ultrasonic wave velocity was determined by finding the transmission time, by measuring the difference of electrical pulses from the transmitting circuit and the receiver circuit [10]. The ultrasonic wave velocity was analyzed using digital oscilloscope output data calculated using equations 1.

$$V = \frac{s}{t} \quad (1)$$

Where: V = velocity of ultrasonic waves (m/s)
s = distance between transmitter and receiver (m)
t = wave transmission time (s)

2.3.2. Coefficient of Attenuation.
The coefficient of attenuation shows the amount of energy loss in the ultrasonic wave after passing through the medium which is influenced by the type of medium passed. The cause of the loss of energy is the occurrence of absorption and other wave events such as reflection, refraction, diffraction, and scattering by the medium. The coefficient of attenuation is measured by observing the decrease of amplitude of the wave after passing through the medium and can be described as a function of the distance traveled by the wave [2].

The attenuation coefficient can be determined by converting the signal voltage sent and received after a certain distance [3]. According to [7], the magnitude of the attenuation coefficient is inversely proportional to the wave velocity. The more types of substances contained in the fruit, the more the interactions of the wave so that the coefficient value of attenuation is greater [10]. According [3], coefficient of attenuation on starfruit is directly proportional to the level of hardness and total fruit acid, and inversely proportional to total soluble solid content. According to Mizrach *et al.* (1989) *in* [11], the magnitude of the ultrasonic wave attenuation can be calculated according to equation 2.

$$\alpha = \frac{1}{x}\left[\ln \frac{A0}{Ax}\right] \quad (2)$$

Where: A0 = Initial amplitude (volt)
Ax = Amplitude after traveling x distance (volt)
X = Distance travelled wave (m)
α = coefficient of attenuation (Np / m)

2.3.3. Zero Moment Power (Mo).
The value of zero moment power (Mo) serves to know the amount of energy transmitted on the material or medium that the wave transmitted. Mo is an area under power spectral density, which can be calculated by performing Fast Fourier Transform (FFT) using Matlab program in the form of amplitude and time relationships between power spectral density and frequency [12].

2.4. Model Development and Validation
The model development was arranged based on the relationship between ultrasonic wave characteristic and purple sweet potato condition. The model for estimating the damage of tubers as a result of the pest was determined by calculating the mean value between the average ultrasonic wave velocity, the coefficient of attenuation and the zero moment power on the healthy and the tubers attacked by the lanas. Based on the mean value, then it can be determined the limit of velocity, the coefficient of attenuation and Mo to differentiate between healthy and pest attacked of sweet potatoes.

The prediction of pest attacks based on a combination of ultrasonic wave characteristics was performed when the equation model constructed from each of the ultrasonic wave characteristic parameters has a good result to estimate the damage to tubers attacked by lanas pest. Model making was done by discriminant analysis to classify an individual or observation into a free and independent group based on a number of explanatory variables.

Validation aims to test the accuracy of the model equations that have been developed by calculating the success rate of tubers selection. Then for fitting the determining model were the success rate of sorting. The greater the success rate of the sorting, the more established the valid model. The rate of sorting errors was calculated using the confusion matrix or sorting error table by the apparent error rate (APER) method.

2.5. Discriminant Analysis
The basic model of discriminant analysis is denoted by the symbol of 'd', and the linear discriminant analysis model of the various independent variables is shown in equation 3.

$$d = b_0 + b_1x_1 + b_2x_2 + b_3x_3 + ... + b_nx_n \quad (3)$$

Where: d = discriminant score
b = discriminant coefficient or weight (0,1,2 ,, n)
x = ultrasonic characteristics (1, 2,, n)

The discriminant analysis procedure starts from the normal multivariate test for the significance test of the discriminant variable which is then followed by the equation-covariant variance matrix test. If the equation test of the variant-covariance matrix is not met then the quadratic discriminant function can be used. Evaluation of mean vector values between groups can be done by testing the hypothesis using multivariate variance analysis (MANOVA) with statistical tests that can be used such as Pillai's Trace, Wilk's Lambda, Hotelling's Trace and Roy's Largest Root.

3. Results and discussion

3.1. Damage of Tubers Due To Lanas Pest Attack
The grouping of the health and damage of purple sweet potato based on the physical characteristics of the tubers is still quite difficult to do accurately. The destructive test results on 45 damage samples and 60 healthy purple sweet potatoes are shown in table 1. Adult lanas pests attack the epidermis of the tubers and make a hole to lay its eggs. The eggs then hatch into larvae and create a hole in the tuber's flesh before turning into pupa and imago (figure 1).

Table 1 Results of manually sorting of purple sweet potato varieties of Ayamurarasaki

Tuber Condition	Healthy tubers	Damage tubers
Black spot	0	45
Hole	3	30
Lanas larvae	0	16
Healthy	57	0

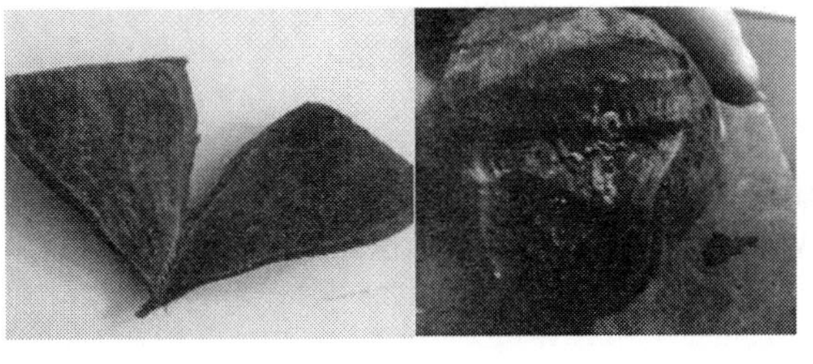

(a) (b)
Figure 1. (a) Healthy purple sweet potato and (b) purple sweet potatoes attacked by lanas

Hardness in the damaged purple sweet potato varieties of Ayamurasaki by lanas pests was found higher than the healthy tubers. Hardness on purple sweet potatoes that was attacked by lanas about 29.96 kgf/cm^2 to 46.73 kgf/cm^2, whereas in healthy tubers was about 30.27 kgf/cm^2 to 42.31 kgf/cm^2, and statistically significantly different between them at the 5% test level, as can be seen in table 2.

Table 2. Average hardness of healthy and damaged purple sweet potatoes

Characteristic	Healthy tuber	Damage tubers
Hardness (kgf/cm^2)	$37.74^a \pm 3.01$	$39.82^b \pm 3.40$

The numbers on the same line followed by the same letter are not significantly different at the 5% test level.

The purple sweet potato responds to the pest attack by producing toxic terpene compounds that cause tubers can not be consumed even at low concentrations and physical damage. The occurrence of malformation (abnormal growth of tissue or organs), thickening and cracks in sweet potatoes attacked by pests [13]. According [6], hole formed by the pest attack will dry and cause the decrease in the elasticity of the tuber tissue. Moreover, if the pest attack continues, the tuber will rot and the water content will increases.

3.2. Transmission Characteristics of Ultrasonic Waves

The ultrasonic wave acoustic properties used to predict damage to purple sweet potato due to pest attack are wave velocity, coefficient of attenuation and zero moment power. The nature of the wave is commonly used in the determination of the quality of agricultural products because it is influenced by the medium passed, and is not damaging medium passed.

Based on the measurement of the acoustic properties of the waves in the healthy and attacked by lanas pest sweet potato, it was found that the ultrasonic wave velocity value and Mo were lower in the healthy tubers than the tuber that attacked by lanas, while will have opposite effect for the coefficient of attenuation. Based on the result of analysis of variance with 95% confidence level, it is known that wave velocity, coefficient of attenuation, and Mo have p-value $< \alpha$ (0.05). It shows that there is a significant difference in the acoustic nature of the healthy purple sweet potato and pest attacked ones. The average value of ultrasonic wave characteristics in purple sweet potatoes is shown in table 3.

Table 3. Average value of ultrasonic wave transmission characteristics in purple sweet potatoes

Characteristics	Healthy tubers	Damage tubers
Wave Velocity (m/s)	$235.30^a \pm 4.68$	$246.55^b \pm 5.50$
Attenuation (Np/m)	$42.41^a \pm 4.12$	$39.83^b \pm 4.70$
Mo	$18.20^a \pm 2.83$	$19.34^b \pm 2.80$

The numbers on the same line followed by the same letter are not significantly different at the 5% test level.

The ultrasonic wave velocity is influenced by the internal condition and modulus elasticity of material. Based on statement of [14], [2], [9] and [10], the hardness of the material will be directly proportional to the wave velocity. The wave velocity of lanas attacked purple sweet potatoes was found higher than the healthy tubers. It is because the hardness of damage tubers is also higher due to the response of tubers that attacked by lanas would cause the thickening of the tissue. Also, ultrasonic wave velocity may also be influenced by water content. A study of ultrasonic evaluation for avocado maturity suggested that greater moisture content causing greater turgor pressure will result in greater ultrasonic wave velocity or positively correlated [15].

The greater attenuation coefficient in the healthy purple sweet potato is caused by its lower hardness compared to lanas attacked tubers. According [14], the amount of energy absorption that occurs depends on the physical characteristics of the wave. The process of wave absorption will often occur in solid medium, so that on raw pineapple fruit with low maturity and high hardness, attenuation coefficient value is smaller than pineapple fruit that has been mature. According to [4], coefficient of attenuation has negative correlation relationship (r = -0.81) with dragon fruit hardness, that is with increasing hardness tend to decrease the coefficient of attenuation.

Based on the coefficient attenuation equation by Mizrach (1989) *in* [11], the distance or diameter of the fruit is inversely proportional to the magnitude of coefficient of attenuation. The larger the diameter of the tuber that means longer ultrasonic wave transmission distance, the coefficient of attenuation will be smaller. The average diameter of healthy purple sweet potato was smaller than that of pests attacked ones, but not significantly different at the 5% test level. The average diameter of healthy sweet purple potato was 4.12 cm, whereas the lanas attacked tubers was 4.30 cm.

Zero moment power (Mo) shows the amount of signal energy transmitted on a medium. The greater the value of Mo in ultrasonic wave transmission then the transmitted energy the greater. The decrease in Mo value is directly proportional to the decrease in average wave magnitude [16]. Mo value obtained on healthy purple sweet potatoes is lower than that of pests attacked tuber. [2] mentioned that the value of Mo decreased along with the older durian fruit with increasingly declining its hardness. [11] also mentioned that the value of Mo tends to decrease with increasing maturity of bananas with a correlation coefficient R2 = 0.7431, the higher the Mo value then the fruit has a high hardness or still an-matured. The prediction of maturity of golden Apollo melons using audio sonic waves resulted in decreasing average Mo values with increasing age of harvest, i.e., from 142.67 ± 53.13 at harvest age 46 DAH to 51.52 ± 14.35 at harvest age of 67 DAH [16].

3.3. Prediction Model of the Damage Purple Sweet Potato Based on Wave Acoustic Characteristics

3.3.1. Wave Velocity.

The prediction model was developed using 2/3 of the total samples, i.e., 40 samples of healthy tubers and 30 samples of lanas attacked sweet potatoes. The ultrasonic wave velocity in lanas attacked purple sweet potatoes was found higher than the healthy purple sweet potato. The wave velocities of the tubers which was attacked by the lanas pest was arranged from 236.26 ms-1 to 261.10 ms-1, while in the healthy sweet purple ranges from 227.12 ms-1 to 245.49 ms-1 (figure 2).

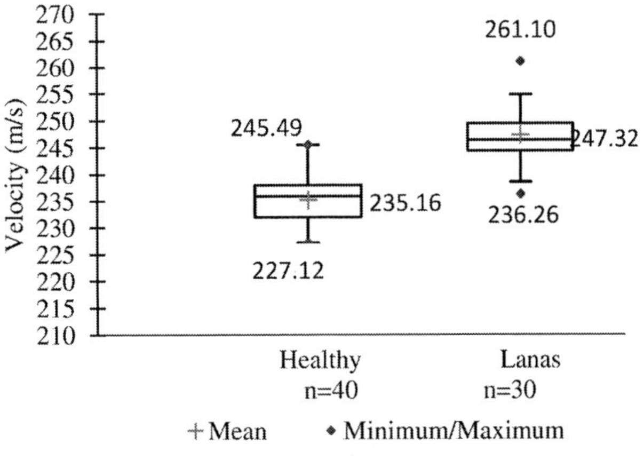

Figure 2. Ultrasonic wave velocity on healthy and lanas attacked sweet purple potato

Based on the value of wave velocity of each group, it was obtained that the threshold value of wave velocity on healthy purple sweet potato and attacked by lanas is 241.24 ms-1. The tuber that has wave velocity smaller than 241.24 ms-1 were grouped into the category of healthy purple yam, while purple sweet potato with wave speed greater than 241.24 ms-1 were grouped into the category of lanas pests attacked tubers. Mathematically, the limit of tubers damage due to pest attack by ultrasonic wave velocity is expressed in equations 4 and 5.

$$V \geq 241.24 \rightarrow L \qquad (4)$$

$$V < 241.24 \rightarrow S \quad (5)$$

Where V is the ultrasonic wave velocity (ms-1), S is a healthy tuber and L is the lanas attacked tubers.

3.3.2. Coefficient of Attenuation.
Sweet purple potato that was attacked by lanas pests was found having a lower average attenuation coefficient compared to healthy tubers. The coefficient of attenuation on healthy purple sweet potatoes ranged from 35.63 Np/m to 54.28 Np/m, whereas in the pests attacked tubers the coefficient of attenuation ranged from 32.18 Np/m to 50.75 Np/m (figure 3).

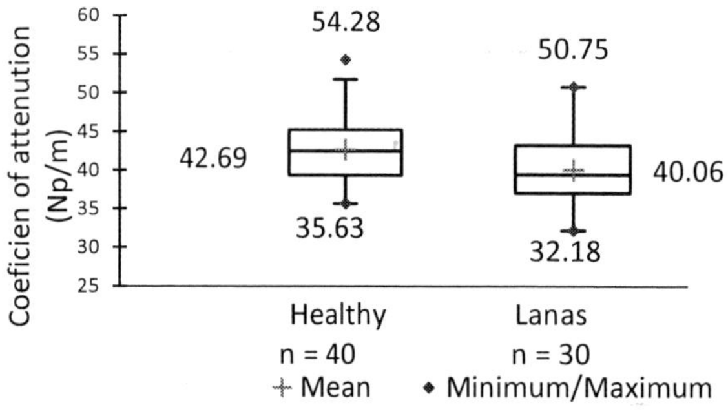

Figure 3. Average coefficient of attenuation on healthy and pest attacked sweet potato

The threshold value of attenuation coefficient between the healthy and lanas pests attacked purple sweet potatoes was 41.38 Np / m. Sweet potato having attenuation coefficient greater than 41.38 Np/m belongs to the category of healthy sweet potato, and which has attenuation coefficient value less than 41.38 Np/m including in the category of sweet potato attacked by pests. Mathematically, the limits of damage of purple potato due to pest attack based on the coefficient of attenuation can be expressed in equation 6 and 7.

$$\alpha \geq 41.38 \rightarrow S \quad (6)$$

$$\alpha < 41.38 \rightarrow L \quad (7)$$

Where α is the coefficient of attenuation (Np/m), S is the healthy tuber and L is tuber attacked by sweet potato pest.

3.3.3. Zero Moment Power (Mo).
In addition to the wave velocity and coefficient of attenuation, the value of zero moment power (Mo) is also used to estimate the damage purple sweet potato due to lanas pest attack. Healthy sweet purple potato has a Mo value ranging from 13.73 to 24.00, while Mo in pest attacked tubers ranged from 13.91 to 24.68 (figure 4). Mo value in sweet purple potato was known from the measurement of ultrasonic wave characteristics in terms of amplitude and time relationship, i.e., by performing FFT.

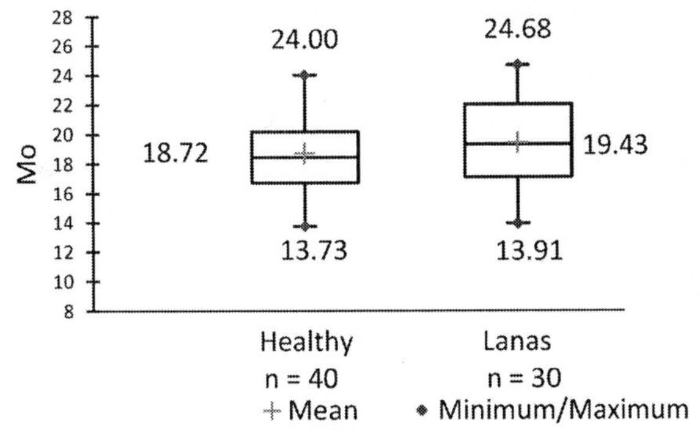

Figure 4. Average Mo on healthy and pests attacked sweet potatoes

The threshold value used to determine the limit of damage between healthy and pests attacked purple sweet potato was 19.08. The tubers have Mo value greater than 19.08 belongs to the category of lanas attacked sweet potato, and that have Mo value less than the limit was a healthy purple sweet potato. Mathematically, the damage to purple sweet potatoes due to pest attack can be written as equation 8 and 9.

$$Mo \geq 19.08 \rightarrow L \quad (8)$$

$$Mo < 19.08 \rightarrow S \quad (9)$$

With Mo is the value of zero moment power, S is a healthy purple sweet potato and L is purple sweet potato that attacked by lanas pest.

3.4. Model Validation

3.4.1. Validation of Estimation Model Based on Ultrasonic Wave Velocity.

Based on equations 4 and 5, validation of the model to test the accuracy of the model equations has been established by calculating the success rate of tubers selection. The distribution graph of ultrasonic wave velocity validation data for model testing can be seen in figure 5.

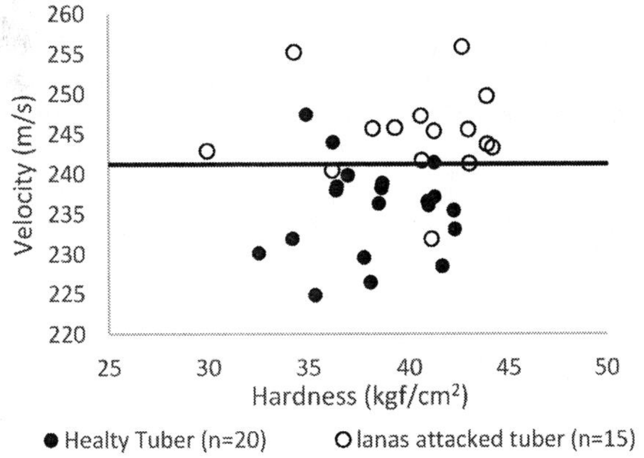

Figure 5. Validation of damage estimation model based on wave velocity

Based on the data of ultrasonic wave velocity validation on healthy and lanas attacked purple sweet potato, there were 3 healthy sweet potatoes considered into the group of lanas attacked potatoes, and 2 attacked sweet potatoes classified into healthy sweet potato. An error in grouping of healthy and lanas pests attacked of purple sweet potato is presented in table 4.

Table 4. Percentage of sorting errors based on wave velocity

From/to	Sorting Results			
	Lanas	Healthy	Total	Error (%)
Healthy	3	17	20	15.00
Lanas	13	2	15	13.33

Based on equations 5 and 6, it was found that the prediction model of purple sweet potato damage due to pest attack can be categorized well with the percentage of success 85.71% and the value of APER of 14.29%.

3.4.2. Validation of Estimation Model Based on Attenuation Coefficient.
The model test using validation data of equations 6 and 7 can be seen in figure 6.

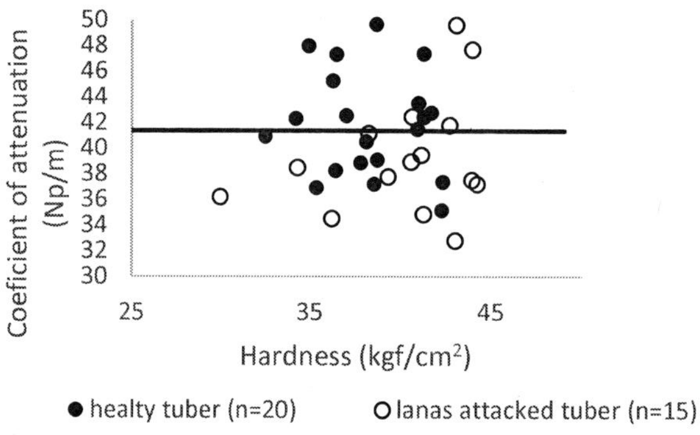

Figure 6. Validation of attenuation coefficient of healthy and lanas attacked of sweet purple potato

The result of the validation of the estimation model of tubers damage based on the coefficient of attenuation shows that as many as 9 healthy purple sweet potatoes are below the borderline or into the damage sweet potato group, and there were 4 damage tubers attacked by the pests that fall into the category of healthy tubers (table 5).

Table 5. Percentage of sorting error based on the coefficient of attenuation

From/to	Sorting Results			
	Lanas	Healthy	Total	Error (%)
Healthy	9	11	20	45.00
Lanas	11	4	15	26.67

The purple sweet potato's grouping based on the attenuation coefficient has a fairly high error compared to the clustering based on the ultrasonic wave velocity. Successful in sorting of purple sweet potato was 62.86% with APER value of 37.14%.

3.4.3. Validation of Estimation Model Based on Zero Moment Power.
Based on equations 8 and 9 concerning the Mo value limit to estimate the damage of the purple sweet potato obtained the model validation results as in figure 7.

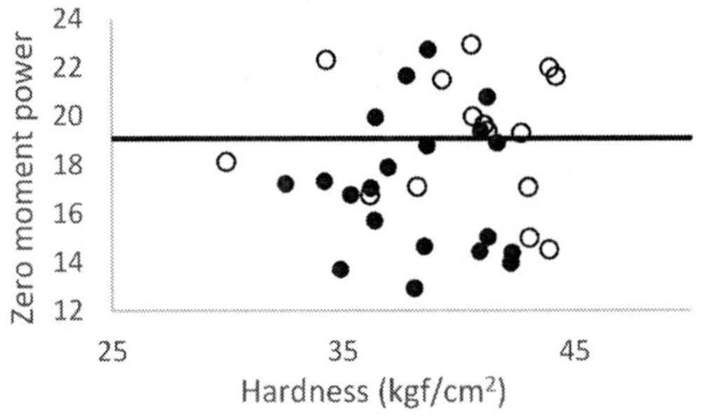

Figure 7. Validation of the damage purple sweet potato estimation model using Mo

The estimation of purple sweet potato damage due to pest attack using Mo value has an unfavorable success compared to grouping using wave velocity. In the estimation value of Mo 19.08, there was an error of grouping of healthy purple sweet potatoes by 25%, and tubers attacked by lanas pest was 40 %. The percentage of success of tubers sorting at Mo limit value was 68.57% with error (APER) of 31.43% (table 6).

Table 6. Percentage of miscellaneous purple potato error based on Mo

From/ to	Sorting Results			
	Lanas	Healthy	Total	Error (%)
Healthy	5	15	20	25.00
Lanas	9	6	15	40.00

3.5. Estimation Model of Lanas Attack Based on Combination of Ultrasonic Wave Characteristics
The grouping of purple sweet potato using a combination of ultrasonic wave characteristics was performed using discriminant analysis. Discriminant analysis is a method of knowing which variables differentiate a group from other groups within a population and can be used to clarify data based on differences in the characteristic data [17].

In determining a model, multi-collinearity testing is required to know that whether an independent variable in a model can be used in the formation of a discriminant function. The multicollinearity test results showed that the model did not experience multi-collinearity because the value of variance inflation factor (VIF) was less than 10 and the tolerance value was more than 0.1 (table 7), so the ultrasonic wave velocity, coefficient of attenuation and zero moment power had no linear relationship between variables and cannot be used as the predictor variable for the formation of discriminant function in estimating lanas pest attack on purple sweet potato.

Table 7. Multi-collinearity test among variable estimators

Statistic	Wave Velocity (m/s)	Attenuation Coefficient (Np/m)	Mo
Tolerance	0.880	0.886	0.99
VIF	1.136	1.129	1.00

In addition, discriminant analysis testing must satisfy several other assumptions, namely multivariate normal distributed data and have the same covariance matrix [18]. The normal multivariate test using the Chi-square plot (figure 8) was performed using SPSS version 15.0 with a significance value of 0.05. There is a high correlation between the distance of the mahalanobis with the Chi-square of 0.992, and the distribution of data tends to form a straight line with more than 50% (54.28%) of the value of the mahalanobis distance less than the value of the Chi-square. The significance value greater than 0.05 indicates that the data of the three variables are normally distributed.

The covariant matrix equation test was performed using Box's test of the XL-stat software to determine the homogeneity of the data used (table 8). The significance value> 0.05 indicates the covariant matrix between groups is homogeneous. The data condition used to predict pest attack on purple yam is multivariate and homogeneous normal distribution, so the discriminant function constructed is linear discriminant function.

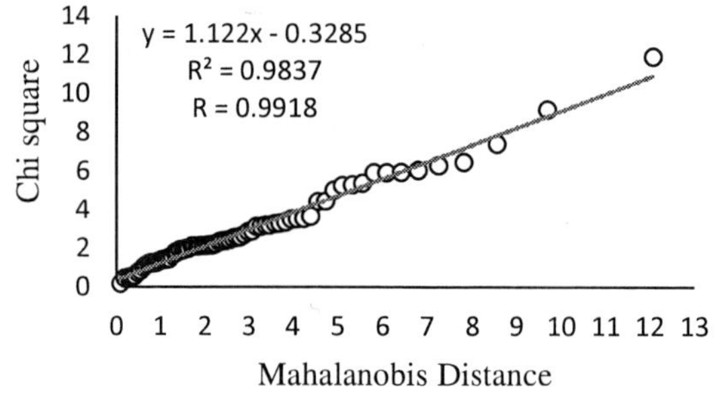

Figure 8. Quartile plot of Chi-square test of multivariate normality

Table 8. Result of equality test of covariance matrix

Test Results	
-2Log(M)	9.481
F (Observed value)	1.502
F (Critical value)	2.099
DF1	6
DF2	27010
p-value	0.173
Alpha	0.05

An equivalent vector equation test was conducted to estimate the variables that could distinguish the healthy purple sweet potato group from attacked pests on the other site. Significance value (p-value) at wave velocity and coefficient of attenuation below 0.05, so it is considered to distinguish well between the healthy purple sweet potato and attacked by the lanas. While Mo has a significance value above 0.05, it means that singularly cannot classify well between healthy and damaged tubers.

After several assumptions or conditions in the preparation of discriminant analysis equations were sufficiently enough, the linear discriminant score function for the grouping between healthy and damage purple sweet potato were established which is shown in equation 10. The limiting value used is the cutting score of 0.3778. The purple sweet potato discriminant score greater than 0.3778 will fall into the category of damage tubers, while the lower ones will fall into the group of healthy tubers. Based on the result of observation data analysis of both categories of purple sweet potatoes, then continued by made grouping based on proximity observation data to main values for each group.

$$d = -51.273 + 0.213X1 - 0.008X2 + 0.041X3 \quad (10)$$

Where: d = discriminant score for purple sweet potato
X1 = ultrasonic wave velocity (m / s)
X2 = coefficient of attenuation (Np / m)
X3 = zero moment power (Mo)

Based on the validation done on the purple sweet potato grouping, there were 3 of 15 tubers attacked by pests into the category of healthy purple sweet potatoes, and 2 of the 20 healthy tubers fall into the category of lanas attacked purple sweet potato (table 9). It was found that the grouping done has a success rate of 85.71% with an APER value of 14.29%.

Table 9. Validation of purple sweet potato grouping using discriminant analysis based on a combination of ultrasonic wave characteristics

From\to	Lanas	Healthy	Total	Error (%)
Lanas	12	3	15	20.00
Healthy	2	18	20	10.00
Total	14	21	35	14.29

4. Conclusions and suggestions

An average ultrasonic wave velocity and the zero moment power (Mo) value were higher in purple sweet potato which attacked by lanas pests compare with healthy tubers, while an opposite effect was found for the average coefficient of attenuation. The result of discriminant analysis shows that the wave velocity, the coefficient of attenuation and Mo can be used to classify the healthy tubers and lanas pests attacked tubers, with the wave velocity as the dominant differentiator factor. It was found that success rate of the model was 85.71% with 14.29% of APER value.

References

[1] Nonci N 2005 Bioecology and control of the beetle *Cylas formicarius* Fabricus (Cleoptera: Curculionidae) *Journal of Agricultural Research* **24**(2): 63-69.
[2] Haryanto B, Budiastra I Wand Purwadaria H 1999 Non-Destructive Quality Evaluation of Fruit with Ultrasonic Waves *Agrimedia* 5(1)
[3] Efriyanti N D 2006 Estimate the maturity level of sweet star fruit using ultrasonic waves (in Indonesian) (Bogor: Graduate School of Bogor Agricultural University)
[4] Djamila S 2010 Evaluation of dragon fruit quality non-destructively by ultrasonic method (in Indonesian) (Bogor: Graduate School of Bogor Agricultural University)
[5] Warji 2008 Estimation of arumanis mangoes damage caused by fruit flies using ultrasonic (in Indonesian) (Bogor: Graduate School of Bogor Agricultural University)
[6] Sutrisno A 2016 Detection of Cilembu sweet potato damage caused by lanas pest attack using ultrasonic wave (in Indonesian) (Bogor: Graduate School of Bogor Agricultural University)
[7] Nasution D 2006 Development of a non-destructive mangosteen fruits evaluation system with ultrasonic waves (in Indonesian) (Bogor: Graduate School of Bogor Agricultural University)
[8] Asmara V T 2014 Improved quality of purple sweet potato paste (*Ipomea batatas* L.) varieties of Ayamuurasaki through enzymatic hydrolysis process (in Indonesian) (Bogor: Graduate School of Bogor Agricultural University)
[9] Juansah J 2005 Design of ultrasound wave measurement system for grading of mangosteen (in Indonesian) (Bogor: Graduate School of Bogor Agricultural University)
[10] Sitompul Y D 2011 Characteristics of ultrasonic waves to detect the quality of Japanese cucumbers (in Indonesian) (Bogor: Graduate School of Bogor Agricultural University)

IOP Conf. Series: Earth and Environmental Science **147** (2018) 012002 doi:10.1088/1755-1315/147/1/012002

[11] Soeseno A 2007 Study of ultrasonic wave characteristics for banana maturity detection (in Indonesian) (Bogor: Graduate School of Bogor Agricultural University)
[12] Cheng Y and Haugh C G 1994 Detecting hollow heart in potatoes using ultrasound *Trans. ASAE* **37**(1) : 217-222.
[13] Ames T, Smit N E J M, Braun AR, O'Sullivan J N and Skoglun L G 1997 *Sweetpotato: Major Pests, Diseases, and Nutritional Disorder* (Lima (PE): International Potato Center (CIP))
[14] Luketsi W P 2016 Determination of maturity level of fresh pineapple fruit non-destructively by ultrasonic method (in Indonesian) (Bogor: Graduate School of Bogor Agricultural University)
[15] Self G K, Ordozgoiti E, Povey M J W and Wainwright H 1994 Ultrasonic evaluation of ripening avocado flesh *Postharvest Biology and Technology* **4** 111-116.
[16] Agusta W 2016 Apollo Melon Fruit Maturity Detection Using Parameters of Sound Signal (in Indonesian) (Bogor: Graduate School of Bogor Agricultural University)
[17] Maria F, Dasari D and Heryanto N 2013 Discriminant analysis of student behavior in fast food (Case Study: Students at UPI, ITB, UNLA and UIN Sunan Gunung Djati) *Journal of EurekaMatika* **1**(1): 107-116.
[18] Cheng Y and Haugh C G 1994 Detecting hollow heart in potatoes using ultrasound *Trans. ASAE* **37**(1):217-222.

Visual Method for Detecting Contaminant on Dried Nutmeg Using Fluorescence Imaging

S A Dahlan[1], U Ahmad[1] and I D M Subrata[1]

[1] Department of Mechanical and Biosystem Engineering, Faculty of Agricultural Engineering and Technology, Bogor Agriculture University, Indonesia

E-mail: sakinahyanidahlan@yahoo.co.id

Abstract. Traditional practice of nutmeg sun-drying causes some fungi such as *Aspergillus flavus* to grow. One of the secondary metabolites of *A. flavus* named aflatoxin (AFs) is known to be carcinogenic, so the dried nutmeg kernel must be aflatoxin-free in the trading. Aflatoxin detection requires time and costly, make it difficult to conduct at the farmers level. This study aims to develop a simple and low-cost method to detect aflatoxin at the farmer level. Fresh nutmeg seeds were dried in two ways; sundried everyday (continuous), and sundried every two days (intermittent), both for around 18 days. The dried nutmeg seeds are then stored in a rice sack under normal conditions until the fungi grow, then they were opened and the images of kernels were captured using a CCD camera, with normal light and UV light sources. Visual observation on images captured in normal light source was able to detect the presence of fungi on dried kernels, by 28.0% for continuous and 26.2% for intermittent sun-drying. Visual observation on images captured in UV light source was able to detect the presence of aflatoxin on dried kernels, indicated by blue luminance on kernel, by 10.4% and 13.4% for continuous and intermittent sun-drying.

1. Introduction

Indonesia is one of the world's largest nutmeg exporting countries with the largest export destination countries are Vietnam (22%), Netherlands(16%), the United States (13%), India (9.4%), and Germany (6.8%) [1]. In some periods, the nutmeg (*Myristica fragrans* Houtt) exported by Indonesia to the European Union was rejected due to the contamination of *Aspergillus flavus* (*A. flavus*) fungus. This type of fungus is known to release secondary metabolites known as aflatoxin that can cause liver cancer in humans and livestock when consumed in certain quantities. Some other fungi that produce aflatoxin are *A. parasiticus*, *A. nomius*, and *A. tamarii*. In addition to nutmeg seeds, *A. flavus* can also be found on other products such as corn, peanuts, milk and derivative products [2]. Some countries in the European Union apply strict regulations to the limits of aflatoxin content in imported foods of 3 to 5 ppb [3], the US Food and Drug Administration (FDA) sets a limit of 20 ppb for aflatoxin content in food and feed [4] , while Indonesia categorizes nutmeg as a spice product with the aflatoxin content limit of 15 ppb [5]. Fungal contamination may develop from improper postharvest handling especially drying and storage stages. To reduce contamination, drying should be done properly, for example on a shelf with a distance of about 1 meter above the ground to avoid direct contact with the soil that can be a source of fungus spread. Storage should be done in a fairly dry place and packed with sacks or closed cans [6].

Several methods have been developed to measure aflatoxin levels such as high-performance liquid chromatography (HPLC), liquid chromatography-mass spectrometry (LC-MS), and enzyme-liked immune assay (ELISA) [7] which require time, costly, and destructively. So a method that can detect aflatoxin levels quickly, does not require high cost, non-destructively and easily to apply especially at farmers level, is needed. From several literatures, aflatoxin is known to fluoresce with blue and green luminescence intensities under UV light with emission wavelengths of 420 - 450 nm and excitation wavelengths of 350 to 370 nm [2]. The objective of this research was to develop a contaminants detection nutmeg kernels visually through images of nutmeg kernels captured using a CCD camera with normal light source to detect a presence of fungi and UV light source to detect aflatoxin on dried nutmeg kernels.

2. Materials and methods

The material used in this research were fresh nutmeg seeds obtained from farmers in Siau Island, Sitaro District, North Sulawesi Province, then sent directly to Bogor by sea carrier. Equipment used in the experiment includes image recording device consisting of a CCD camera (charged couple device), a set of PC (personal computer), camera buffer, board to place the samples, four pieces ordinary Philips 5 watts lamp as a normal light source, and four pieces UV lamps SANKO DENKI brand with wavelength of 352 nm as the source of fluorescent light, as well as a set of tools for measurement of moisture content such as oven, analytical scales, cups, and desiccators. Analysis of moisture content was done based on AOAC standard (1999).

Research begins with sun-drying nutmeg seeds in two different methods, daily sun-drying (continuous sun-drying) and sun-drying every two days (intermittent sun-drying) for 18 days. Dried nutmeg seeds then were stored in normal condition (room temperature was 25 °C) and packed in wooden type plastic rice sack until the fungi grow. The dried seeds then were cracked and the images of kernels were captured using a CCD camera with normal light and UV light sources. Visual observation on images was conducted in the room and chemical analysis for aflatoxin level was conducted by HPLC method in a laboratory at SEAMO BIOTROP.

3. Results and discussion

3.1. Nutmeg seed drying

The fresh nutmeg fruits were harvested by farmers in Siau Island, Sitaro District in North Sulawesi and opened to obtain the fresh seeds, the packed and shipped to Bogor directly. It took eight days to reach the laboratory, with an initial moisture content of 33.3%. All the nutmeg kernels were sundried with daily temperature as shown in figure 1. Drying of nutmeg seeds is usually done to produce the aroma of nutmeg kernels.

Temperature and relative humidity vary according to the air condition during sun-drying of nutmeg seeds. The occurrence of variations or fluctuations in air temperature was normal and assumed to be last for eight hours period when the sun-drying took place. This fluctuation was closely related to the process of energy exchange that takes place in the atmosphere [8]. The day 1 and day 3 were the lowest temperature of 30 °C during sun-drying. The highest temperature during the drying process was 41 °C, occurred on the day 8. In the sun-drying of nutmeg seeds, the recommended temperature for maintaining high-quality nutmeg kernels should be not more than 45 °C. High drying temperatures can cause crack or wrinkle on nutmegs kernels due to the melting of fat content. Besides causing wrinkled on kernels, high drying temperatures can reduce the distinctive aroma of nutmeg kernels [6], thereby decreasing the quality of nutmeg kernels in trading.

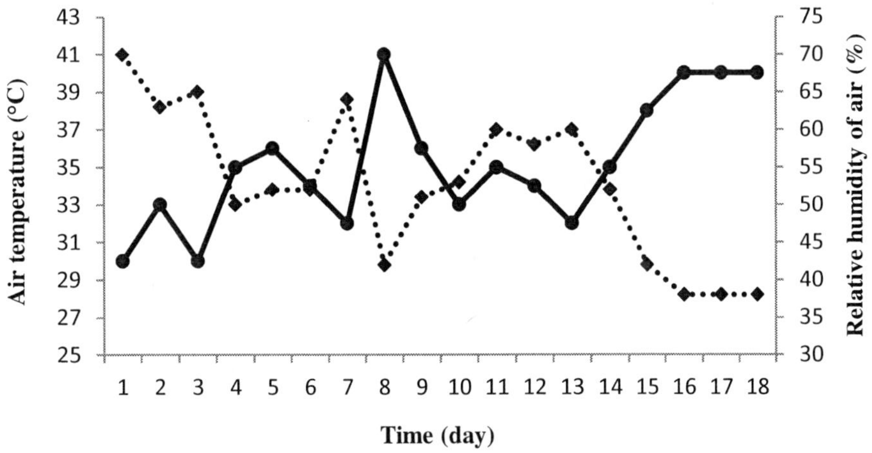

Figure 1. Temperature (●) and relative humidity (♦) of air during drying

The initial moisture content of nutmeg seeds was 33.3%. In general, the chart of changes in the moisture content of nutmeg seed with continuous sun-drying and intermittent sun-drying show the same pattern, which increased after the first day of sun-drying and then tends to decrease until the end of sun-drying process. figure 2 shows changes in the moisture content of nutmeg seeds that were sundried everyday. Sun-drying was discontinued at day 17 because the water content had reached 13.5% and would not decrease anymore because it reached the equilibrium moisture content, known from constant weight of nutmeg seeds. The moisture content of nutmeg seeds which were sundried every two days decreased until the sun-drying was stopped when it reached 10.7% (figure 3) at day 16 of drying, when constant weight of nutmeg seeds was observed. This method takes a shorter time compared to daily sun-drying. This is related to the characteristics of the material that is less exposed to high temperatures due to drying then the evaporation will become faster when the material is dried. The moisture content of nutmeg seeds at the request of the European Spices Association (ESA) is 12% wet basis. Meanwhile, according to SNI (No. SNI 01-0006-1993) the maximum moisture content of nutmeg seeds on nut quality standard requirements of nutmeg is 10% [6].

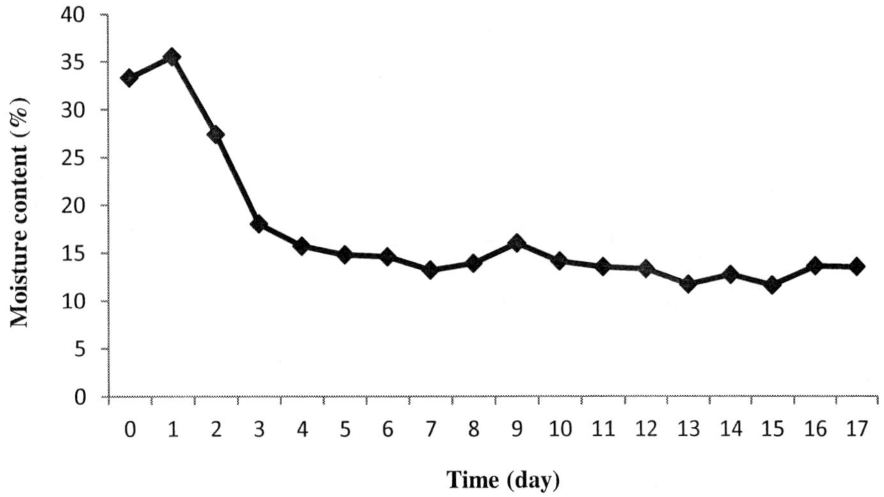

Figure 2. Changes in the moisture content of nutmeg seeds sundried every day (continuous sun-drying)

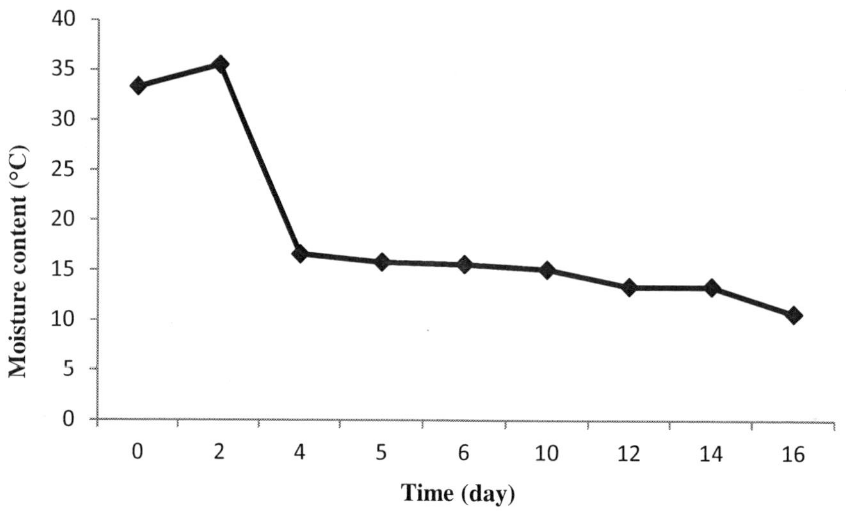

Figure 3. Changes in moisture content of dried nutmeg seed every two days (intermittent drying)

3.2. Visual appearance of nutmeg seeds with normal light source

Figure 4 is a visual appearance of the fungus on dry nutmeg kernels. Nutmeg kernels are the original Indonesian herbs and are highly susceptible to fungal infestations. The fungus can spread through soil and air by two main factors that support contamination such as temperature and humidity [3]. Through visual observation, total fungus infestation on nutmeg seeds with continuous sun-drying reached 28% with 3.2% high infestation, 7.2% medium infestation, and 17,6% low infestation. The nutmeg seed dried every two days resulted in a total infestation of 26.2% with high infested seeds of 12.2%, medium infestation of 4.3%, and low infested of 9.8% (table 1). The maximum number of fungi in the general requirements of nutmeg quality according to SNI (No. SNI 01-0006-1993) is 8% [6].

Daily sun-drying in the open air allows more fungi to contaminate nutmeg seeds compared with nutmegs that are dried in every two days, but contamination does not develop and spread because the nutmeg was exposed to higher temperatures longer as well as lower humidity does not support the growth and development of the fungus. While the sun-drying done every two days allows the fungus infestation to grow and spread faster so that more infestations found in high category. This phenomenon occurred because when nutmeg seeds were not dried then the nutmeg directly interact with the higher relative humidity and lower temperature of the room which is very supportive for fungus growth.

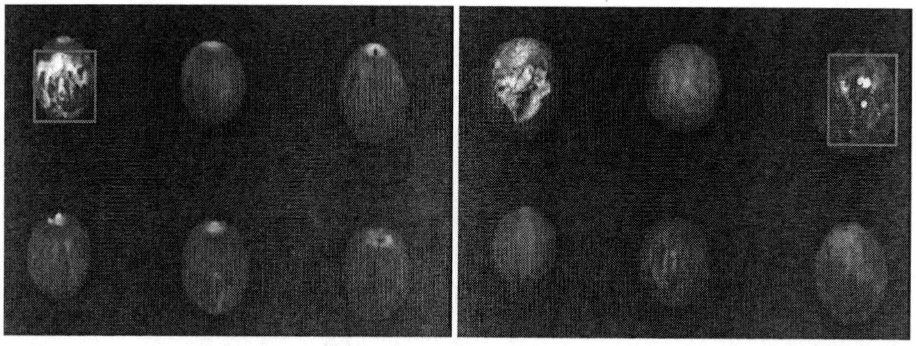

Figure 4. Image of nutmeg seeds infested by fungus observed using a CCD camera with normal light source

Table 1. Presence of fungus on nutmeg kernels observed on images captured using CCD camera with normal light source

Sample	N	n (%)	Infestation (%)		
			low	medium	High
Continuous Drying	125	35 (28,0%)	22 (17,6%)	9 (7,2%)	4 (3,2%)
Intermittent Drying	164	43 (26,2%)	16 (9,8%)	7 (4,3%)	20 (12,2%)

N=total number of nutmeg kernels; n=number of contaminated nutmeg kernels

3.3. Visual appearance of nutmeg seeds with UV light sources

The presence of aflatoxin observed visually on nutmeg kernels were characterized by blue luminescence on the monitor screen (figure 5). The high percentage of *A. flavus* contamination in dried nutmeg kernels indicates the risk of aflatoxin formation. The amount of *A. flavus* contamination in the dried food can be affected by the conditions and the duration of drying and storage. Drying and storage conditions play an important role for growth of *A. flavus* during storage [9]. The results (table 1) showed that overall the total nutmeg kernels contained aflatoxin by 10.4% and 13.4%, respectively, for nutmegs dried everyday and dried every two days. The percentage of aflatoxin on continuous sun-drying nutmeg were 8.0% for the low category, 0.8% medium category and 3.2% high category. The percentage of aflatoxin on nutmeg kernels with intermittent sun-drying were 6,09% for low category, 3.05% medium category, and 4.27% high category using the same sample as in table 1.

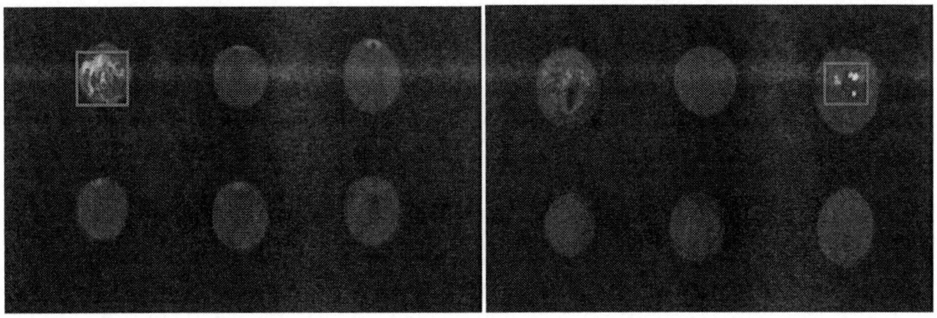

Figure 5. Luminescence of aflatoxin on nutmeg seed using CCD camera with UV light source

Table 2. Presentation of luminescence on nutmeg using CCD camera with UV light sources

Sample	N	n (%)	Infestation (%)		
			low	medium	High
Continuous Drying	125	13 (10,4%)	10 (8,0%)	1 (0,8%)	2 (1,6%)
Intermittent Drying	164	22 (13,4%)	10 (6,1%)	5 (3,1%)	7 (4,3%)

N=total number of nutmeg kernels; n=number of contaminated nutmeg kernels

In general, more than 10 types of aflatoxin have been identified, but only aflatoxin B1, B2, G1, and G2 that have important roles due to their toxicity and are often found in food and feed products [2]. Several methods to detect aflatoxin have been developed include high-performance liquid chromatography (HPLC), liquid chromatography-mass spectrometry (LC-MS), and enzyme-liked immune assay (ELISA) [7]. Nutmeg kernel samples with continuous sun-drying method and intermittent sun-drying have been tested for its aflatoxin content of B1, B2, G1, and G2 by HPLC (High Performance Liquid Chromatographic) method and overall aflatoxin contamination is less than the limit that can be detected by the tools of <0.45 ppb for aflatoxin B1 (AFB1), <0.43 ppb AFB2, <0.48 ppb AFG1, and <0.54 ppb AFG2 (table 3). However, in visual observation through camera with UV light source, aflatoxin can be seen by the presence of blue luminescence in the area of the fungus on the nutmeg kernels. Based on physical features, aflatoxin compounds when exposed to high-energy light will emit fluorescence light with specific wavelengths [10].

Table 3. Total aflatoxin content in nutmeg kernels by ELISA method

Sample	Aflatoxin content (ppb)			
	B1	B2	G1	G2
Continous Drying	<0,45	<0,43	<0,48	<0,54
Intermittent Drying	<0,45	<0,43	<0,48	<0,54

4. Conclusions

Daily drying (continuous drying) and drying in every two days (intermittent drying) of nutmeg seeds in ambient temperature varying between 30-41 °C and 38-70% RH result in different final nutmeg seeds moisture content. In continuous drying, the nutmeg seeds could reach a moisture content of 13.5% within 16 days while in intermittent drying, the nutmeg seeds could reach a moisture content of 10.7% within the same drying time.

CCD camera with normal light could only produced images of nutmeg kernels with the fungus, while the same camera with UV light as a source for fluorescence effect could detect the existence of aflatoxin indicating by blue luminescence on the images produced.

References

[1] The Observatory of Eonomic Complexity 2016 [internet].[downloaded July 2017]. Available at http://atlas.media.mit.edu/en/visualize/tree_map/hs92/export/idn/show/0908/2016/

[2] Tabata S 2011 *Mycotoxins: Aflatoxins and Related Compounds* (Japan: Tokyo Metropolitan Institute of Public Health)

[3] Bhatnagar D, Ehrlich K C, Moore G G and Payne G A 2014 *Aspergillus flavus* Encyclopedia of Food Microbiology vol 1

[4] Mathur-Bhatnagar P, Sunkara S, Panwar-Bhatnagar M, Waliyar F and Sharma K K 2015 *Plant Science* **234** 119-132

[5] Citanirmala N M V 2016 Study of Applied Implementation of Minister of Agriculture Number 53 Year 2012 for Control of Aflatoxin in Nutmeg (in Indonesian) (Bogor: Graduate School of Bogor Agricultural University)

[6] Natawidjaya H, Ametung M U, Nurjannah N, Nuraini and Didu N 2012 Technical Guidelines for Postharvest Handling Nutmeg (in Indonesian) (Jakarta: Direktorat Jenderal Perkebunan)

[7] Rodriguez M C, Sáchez G H, Sobrero M S, Schenone A V and Marsili N R 2013 *Microchem. J.* **110** 480-484

[8] Lakitan B 1994 Climatology Basics (in Indonesian) (Jakarta: Raja Grafindo Persada)

[9] Pereira M M G, Carvalho E P and Prado G 2002 *B. Do Ceppa.* **20** 141-156

[10] Ferdi 2010 Use of Flourescence Spectroscopy Method for Fungal Detection of Aspergillus flavus Link. on the Corn Seed (in Indonesian). (Bogor: Graduate School of Bogor Agricultural University)

IOP Conf. Series: Earth and Environmental Science **147** (2018) 012004 doi:10.1088/1755-1315/147/1/012004

Prediction of Caffeine Content in Java Preanger Coffee Beans by NIR Spectroscopy Using PLS and MLR Method

I W Budiastra[1], Sutrisno[1], S Widyotomo[2] and P C Ayu[1]

[1] Department of Mechanical and Biosystem Engineering, Bogor Agricultural University, Indonesia
[2] Indonesian Coffee and Cocoa Research Institute, Jember, Indonesia

E-mail: wbudiastra@gmail.com

Abstract. Caffeine is one of important components in coffee that contributes to the coffee beverages flavor. Caffeine concentration in coffee bean is usually determined by chemical method which is time consuming and destructive method. A nondestructive method using NIR spectroscopy was successfully applied to determine the caffeine concentration of Arabica gayo coffee bean. In this study, NIR Spectroscopy was assessed to determine the caffeine concentration of java preanger coffee bean. A hundred samples, each consist of 96 g coffee beans were prepared for reflectance and chemical measurement. Reflectance of the sample was measured by FT-NIR spectrometer in the wavelength of 1000-2500 nm (10000-4000 cm-1) followed by determination of caffeine content using LCMS method. Calibration of NIR spectra and the caffeine content was carried out using PLS and MLR methods. Several spectra data processing was conducted to increase the accuracy of prediction. The result of the study showed that caffeine content could be determined by PLS model using 7 factors and spectra data processing of combination of the first derivative and MSC of spectra absorbance (r= 0.946; CV= 1.54 %; RPD= 2.28). A lower accuracy was obtained by MLR model consisted of three caffeine and other four absorption wavelengths (r= 0.683; CV= 3.31%; RPD= 1.18).

1. Introduction

Coffee is one of important commodities for Indonesia. It contributes to Indonesia economy and community welfare. Indonesia is rich in arabica coffee, namely arabica java preanger, arabica gayo, mandheling, toraja, and kintamani. These coffee have a unique taste and flavor which influenced by its chemical composition such as carbohydrate, protein, caffeine, trigonelline and chlorogenic acid.

Caffeine is one of the important chemical components of coffee. Caffeine has stimulant properties on our central nervous system and becomes most widely consumed psychoactive substance in the world [1]. Caffeine concentration in coffee bean is usually determined by chemical method which is time consuming and destructive method. Some efforts to find a fast and nondestructive method to determine caffeine content of coffee have been done; one of them is using NIR spectroscopy. However, most of the research was carried out in ground coffee beans [2; 3]. A few researches on NIR spectroscopy to determine caffeine content of coffee bean. Rosita et.al. 2016 successfully applied NIR spectroscopy and PLS to determine the caffeine concentration of Arabica gayo coffee bean. In this research, NIR spectroscopy was assessed to determine caffeine content in java preanger green coffee bean using PLS and MLR method. The inclusion of MLR method in this research aimed to develop a fastest detection and sorting system for coffee beans in industry in the near future, since MLR method can reduce wavelength numbers and thus reduce processing time compared to PLS method.

2. Materials and methods

2.1. Material and Apparatus

A batch of Arabica java preanger green coffee bean (water content 12–14%) was obtained from Indonesian Coffee and Cocoa Research Institute (ICCRI) Jember, Indonesia. Apparatus used in this study were NIR spectrometer type NIRFlex N-500 (BUCHI Labortechnic AG. Switzerland), LC-MS 2020 (Shimadzu, Kyoto, Japan), computer equipped with unscrambler software v 10.3 (CAMO, Norway) and Microsoft Excel, digital scales, petri dish, and grinder.

2.2. NIRS Measurement

A sample of coffee bean was weighed about 96 g for NIR measurement. The sample was put in a petri dish and arranged into 4 layers [4]. Measurement of NIR spectra was conducted using NIR spectrometer in the wavelength of 10000 – 4000 cm^{-1} (1000 – 2500 nm) with 4/cm interval and scan speed of 3 scans/s (Temperature around 22 – 25 °C). The number of samples for NIR testing was 100 samples.

2.3. Chemical Analysis

After NIR spectra measurement, the sample was subjected to chemical analysis. Each sample was grounded using grinder; about 0.2 g of ground coffee was used for caffeine analysis. Caffeine standard, formic acid, acetonitrile, and distilled water were used in chemical analysis. Determination of caffeine by chemical method was conducted using LCMS procedure. To express the amount of caffeine on a dry weight basis (db.), the water content was first determined using AOAC method [5]. Afterwards, samples of 0.2 g of ground coffee were extracted before LC-MS analysis. The LC equipment comprised ESI-APCI combination source (DUIS) and C-18 water was used as column in LCMS analysis with temperature of 40°C. Subsequently, mobile phase were divided into mobile phase A (40% of water/formic acid) and mobile phase B (60% of acetonitrile). The flow rate was 0.2 ml/min, and volume of sample injected was 2 μm. Determination of caffeine content was achieved by comparing peak area measurement and the calibration graph of caffeine standard with correlation coefficient of 0.999.

2.4. NIR Data Processing and Analysis

The reflectance spectra of samples, consisted of 100 reflectance observation, was transformed into absorbance spectra (log (1//R)) for a linear correlation of NIR values and the caffeine content. In this study, the calibration methods were compared between PLS and MLR method using unscramble software. To obtain the best prediction, several NIR spectra data processing method like first and second derivative, multiple scatter correction (MSC), standard normal variate (SNV) and combination of both first and second derivative with MSC and SNV were performed. Samples were divided into calibration and validation samples that were 67 and 33 samples respectively. Hence, cross-validation was used to test the performance of calibration model. The parameters used to evaluate the capability of model included coefficient correlation (r), standard error (SE), coefficient of variation (CV) and ratio of standard deviation to SE (RPD).

3. Results and discussion

3.1. NIR Spectra and Caffeine Concentration of Java Preanger Coffee Beans

Figure 1 shows the original absorbance spectra of the lowest and highest caffeine concentration in Arabica java preanger green coffee bean. The spectra shape was influenced by several aspects, in particular major chemical concentration (water, carbohydrate, etc.) and the particle size. The peak of spectra indicated in 1450 nm and 1940 nm that due to water absorption. The peak in 1210 nm and 1780 nm indicate carbohydrate and fat absorption. The caffeine absorption 1128, 1298 and 1672 nm could not be observed in original NIR absorption spectra. Therefore, to analyze the caffeine concentration, the data processing such as derivative method may be required. The difference in absorption between the highest and the lowest caffeine content is very small indicated that determination of caffeine will be difficult.

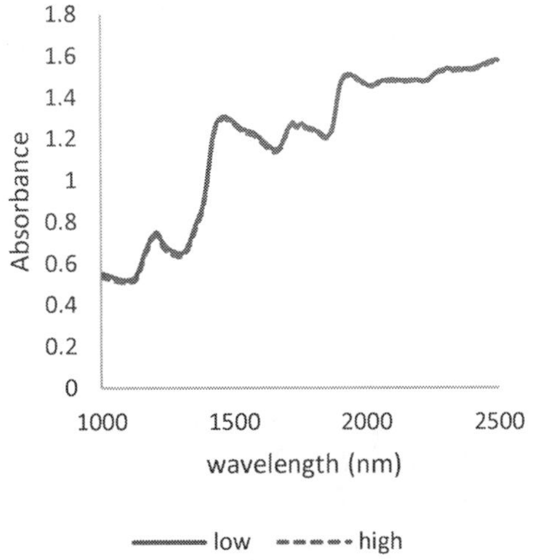

Figure 1. NIR absorbance spectra of the lowest and highest caffeine concentration in java preanger coffee bean

The caffeine concentration in java Preanger coffee bean was low around 1.18 – 1.41 % (table 1). This concentration is higher than concentration of Arabica gayo coffee bean which around 0.194–0.328% [6]. This is due to the difference of origin place, variety, and complex reaction that happened along the processing stage. Also, the water loss in materials also influences the caffeine concentration [3].

Table 1. Caffeine concentration in java preanger coffee beans

Process	Range (%)	Mean (%)	Standard deviation (%)
Calibration	1.18 – 1.41	1.31	0.056
Validation	1.21 – 1.40	1.32	0.047

3.2. Results of calibration and validation

The best calibration and validation result of caffeine using PLS and MLR method is shown in table 2 and 3, respectively. In PLS method, the best calibration model was achieved using combination of first derivative and MSC data processing and 7 PLS factors indicated by the high correlation coefficient (0.946), high RPD (2.28) and low CV (1.54 %). The first derivative was used to enhance the spectral resolution since changes in the gradient are examined so that this data processing suitable for complex spectra [7]. Also, MSC was performed to remove scatter and baseline shift in the obtained spectra [8], afterward the value of error validation decrease which influences RPD and consistency. According to Mouazen et al. [9], RPD value must be higher than 1.5 so that the calibration model can be used.

However, in a study on predicting caffeine concentration in Arabica gayo green coffee bean using NIRS [6], the second derivative method was selected as the best calibration model using PLS method indicated by high correlation coefficient (0.97), low SEC and SEP (0.005 and 0.007%) and low coefficient of variation (1.76%). There is a difference between error value in this research and [6] due to the difference of caffeine concentration, particle size of sample, and other properties in sample. The caffeine concentration in java preanger was around 1.18 – 1.41% while in gayo coffee was around 0.194 – 0.328%. Another properties concentration in sample such as carbohydrate, protein, lipid, and water are also different. This is due to the difference of origin place, variety, and complex reaction that happened along the processing stage [3]. Higher concentration will show a higher absorbance value in the spectra. Moreover, the size of gayo and java preanger coffee is also different. The size of java preanger coffee is bigger than gayo coffee; a bigger size will show bigger absorbance value in the spectra.

Table 2. Result of calibration and validation for caffeine using PLS method

Data Processing	Factors	r	SEC (%)	SEP (%)	CV (%)	RPD	Consistency (%)
Original	7	0,582	0,045	0,042	3,18	1,11	106,76
1st derivative	6	0,943	0,018	0,023	1,72	2,05	81,02
2nd derivative	5	0,953	0,017	0,023	1,7	2,06	74,62
1st derivative ; MSC	7	0,946	0,018	0,0206	1,54	2,28	87,57
2nd derivative ; MSC	4	0,942	0,019	0,023	1,71	2,06	82,33
1st derivative ; SNV	7	0,947	0,018	0,0207	1,55	2,27	87,09
2nd derivative ; SNV	5	0,958	0,016	0,022	1,65	2,13	72,69

On the other hand, MLR method could not give the accurate and robustness prediction (table 3). Increasing variables input of absorption in the number of selected wavelengths could increase accuracy, but the accuracy is still lower than PLS method. The wavelengths selected for predicting caffeine content using MLR method were a combination of caffeine wavelengths (1128, 1298, 1672 nm) [10] and several wavelengths which accommodate the scatter and intercorrelation effects between chemical compounds [11].

The MLR calibration method also produces the models with lower accuracies in predicting the caffeine content of Java Preanger coffee beans, even with variables input of processed absorption spectra using dg1, dg2 and normalization (table 4).

Table 3. Result of calibration and validation for caffeine using MLR method

Combination of selected wavelength	r	SEC (%)	SEP (%)	CV (%)	RPD	Consistency (%)
1128	0,119	0,0535	0,0501	3,81	1,03	106,79
1128, 1066	0,385	0,0497	0,0517	3,93	0,99	96,09
1128, 1066, 1940	0,388	0,0496	0,0524	3,98	0,98	94,70
1128, 1066, 1940, 1298	0,508	0,0462	0,0516	3,92	0,99	89,60
1128, 1066, 1940, 1298, 1672	0,423	0,0488	0,0533	4,05	0,97	91,55
1128, 1066, 1940, 1298, 1672, 1959	0,508	0,0464	0,0495	3,77	1,04	93,62
1128, 1066, 1940, 1298, 1672, 1959, 1032	0,683	0,0393	0,0435	3,31	1,18	90,36
1128, 1066, 1940, 1298, 1672, 1959, 1032, 1125	0,683	0,0393	0,0433	3,29	1,19	90,73
1128, 1066, 1940, 1298, 1672, 1959, 1032, 1125,1120	0,685	0,0392	0,0437	3,32	1,18	89,76
1128, 1066, 1940, 1298, 1672, 1959, 1032,1125,1120,1063, 1060,1053	0,695	0,0382	0,0425	3,23	1,21	89,98
1128, 1066, 1940, 1298, 1672, 1959,1032,1125,1120,1063,1060, 1053,1040,1029	0,709	0,0379	0,0418	3,18	1,23	90,77
1128, 1066, 1940, 1298, 1672, 1959,1032,1125,1120,1063,1060, 1053,1040,1029,1114,1096,1088, 1074	0,761	0,0380	0,0349	2,66	1,47	108,71
1128, 1066, 1940, 1298, 1672, 1959,1032,1125,1120,1063,1060, 1053,1040,1029,1114,1096,1088, 1074,1018,1015,1010,1007,1006, 1020,1019	0,844	0,0289	0,0338	2,57	1,52	85,45
1006.1007,1010,1015,1018,1019, 1020,1026,1029,1032,1039,1052, 1060,1063,1066,1075,1083,1088, 1089,1096,1106,1111,1114,1120, 1125	0,908	0,0233	0,0255	1,93	1,84	91,32

IOP Conf. Series: Earth and Environmental Science **147** (2018) 012004 doi:10.1088/1755-1315/147/1/012004

Table 4. Result of calibration and validation for caffeine using NIR data processing and MLR method

Data processing	r	SEC (%)	SEP (%)	CV (%)	RPD	Consistency (%)
dg1	0,600048	0,043065	0,044899	3,41	1,14	95,916
dg2	0,354821	0,050331	0,051564	3,92	1,00	97,610
n01	0,680787	0,039432	0,044899	3,41	1,14	87,825

Fig. 2 and Fig. 3 showed the scatter plot of reference and prediction values of caffeine using PLS and MLR method (using 7 selected wavelengths), respectively. figure 2 showed a high correlation and a lower SEC between caffeine predicted using NIRS and caffeine reference using LCMS, indicated by r-value > 0.9 and SEC < 0.023%. The results indicate that the obtained calibration model using PLS method could predict caffeine concentration in java preanger green coffee bean. On the other hand, the model developed by MLR method gives a lower accuracy in predicting the caffeine concentration of java preanger coffee beans.

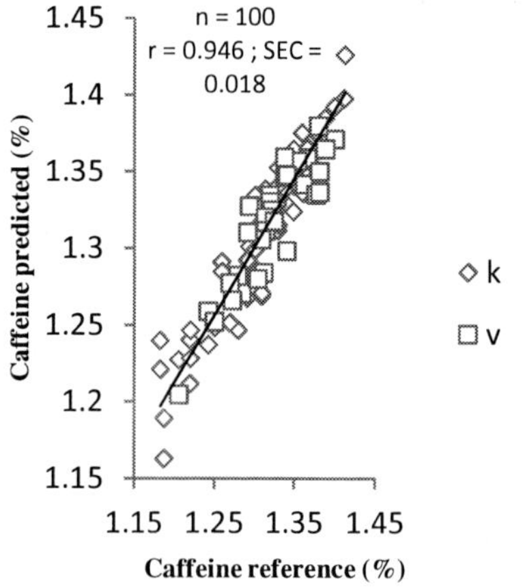

Figure 2. Plots of caffeine referenced vs. predicted using PLS method

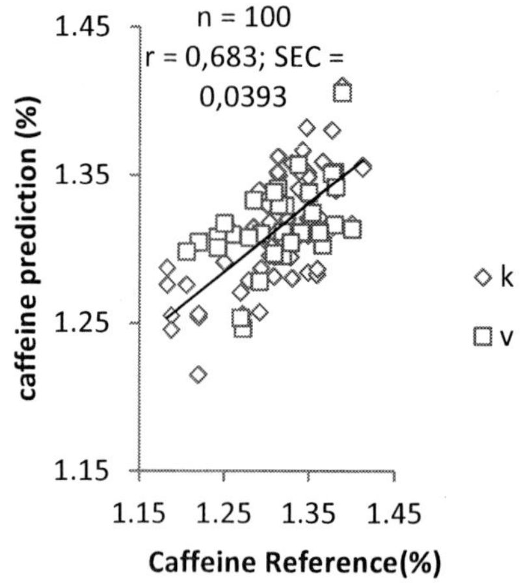

Figure 3. Plots of caffeine referenced vs. predicted using MLR method

4. Conclusion

The best calibration model of PLS for good prediction of the caffeine concentration of java preanger coffee bean was obtained by combination of first derivative and MSC data processing and using 7 PLS factors (r= 0.946, CV= 1.54%, RPD= 2.28 and consistency 87.57%). The accuracy of the PLS model is higher and better than the MLR model. This study showed that NIR spectroscopy and the PLS model can be used to determine the caffeine concentration in Arabica java preanger green coffee bean.

Acknowledgement

We would like to thanks the Directorate of Research, Ministry of Research, Technology and Higher Education for supporting this research (STRANAS Research Grant of FY 2017).

References

[1] Campo GD, Borrego I, Caracena R and Zuriarrain J 2010 *Talanta* **81** 367-371
[2] Huck CW, Guggenbichler W and Bonn GK 2005 *Analytica Chimica Acta* **538** 195-203
[3] Zhang X, Li W, Yin B, Chen W, Kelly DP, Wang X, Zheng K and Du Y 2013 *Molecular and Biomolecular Spectroscopy* **114** 350-356
[4] Rosita R 2016 *Non-destructive Determination of Arabika Gayo Coffee Chemical Composition with Near Infrared Spectroscopy* (in Indonesian) (Bogor: Graduate School of Bogor Agricultural University)
[5] [AOAC] Association of Official Analytical Chemyst (USA) 2005 *Official Method of Analysis of the Association of Offial Analytical of Chemyst* (Arlington: The association of official analytical chemyst, Inc)
[6] Rosita R, Budiastra I W and Sutrisno 2016 *JTEP Jurnal Keteknikan Pertanian* **4** 179-186.
[7] Stuart B 2004 *Infrared Spectroscopy; Fundamentals and Applications* (John Wiley & Sons, Ltd)
[8] Shan J, Suzuki T, Ogawa Y, Kondo N 2015 *Engineering in Agriculture, Environment and Food* **8** 195-199.
[9] Mouazen A M, W Saeys, J Xing, J De Baerdemaeker and H Ramon 2005 *J Near Infrared Spectroscopy* **13** 87-97.
[10] Ribeiro J S, Ferreira M M C and Salva T J G 2011 *Talanta* **83** 1352-58.
[11] Budiastra I W, Ikeda Y and Nishizu T 1998 *Journal of JSAM* **60** 117 – 127.

A Study of Light Level Effect on the Accuracy of Image Processing-based Tomato Grading

D Prijatna[1], M Muhaemin[1], R P Wulandari[2], T Herwanto[1], M Saukat[1] and W K Sugandi[1]

[1] Department of Agricultural and Bio-system Engineering, Faculty of Agro-Industrial Technology, Universitas Padjadjaran, Jatinangor Campus, Indonesia

E-mail: dedyprijatna@yahoo.com

Abstract. Image processing method has been used in non-destructive tests of agricultural products. Compared to manual method, image processing method may produce more objective and consistent results. Image capturing box installed in currently used tomato grading machine (TEP-4) is equipped with four fluorescence lamps to illuminate the processed tomatoes. Since the performance of any lamp will decrease if its service time has exceeded its lifetime, it is predicted that this will affect tomato classification. The objective of this study was to determine the minimum light levels which affect classification accuracy. This study was conducted by varying light level from minimum and maximum on tomatoes in image capturing boxes and then investigates its effects on image characteristics. Research results showed that light intensity affects two variables which are important for classification, for example, area and color of captured image. Image processing program was able to determine correctly the weight and classification of tomatoes when light level was 30 lx to 140 lx.

1. Introduction

Post-harvest handling is a chain of activities to maintain and achieve good quality agricultural product. One of the activities is grading in which product is classified according to consumer preferences [1]. Currently, most tomato grading is implemented using manual method mostly based on visual appearance using human eyes. It was observed that this method of grading tends to be subjective [2] and in long working hours may yield inconsistent results [3].

Image processing has long been used in industry mostly for quality control and automation. Although its application in agriculture came later, it is now used in production as well as post-harvest handling. In agricultural production, image processing is used for detecting plant condition and identifying pest [4]; [5]. In post-harvest processing, image processing is applied in the non-destructive test to evaluate the quality of agricultural products such as fruits and vegetables [6]; [7].

Department of Agricultural Engineering, Universitas Padjadjaran (UNPAD) has long been developing automatic tomato grading machine based on image processing to replace human sensory perception. The first prototype of the machine has been capable of using image processing to accurately classify tomatoes according to grading criteria given by a commercial tomato supplier. Nevertheless, their capacity was less than that of human [8]. This prototype has been further developed to improved mostly on its capacity. The last prototype has a capacity of 6000 tomatoes per hour [9].

IOP Conf. Series: Earth and Environmental Science **147** (2018) 012005 doi:10.1088/1755-1315/147/1/012005

In any of the grading machines, image of tomatoes is captured in a box furnished with digital camera and fluorescence lamps. This type of lamp is known that it will decrease its lighting ability with time. Because the quality of image is highly dependent on light [10], it is predicted reduced light intensity will affect the accuracy of tomato grading. Therefore, a study of the effect of light level on the accuracy of grading is necessary.

2. Materials and methods

This research was conducted in the Laboratory of Farm Machinery, Department of Agricultural Engineering, Universitas Padjadjaran from June 2015 to October 2016.The outline of steps of this research is given in figure 1.

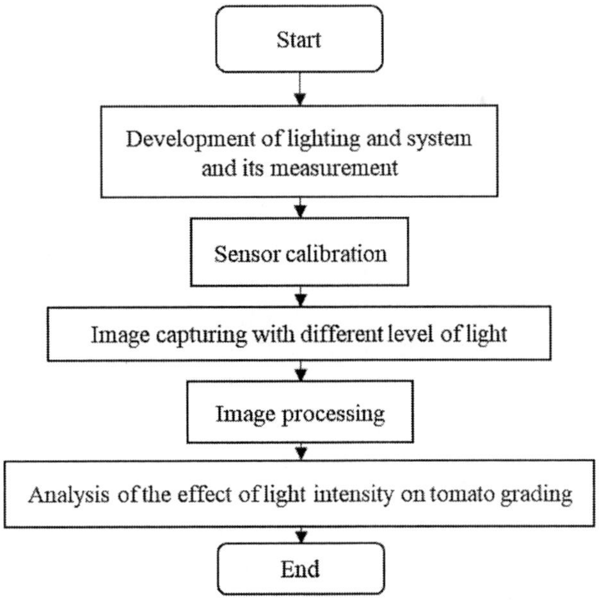

Figure 1. Research steps

2.1. *Construction of variable-intensity lighting system and sensor calibration*

In previously available image capturing box, the intensity of light could not be varied widely for experiment. Therefore, a new lighting system was developed using high powered LED. To ensure a ripple-free DC current which powered the LED, AC power was converted to DC using step-down transformer, rectifier and capacitor. Four light sensors BH1750 were arranged in four corner of the box and used to measure the emitted light. The data from the sensor were read by a microcontroller and then send to computer (figure 2). Before using the sensor to measure light intensity in image capturing box, the sensors were calibrated using lux meter *Lutron* EM-900. Calibration processes were conducted one sensor at a time at its respective position.

2.2. *Image Capturing and Processing*

Tomatoes were placed on conveyor and then move manually until they were located near the center of image capturing box. Light intensity was set and image was taken. For this image capturing event, light intensity was varied from 0 to 154 lx. Images were then processed using visual basic program used by the machine to predict tomato weight and calculate normalized green color of tomato. This two values were then used to determine tomato grade. In this respect, normal light was used as reference.

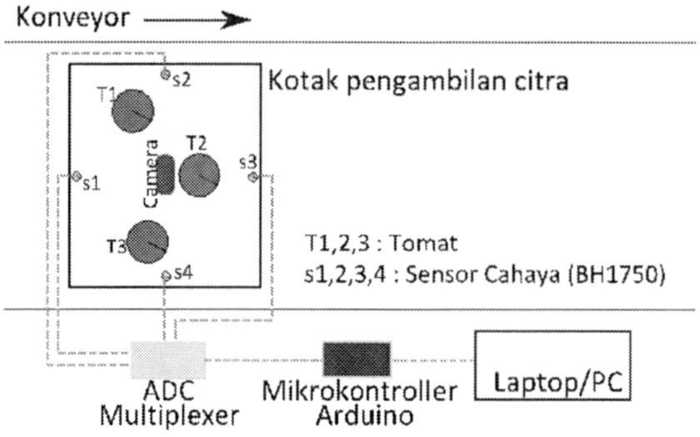

(a)

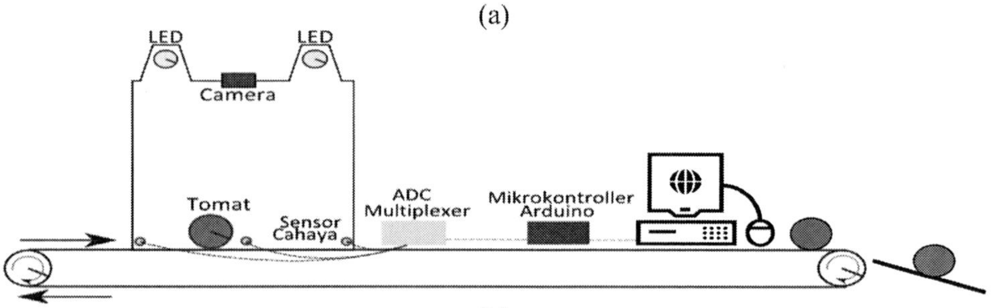

(b)

Figure 2. Experimental setup
(a) Top view (b) Front view

3. Results and discussion

3.1. *Sensor Calibration and distribution of light levels*

Results of sensor calibration are given in figure 3. Within the range of experiment, each sensor has similar type of curve but slightly different value. All regression equations are linear and all of coefficients of determination were larger than 0.999 which means very accurate.

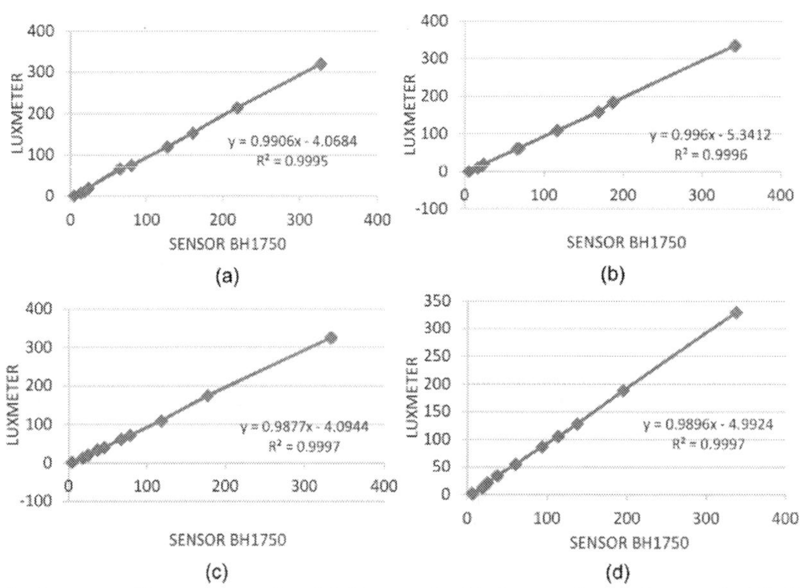

Figure 3. Sensor calibration results: (a) Sensor 1, (b) Sensor 2, (c) Sensor 3, (d) Sensor 4.

Before used to capture tomato image, distribution of light in the box was measured by calculating luminance value from the RGB of conveyor image without tomato (figure 4). Distributions of luminance, as well as its histogram, are given in figure 5. Based on the figure, it can be concluded that light was evenly distributed in the box and image taken within it would not be distorted due to its light distribution.

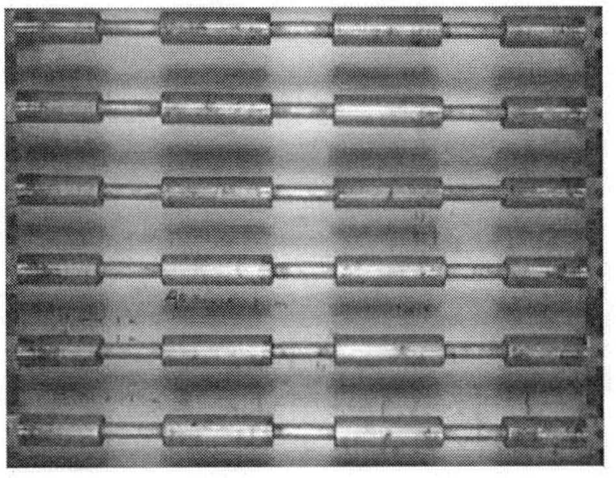

Figure 4. Conveyor without tomato

3.2. *Effect of Light Intensity on Tomato Grading*

The effect of light intensity of tomato image area is shown in figure 6. It can be seen from the figure that as the intensity increase, the image area also increase until about 30 lx. After reaching this value, then the image area is nearly constant at about 2200 pixel. Further checking on the image processing pipeline showed that when light intensity was low some parts of tomato edge look darker and image processing program instead of identifying it as tomato, identify it as part of background. Consequently, calculated tomato image area was lower than expected.

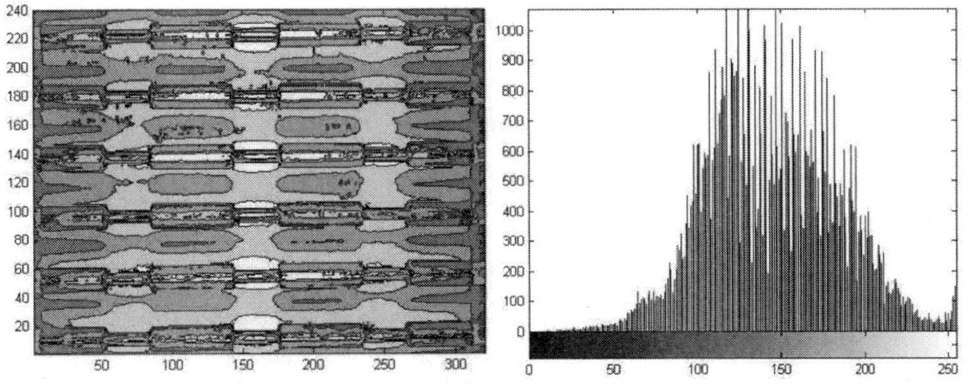

Figure 5. Luminance value of conveyor image

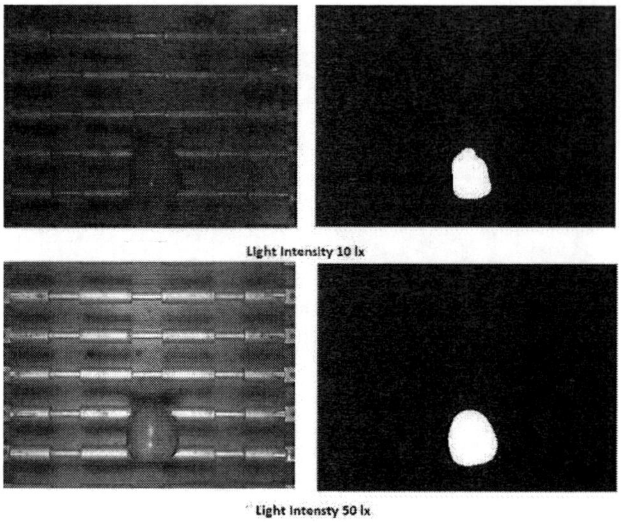

Figure 6. Captured image in light intensity of 10 lx (top), and 50 lx (bottom)

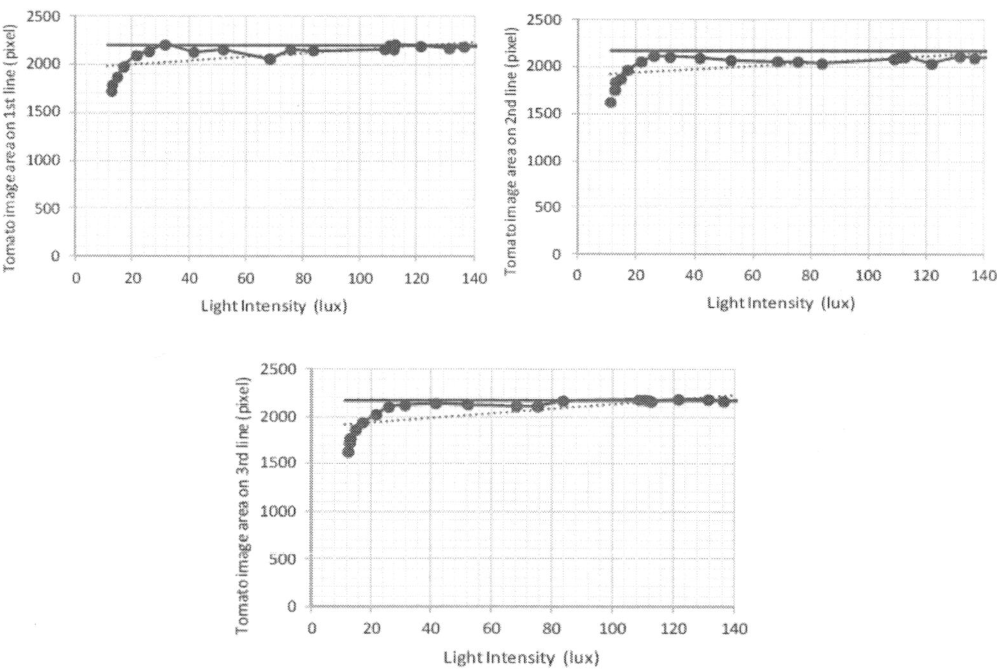

Figure 7. The effects on light intensity on image area

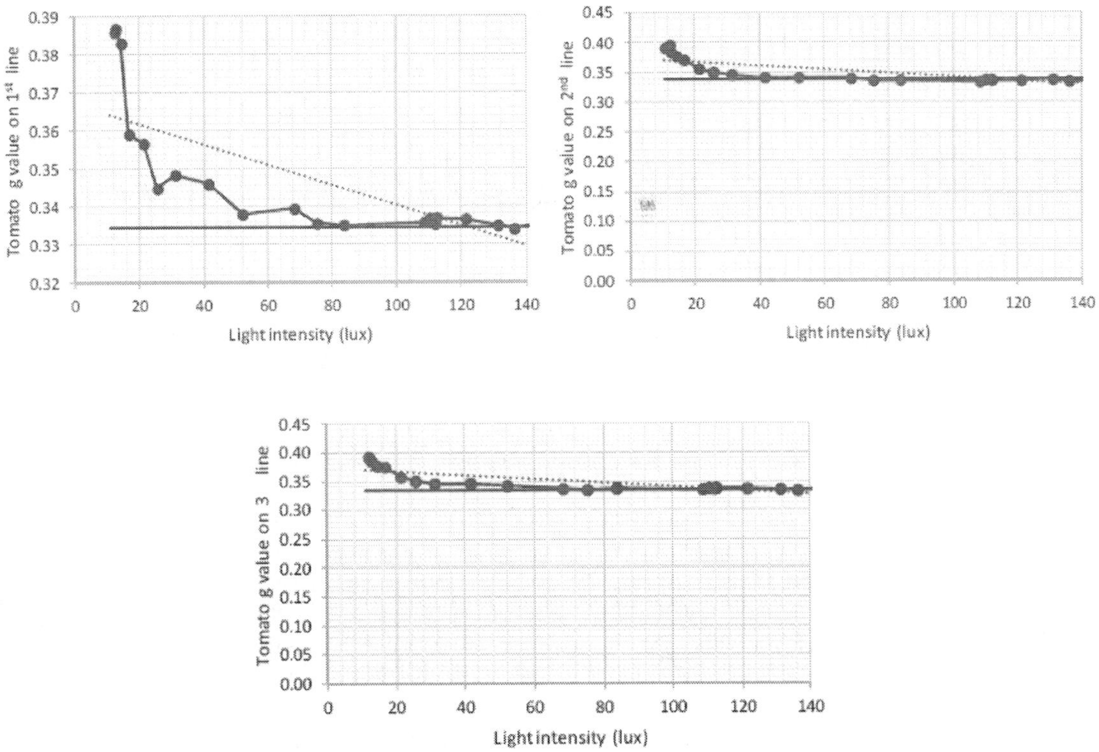

Figure 8. The effect of light intensity on predicted tomato weight

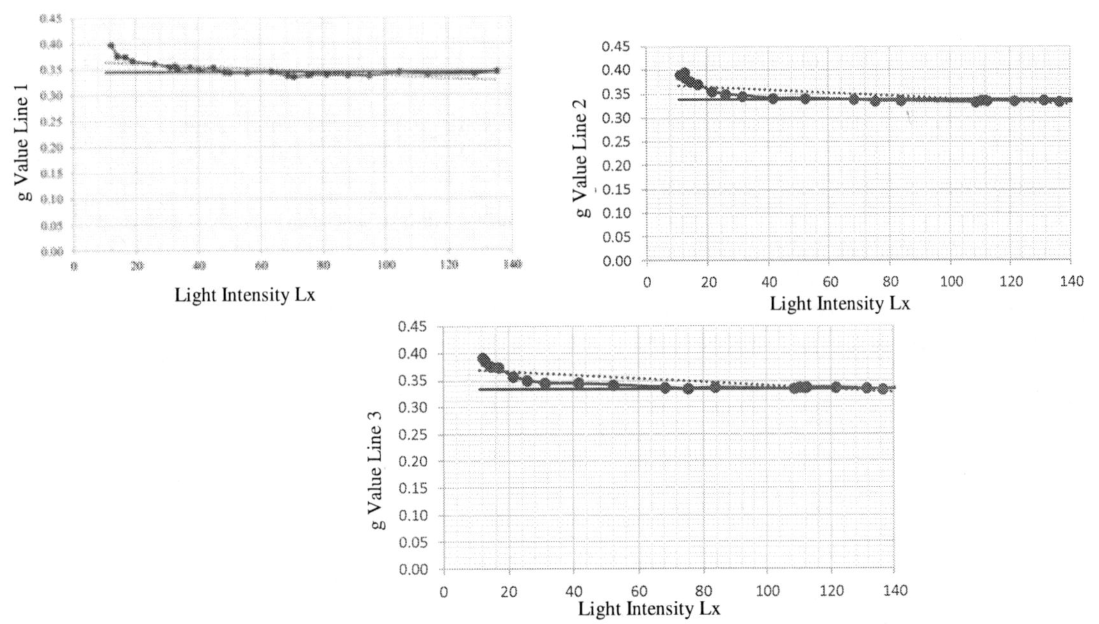

Figure 9. The effects of light intensity on normalized green

Table1. Light intensity levels at various criteria (lx)

	Tomato 1			Tomato 2			Tomato 3			Ave-rage
	Line 1	Line 2	Line 3	Line 1	Line 2	Line 3	Line 1	Line 2	Line 3	
Some tomatoes can be detected	-	10,97	11,98	-	11,22	11,22	-	-	11,72	-
All tomatoes can be detected	12,48	12,48	12,48	12,23	12,23	12,23	12,23	12,23	12,23	12,31
Tomato image area reach reference value	21,31	21,31	21,31	14,50	14,50	14,50	19,04	19,04	26,10	17,95
Normalized green value reach reference value	52,08	68,22	68,22	44,51	50,32	55,37	33,42	55,86	55,86	53,76

Similar results were obtained for tomato weights (figure 7). The weight of tomato become nearly constant after reaching the value of intensity of 30 lx. This can be understood since the weight of tomato was calculated from tomato image area. The bigger the image area, the heavier the tomato.

The maturity of tomato can be identified by its color. Mature tomato display higher value of red and lower value of green. Previous study had showed that normalized green (g) represent the maturity better than red or blue [11]. Light level also affects color perception of tomato. Value of normalized green in various light intensity are given in figure 8. It is clear that light intensity affects the value of g. At low light intensity, the g value was high and then decrease until light intensity reached a value between 30 to 40 lx. After this light intensity value, then the value of g remain constant.

Currently used program can detect tomato at minimum light level of 11 lx (table 1). At this light level, not all tomatoes can be detected. When the light level was increased to 12.3 lx, all tomato can be detected. Nevertheless, the value of tomato variables pertinent to grading was not right yet. The value of image area close to its reference value started to occur when minimum light level was 18 lx. Similar things occurred to the value of normalized green was at light level of 53.8 lx. These data showed that the light levels affect color perception more than image area. Since grading decision was based on image area and normalized green values, the accurate grading of tomato start to occur when light level of 54 lx. Below this light level, program will read tomato image as background image

4. Conclusions

Based on findings above, the following conclusions can be drawn:

a. Light levels affect the result of both predicted weight and perceived color of processed tomatoes.
b. The light levels affect color perception more than tomato image area
c. For accurate grading of tomatoes, minimum light level was 54 lx. Below this light level, program tend to classify tomatoes incorrectly.

Acknowledgement

The authors expressed sincere gratitude to Directorate General for Higher Education for providing fund for this research

References

[1] Zain S, U Suhadi, Sawitri and U Ibrahim 2005 *Postharvest Agriculture product Engineering* (in Indonesian) (Bandung: Giratuma)

[2] Sun D W and T Brosnan 2003 Improving quality inspection of food products by computervision: A review. *Journal of Food Engineering* **61**:3-16.

[3] Francis F J 1980 Colour quality evaluation of horticultural crops. *HortScience* **15**(1):14-15

[4] Sunagar V B, P A Kattimani, V A Padasali and N V Hiremath 2014 *Estimation of nitrogen content inleaves using image processing*. Proceedings of International Conference on

Advances in Engineering and Technology (Goa) (India) pp 25-28

[5] Patil J K and R Kumar 2011 Advances in image processing for detection of plant diseases *Journal of Advanced Bioinformatics Applications and Research* **2**(2):135-141

[6] Ahmad U, I D M Subrata and S Gunayanti 2004 Manggo Grading Based on visuality using Image Processing (in Indonesian) *Jurnal Keteknikan Pertanian.* **18**(1)

[7] Barrios A G, R A B Lopez, E R Garcia, M T Ayala and G M S Zarazua 2011 Tomato quality evaluation with image processing: A review *African Journal of Agricultural Research* **6**(14):3333-39.

[8] Muhaemin M, T Herwanto, D Prijatna, M.Saukat and W K Sugandhi 2012 *Development of Model Tomato Grading Machine Based on Visual Evaluation* (In Indonesian) Proceeding Seminar Nasional PERTETA (Bali) (Indonesia)

[9] T Herwanto, Muhaemin M, D Prijatna, M Saukat and W K Sugandhi 2014 Scale-up Tomato Grading Machine Prototype Based on Visual Evaluation (In Indonesian) Laporan Penelitian. LPPM Universitas Padjadjaran.

[10] Murinto, K. 2009. Retinex Method Implementation for Image Enlightenment (In Indonesian) *Jurnal Informatika Fakultas Teknologi Industri Universitas Ahmad Dahlan Yogyakarta* **3**(2)

Mechanical Damage Detection of Indonesia Local Citrus Based on Fluorescence Imaging

T H Siregar[1], U Ahmad[1], Sutrisno[1] and A Maddu[2]

[1] Department of Mechanical and Biosystem Engineering, Faculty of Agricultural Engineering and Technology, Bogor Agriculture University, Indonesia
[2] Department of Physics, Faculty of Natural Sciences and Mathematic, Bogor Agriculture University, Indonesia

E-mail: tikahafzara@gmail.com

Abstract. Citrus experienced physical damage in peel will produce essential oils that contain polymethoxylated flavone. Polymethoxylated flavone is fluorescence substance; thus can be detected by fluorescence imaging. This study aims to study the fluorescence spectra characteristic and to determine the damage region in citrus peel based on fluorescence image. Pulung citrus from Batu district, East Java, as a famous citrus production area in Indonesia, was used in the experiment. It was observed that the image processing could detect the mechanical damage region. Fluorescence imaging can be used to classify the citrus into two categories, sound and defect citruses.

1. Introduction

Indonesia has more than 19 varieties of citruses concentrated in some islands [1], and one superior variety is a Keprok Pulung citrus, which is famous in the country. However, the citrus is subject to damage during handling and transportation. The damaged fruit should be separated as soon as possible to prevent spreading rotten started from the damaged ones. Until now, sorting method of the damaged fruit is manually conducted which is difficult especially at the early stage of damage.

In Indonesia, research on substances flavonoids in citrus peels is benefited to pharmaceutical, with a target of the drug product for anti-proliferative in cancer cells [2]. Until recently, there has been no research on the utilization technology of fluorescence spectroscopy in the field of agricultural engineering in Indonesia. Fluorescence detection method can be applied to detect damaged fruit because the essential oil in citrus peel contains polymethoxylated flavonoid, which is known as a glowing substance under UV light. Recent studies have identified the fluorescence substances in Mandarin [3], and conducted an investigation of wave excitation for fluorescence emission from orange peel that is exposed to UV light [4]. From this research it has been known to fluorescence wavelengths of 15 varieties of oranges grown in Japan are the 300-700 nm. Furthermore, spectra data was used to support the development of machine vision to detect the damage of fruit. Hyperspectral imaging could be used to detect the bruise in apple [5], hyperspectral imaging used for safety inspection of food and agricultural products [6][7]. The study used machine vision to detect the damage of citrus fruit [8][9]. Another research studied the pattern of fluorescence associated with citrus peel defects and had separated the citrus defect in some categories [10].

Fluorescence wavelength is derived from the substance available in citrus peel. Therefore, fluorescence spectroscopy characteristic of the essential oil in citrus peel provide information for the proper lamp used to excite the citrus. The fluorescence image is collected by image acquisition system. The mechanical damage in citrus peel detects from the fluorescence imaging. This study aims to study the fluorescence characteristic of Indonesia local citrus. This study will develop the image acquisition system with proper light to get the fluorescence image. Furthermore, the fluorescence characteristic and the image acquisition system can be used to develop a non-destructive evaluation system for damaged citrus at the early stage which is unseen to human eyes.

2. Materials and methods

2.1. Sample preparation

This experiment used Keprok Pulung citrus (a citrus variety grown in Batu District, East Java Province) as a representative of Indonesian citrus (figure 1). Three groups of samples were prepared, a sound group and two defect groups of scratched fruits caused by swapping rough surface sandpaper (No. 100) and bruised fruits caused by pressing the citrus with rheometer in 19,61 N. The three groups of samples then stored in a room with air conditioning set to 25°C.

Some peel was taken from the sample for extraction. Procedure to prepare peel extract is as follows; citrus peel by weight of 3.1 g was mashed and mixed with 10 mL of pure n-hexane for spectroscopy as a solvent. The extract was stirred and kept for one night at the same temperature as the fruit stored. Citrus peel extract was later used for data retrieval using fluorescence spectrophotometer for absorption, excitation, and fluorescence wavelength observation. Citrus peel extract was placed in quartz cell, and spectra were measured. First, the absorbance spectra were recorded. From spectra curve, peak of absorbance spectra were taken as excitation wavelength for measuring itself. After determining the best excitation wavelength, the fluorescence spectra were measured. Absorbance spectra were recorded on a UV-Vis Spectrophotometer (Ocean Optics, USA). Fluorescence spectra were recorded with USB Fluorescence spectrophotometer (Ocean Optics, USA). Excitation light for spectra analysis was provided by UVGL-58 UV lamp with wavelength at 365 nm.

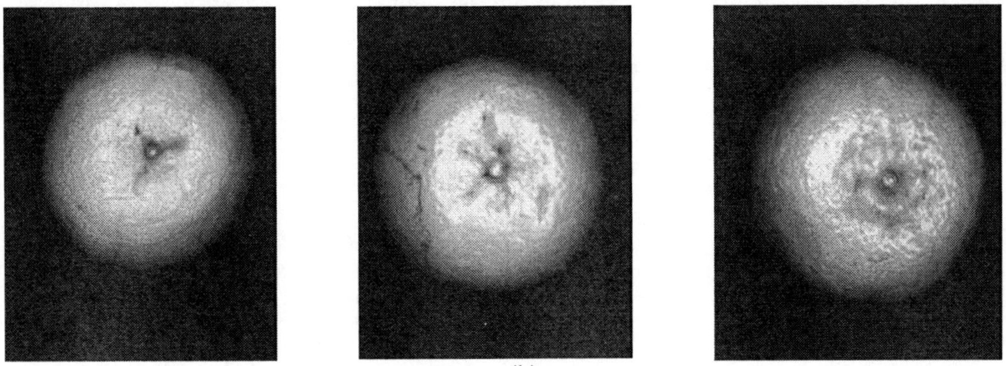

(a) (b) (c)

Figure 1. Keprok Pulung Citrus after treatment; (a) sound, (b) scratched, (c) bruised

2.2. Fluorescence image acquisition

Fluorescence images were captured by an image acquisition system that consisted of 4 UV lamps (Sankyo Denky) 15 watt for illuminating the sample with 352 nm and VGA camera (The Image Source) with resolution 744 x 480 pixel. Segmentation process was done to recognize the damage region in citrus. The image acquisition system captured image in red, green, blue (RGB) image. RGB image was less suitable for object recognition. The RGB image was transformed to hue saturation intensity (HSI) image. HSI combined information both color and grayscale of an image. Hue indicated the type of color or the color pattern where the color was found in the color spectra. It was easier to identify objects with different hue, by giving grey level in hue image. The image processing was conducted by using software Matlab R2013a. The damage region in citrus peel was determined by image processing program. The RGB image was transformed into HSI image. Then segmenting the hue image for a grey level between 150 and 160. The image processing shows in figure 2.

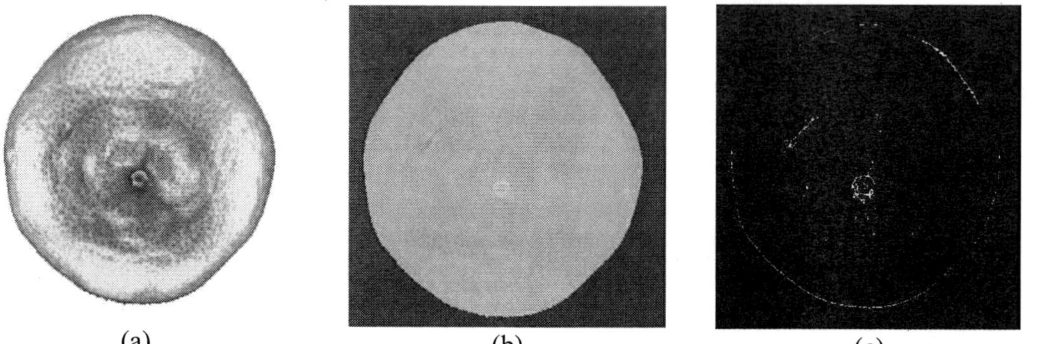

(a) (b) (c)

Figure 2. Image Processing ; (a) RGB image, (b) HSI image, (c) segmentation the damage region

3. Results and discussion

3.1. Characteristic of absorbance and fluorescence spectroscopy from Indonesia citrus.

Two dominant peaks were found in the absorbance of the citrus fruit (figure 3). The primary peak appeared at 330 nm to 350 nm. Another peak appeared at 420 nm to 450 nm. It means that there are two substances contain in citrus peel extract. The first peak increased from 310 nm to 330 nm and decreased gradually until 380 nm. The previous study identified that heptamethoxy flavone was one of the major fluorescence substances in rotten citrus fruits and the excitation and fluorescent wavelengths of one of the substances were between 360 nm to 370 nm and between 530 nm to 550 nm, respectively [3]. This information used to select the proper lamp to excite the sample. In this study, the lamp to excite the sample was 365 nm. It was reported that the excitation wavelength of 365 nm provided better fluorescence spectra of eight varieties japanese citrus [11].

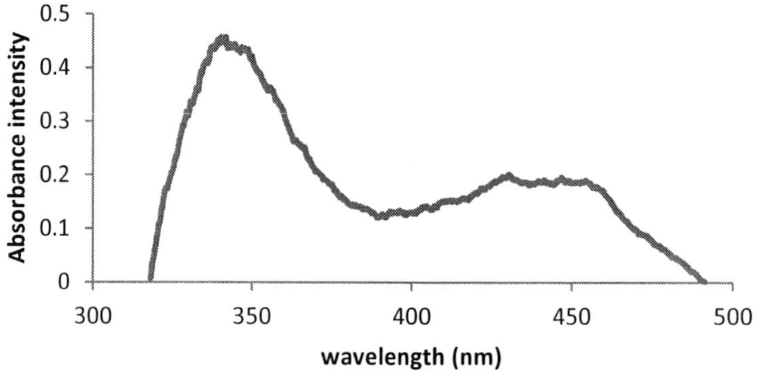

Figure 3. Absorbance spectra

Figure 4 shows the fluorescence spectra of citrus peel extract. The first peak appeared in 500 nm to 520 nm and second was in 740 nm to 750 nm. Mandarin citrus has fluorescence substance that was excited with light from 320 nm to 390 nm and emits the fluorescence of 520 to 570 nm [3]. The fluorescence substance in Mandarin citrus was polyethoxylated flavone. From this literature, the

IOP Conf. Series: Earth and Environmental Science **147** (2018) 012006 doi:10.1088/1755-1315/147/1/012006

fluorescence substance from Indonesia citrus supposed to be the same as the mandarin citrus. The first peak at 500 nm to 520 nm should be polymethoxylated flavone. Polymethoxylated flavone could dissolve in n-hexane. This substance will release in the citrus peel when citrus got damage. From this study, the fluorescence method can be applied to be damage detection method of Indonesia citrus.

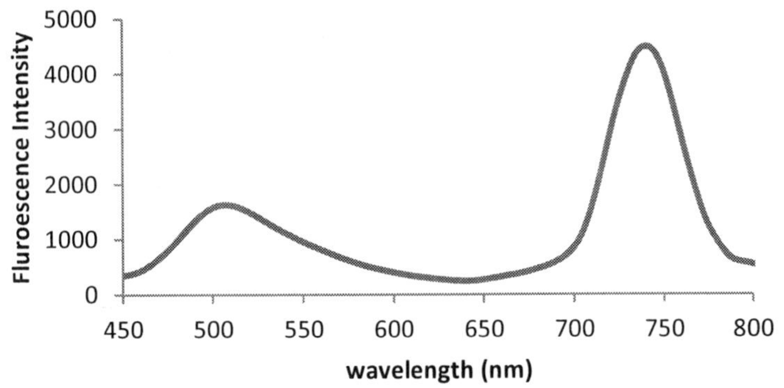

Figure 4. Fluorescence spectra

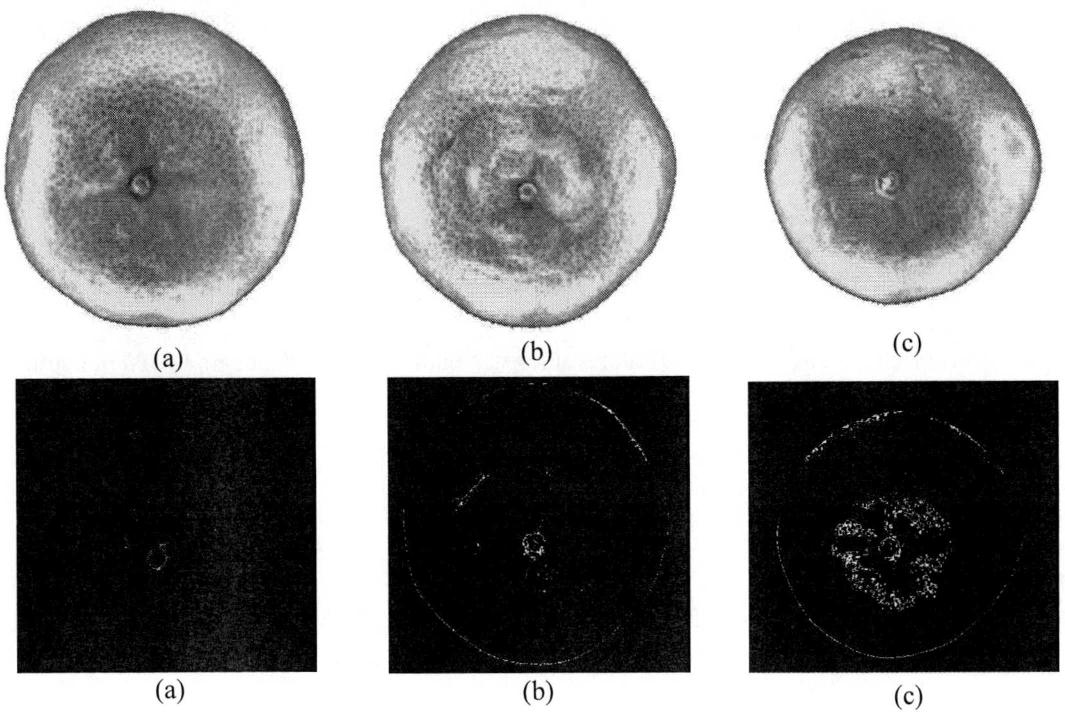

Figure 5. Fluorescence images on the top and process images on the bottom; (a) sound, (b) scratched, (c) bruised

3.2. Fluorescence image processing

Mechanical damage in citrus peel was marked by fluorescence substance that glows under UV light excitation. Based on fluorescence spectra data, the fluorescence wavelength was 500 nm to 520 nm when excited by 365 nm. The fluorescence wavelength was longer than the UV light wavelength. There was the different color region between the sound part and damaged part in citrus peel. Figure 5 shows the image processing program detected the damage region in citrus peel. White spot associated with the mechanical damage in citrus peel.

Damaged regions in citrus peel had low contrast compare to sound peel. Green color emits because of 352 nm UV light excitation with high intensity could be observed from fluorescence image with low intensity dim. Blue color emits from 352 nm UV light with high intensity effected the green color from fluorescence image with low intensity dim. Figure 6 shows the citrus that marked with green dots as damage region by the program. The region inside black line was the artificial damage region. The image processing program cannot detect the region of the damage peel accurately. Figure 6a should be clear of green dots. But image processing program detected the stem part of citrus as the damage region. The stem part was detected as the damage region for each treatment. The percentage of damage part counted higher because the stem part was recognized as the damage region. Needs to improve the image processing program to separate the stem part from the damage region.

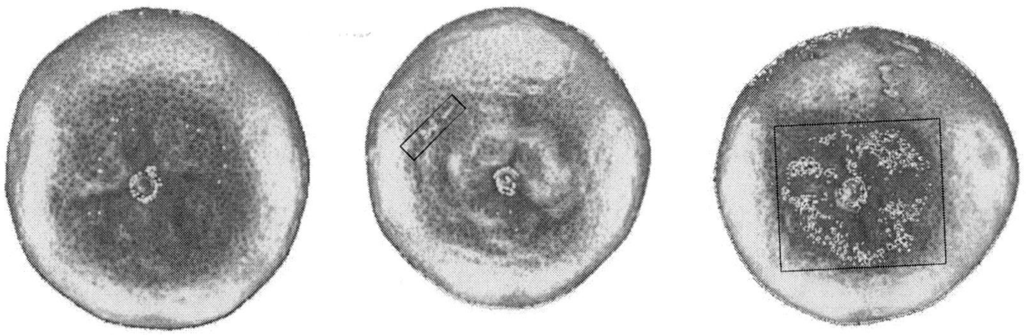

(a) (b) (c)

Figure 6. Fluorescence images of citrus marked with damage region for each treatment; (a) sound, (b) scratched, (c) bruised

4. Conclusions

Fluorescence substance in Keprok Pulung citrus has fluorescence wavelength in 500 nm until 520 nm when excited with 365 nm. Fluorescence substance produced by mechanical damage in Keprok Pulung citrus peel can be detected using fluorescence imaging based on machine vision system. The image acquisition set up can detect artificial mechanical damage such as scratched and bruised damage in Keprok Pulung citrus, but recognized the stem part as the damaged region. It requires to develop a better image acquisition system and to improve the image processing program.

References

[1] Agisimanto D C, Martasari A and Supriyanto 2007 Primer different of RAPD and ISSR in Identification of Siam Orange Genetic Relationship (*Citrus suhuniensis* L.Tan) (In Indonesian) *Jurnal Holtikultura* **17**(2): 101-110

[2] Redha A 2010 Flavonoid : Structure, Antioksidatif and Role in Biologic System (In Indonesian) *Jurnal Belian* **9**(2):196-202

[3] Kondo N M.,Kuramoto H, Shimizu Y, Ogawa M, Kurita T, Nishizu V K, Chong K and Yamamoto 2009 Identification of fluorescent substance in Mandarin orange skin for machine vision system to detect rotten citrus fruits *EAEF Research Paper* **2**(2): 54-59

[4] Momin M A, Kondo N, Kuramoto M, Ogawa Y, Yamamoto K and Shiigi T 2012 Investigation of excitation wavelength for fluorescence emission of Citrus peels based on UV-VIS spectra *EAEF Research Paper* **5**(4):126-132

[5] Lu R 2003 Detection of bruises on apples using near-infrared hyperspectral imaging *American*

Society of Agricultural Engineers ASAE **46**(2):1-8
[6] Blasco J, Aleixos N, Gomez J and Molto E 2007 Citrus sorting by identification of the most common defects using multispectral computer vision *Journal of Food Engineering* **83** (2007) 384-393
[7] Lu R and Chen Y R 2012 Hyperspectral imaging for safety inspection of food and agricultural products *Proc. SPIE* **3544**
[8] Garcia F L, Garcia G A, Blasco J, Aleixos N and Valiente J M 2010 Automatic detection of skin defects in citrus fruits using a multivariate image analysis approach *Computers and Electronics in agriculture Journal* 2010: 02001
[9] Slaughter D C, Obenland D M, Thompson J F, Arpaia M L and Margosan D 2007 Non-destructive freeze damage detection in oranges using machine vision and ultraviolet fluorescence *Journal of Postharvest Biology and Technology* **48** 341-346
[10] Momin M A, Kondo N, Ogawa Y, Ido K and Ninomiya K 2013 Patterns of fluorescence associated with citrus peel defects Engineering in Agriculture *EAEF Research Paper* **6**(2): 54-60
[11] Momin M A, Kondo N, Kuramoto M, Ogawa Y and Shiigi T 2015 Study on excitation and fluorescence spectras of Japanese citruses to construct machine vision systems for acquiring fluorescent images *Proc. Of SPIE* **8027**

IOP Conf. Series: Earth and Environmental Science **147** (2018) 012007 doi:10.1088/1755-1315/147/1/012007

Potency of Purple Sweet Potato's Anthocyanin as Biosensor for Detection of Chemicals in Food Products

A Wulandari[1], TC Sunarti[1], F Fahma[1] and E Noor[1]

[1] Study Program of Agroindustrial Technology, Department of Agroindustrial Technology, Bogor Agricultural University, Indonesia

Email: titi-cs@apps.ipb.ac.id

Abstract. Bioactive compounds such as anthocyanin are a natural ingredient that produces color with typical specificity. Anthocyanin from *Ayamurasaki* purple sweet potato (*Ipomoea batatas* L.) was extracted in ethanol and used as crude anthocyanin extracts. The color of bioactive anthocyanin can be used as a biosensor to detect chemical of food products because it provides a unique color change. However, the each bioactive has a particular sensitivity and selectivity to a specific chemical, so it is necessary to select and test the selectivity. Six chemicals, which were sodium nitrite, sodium benzoate, sodium cyclamate (food additives), formalin, borax (illegal food preservatives), and residue fertilizer (urea) were tested and observed for its color change. The results showed that the bioactive anthocyanin of purple sweet potato with the concentration of ± 42.65 ppm had better selectivity and sensitivity to sodium nitrite with a detection limit of 100 ppm, where the color change response time ranged from 15-20 minutes. The selectivity and sensitivity of this bioactive can be used as the basic information for the development of biosensor.

1. Introduction

Food safety is an absolute must for foods that are consumed by a human to survive. Mostly found in the excessive amount of chemicals in Indonesian food products were food additives such as sodium benzoate, sodium nitrite, sodium cyclamate, illegal preservatives such as borax and formalin [1], and residue fertilizer as nitrate [2]. Sodium benzoate and sodium nitrite were used for inhibiting the growth of microorganisms like bacteria, mold, and insects [3], sodium cyclamate was mostly used as an artificial sweetener with 30-50 times degree of sweetness compared to sugar [4]. The residue from urea fertilizer was found mainly in the products of fruits and vegetables as nitrate [2]. Those chemicals accumulated in the human body might result in various effects, from cancer, asthma, allergy, skin rash, vasodilator effect, and the generation of carcinogenic nitrosamine and methemoglobin [4-6]. Formalin and borax are highly toxic and carcinogenic chemicals that are not allowed by the government to be used in food products [7, 8]. Thus, it is necessary to search for the way to detect those chemicals. General methods to identify the presence of those chemicals are spectroscopy, spectrophotometry, chromatography like gas chromatography (GC), High-Performance Liquid Chromatography (HPLC), and electrochemical methods [9-12]. However, these methods had disadvantages of requiring high technology and specific experts to operate, restricted to lab-scale, and being multi-step and thus having a long time process [12-14] making it unreliable for routine and economical analysis. Owing to those facts, it is necessary to develop a practical, simple and low-cost chemical sensor for food products with appropriate sensitivity, stability, and selectivity.

The biosensor is a useful device for analyzing bio-molecular interaction in the clinical, biochemical and environmental field. It has the advantages of portable, affordable, easy to use and selective and thus it is widely utilized in food industries to detect harmful contaminants such as metals, pesticide, mycotoxin or pathogenic organisms [15]. One of the biosensors is colorimetric, which works based on color change

In this research, the biosensor for chemicals in food products was developed based on anthocyanin from purple sweet potato. This bioactive is a natural compound which generates colors with particular specificity. Anthocyanin is a bioactive in flavonoid group which has varying color from red to blue and purple and is found in tubers, fruits, flowers, and vegetables with color intensity based on its pH [16-18]. Anthocyanin is water-soluble and is safe to be consumed. Thus it is often used as natural coloring agent [19]. The pigment of anthocyanin is degradable by the exposure of light, pH, temperature, oxygen, ascorbic acid and sugar [20]. Anthocyanin is composed of anthocyanidin components consisting pelargonidin, peonidin, malvidin, petunidin, delphinidin and cyanidin [21].The components in each plant differ depending on its type [22, 23] while the main components of anthocyanin in the purple sweet potato were cyanidin and peonidin [22].

The anthocyanin was extracted from *Ayamurasaki* purple sweet potato (*Ipomoea batatas* L.) in ethanol and was used as the crude anthocyanin extract. The color of this bioactive can be utilized as the biosensor to detect chemical in food products because the color changes significantly when interacting with a particular chemical compound. However, anthocyanin has a particular sensitivity and selectivity to each chemical. Therefore, it is required to conduct tests to investigate the sensitivity and selectivity to particular chemicals. This research aims to investigate the selectivity and sensitivity of anthocyanin against various chemicals in food products (sodium benzoate, sodium nitrite, sodium cyclamate, urea fertilizer, formalin, and borax).

2. Materials and methods

2.1. Materials

The materials in this work were *Ayamurasaki* purple sweet potatoes (*Ipomoea batatas* L.) which were obtained in the local market in Cikarawang, Bogor. Acetic acid, formalin, borax (sodium tetraborate) and urea were from Merck (Darmstadt, Germany); distilled water, ethanol 96%, sodium benzoate, sodium cyclamate were obtained from "*Setia Guna*" chemical shop in Bogor. Sodium nitrite was from "*Brataco*" chemical shop in Bogor, Indonesia. The blender (Philips), centrifuge (IEC Clinical Centrifuge, USA), vacuum filter (Value, 2 Stage Vacuum Pump, VE 2100 N), filter paper, water bath, pH meter (Beckman), spectrophotometer UV-VIS (Thermo Scientific, Genesys 10S, Germany), digital camera (Canon, Power Shot A2300 HD) and glass vessels were used for analysis.

2.2. Sample preparation and anthocyanin extraction from purple sweet potato

The purple sweet potato was washed, peeled, and cut with ± 3 cm length. The sample then was blanched for 7 minutes, mashed and frozen at -27°C. The frozen sample with 350 g was crushed with blender for 3 minutes with an addition of 700 mL solvent (solvent: sample = 1:2). The solvent consisted of ethanol 96%, acetic acid, and distilled water with a ratio of 25:1:5, respectively. Next, the extract was drained with fabric. The filtrate was heated in water bath at 50 °C to vaporize ethanol to obtain a more concentrated filtrate. The pigment filtrate then was filtered using vacuum filter with filter paper. The filtrate was centrifuged with 3000 rpm for 15 minutes. After that, the supernatant was kept in a dark glass bottle in a refrigerator for further analysis.

2.3. pH measurement of chemicals

Chemicals (sodium cyclamate, sodium benzoate, sodium nitrite, urea, formalin, and borax) was prepared each as a solution with 10% concentration and was measured by pH meter.

2.4. Selectivity test of purple sweet potato's anthocyanin against chemicals

Anthocyanin has a particular sensitivity and selectivity to specific chemicals. In this research, anthocyanin was tested for sodium cyclamate, borax, sodium nitrite, urea, formalin, and borax. This test aimed to investigate the chemicals which elicited a significant color change after the reaction with anthocyanin, thus providing vital information for further development of biosensor.

Crude anthocyanin extract with a concentration of 42.65 ppm was added with each chemical solution (sodium cyclamate, sodium benzoate, sodium nitrite, urea, formalin, and borax) in a high concentration of 1% and 10%, with a ratio of 1:4 (anthocyanin: chemical). After 15 minutes, the color change was observed, and the absorbance was measured using spectrophotometer UV-VIS in the range of 400-760 nm.

2.5. Sensitivity test of purple sweet potato's anthocyanin to selected chemicals

Chemicals from previous test which reacted to anthocyanin underwent further analysis with broader range of concentration: 10, 25, 50, 100, 250, 500, 1,000, 2,500, 5,000, and 10,000 ppm, by reacting anthocyanin solution (42.65 ppm) with each selected chemical in the concentration of 10, 25, 50, 100, 250, 500, 1,000, 2,500, 5,000, and 10,000 ppm, with a ratio of 1:4 (anthocyanin: selected chemical). The color change was recorded with a digital camera for ± 30 minutes, and the absorbance was measured with the spectrophotometer.

3. Results and discussion

3.1. pH measurement of chemicals

Table 1 shows the result of pH measurement of chemicals. Sodium nitrite, sodium benzoate, urea, and borax had a pH ranging from 7 – 9, whereas sodium cyclamate tended to be neutral and formalin was acidic. This characterization was conducted to identify the sensitivity of anthocyanin to the chemicals with a particular pH value with a color change from red to purple, blue, green and yellow, which was also reported in [24]. Anthocyanin colored red at pH of 1 – 2, blue at pH of 6.5 – 9, and yellow at pH > 9 [24]. With these results, we can predict the chemical color change when reacting with anthocyanin and then be matched against the observed results of the anthocyanin's sensitivity to the chemicals.

Table 1. pH of chemicals

Chemicals	pH
Sodium cyclamate	6.532
Sodium nitrite	8.986
Sodium benzoate	7.802
Urea	8.052
Borax	9.342
Formalin	3.004

3.2. Selectivity of anthocyanin purple sweet potato's against chemicals

3.2.1. Color characteristics of purple sweet potato's anthocyanin to chemicals. Figure 1 displays the result of the test where it can be seen that anthocyanin has a good selectivity against sodium nitrite to the concentration of 1% (10,000 ppm), indicated by the significant color change from reddish to yellowish. The color change into yellowish was owing to the nucleophilic substitution of –nucleophiles (nitrate ion) at the C-4 position in the flavylium structure of anthocyanin and thus forming a complex of anthocyanin-nitrite.

Figure 1. Purple sweet potato's anthocyanin selectivity to chemicals(1%) (KL: control/anthocyanin, NS: sodium cyclamate, NB: sodium benzoate, NN: sodium nitrite, UR: urea, FN: formalin, BX: borax.

The formation of anthocyanin-nitrite induced flavylium cation (red) to transform into chalcone (yellow) [25, 26]. Meanwhile, other chemicals (sodium benzoate, cyclamate, urea, formalin, and borax) did not react well with anthocyanin as there was no significant change in color; they were relatively stable at reddish color. Based on pH measurement, sodium nitrite had a pH of 8.9. Anthocyanin of red sweet potato had a bluish color in pH of 8-9 and had a color change into yellowish when in pH of 11.7 [27]; thus we could say that the change occurred not only due to the pH change, but also a reaction occurred between nitrite ion and flavylium ion in anthocyanin. We concluded that anthocyanin of purple sweet potato had an excellent sensitivity to sodium nitrite.

Flavonoid from *Schisandra chinensis* fruit powder could enhance reddish color in sausage, indicated by the increase of a* (redness) in L*a*b system; however the addition of nitrite as preservative caused sausage color to fade and turned into yellowish, indicated by the increase of b* (yellowness) [28]. This phenomenon showed that nitrite was able to alter the structure of anthocyanin and fade its color into yellowish.

However, in the test of selectivity of anthocyanin to chemicals at 10% concentration each, there was an interaction to borax as well. Figure 2 shows the result of the anthocyanin-borax interaction, resulting in a bluish solution (from reddish).

Figure 2. Purple sweet potato's anthocyanin selectivity against chemicals (10%) (KL: control/anthocyanin, NS: sodium cyclamate, NB: sodium benzoate, NN: sodium nitrite, UR: urea, FN: formalin, BX: borax.

Anthocyanin of purple sweet potato turned into bluish after reacting with borax (10%), most likely due to flavylium ion which gave reddish color turned into quinonoid compound by the presence of borax substituting the position of C-4 [24-26]. Despite that, the chemical contaminant selected for further analysis was sodium nitrite for its better (significant) sensitivity.

3.2.2. Spectrum characterization of purple sweet potato's anthocyanin against chemicals by spectrophotometer UV-VIS in 400-760 nm. Spectrophotometry showed that anthocyanin (KL), sodium cyclamate (NS), benzoate (NB), formalin (FN) and borax (BX) had a peak at 540 nm while urea had a peak at 520 nm. Figure 3 shows the details of spectrophotometry result.

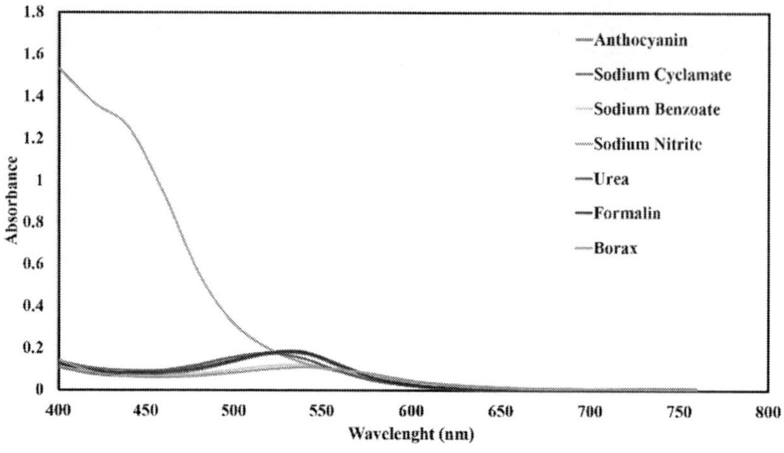

Figure 3. Spectrum of the samples (anthocyanin-1% chemicals) in λ 400-760 nm

This result describes insignificant color change of the anthocyanin-chemicals (NS, NB, FN, BX, and UR) reaction, where the colors were stable at reddish color. At the wavelength of 520 and 540 nm, the visible complementary color of red was reflected, showing there was no reaction between purple sweet potato's anthocyanin and NS, NB, FN, BX, and UR. In contrast, the interaction of purple sweet potato's anthocyanin and sodium nitrite (NN) resulted in the peak at the wavelength of 440 nm. There was a hypsochromic shift from 540 nm (anthocyanin) to 440 nm (anthocyanin-nitrite), indicating a color change after anthocyanin reacted with sodium nitrite, which was pinkish-red to yellow.

The absorbance of each solution (anthocyanin-chemicals) at 440 nm clearly explained that only anthocyanin-sodium nitrite had a peak (figure 4), resulting in a color change to yellowish after anthocyanin reacted with sodium nitrite. This is because there is a change of flavylium ions from anthocyanin (red) to form chalcone (yellow) [25, 26]. In a case of lead (Pb) detection by using curcumin, the spectrophotometry result showed the curcumin-lead had a peak at specific wavelength whereas the solution of curcumin and other metal ions had not [14].

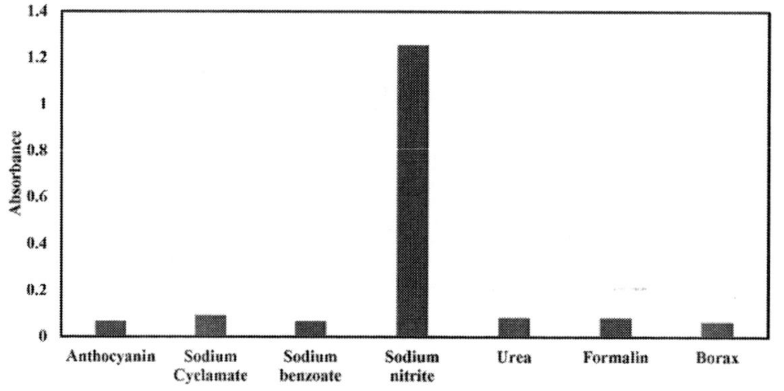

Figure 4. The spectrum of samples (anthocyanin-1% chemicals) at 440 nm

As described in figure 4, we concluded that anthocyanin of purple sweet potato had an excellent selectivity to sodium nitrite and was not influenced by the presence of other chemicals (sodium benzoate, cyclamate, urea, borax, and formalin). Thus, sodium nitrite was selected as the sample for sensitivity test with a concentration range of 1% to 10%.

3.3. *Sensitivity test of purple sweet potato's anthocyanin to sodium nitrite*

3.3.1. Color characterization of purple sweet potato's anthocyanin against sodium nitrite. The sensitivity test was undergone to anthocyanin against sodium nitrite in the concentration of 10–10,000 ppm. Figure 5 exhibits the result that there is color degradation from pinkish-red color to pale orange to bold yellow (from the addition of 0 – 10,000 ppm of sodium nitrite/NN). In the sample of anthocyanin-sodium nitrite (100 ppm), the color started to change into pale orange. A significant change of color into yellowish was revealed in the sample with 500 ppm of NN.

Figure 5. Color sensitivity of purple sweet potato's anthocyanin against sodium nitrite in various concentrations (ppm)

This result meant that the limit of detection (LOD) of purple sweet potato's anthocyanin to detect the presence of sodium nitrite was 100 ppm with the anthocyanin concentration of ± 42.65 ppm. The duration until the color change differed for each sample with different concentration. The higher the concentration of NN was, the change was faster. The response time rate of anthocyanin to change color after reacting with sodium nitrite was 15 – 20 minutes.

In a test of the colorimetric sensor using nanoparticle of Ag@Au to detect sodium nitrite, it could detect sodium nitrite with a LOD of 1 μM while it had a good selectivity against other ions as well [29]. The sensor showed a change from purplish to reddish with the presence of nitrite [29]. Compared to that case, this research had lower sensitivity, most probably due to the use of nanoparticle in that other research. The use of nanoparticle might enhance the absorbance of nitrite. Nonetheless, this research has the advantage of using an eco-friendly bioactive – which was anthocyanin, and is affordable as well, for the sensor.

3.3.2. Spectrum characterization of purple sweet potato's anthocyanin-sodium nitrite by using spectrophotometer UV-VIS in the wavelength of 400-760 nm. Spectrophotometry result exhibited that the solution of anthocyanin-nitrite in sodium nitrite (NN) concentration of 0-250 ppm had a peak at 540 nm, whereas the solution with an NN concentration of 500-10,000 ppm had a peak at 440 nm, indicating a hypsochromic shift from 540 nm (NN concentration of 0-250 ppm) to 440 nm (NN concentration of 500-10,000 ppm). This phenomenon can be observed in figure 6.

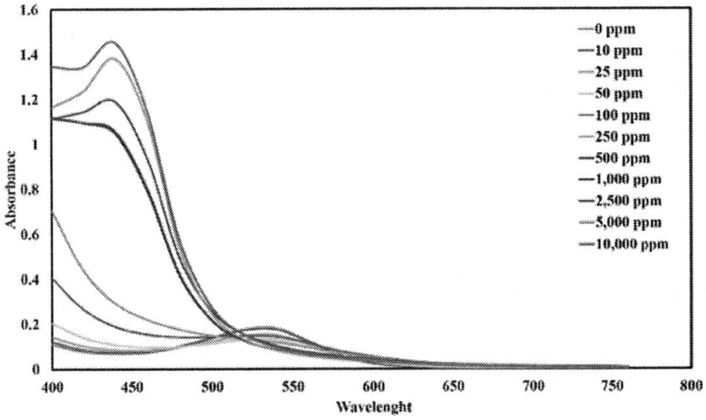

Figure 6. The spectrum of purple sweet potato's anthocyanin against sodium nitrite in various concentrations (0-10,000 ppm), at 400 - 760 nm

The shifting of wavelength indicated there was a color change from pinkish-red to yellowish. Beside hypsochromic shifting, a hyperchromic shift occurred from the solution with sodium nitrite concentration of 0-250 ppm (540 nm) to the solution with 500-10,000 ppm sodium nitrite (440 nm). Hyperchromic, the shifting from lower concentration to a higher concentration, was due to the color change of anthocyanin-nitrite from pinkish-red to bold yellow.

Figure 7 displays the peak drop of anthocyanin-nitrite along with the increase of concentration (from the concentration of 0-250 nm) at the peak of 540 nm. This drop indicated there was a paling of reddish color of anthocyanin at 540 nm. This phenomenon was also reported in the case of peak drop of Ag@AuNP at 526 nm in the solution with various nitrite concentrations of 0.1-100 μM [29] and as well in the case of the curcumin-lead peak at 456 nm in the various lead concentration of 10-100 μM [14].

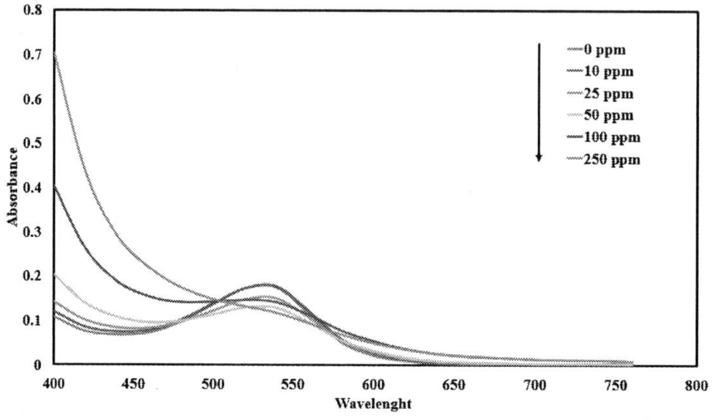

Figure 7. The spectrum of purple sweet potato's anthocyanin against sodium nitrite in various concentrations (0-250 ppm), at 400-760 nm

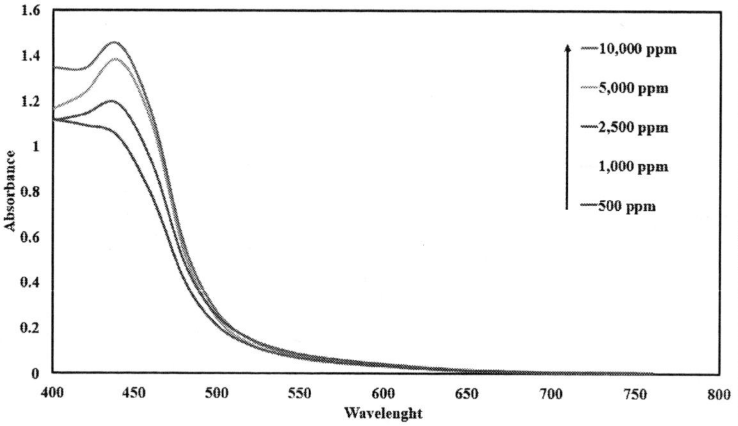

Figure 8. The spectrum of purple sweet potato's anthocyanin against sodium nitrite in various concentrations (500-10,000 ppm), at 400-760 nm

The increase of absorbance to 440 nm (λ = 400-760 nm) occurred in the anthocyanin-nitrite solution with various sodium nitrite concentrations of 500-10,000 ppm. The absorbance increased along with the increase of sodium nitrite concentration thus the yellow color intensified. In the case of the sensor to identify formaldehyde using anthocyanin reaction with hydroxylamine sulfate, there was also an increase of spectrum absorbance along with the increase of formaldehyde concentration (0.0-6.4 ppm) at 506 nm [30].

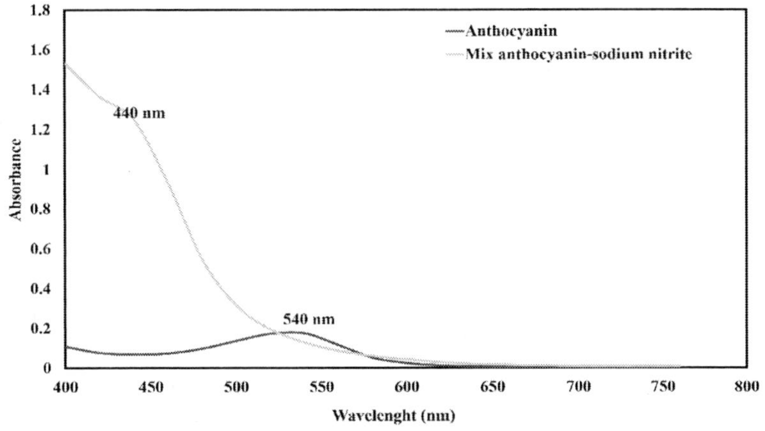

Figure 9. The spectrum of anthocyanin (control) and anthocyanin-nitrite (10,000 ppm) at 400-760 nm

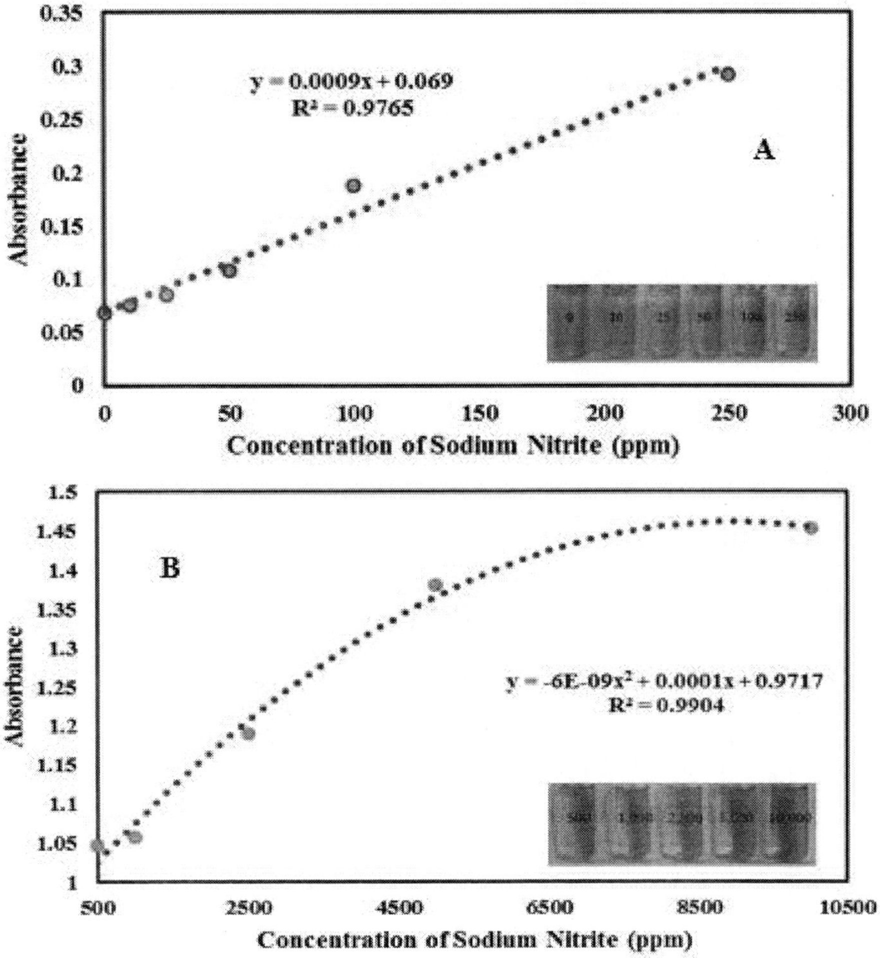

Figure 10. The calibration plot of absorbance of anthocyanin-nitrite at 440 nm and nitrite concentration range of 0-250 ppm (Fig. 10A) and 500-10000 ppm (Fig. 10B).

Figure 9 shows the comparison of the absorbance of anthocyanin and anthocyanin-nitrite. Anthocyanin had a peak at 540 nm while anthocyanin-nitrite had the peak at 440 nm, explaining that the color of the reddish color of anthocyanin changed to the yellowish color of anthocyanin-nitrite after anthocyanin reacted with sodium nitrite.

3.4. Analytical performance

Figure 10 was a plot calibration between absorbance of anthocyanin-nitrite at 440 nm and nitrite concentration range of 0-250 ppm (Fig. 10A) and 500-10000 ppm (Fig. 10B). The relationship of both produced a linear equation with the value of the equation of $y = 0.0009x + 0.069$ and the correlation coefficient of 0.9765 (Fig.10A) showing that there was an increase in absorbance as the concentration of sodium nitrite increases. Figure 10B shows the correlation curve between the absorbance of anthocyanin-nitrite and the concentration of nitrite form a polynomial line with a correlation coefficient of 0.9904, indicating that an increase in absorbance as the sodium nitrite concentration increases to the optimum point until constant.

4. Conclusions

Purple sweet potato's anthocyanin of ± 42.65 ppm had excellent sensitivity and selectivity to sodium nitrite. Anthocyanin could detect sodium nitrite with a limit of detection of 100 ppm. The color change response time of the reaction of anthocyanin with sodium nitrite ranged from 15 to 20 minutes. These properties of selectivity and sensitivity were the primary considerations of the development of biosensor. Hence, anthocyanin extracted from *Ayamurasaki* purple sweet potatoes has a great potential to be implemented into an eco-friendly, easy to use, affordable, and has a proper sensitivity and selectivity biosensor development for sodium nitrite detection.

Acknowledgement

This research especially thank the Direktorat Jenderal Penguatan Riset dan Pengembangan Kementerian Riset, Teknologi dan Pendidikan Tinggi Lembaga Penelitian dan Pengabdian Kepada Masyarakat, Indonesian Government for the funding and support to this research, with is stated at 1105/IT3.11/LT/2017.

References

[1] National Agency for Drug and Food Control 2016 Annual report of result of security and quality supervision food products [in Indonesian]. http://www.pom.go.id/new/admin/dat/20171127/laptah2016.pdf [2018-01-07]

[2] Cintya H, Silalahi J, Putra E D L and Satria D 2016 Analysis of Nitrite and Nitrate in Vegetables in Medan City *Der. Pharma. Chemica* **8** 47-52

[3] Sharma S 2015 Food Preservatives and their harmful effects *IJSRP* **5** 1-2

[4] Pandey RM and Upadhyay SK 2012 *Food Additive* ed Yehia El-Samragy (Europe : *InTech*) chapter 1 pp 1-30 http://www.intechopen.com/books/food-additive/food-additive [2017-05-05]

[5] Mitacek E J, Brunnemann K D, Suttajit M, Caplan LS, Gagna C E, Bhothisuwan K, Siriamornpun S, Hummel CF, Ohshima H, Roy R and Martin N 2008 Geographic distribution of liver and stomach cancers in Thailand in relation to estimated dietary intake of nitrate, nitrite, and nitrosodimethylamine *Nutr. Cancer* **60** 196–203

[6] Marco A, Navarro J L and Flores M 2006 The influence of nitrite and nitrate on microbial, chemical and sensory parameters of slow dry fermented sausage *Meat. Sci.* **73** 660–673

[7] Dengbin Yu , Daming Yong and Shaojun Dong 2013 Toxicity detection of sodium nitrite, borax and aluminum potassium sulfate using electrochemical method *J Environ Sci.* **25** 785-90

[8] IARC (International Agency for Research on Cancer) 2006 Monographs on the evaluation of carcinogenic risks to humans, vol. 88. Formaldehyde, 2-butoxyethanol and 1-tert-butoxy-2-propanol http://inchem.org/documents/iarc/vol88/volume88.pdf [2017-08-19]

[9] Tang J H, Wang X M, Sheng G Y and Fu J M 2005 The analytical research progress of formaldehyde and carbonyl compound in air *Analytical Chemistry* **33** 134-140 (in Chinese)
[10] Davis J, Compton R G 2000 Sonoelectrochemically enhanced nitrite detection *Analytica Chimica Acta*. **404** 241-7
[11] Davis J, Moorcroft M J, Wilkins SJ, Compton R G and Cardosi M F 2000 Electrochemical detection of nitrate and nitrite at a copper modified electrode *Analyst*. **125** 737-742
[12] Chiou J, Arthur H H L, Hang W L and Wing-tak W 2015 Rapid testing methods for food contaminants and toxicants *Journal of Integrative Agriculture* **14** 2243-64
[13] Zhao Z and Cai X 1988 Determination of trace nitrites by catalytic polarography in ferrous iron thiocyanate medium *Journal of Electroanalytical Chemistry and Interfacial Electrochemistry* **252** 361-370
[14] Raj S and Dhesingh RS 2016 Curcumin based biocompatible nanofibers for lead ion detection *Sensors and Actuators* B **226** 318-25
[15] Santoro K and Ricciardi C 2016 *Biosensors*. Politecnico di Torino– LATEMAR Unit (Torino, Italy)
[16] Torskangerpoll K and Andersen OM 2005 Colour stability of anthocyanins in aqueous solutions at various pH values *Food Chem.* **89** 427-40
[17] Burdulis D, Sarkinas A, Jasutiené I, Stackevicené E, Nikolajevas L and Janulis V 2009 Comparative study of anthocyanin composition, antimicrobial and antioxidant activity in bilberry (*Vaccinium myrtillus* L.) and blueberry (*Vaccinium corymbosum* L.) fruits. *Acta Pol.Pharm.* **66** 399-408
[18] Jensen M B, Bergamo C A, Payet R M, Liu X and Konczak I 2011 Influence of copigment derived from Tasmannia pepper leaf on Davidson's plum anthocyanins *J.Food.Sci*. **76** C447–53
[19] Chisté R C, Lopes A S and Faria L J G 2010 Thermal and light degradation kinetics of anthocyanin extracts from mangosteen peel (*Garcinia mangostana* L.) *Int.J.Food.Sci Tech*. **45** 1902-08
[20] Castañeda-Ovando A, Pacheco-Hernández M L, Páez-Hernández M E, Rodriguez J A and Galán-Vidal CA 2009 Chemical studies of anthocyanins: A review *Food Chem.* **113** 859-871
[21] Harborne JB 1967 *Comparative biochemistry of the flavonoids* (New York:Academic Press)
[22] Montilla EC, Hillebrand S, Butschbach D, Baldermann S, Watanabe N and Winterhalter P 2010 Preparative isolation of anthocyanins from Japanese purple sweet potato (*Ipomoea batatas* L.) varieties by high-speed countercurrent chromatography *J. Agric. Food Chem.* **58** 9899-9904
[23] Terahara N, Saito N, Honda T, Toki K and Osajima Y 1990 Structure of ternatin A1, the largest ternatin in the major blue anthocyanins from *Clitoria ternatea* flowers *Tetrahedron Lett.* **31** 2921-24
[24] Março P H, Poppi R J, Scarminio I S and Tauler R 2011 Investigation of the pH effect and UV radiation on kinetic degradation of anthocyanin mixtures extracted from *Hibiscus acetosella* *Food Chem.* **125** 1020-27
[25] Fichtner C Grigoriy R and Herbert M 2001 Kinetics of the Reactions of Flavylium Ions with π-Nucleophiles *Eur. J. Org. Chem.* **23** 4451-56
[26] Galán-Vidal C A, Castañeda-Ovando A, Páez-Hernández M E and Contreras-López E 2014 Determination of Nitritees in Commercial Sausages by Anthocyanins Degradation. Experimental Design and Optimization *J. Mex. Chem. Soc.* **58** 180-184
[27] Cevallos-Casals B A and Cisneros-Zevallos L 2004 Stability of anthocyanin-based aqueous extracts of Andean purple corn and red-fleshed sweet potato compared to synthetic and natural colorants *Food Chem.* **86** 69–77
[28] Jin S K and Park J H 2013 Effect of the addition of *Schisandra chinensis* powder on the physico-chemical characteristics of sausage *Asian-Australas.J.Anim.Sci.* **26** 1753-61
[29] Li T, Li Y, Zhang Y, Dong C, Shen Z and Wu A 2015 A colorimetric nitritee detection system with excellent selectivity and high sensitivity based on Ag@Au nanoparticles *Analyst* **140** 1076-81

[30] Meng Q, Han T , Wang G , Zheng N , Cao C and Xie S 2014 Preparation of a natural dye doped Ormosil coating for the detection of formaldehyde in the optical gas sensor *Sensors and Actuators* B **196** 238-244

Estimation of Sensory Analysis Cupping Test Arabica Coffee Using NIR Spectroscopy

Safrizal[1,2], Sutrisno[3], Lilik P E N[3], U Ahmad[3] and Samsudin[4]

[1] Postgraduate Student of Bogor Agricultural University (IPB), Indonesia
[2] Syiah Kuala University, Indonesia
[3] Department of Mechanical and Biosystem Engineering, Bogor Agricultural University (IPB), Indonesia
[4] Researcher of BALITTRI Bogor, Indonesia

E-mail: rizal_oct@yahoo.com

Abstract. Flavors have become the most important coffee quality parameters now day, many coffee consuming countries require certain taste scores for the coffee to be ordered, the currently used cupping method of appraisal is the method designed by The Specialty Coffee Association Of America (SCAA), from several previous studies was found that Near-Infrared Spectroscopy (NIRS) can be used to detect chemical composition of certain materials including those associated with flavor so it is possible also to be applied to coffee powder. The aim of this research is to get correlation between NIRS spectrum with cupping scoring by tester, then look at the possibility of testing coffee taste sensors using NIRS spectrum. The coffee samples were taken from various places, altitudes and postharvest handling methods, then the samples were prepared following the SCAA protocol, for sensory analysis was done in two ways, with the expert tester and with the NIRS test. The calibration between both found that Without pretreatment using PLS get RMSE cross validation 6.14, using Multiplicative Scatter Correction spectra obtained RMSE cross validation 5.43, the best RMSE cross-validation was 1.73 achieved by de-trending correction, NIRS can be used to predict the score of cupping.

1. Background

Coffee has become a popular drink and consumed all over the world, just as coffee grapes have a distinctive flavor for each producing area so as to have their own fans for a particular taste. The ordering of coffee by the current consumer country has been based on the quality of the taste so that the better it feels the greater the market demand and the expensive.

The method of determining flavor quality commonly used in many countries is the SCAA method, the grade of quality tastes differentiated by direct tasting coffee that has been brewed by certified testers, its tasting method has been standardized by SCAA to eliminate other factors that influence taste, its method called cupping. Cupping scores are strongly influenced by the geographic place where coffee is grown, fertilization during planting and protective plant density resulting in the chemical composition of the resulting coffee beans. Research on several agricultural commodities has been found that chemical composition some agricultural commodities can be detected by using NIRS, in cocoa [14], Gambier [13]. The same approach is also expected to be applied to coffee cupping that will replace tasting method using the human sensory organ, the grade of coffee cupping is classified as table 1.

IOP Conf. Series: Earth and Environmental Science **147** (2018) 012008 doi:10.1088/1755-1315/147/1/012008

Table 1. Classification of quality issued by SCAA

Total Score	Quality Classification	
90 – 100	Outstanding	Specialty
85 – 89.99	Excellent	
80 – 84.99	Very good	
< 80	Below Specialty Quality	Not Specialty

The aim of this research is to get correlation between NIRS spectra with cupping scoring by tester, then look at the possibility of testing coffee taste sensors using NIRS spectra

2. Methods

2.1. Sample preparation
Coffee beans taken directly from the Gayo highlands by random sampling of 50 samples in 50 locations with uniform height of planting and varieties so as to obtain samples with varying cupping quality, then the sample of coffee is prepared, Preparation of coffee samples begins with roast following the procedures issued by SCAA [11] include :

- The sample should be roasted within 24 hours of cupping and allowed to rest for at least 8 hours.
- The roast profile should be a light to light-medium roast, measured via the M- Basic (Gourmet) Agtron scale of approximately 58 on whole bean and 63 on the ground, +/- 1 point (55-60 on the standard scale or Agtron/SCAA Roast tile #55).
- The roast should be completed in no less than 8 minutes and no more than 12 minutes. Scorching or tipping should not be apparent.
- The sample should be immediately air-cooled (no water quenching). When they reach room temperature (app. 75° F or 20° C), completed samples should then be stored in airtight containers or non-permeable bags until cupping to minimize exposure to air and prevent contamination.
- Samples should be stored in a cool dark place, but not refrigerated or frozen.

The ready-made coffee samples were then divided into two, each 100 grams, partly used to perform cupping sensory analysis while others were used for their NIRS testing as shown in figure 1.

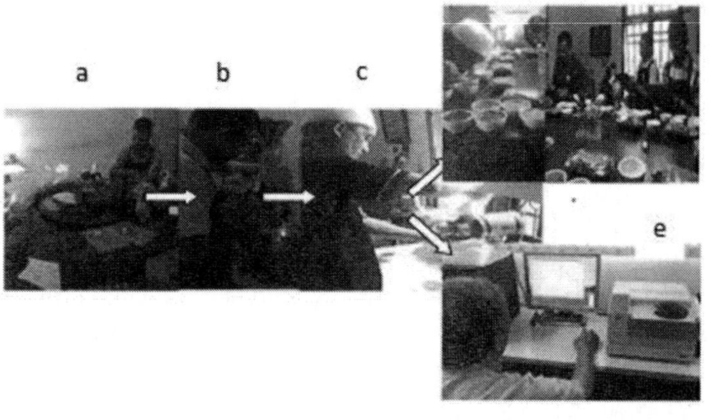

Figure 1. The flow of a coffee sample process is done by a. labeling, b roasting, c grinding, sensory analysis done by d. expert cupper and e. with NIRS

2.2. Cupping score analysis

Before the analysis of sensory roasted coffee first milled maximum of 30 minutes before brewing, the size of the coffee powder used for cupping test is to pass through a 20 mesh screen size. The number of cups used for cupping analysis of each sample is 5 cups.

The analysis of cupping score is done by involving 5 experts cupper(the name for coffee drinkers), in the test procedure issued by SCAA [11] the cupping activity or taste test of a taster has to face several cups of coffee and taste it in certain ways. One of the purposes of cupping is to know the difference between the various cups of coffee based on the quality of taste and aroma. The aroma can be detected in two ways, namely smell the aroma when coffee is still a powder and shortly after brewed with hot water. When coffee is still a powder, can also smell the level of acidity (acidity). Next, by tasting the coffee, cupper can know the flavor or taste of coffee varies. The taste of each coffee is certainly different from each other, starting only felt one character that stands out in a complex sense. All that happens because of several factors such as the genetic coffee tree, the process of planting, postharvest process, until the roaster process. Then the body, the viscosity with different weights and the last after taste, the impression of the taste of the coffee whether it will quickly disappear or not. Through cupping can be seen that each original coffee has the characteristics and uniqueness according to the origin of the coffee area.

2.3. Measurement of the NIRS spectra

The measurement of the spectra of the coffee powder is done by using a solid Buchi NIRFlex N-500 spectrophotometer by placing the sample into a petri dish, as shown in Fig. 1 next the sample is scanned three times at different positions. Spectrophotometer operation and near-infrared spectrum data collection were performed with NIR ware 1.2 software [3] at room temperature. The spectrum data was collected by reflectance measurements of the samples with a 0.4 nm data interval on a scale of 1000-2500 nm as was done by Scholz [12]. The spectra generated by each sample is averaged which is then used to construct the calibration model and stored as the reflectance value (R).

The measurement principle of the Buchi NIRFlex N-500 solid spectrophotometer as shown in figure 2 of the manual book. The process of measuring the spectrum begins by emitting a wave or NIR ray to optical 1 and then proceeding to mirror 1, light barrier and forwarded to the sample perpendicularly. When the light has reached the sample, the molecular bonds in the material will vibrate at the same frequency as the transmitted NIR wave, then wave absorption occurs. Waves absorbed by some materials are reflected back and transmitted to mirror 2, next to optics 2 to be forwarded to the detector. The detector will record the resulting spectrum which is then forwarded to the computer for chemometric acquisition and analysis.

IOP Conf. Series: Earth and Environmental Science **147** (2018) 012008 doi:10.1088/1755-1315/147/1/012008

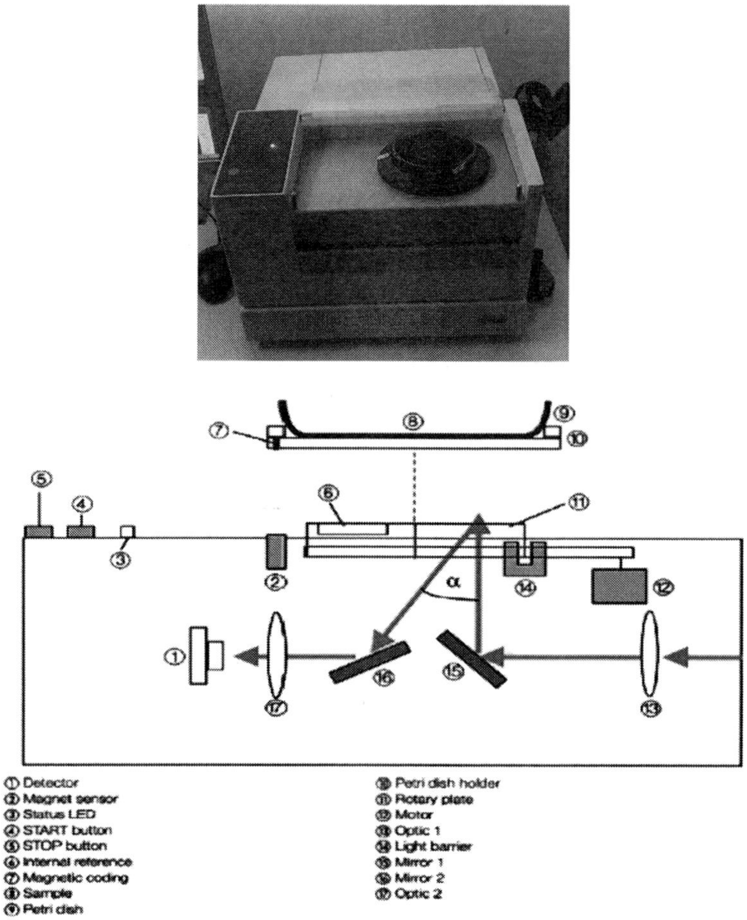

Figure 2. Measurement of near infrared spectra using spectrophotometer and functional principle of spectrophotometer (Buchi Labortechnik AG, 2007)

2.4. *NIRS data pretreatment*

The uses The Unscrambler®X version 10.3 The spectra is corrected with two data pretreatments, namely the normalization represented by Multiplicative Scatter Correction (MSC) and the transformation group represented by De-Trending (DT).

Pretreated spectra were calibrated using Partial Least Squares (PLS) which will be compared with the results of cupping test conducted by expert cupper. PLS is one of the most popular methods for multivariate calibration of NIR spectral data. PLS retrieve (NIR spectral data) X and Y (attribute quality desired) matrix consider when developing model to find latent variable in X which will best predict latent variable in Y. PLS maximize covariance between X and Y. In this case, the convergence of system for minimum residual errors is often achieved in fewer factors. PLS also leads to a reduction in the number of latent variables [6] [4].

3. Results

3.1. *Cupping score*

Cupping is done by certified expert testers so that the value of cupping is tested in accordance with consumer demand. The cupping values obtained by the Gayo arabica coffee sample test vary from 67.75 as the lowest to 85.25, the cupping score above 80 is considered as specialty coffee by SCAA.

3.2. *NIRS spectrum*

NIRS spectrum was then tested for PLS calibration to obtain correlations between NIRS spectra and cupping values, the results obtained can still be corrected with spectral correction using MSC and DT, the baseline spectrum and spectrum correction can be seen in figure 3.

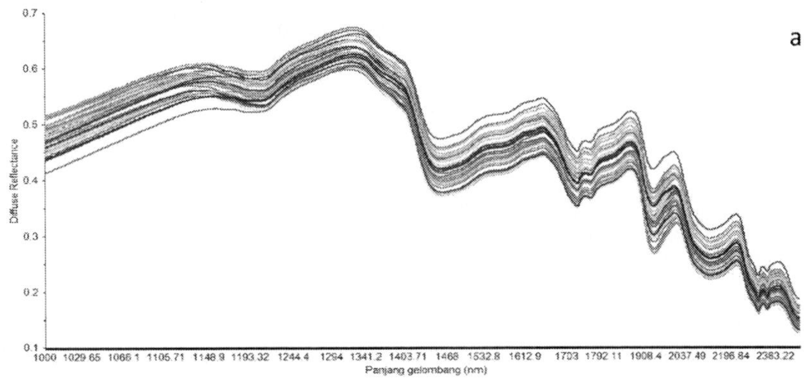

(a)

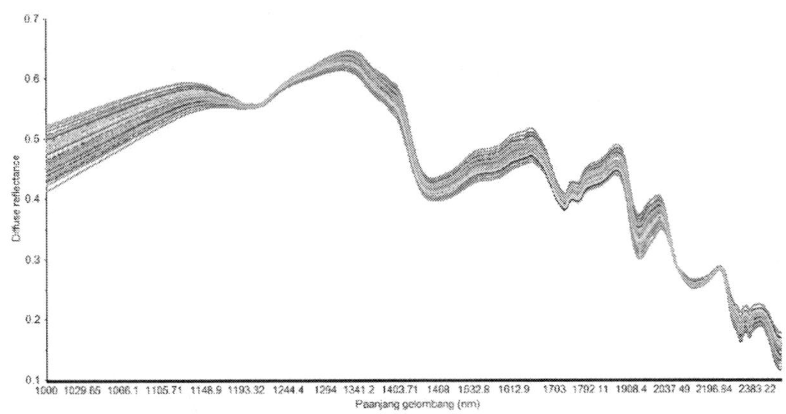

(b)

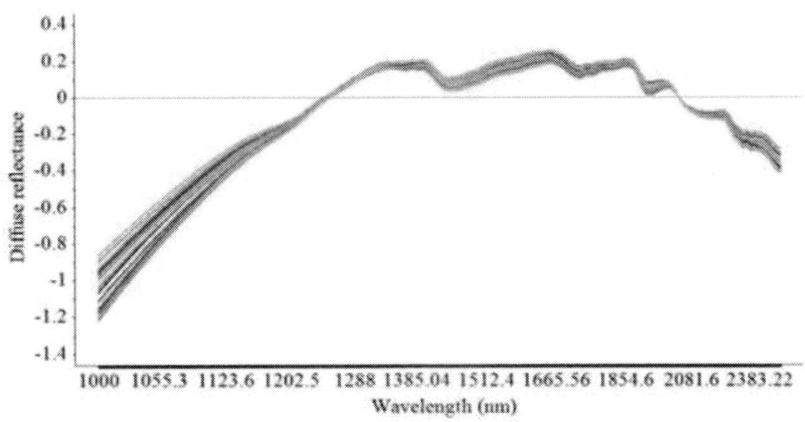

(c)

Figure 3. Absorbance spectra for coffee powder obtained from (a) raw data and, (b) MSC, (c) DT data

3.3. Predict cupping score

The basic spectra of coffee powder show the presence of several chemical components in coffee such as chlorogenic acid, caffeine as shown in figure 3. The dominant wavelength is seen at 1390 nm - 1436nm and in 1917nm - 1962nm after plotted to figure 4 shows close to the acid group component. The cupping score is influenced by the chemical composition present in the coffee powder while the chemical composition present in the coffee powder will affect the absorption of tested NIRS.

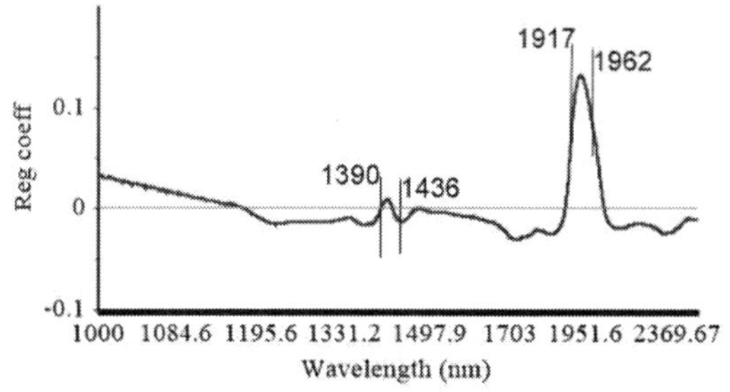

Figure 4. Loading plot analysis results on coffee

The result of prediction of PLS without pretreatment yields good r value that is 0.68. While the RMSEC and SEC errors are small, well below the SD data value. Furthermore, if viewed from the value of RPD, it is still above two. According to Nicolai [9] the value of RPD within 2-3 indicates that the model includes good model performance.

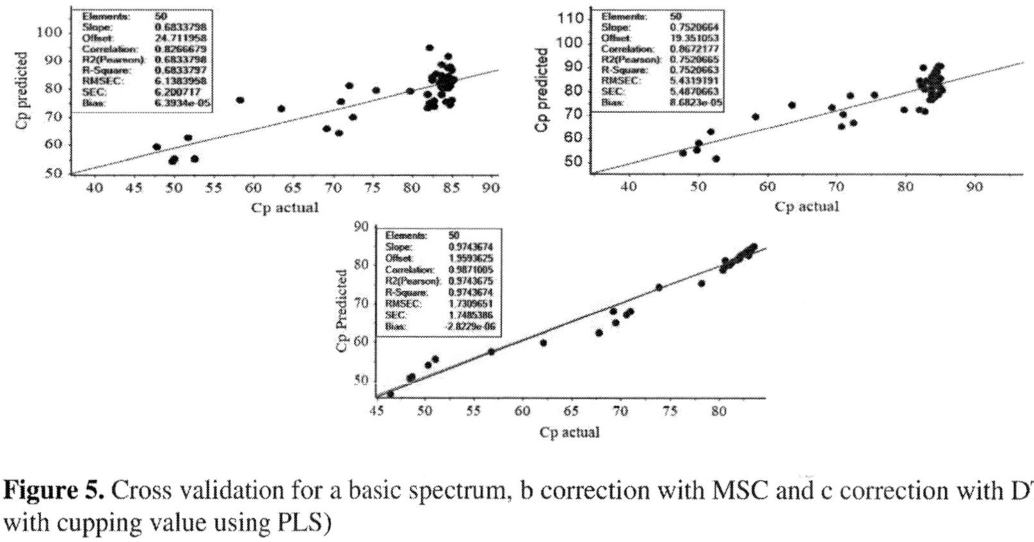

Figure 5. Cross validation for a basic spectrum, b correction with MSC and c correction with DT with cupping value using PLS)

The prediction of cupping score better results with corrected spectrum as shown in figure 5 cross-validation made on the basic spectrum get the value of R2 0.68 while after corrected its spectrum using MSC yield value R2 0.752 while using DT get better 0.97. For the RMSEC value for using the basic spectrum of 6.14, the smaller the MSC correction used to obtain the value of 5.43 and with the correction DT value of 1.73 means that the prediction of the score of gayo coffee cupping better results using NIRS spectrum that has been corrected with DT.

4. Conclusions

Cupping analysis can be done by reducing human involvement in sensory analysis by using NIRS spectra, the best result from cupping analysis using NIRS obtained on the corrected spectra using De-Trending (DT) and PLS calibration, it can be seen from the result of cross-validation resulting in very small RMSEC 1.73

References

[1] Alessandrini L, Romani S, Pinnavaia G and Rosa M D 2008 Near Infrared Spectroscopy: An Analytical Tool to Predict Coffee Roasting Degree *Analytica Chimica Acta* **625** pp 95-100

[2] Badan Standardisasi Nasional 2008 Coffee bean National Standard of Indonesia (SNI) No. 01-2907-2008 (in Indonesian) (Jakarta)

[3] Büchi Labortechnik A G 2007 *Operation manual NIRFLEX®* N-500 accessed at http//www.buchi.com. [1 November 2015]

[4] CAMO 2012 *Method Reference The Unscrambler® X 10.2*

[5] Cen H and He Y 2007 Theory and Application of Near Infrared Reflectance Spectroscopy in Determination of Food quality *Trends Food Sci Technol* **18**:72-83

[6] Felizardo P, Baptista P, Menezes J C and Correia M J N 2007 Multivariate Near Infrared Spectroscopy Models for Predicting Methanol and Water Content in Biodiesel *Analytica Chimica Acta* **595**: 107-113

[7] [ICO] International Coffee Organization 2009 *Statistics. Breakdown of Exports of Green Arabica and Green Robusta of Countries Exporting Significant Volumes of Both Types of Coffee*. www.ico.org (diakses 21 January 2015)

[8] [ITC] International Trade Centre 2011 *The Coffee Exporter's Guide*, Third Edition ITC, Palais des Nations (Geneva) (*www.intracen.org*)

[9] Nicolai B M, K Beullens, E Bobelyn, A Peirs, W Saeys, K I Theron and J Lamertyn 2007 Nondestructive Measurement of Fruit and Vege Tabel Quality by Means of NIR Spectroscopy: a Review *Postharvest Biology and Technology* **46** : 99-118

[10] Nicoli M C, Anese M, Manzocco L, Lerici C R 1997 Antioxidant properties of coffee brews in relation to the roasting degree *Lebensm. Wissens. Tech.* **30** 292–297

[11] Novaes Moreira F J, Oigman S S, De Souza F R O M, Neto Aquino F R 2014 New Approaches On The Analyses Of Thermolabile Coffee Diterpenes by Gas Chromatography and Its Relationship With Cup Quality *Talanta Journal*

[12] SCAA 2014 *SCAA Protocol Cupping Specialty Coffee*. https://www.scaa.org/?page=resources&d=cupping-protocols (accessed 15 November 2015)

[13] Scholz M B S, Pagiatto N F, Kitzberger C S G, Pereira L F P, Davrieux F, Charmetant P, Leroy T 2013 *Validation of Near-infrared Spectroscopy For the Quantification of Cafestoland Kahweol in Green Coffee*

[14] Suryani A 2014 Development of NIR Spectroscopy Method for Measurement of Catechin Content and Water Content of Gambir (*Uncaria gambir* Roxb.) in Nondestructive (in Indonesian) (Bogor: Graduate School of Bogor Agricultural University)

[15] Zulfahrizal 2014 Development of Nondestructive Measurement Methods To Determine Quality And Fermentation Of Whole Cocoa Beans Using NIR Spectroscopy (in Indonesian) (Bogor: Graduate School of Bogor Agricultural University)

Preparation of Multi-layered Microcapsules from Nanofibrils of Soy Protein Isolate using Layer-by-Layer Adsorption Method

N. Purwanti[1,2], Warji[3], S.S. Mardjan[1], S. Yuliani[4] and K. Schroën[5]

[1] Department of Mechanical and Biosystem Engineering, Faculty of Agricultural Engineering and Technology, Bogor Agricultural University, Bogor
[2] Southeast Asia Food and Agriculture Science and Technology Center (SEAFAST Center), Bogor Agricultural University, Bogor
[3] Department of Agricultural Engineering, Faculty of Agricultural, Lampung University, Lampung
[4] Indonesian Center for Agricultural Postharvest Research and Development (ICAPRD), Ministry of Agriculture of Republic of Indonesia, Bogor
[5] Food Process Engineering Group, Wageningen University and Research, Bornse Weilanden 9, 6708 WG Wageningen, The Netherlands.

E-mail: nanik_purwanti@ipb.ac.id.

Abstract. Nanofibrils were formed from soy protein isolate (SPI), which was isolated from black soybean, and commercial SPI by heating 2% w/w SPI suspensions at pH 2 and 80°C for 16 h. Nanofibrils from both types of SPI have branches and curvy structures with diameters between 10 and 20 nm. These fibrils, together with high methoxyl pectin (HMP), were used to make up to seven layered microcapsules at pH 3.5. Initially, negatively charged HMP (0.1 %w/w) was adsorbed onto oil droplets stabilized by unheated SPI (0.1 %w/w) which was positively charged. Onto this base, following layers were added by alternately adsorbing positively charged SPI nanofibrils and negatively charged HMP until the desired numbers of layers were achieved. Both types of SPI resulted in similar morphology of microcapsules. The microcapsules had a diameter between 12 and 18 μm with shell thickness of around1.7 μm and a final charge of +3.5 mV. This research showed that SPI nanofibrils can be used effectively as building blocks for constructing multilayer microcapsules that have much thicker shells for the same number of layers as examples from literature.

1. Introduction

Whey protein isolate (WPI) originates from milk and consists of mixed proteins, i.e., β-lactoglobulin, α-lactalbumin, bovine serum albumin and immunoglobulin [1]. WPI is pretty much the golden standard ingredient in food, and well-known for its functionalities such as foaming, emulsifying, thickening/gelling and texturizing properties [2]. Since proteins of animal origin are much less sustainable than their plant counterparts, many researchers are looking for plant-based proteins that have similar, or is possibly even better properties than whey protein isolate. A good candidate is soy protein isolate (SPI) that also consists of mixed proteins, i.e., soy glycinin and β-conglycinin [3].

Both WPI and SPI can form amyloid fibrils at acidic conditions under heating. For WPI it was reported that these fibrils, could be used very efficiently as building blocks for microcapsules prepared by the layer-by-layer adsorption method [4] and [5]. SPI forms branched and curvy structures of fibrils instead of the long and straight WPI fibrils. The WPI fibrils strengthened the capsules to a large extent due to their persistence length, and it is expected that the branched SPI fibrils could even be better building blocks due to their persistence length that is now related to surface coverage. The potential of SPI nanofibrils as building blocks of multilayer microcapsules using layer-by-layer adsorption method has been indicated by Warji et al. [5], who isolated SPI from yellow soybean. The protein content of both yellow and black soybeans, particularly from superior cultivars, is about 40% or higher than this, but it is certainly much higher than the protein content of imported soybeans. The current study specifically elucidates the details on preparation and physical properties of the microcapsules reinforced by SPI nanofibrils, in which the SPI is isolated from local black soybean. Nanofibrils prepared from commercial SPI were also used in this study as a reference.

2. Materials and methods

2.1. *Materials and chemicals*

Local black soybean was obtained from farmers in Yogyakarta. Commercial soy protein isolate (Newpro 90 #20140319NP) was kindly provided by PT. Pintu Mas Mulia Kimia, Indonesia. High methoxyl pectin (HMP) with a degree of methoxylation about 70-75% was obtained from Sigma Aldrich, USA (CAS. No. 9000-69-5). Other chemicals used in synthesis were n-hexane (AR, Cat. No. A-1045, Smart-Lab, Indonesia), 2-Mercaptoethanol for synthesis (Cat. No. 805740, Merck, Germany), Tris(hydroxymethyl) aminomethane (Cat. No. 108382, Merck, Germany), n-hexadecane (Cat.No. 820633, Merck, Germany) and HCl 37% (Cat.No. 1000317, Merck, Germany). Double distilled water was used to disperse the proteins.

2.2. *Protein isolation*

Protein from soybean was isolated using adapted methods from [3] and [6]. In brief, soybean meal (SBM), which was prepared from black soybean, was defatted five times with hexane at room temperature (SBM/hexane ratio of 1:5). The mixture was filtered using filter papers (KRTS60, Indonesian) and then, the defatted SBM was left to dry at room temperature. The protein was isolated by suspending defatted SBM in 30 mM Tris-HCl buffer pH 8 that contained 10 mM 2-mercaptoethanol (145 g of defatted SBM in 1.5 L of buffer). The suspension was stirred at room temperature for 1.5 h followed by centrifugation at 19,000 × *g*, 20 °C for 30 min (Beckman, Model J2-21 Ultracentrifuge, USA) and filtration (Whatman, Schleicher & Schuell, type 595, Cat. No.1001125). The supernatant was set to pH 4.8 using 6 M HCl solution to induce protein precipitation while stirring for 1 h. The resulting suspension was centrifuged at 19,000 × *g*, 20 °C for 30 min and the precipitates ware harvested. The precipitates were washed twice using double distilled water and centrifuged after each washing step. The precipitate was then re-suspended in double distilled water and set to pH 8 using 2 M Tris solution. The suspension was stirred overnight; its pH was re-adjusted again to pH 8 and then, freeze-dried at 200 millitorrs and -25 °C for at least 24 h (depends on the sample size) to obtain SPI powder.

2.3. *Preparation of protein nanofibrils*
Nanofibrils were prepared following the method of [5]. From SPI, originating from black soybean and the commercial one, suspensions were prepared by dispersing 2 %w/w of the proteins in double distilled water. The protein suspensions were stirred overnight to complete hydration and then, the pH was set to 2.0 using 6 M HCl solution. The protein suspensions at this pH were then heated in a water bath at 80 °C (internal temperature) while stirring for 16 h.

2.4. *Microcapsules preparation using layer-by-layer (LbL) adsorption method*
Multilayer microcapsules made from SPI nanofibrils and HMP were prepared using LbL adsorption method following [5] that adopted the method of [7]. HMP solution and unheated SPI suspension were prepared by dispersing 0.1% w/w HMP and SPI in 25 mM sodium chloride solution at pH 3.5. The suspension/solution was centrifuged at 925 xg in a Kokusan centrifuge (H-103N Series, Japan) for 30 min to precipitate undissolved materials and then, filtered through 0.45 µm cellulose acetate syringe filter (25CS045AS, DISMIC-25CS, ADVATEC®, Japan). Oil-in-water emulsions were prepared as the template for the capsules. Hexadecane (1% w/w) was dispersed in 0.1 %w/w unheated SPI solution pH 3.5 using a rotor-stator dispersion tool (T25 Ultra-Turrax®, Ika, Germany) equipped with S25N – 25F dispersing element (IKA, Germany) at 9500 rpm for 1 min. The emulsion droplets had positive charges because pH of the protein was below its isoelectric point. The droplets were separated from the remaining protein solution using centrifugation at 70 x and then dispersed in HMP solution that was set at pH 3.5. A layer with negative charges was formed on top of positively charged emulsion droplets because HMP carries negative charges at this pH. These bi-layered droplets were harvested by centrifugation and then dispersed in positively-charged solution of protein fibrils. The protein nanofibrils deposited onto the droplets as the third layer. Subsequently, additional layers of HMP and protein nanofibrils were deposited onto the droplets by repeating the same procedures until seven layers were formed around the oil droplet. The final harvested droplets were freeze-dried at 200 millitorrs and -25 °C for 24 hours to evacuate hexadecane; therefore, dried hollow microcapsules were obtained.

2.5. *Transmission Electron Microscopy (TEM)*
TEM method was adapted from [8] to observe the morphology of the protein nanofibrils. The fibrils were diluted to 0.05 wt % in HCl solution of pH 2.0. A drop of the diluted sample was deposited onto a 5- nm thick carbon support film on a copper grid (400 mesh). The excess was removed after 15 s using a piece of filter paper. Electron micrographs were made using TEM (G^2F20, FEI Technai™, USA) operated at 120 kV.

2.6. *Scanning Electron Microscopy (SEM)*
SEM (EVO MA 10, Zeiss, Germany) was used to observe the morphology of the microcapsules using the adapted method of [4]. The freeze-dried microcapsules were placed onto brass holders with double- sided sticky carbon tape. The microcapsules were attached to the sticky layer with pressurized air. The microcapsules were sputter coated with gold and the analysis was performed at room temperature at a working distance of 15 mm using electron high tension (EHT) detection at 17 kV.

2.7. *Zeta potential of nanofibrils*
Charges of the proteins, HMP and nanofibrils were measured in terms of ζ-potential using ZetaSizer Nano ZS (Malvern Instruments, UK). Five to ten measurements were performed for each sample.

3. Results and discussion

3.1. *SPI and nanofibrils*

Protein from local black soybean had a protein content of 82.9% on wet basis or 87.0% on dry basis with yield of isolation around 54.7% (calculated based on the protein content in the beans). The protein content of SPI obtained in this research is close to the one indicated by Codex Standard [9], i.e., 90% on dry basis. No particular structures were observed (figure 1a and c) in unheated SPI suspensions, but after heating 2% w/w SPI suspensions at the conditions described previously, protein fibrils with branched and curvy structures with diameter between 10 and 20 nm were obtained (figure 1b and 1d). There was no obvious difference between fibrils made of SPI isolated from black soybean and of commercial SPI. During heating above denaturation temperature, polypeptides broke down and the low pH stimulated formation of new bonds, which is a process known as fibril configuration [3], [10], [11] and [12].

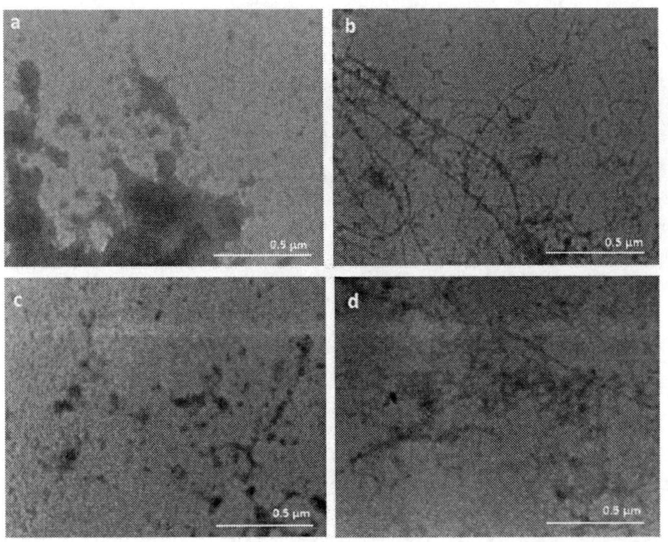

Figure 1. TEM images of (a) SPI of local black soybean, (b) SPI nanofibrils of local black soybean, (c) commercial SPI, and (d) nanofibrils of commercial SPI.

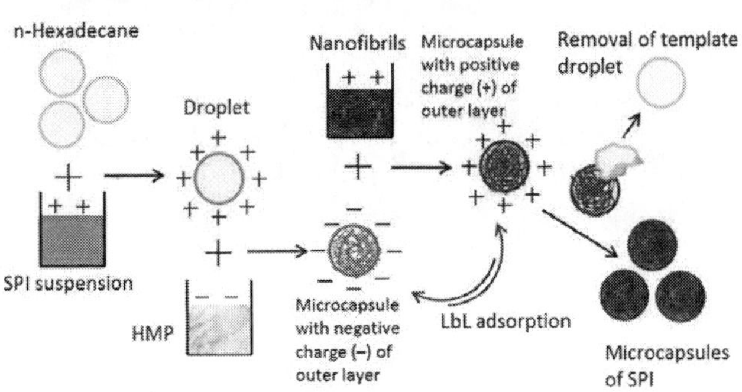

Figure 2. Schematic overview of layer-by-layer adsorption method.

IOP Conf. Series: Earth and Environmental Science **147** (2018) 012009 doi:10.1088/1755-1315/147/1/012009

3.2. *Multilayer microcapsules*

In figure 2, microcapsules assembly using the LbL adsorption method is illustrated. The images of microcapsules of local black soybean shown in figure 3a were similar to those made of commercial SPI (figure 3b), although the first ones seem to be smoother. The microcapsules were spherical, compact and hollow, and due to the harsh conditions during freeze-drying they opened up which allowed us to measure shell thickness as discussed later. It is expected that the cavities of these microcapsules are suitable to deliver e.g. functional food materials, or drugs, to a specific location along gastrointestinal tract.

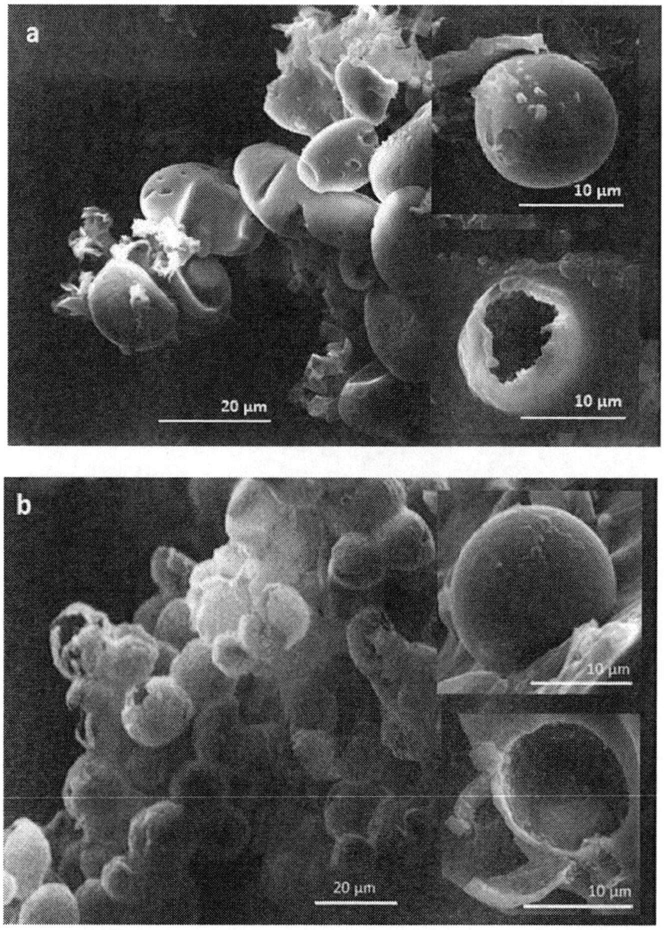

Figure 3. SEM images of microcapsules constructed by (a) SPI nanofibrils of local black soybean and (b) nanofibrils of commercial SPI.

The charge of the microcapsules after addition of each layer was as indicated in figure 4. The first layer, which was unheated SPI solution, was positively charged. The second layer, which was HMP, was negatively charged. Following odd-numbered layers corresponded to the layers constructed by SPI nanofibrils which were positively charged; even numbered layers corresponded to HMP. The final charges of the microcapsules were about +3.5 mV, both for the ones made of SPI from local black soybean and commercial SPI.

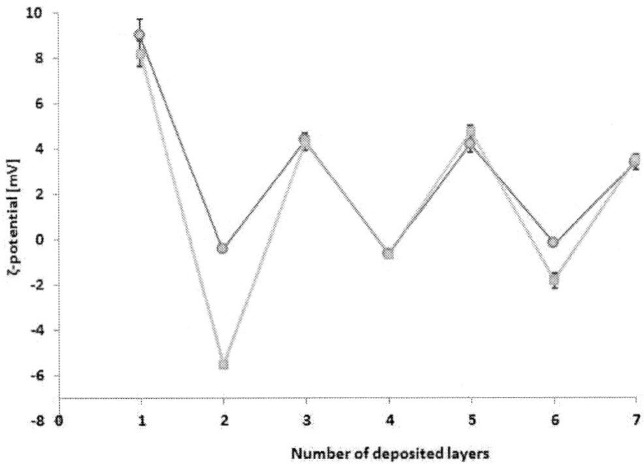

Figure 4. Zeta potential of multilayer microcapsules constructed by (—▪—) SPI nanofibrils of local black soybean and (—○—) nanofibrils of commercial SPI.

The microcapsules had average diameters of 14.33 μm and 16.74 μm for the ones made with SPI of local black soybean and commercial SPI, respectively (figure 5). These diameters were similar to the ones reported by [4] and [5] for seven layers microcapsules reinforced by WPI nanofibrils and SPI nanofibrils (from yellow soybean), respectively. The microcapsules made from nanofibrils of SPI of local black soybean and commercial SPI also had a similar thickness, i.e., 1.13 and 1.42 μm, respectively, but these values were much higher than the thickness of seven-layer microcapsules reinforced by WPI nanofibrils, which was about 250 nm, as reported by [4]. This particular difference is suggested as the result of different structures of the protein fibrils and thickness of the fibrils; with soy protein, thicker layers can be built using fewer layers, which makes the production process simpler.

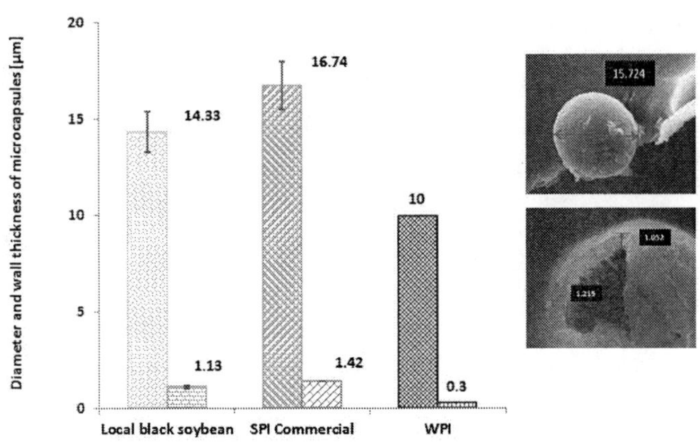

Figure 5. ( ) denotes diameter of microcapsules constructed with SPI nanofibrils of local black soybean, ( ) same but for commercial SPI, ( ) thickness of microcapsules constructed with SPI nanofibrils of local black soybean and ( ) same for nanofibrils of commercial SPI. ( ) denotes diameter and ( ) thickness of WPI microcapsules [7].

4. Conclusions

SPI, which was isolated from local black soybean, and commercial SPI can be converted into fibrillar structures that were curvy and branched by heating the suspension at pH 2 and 80 °C for 16 h. These structures successfully constructed multilayer microcapsules together with HMP using LbL adsorption method, making use of self-assembly of two polymers having opposite charges. Hollow microcapsules with similar diameter, thickness, and final charges were obtained using both types of SPI; compared to microcapsules made of WPI, the soy microcapsules reached higher thickness at a lower number of added layers, therewith making preparation simpler.

Acknowledgments

The research was financially supported by International Research Collaboration and Scientific Publication with contract number 011/SP2H/LT/DRPM/IV/2017, date April, 20th 2017. The authors also thank PT DKSH Indonesia for supporting ZetaSizer NanoZS (Malvern Instruments, UK).

References

[1] Purwanti N, Fox M, Schroen K and de Jong P 2009 The microheater: A new tool for rapid determination of food kinetics *Journal of Food Engineering* **91** 78
[2] Purwanti N, Van der Goot A J, Boom R M and Vereijken J M *2010* New directions towards structure formation and stability of protein-rich foods from globular proteins Trends in Food Science & Technology **21** 85
[3] Akkermans C, van der Goot A J, Venema P, Gruppen J M and Boom R M 2007 Micrometer-sized fibrillar protein aggregates from soy glycinin and soy protein isolate *Journal of Agricultural and Food Chemistry* **5** 9877
[4] Rossier-Miranda F J, Schroën K and Boom R 2010 Mechanical characterization and pH response of fibril-reinforced microcapsules prepared by layer-by-layer adsorption Langmuir **26** 19106
[5] Warji, Mardjan S S, Yuliani S and Purwanti N 2017 Characterization of nanofibrils from soy protein and their potential applications for food thickener and building blocks of microcapsules *International Journal of Food Properties* **20** s1121
[6] Kuipers B J H, van Koningsveld G A, Alting A C, Driehuis F, Gruppen H and Voragen A G J 2005 Enzymatic hydrolysis as a means of expanding the cold gelation conditions of soy proteins *Journal Agricultural Food Chemistry* **53** 1031
[7] Sagis L M C, de Ruiter R, Rossier-Miranda F J, de Ruiter J, Schroën K, van Aelst A C, Kieft H, Boom R and van der Linden E 2008 Polymer microcapsules with a fiber-reinforced nanocomposite shell Langmuir **24** 1608
[8] Bolder S G, Hendrickx H, Sagis L M C and van der Linden E 2006 Fibril assemblies in aqueous whey protein mixtures *Journal of Agricultural and Food Chemistry* **54** 4229
[9] Codex Standard 175-1989. 2016. Codex General Standard for Soy Protein Product; www.fao.org/input/download/standards/325/CXS_175e.pdf.
[10] Arnaudov L N, de Vries R, Ippel H and van Mierlo C P M 2003 Multiple steps during the formation of β-lactoglobulin fibril Biomacromolecules **4** 1614
[11] Mishra R, Sorgjerd K, Nystrom S, Nordigarden A, Yu, Y C and Hammarstrom P 2007 Lysozyme amyloidogenesis is accelerated by specific nicking and fragmentation but decelerated by intact protein binding and conversion J. Mol. Biol. 366 1029
[12] van der Linden E and Venema P 2007 Self-assembly and aggregation of proteins Current Opinion in Colloid & Interface Science **12** 158.

The Classification of Ground Roasted Decaffeinated Coffee Using UV-VIS Spectroscopy and SIMCA Method

M Yulia[1], A R Asnaning[1], and D Suhandy[2]

[1] Department of Agricultural Technology, Lampung State Polytechnic, Indonesia.
[2] Department of Agricultural Engineering, The University of Lampung, Indonesia

E-mail: meinilwitayulia@polinela.ac.id

Abstract. In this work, an investigation on the classification between decaffeinated and non-decaffeinated coffee samples using UV-VIS spectroscopy and SIMCA method was investigated. Total 200 samples of ground roasted coffee were used (100 samples for decaffeinated coffee and 100 samples for non-decaffeinated coffee). After extraction and dilution, the spectra of coffee samples solution were acquired using a UV-VIS spectrometer (Genesys™ 10S UV-VIS, Thermo Scientific, USA) in the range of 190-1100 nm. The multivariate analyses of the spectra were performed using principal component analysis (PCA) and soft independent modeling of class analogy (SIMCA). The SIMCA model showed that the classification between decaffeinated and non-decaffeinated coffee samples was detected with 100% sensitivity and specificity.

1. Introduction

Coffee is the most frequently consumed functional food worldwide [1]. Caffeine is one of the functional ingredients in coffee and one of the main alkaloid in coffee beans, accounting for 1 to 4% (dry basis), with large variation within cultivars and among them [2-4]. Caffeine contents are strongly related to the quality of coffee beverages because it contributes to its bitterness [5].

It has been reported that caffeine gives benefits to our health. Caffeine was reported well for increasing alertness, through stimulation of the central nervous system, rising blood circulation and respiration, is probably the main reason for coffee popularity [2, 6]. Other possible benefits of caffeine include mood enhancement, better exercise performance and reaction time, and reduction of symptoms associated with Parkinson's disease and tremors [7]. For more detailed report on positive effects of moderate consumption of caffeine has been published [8].

However, caffeine has also some negative effects such as sleeplessness and mild addiction, which has prompted development of a decaffeinated coffee industry (estimated for around 10–15% of the total amount of coffee consumed in the world). High doses of caffeine also cause anxiety, restlessness, tension, nervousness, and psychomotor agitation [9], while long-term use of this alkaloid may increase the risk of cardiovascular diseases, with individual differences in caffeine response, probably related to genetic factors [10].

The adverse side effects of caffeine [11] have increased the market for decaffeinated coffee to about 10% of coffee consumption worldwide (www.ncausa.org). Recently, the need for decaffeinated coffee authentication is increasing due to the popularity of decaffeinated coffee and its higher price than that of non-decaffeinated coffee. The purpose of this authentication is to guarantee the purity of decaffeinated coffee and to determine whether it has been mixed with non-decaffeinated coffee in order to satisfy food quality and safety requirements [12].

Several methods have been conducted to determine caffeine content in coffee beverage [13] and in roasted coffee [14-15]. Most of the previously reported studies for caffeine content determination, however, involved the use of high-cost devices such as FT-NIR (Fourier transform near infrared). On the other hand, a relatively low cost analytical based method using UV-VIS spectroscopy which utilizes wavelength region 200-700 nm has been used for coffee authentication in civet coffee [16] and in peaberry coffee [17]. The results were satisfied with coefficient of determinations were more than 0.90 and RPDs were more than 3.0 [16-17].

However, to our knowledge, no studies are available on the application of low-cost analysis based on UV-VIS spectroscopy to discriminate between decaffeinated and non-decaffeinated coffee samples. Therefore, the objective of this study was to investigate the feasibility of using UV-VIS spectroscopy and SIMCA method to classify decaffeinated and non-decaffeinated coffee samples.

2. Materials and Methods

2.1. Samples

In this research, a number of 200 samples of ground roasted decaffeinated and non-decaffeinated coffee was provided (1 gram weight for each sample). Since particle size in coffee powder has a significant influence on the spectra obtained, we sieved all coffee samples through a nest of U. S. standard sieves (mesh number of 40) on a Meinzer II sieve shaker (CSC Scientific Company, Inc. USA) for 10 minutes to obtain a particle size of 420 μm. These experiments were performed at room temperature (around 27-29°C). An aqueous extraction procedure of the coffee samples was conducted based on reference [16-17].

2.2. UV-VIS spectral acquisition

The UV-VIS spectral data of aqueous coffee samples were obtained in the range of 200-700 nm by using a UV-VIS spectrometer (Genesys™ 10S UV-Vis, Thermo Scientific, USA). This spectrometer was equipped with a quartz cell with optical path of 10 mm. The spectral acquisition was done at a spectral resolution of 1 nm at a room temperature. The original spectra (without any preprocessing) were used for further analysis. All spectral acquisitions were conducted at room temperature (around 28°C).

2.3. Chemometric analysis

The chemometric analysis was performed by using software The Unscrambler 9.8 (CAMO, AS, Norway). In order to verify the repeatability of the UV-VIS spectral data, we took 2 measurements for each sample and obtained mean values for further analysis. All data acquired were then classified into decaffeinated or non-decaffeinated class based on result from algorithm of 'soft independent modeling class analogy' (SIMCA). For SIMCA, the samples were randomly divided into two groups: calibration set (70 samples for decaffeinated and non-decaffeinated, respectively) and prediction set (30 samples for decaffeinated and non-decaffeinated, respectively).

3. Results and Discussion

3.1. Spectral data of decaffeinated and non-decaffeinated coffee in the range of 200-700 nm

Figure 1 shows spectral data of 200 samples of decaffeinated and non-decaffeinated coffee in the range of 200-700 nm. In the left part of figure 1, it is difficult to see any difference between the two types of coffee since that all spectra have similarity in shape. For this reason, the average of 100 samples of decaffeinated and 100 samples of non-decaffeinated was calculated and the result is shown in right part of figure 1. It is noted that the overall intensity of decaffeinated and non-decaffeinated coffee samples was very close each other except in the region 280-350 nm. In this part, there is a significant difference between decaffeinated and non-decaffeinated coffee samples. The average non-decaffeinated coffee sample has higher absorbance than that of the average decaffeinated coffee sample. For this reason, the region 280-350 nm may give us a good separation between decaffeinated and non-decaffeinated coffee samples. The wavelength at around 280 nm may refer to the absorbance of caffeine, while 310 and 350 nm are closely related to the absorbance of caffeic acid and chlorogenic acids (CGA), respectively [18-19].

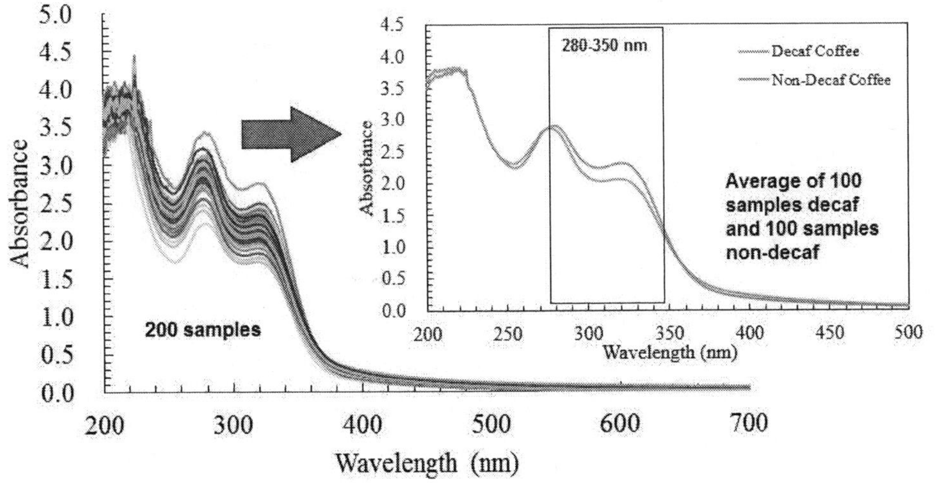

Figure 1. Spectral data of decaffeinated and non-decaffeinated coffee samples in the range of 200-700 nm.

3.2. Principal component analysis (PCA)

Figure 2 shows the result of PCA for all samples (200 samples) in the range of 280-350 nm with full-cross validation method. PCA was performed in order to investigate the presence of outliers and any trend of discrimination between decaffeinated and non-decaffeinated coffee samples. PC1 accounted for 59% of the variation explanation and described most variation in the calibration spectra. PC2 represented 25% of the variation in the spectra. It can be seen that using the two PCs, there is a clear separation between decaffeinated and non-decaffeinated coffee samples. The 95% confidence ellipse is also presented in figure 2 and it is based on Hotelling's T-square statistics. Hotelling's T-square statistic is a measure of the variation of a sample within a PLS model and a high T-square level for a certain sample indicates a high influence of the sample on the model. A number of 25 samples found outside the ellipse (high T-square level) and it was identified as outliers in this study. We decided to remove those potential outliers for further analysis. Thus, the remaining samples were 175 samples and it was used for SIMCA. Figure 3 shows the result of PCA after removing 25 samples (using 175 samples). It can be seen that using two PCs (PC1=98% and PC2=2%), the separation between decaffeinated (red color) and non-decaffeinated (blue color) coffee samples are very clear. The decaffeinated coffee was mostly located in the left part of PC1 (PC1<0) while non-decaffeinated coffee was completely positioned on the right part of PC1 (PC1>0).

Figure 4 shows the loadings plot of first two PCs. Examining the loading plot, the main wavelength region responsible for the separation of the samples were 300-350 nm with the peak located at around 325 nm.

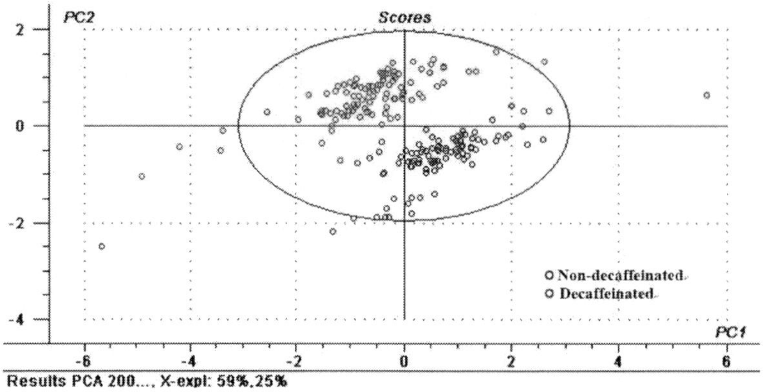

Figure 2. The result of PCA on all samples (200 samples decaffeinated and non-decaffeinated coffee) in the range wavelength of 280-350 nm.

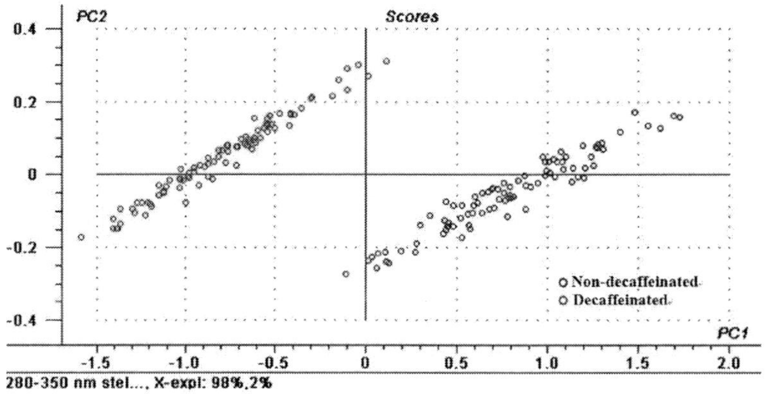

Figure 3. The result of PCA on remaining samples (175 samples decaffeinated and non-decaffeinated coffee) in the range wavelength of 280-350 nm.

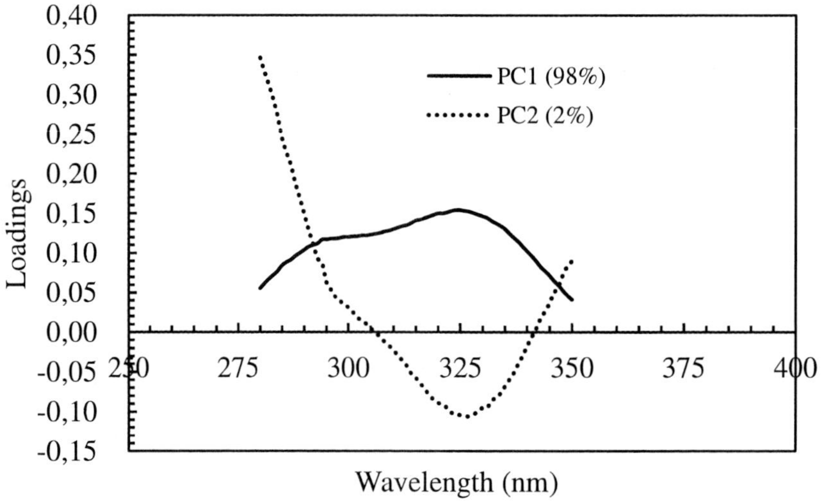

Figure 4. The loading plot of decaffeinated and non-decaffeinated coffee samples in the plane defined by the first two principal components.

3.3. Developing and evaluating SIMCA model for each class

For SIMCA, first, we developed SIMCA model for each class. For this purpose, we performed PCA for the calibration set for each class. SIMCA model decaffeinated coffee was developed using PCA with 60 samples. SIMCA model non-decaffeinated coffee was developed using PCA with 64 samples. The performance of SIMCA model was evaluated using three parameters: accuracy, sensitivity, and specificity using prediction sample set (23 samples for decaffeinated and 28 samples for non-decaffeinated). The accuracy is the proportion of both true positives (23 samples for decaffeinated) and true negatives (28 samples for non-decaffeinated) among the total number of cases examined (total 51 samples). The sensitivity is the SIMCA model ability to correctly classify the samples, relating the predicted samples to being in a class with the samples that actually are in this class. The specificity is the SIMCA model ability to correctly classify the samples, relating the predicted samples to not being in a class with the samples that actually are not in this class [20].Table 1 shows the confusion matrix of the result of classification using the developed SIMCA model. The confusion matrix presented in Table 1 enabled to detect if some samples of a given class are confused with another class. It can be seen that there were no any confusing samples. All prediction samples were correctly identified without confusion (no false positive or false negative) resulted in 100% of accuracy, sensitivity, and specificity.

Table 1. Confusion matrix of prediction samples calculated using developed SIMCA model.

Items	Decaffeinated Classified by using SIMCA model	Non-Decaffeinated Classified by using SIMCA model
Decaffeinated Class (actual)	23	0
Non-Decaffeinated Class (actual)	0	28

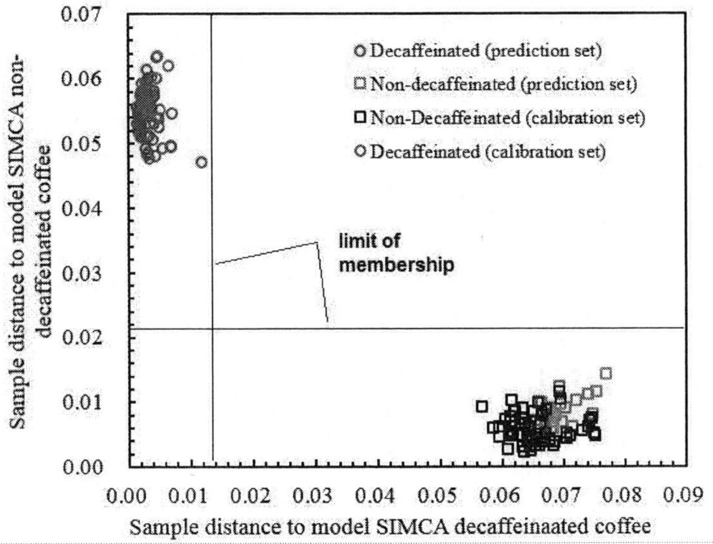

Figure 5. Coomans plot of decaffeinated and non-decaffeinated coffee samples calculated using developed SIMCA model.

The result from the SIMCA analysis was also presented in a plot called the Coomans plot for decaffeinated and non-decaffeinated models, where distances between two classes were plotted against each other in a scores plot (figure 5). For this, first, each limit of membership (for decaffeinated and non-decaffeinated coffee) was calculated. Samples in the prediction set are projected into each SIMCA model and their residual distance calculated. The residual standard deviation was used as a measure of the critical distance for classification (limit of membership). If the residual distance from the model was below their statistical limit, the sample was defined to that class and if the residual variance was higher, then the sample did not belong to that class [16]. The Coomans plots also indicated the large distances between the decaffeinated and non-decaffeinated classes with 99% confidence level.

4. Conclusion

In this research, we propose a rapid differentiation and classification method for decaffeinated and non-decaffeinated coffee using a relatively low-cost analysis method based on UV-VIS spectroscopy and SIMCA method. PCA and SIMCA are commonly used to analyze chemical data by displaying the predominant components to provide accurate evaluation. Within this research, our results confirmed high differentiation capacity of PCA and superior classification potential of SIMCA applied to discrimination between decaffeinated and non-decaffeinated coffee. The developed SIMCA model of decaffeinated and non-decaffeinated coffee could be used well to predict the class of new samples without any confusion. The obtained classification results were very good with 100% rate for accuracy, sensitivity, and specificity. This research has demonstrated that UV-VIS spectroscopy is found to be an efficient instrument for decaffeinated and non-decaffeinated classification and it can be recommended for practice.

Acknowledgement

The present research was supported by Kemenristekdikti Indonesia (Hibah Penelitian Dosen Pemula (PDP) 2017 (Nomor: 065.8/PL15.8/LT/2017)).

References

[1] Bisht S and Sisodia SS 2010 *J Nat Pharm* **1** 58–65.
[2] Belitz H D, Grosch W and Schieberle P 2009 *Food chemistry* (4th ed.). (Heidelberg: Springer).
[3] Dessalegn Y, Labuschagne M T, Osthoff G and Herselman L 2008 *J. Sci. Food Agric.* **88** 1726–30.
[4] Mazzafera P and Silvarolla M B 2010 *Seed Sci. Res.* **20** 163–7.
[5] Farah A, Monteiro M C, Calado V, Franca AS and Trugo L C 2006 *Food Chem.* **98** 373–80.
[6] Reich M S, Dietrich M S, Finlayson A J R, Fischer E F and Martin P R 2008 *Alcohol. Clin. Exp. Res.* **32** 1799–1806.
[7] Heckman M A, Weil J and González de Mejía E 2010 *J. Food Sci.* **75** R77–R87.
[8] Glade M J 2010 *Nutrition* **26** 932–8.
[9] Daly J W and Fredholm B B 1998 *Drug Alcohol Depend.* **51** 199–206.
[10] Yang A, Palmer A A and de Wit H 2010 *Psychopharmacology* **211** 245–527.
[11] James JE 1991 *Caffeine and Health* (Academic: San Diego).
[12] Pizarro C, Esteban-Diez I, Gonzalez-Saiz JM and Forina M 2007 *J Agric. Food Chem.* **55** 7477–88.
[13] Wang D, Min S G, Duan J, Xiong Y M and Li Q Q 2012 *Guang Pu Xue Yu Guang Pu Fen Xi/Spectroscopy and Spectral Analysis* **32** 982–4.
[14] Zhang X, Li W, Yin B, Chen W, Kelly DP and Wang X 2013 *Spectrochim Acta A Mol Biomol Spectrosc.* **114** 350–6.
[15] Garrigues J M, Bouhsain Z, Garrigues S and de la Guardia M 2000 *J Anal. Chem.* **366** 319–22.
[16] Suhandy D and Yulia M 2017 *Int. J. Food Prop.* **20** S331–339.
[17] Suhandy D and Yulia M 2017 *Int. J Food Sci.*
[18] Belay A, Ture K, Redi M and Asfaw A 2008 *Food Chem.* **108** 310–315.
[19] Souto U T C P, Barbosa M F, Dantas H V, Pontes A S, Lyra W S, Diniz P H G D, Araújo M C U and Silva E C 2015 LWT - *Food Sci. Technol.* **63** 1037–41.
[20] Bona E, Marquetti I, Link J V, Makimori G Y F, Arca V C, Lemes A L G, Ferreira J M G, Scholz M B S, Valderrama P and Poppi R J 2017 LWT - *Food Sci. Technol.* **76** 330–336.

The Detection and Quantification of Adulteration in Ground Roasted Asian Palm Civet Coffee Using Near-Infrared Spectroscopy in Tandem with Chemometrics

D Suhandy[1], M Yulia[2], Y Ogawa[3] and N Kondo[3]

[1] Department of Agricultural Engineering, The University of Lampung, Indonesia.
[2] Department of Agricultural Technology, Lampung State Polytechnic, Indonesia
[3] Laboratory of Bio-Sensing Engineering, Kyoto University, Japan.

E-mail: diding.sughandy@fp.unila.ac.id

Abstract. In the present research, an evaluation of using near infrared (NIR) spectroscopy in tandem with full spectrum partial least squares (FS-PLS) regression for quantification of degree of adulteration in civet coffee was conducted. A number of 126 ground roasted coffee samples with degree of adulteration 0-51% were prepared. Spectral data were acquired using a NIR spectrometer equipped with an integrating sphere for diffuse reflectance measurement in the range of 1300-2500 nm. The samples were divided into two groups calibration sample set (84 samples) and prediction sample set (42 samples). The calibration model was developed on original spectra using FS-PLS regression with full-cross validation method. The calibration model exhibited the determination coefficient R^2=0.96 for calibration and R^2=0.92 for validation. The prediction resulted in low root mean square error of prediction (RMSEP) (4.67%) and high ratio prediction to deviation (RPD) (3.75). In conclusion, the degree of adulteration in civet coffee have been quantified successfully by using NIR spectroscopy and FS-PLS regression in a non-destructive, economical, precise, and highly sensitive method, which uses very simple sample preparation.

1. Introduction

Coffee is one of the most popular beverages in the world [1]. Coffee can be regarded as a large global commodity which has one of the greatest socio-economic impacts when compared with the different activities involved in the agricultural trade [2]. Recent data shows that global coffee exports amounted to 9.13 million bags in October 2016, with 0.9% estimated increase in global coffee production in 2015/2016 as compared to 2014/2015 [3].

Up to now, coffee has approximately 70 species [4]. However, in term of popularity and profitability, the two most important species grown in the world are Arabica (*Coffea arabica*) and Robusta (*Coffea canephora*). Arabica coffee accounts for approximately 64% of the world production, with the remaining 36% derived from Robusta coffee [5-6]. Another important type of coffee is luwak coffee or Asian palm civet coffee or kopi luwak (Indonesian words for coffee and palm civet) which is well known as the world's priciest and rarest coffee [7-8].

As one of the most profitable trading products, luwak coffee has been a target for fraud trading by mixing luwak coffee with other cheaper coffee [8]. In order to protect the authenticity of luwak coffee and protect consumer from luwak coffee adulteration, it is very important to develop a robust and easy method for adulteration detection and quantification in luwak coffee [8].

Several methods both in quantitative and qualitative analysis of civet coffee has been conducted such as using gas chromatography (GC) [9] and UV-Vis spectroscopy [8] with acceptable results. However, those techniques have several disadvantages such as time-consuming (need sample preparation such extraction), produce chemical waste, relatively expensive method and difficult for on-line measurement system.

On the other hand, another spectroscopic method based on vibrational techniques, such as NIR (*near infrared*) spectroscopy is particularly prone to the development of method that is fast, reliable, simple to implement and require no pretreatment of the sample. The use of NIR spectroscopy almost always requires its combination with multivariate chemometric tools, providing direct, non-destructive, and low-cost methods, with no consumption of reagents or solvents and in accordance with the principles of green chemistry. To the best of our knowledge, there is no reported studies so far on the use of NIR spectroscopy for quantification of degree of adulteration in civet coffee. Therefore, the objective of this research is to develop and evaluate a PLS calibration model for quantification of degree of adulteration in civet coffee.

2. Materials and methods

2.1. *Samples*

A number of 2 kg ground roasted civet and non-civet coffee samples were collected directly from coffee farmers at Liwa, Lampung, Indonesia. All samples were grinded using home-coffee-grinder (Sayota). Particle sizes were not uniform in the ground coffee powder. Since particle size in coffee powder has a significant influence on the spectra obtained [10], we sieved all coffee samples through a nest of U. S. standard sieves (mesh number of 40) on a Meinzer II sieve shaker (CSC Scientific Company, Inc. USA) for 10 minutes to obtain a particle size of 420 μm. These experiments were performed at room temperature (around 27-29°C). In this study, a total of 126 civet ground roasted coffee samples with different degree of adulteration (0-51%) were prepared for samples. The 126 datasets were randomly split into two subsets: calibration (84 samples) and prediction (42 samples) sets. Calibration set was used for developing PLS calibration model while prediction set was used to evaluate the performance of the developed PLS calibration model. Table 1 shows the characteristics of calibration and prediction sample sets.

Table 1. Characteristics of calibration and prediction sample sets.

Items	Calibration sample set	Prediction sample set
Number of samples	84	42
Range	0 ~ 51.04	0 ~ 51.04
Mean	24.05	24.05
Standard Deviation	17.22	17.32
Unit	%	%

2.2. *Spectral acquisition*

The NIR spectra were acquired using a V-670 UV-Visible-NIR Spectrophotometer (JASCO Co., Japan). This spectrometer was equipped with an integrating sphere beside detector (ISN-723, JASCO Co.) which is connected to the spectrometer by an optical fiber. The V-670 double-beam spectrometer utilizes a unique, single monochromator design covering a wavelength range from 190 to 2500 nm. This spectrometer has two detectors: A PMT detector is provided for the UV/VIS region and a Peltier-cooled PbS detector is employed for the NIR region. For light source, this spectrometer utilizes two types of lamps: Halogen lamp and Deuterium lamp.

The NIR instrument was controlled by a compatible Windows XP system, and Spectra Manager (V-670, JASCO Co.) was used to acquire spectra data. A 4 cm by 4 cm square sample holder was used; with care taken to ensure every sample (weight of approx. 0.2 g) was placed into the sample holder and the sample surface kept flat. Each spectrum was measured from 1300 to 2500 nm, with a bandwidth of 20.0 nm, and a scan speed of 1000 nm/min. All experiments were conducted at room temperature (around 20°C).

2.3. *Multivariate analysis*

Principal component analysis (PCA) was done before developing the partial least squares (PLS) calibration model in order to analyze the relevant and interpretable structure in the data [11]. The PLS regression was used to develop the calibration model for original spectra. The quality of the calibration model was evaluated by analyzing the values of the coefficient of determination between predicted and actual degree of adulteration for the calibration set (R^2_{cal}), root mean square error of calibration (RMSEC), and root mean square error of cross-validation (RMSECV). As mentioned by reference [12], a calibration model with R^2 value greater than 0.91 is considered to be an excellent calibration, while an R^2 value between 0.82 and 0.90 results in good prediction. A small difference between RMSEC and RMSECV value was also important to avoid 'over-fitting' in the calibration model [13].

Prediction set was provided and it was used to evaluate the prediction ability of the model through the value of coefficient of determination between predicted and actual degree of adulteration for the prediction set (R^2_{pred}), the root mean square error of prediction (RMSEP), bias between actual and predicted degree of adulteration and ratio prediction to deviation (RPD) value ($RPD = SD_{validation\ set}/SEP$) [14]. The calibration model should have as high as possible RPD value. The RPD value is desired to be larger than 3 for an acceptable calibration [12]. An RPD value less than 1.5 indicates incorrect predictions and the calibration model cannot be used for further prediction [12, 15]. RMSEP is generally not higher than 1.2 times the RMSEC [16]. PCA and PLS regression were generated using the Unscrambler 9.8 program (CAMO, Oslo, Norway).

3. Results and discussion

3.1. *Spectral data of civet coffee with several degree of adulteration*

Figure 1 shows spectral data of civet coffee samples with three level of degree of adulteration (low, middle and high) in the range of 1300-2500 nm. There are several peaks at 1460 nm, 1740 nm and 1930 nm. The wavelengths in the region 1400-1500 nm may refer to the first overtone of the OH functional group [17-18], of chlorogenic acids, water, alcohols, and especially carbohydrates present in the civet coffee samples [5, 19-20]. It is noted that increasing of degree of adulteration tends to decrease the intensity of absorbance.

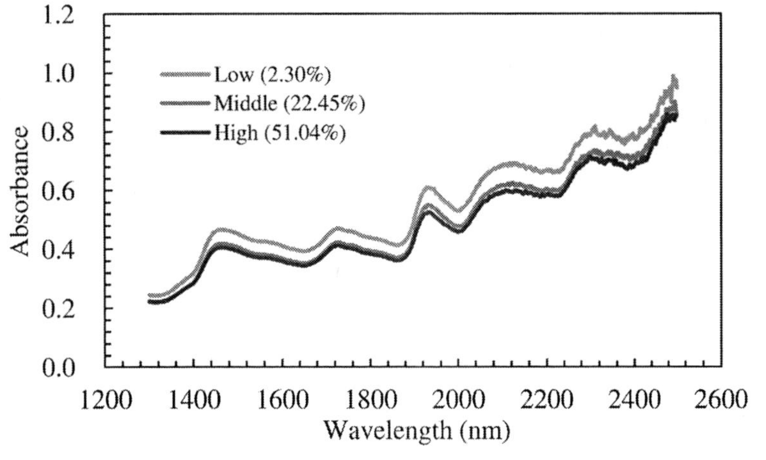

Figure 1. Spectral data of civet coffee samples with different degree of adulteration (low, middle and high) in the range of 1300-2500 nm.

3.2. *PCA*

Prior to developing PLS calibration model, an exploratory analysis using PCA (*principal component analysis*) on all spectral data was performed in order to investigate the presence of outliers and any trend of discrimination among civet coffee samples from different degree of adulteration. The scores plot of the two first principal components (PC1 vs PC2) is presented in figure 2. PC1 accounted for 93% of the variation explanation and described most variation in the calibration spectra. PC2 represented 4% of the variation in the spectra. The first two PCs accumulatively accounted for 97% of the variation among samples. Figure 2 shows that the samples can be clustered into three level of degree of adulteration: low (0-15.60%), middle (20.12-35.54%) and high (40.52-51.04%). It is interesting that in general there is a tendency of an increase in adulterant content in the negative direction of PC1, i.e., PC1 < 0. The civet coffee with 100% purity (0% of degree of adulteration) is found in the positive quadrant of PC1, i.e., PC1 > 0, (figure 2). Therefore, as the adulterant content in the blend increases, the samples are positioned in the direction of the negative quadrant of PC1, i.e., PC1 < 0.

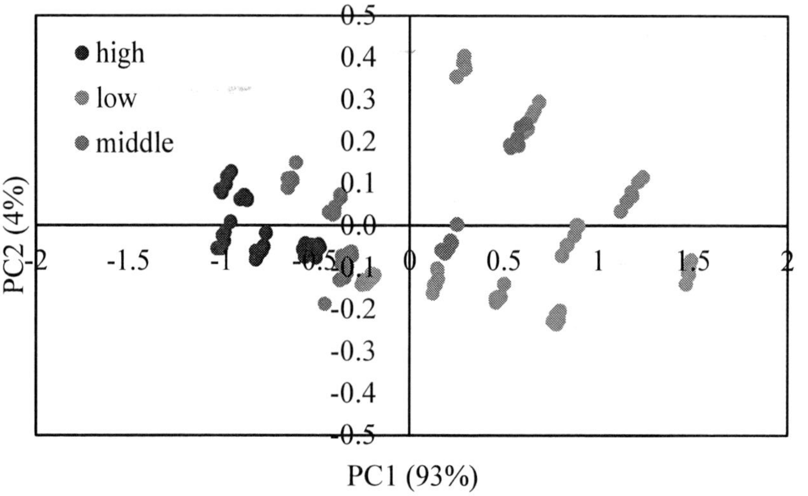

Figure 2. The result of PCA on all samples with different degree of adulteration (low, middle and high).

3.3. *Developing PLS calibration model*

The PLS calibration model using original spectra was developed with PLS1 regression and full-cross validation method. The result is depicted in figure 3. The quality of model is very excellent with very high coefficient of determination (R^2_{cal}=0.96), low RMSEC (3.62%) and low RMSECV (4.83%). The difference between RMSEC and RMSECV is also small and relatively low number of latent variables (LVs) was included in the model. The quality of the developed calibration model was comparable to that of several previously reported studies on the determination of adulteration in civet coffee using other methods [8]. The quantification of adulteration in civet coffee using UV-Vis spectroscopy has been reported with the best calibration model, was observed using Savitzky-Golay smoothing spectra with R^2_{cal}=0.98 [8]. The high R^2 and low values of RMSEC and RMSECV indicate that the degree of adulteration in civet coffee blend could be reliably predicted by the developed PLS calibration model. Although the reference [8] got higher quality of calibration model with higher R^2, the research was conducted using ground roasted civet coffee and involved an extraction process. But, in this present study, a direct measurement of ground roasted civet coffee (no extraction) was conducted, hence, faster measurement was expected.

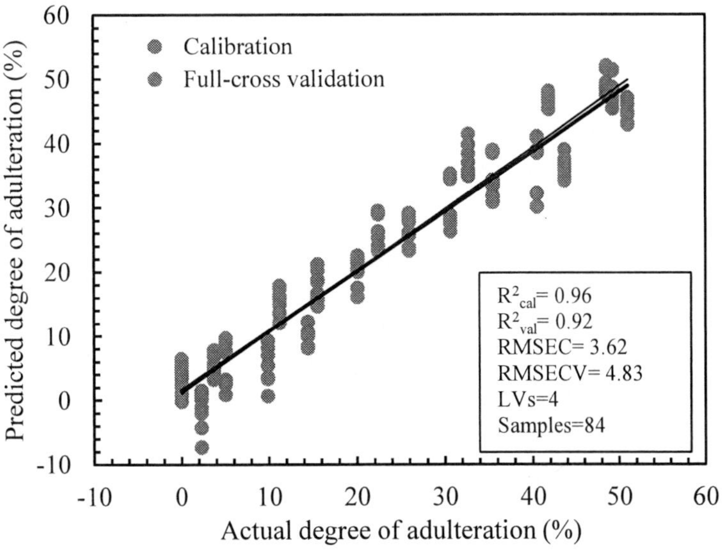

Figure 3. Scatter plot between actual and predicted degree of adulteration in the calibration model.

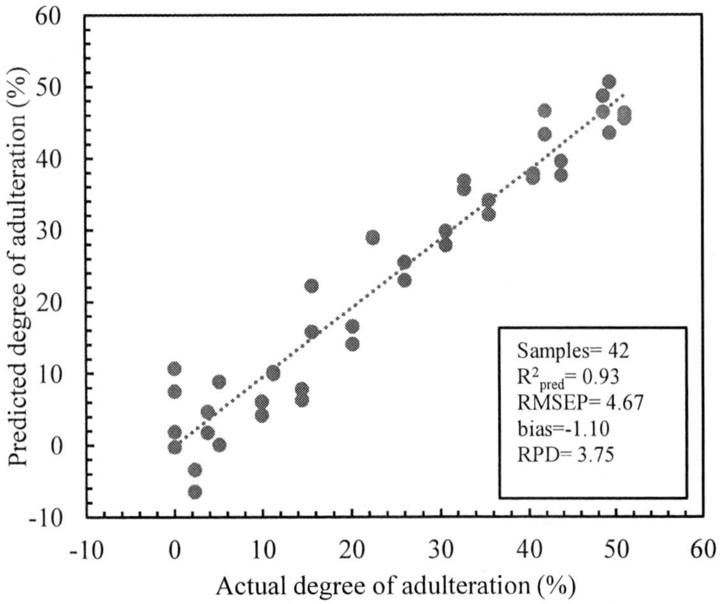

Figure 4. Scatter plot between actual and predicted degree of adulteration in the prediction step.

3.4. *Prediction unknown new samples*

In order to evaluate the performance of the developed PLS calibration model, a prediction of degree of adulteration in unknown new samples of prediction set was performed. For this purpose, a number of 42 samples was provided. The result for prediction was shown in figure 4. The quality of prediction is also very good with high coefficient of determination (R^2_{pred}=0.93), low RMSEP value (4.67%) and low bias (-1.10%). It is noted that the RMSEP value is not exceed 1.2 times than the RMSEC value. The prediction resulted in high RPD value (RPD=3.75). The model prediction results with $R^2 > 0.80$ and RPD in the 2.5–3.0 range was considered reliable and the model could be used as a good prediction model for determination of degree of adulteration in civet coffee. Based on figure 4 and using a method proposed by reference [21], the limit of quantification of this method was 6.98%.

4. Conclusion

In this research, NIR spectroscopy, combined with full spectrum PLS regression model, is confirmed to be an interesting tool for non-destructive quantification of degree of adulteration in civet coffee. The PLS calibration model gives acceptable result with high coefficient of determination both in calibration and validation step. The prediction of unknown new samples using the proposed PLS calibration model resulted in low RMSEP and high RPD value. This result can be potentially applied for non-destructive authentication system for civet coffee.

Acknowledgements

The authors would like to thank Laboratory of Bio-Sensing Engineering of Kyoto University for giving the author permission to use NIR spectrometer and its supporting devices (through JASSO Follow up Research Program 2016). Some part of this research was also funded by Kemenristekdikti Indonesia (Hibah Strategis Nasional/STRANAS 2016–2017 No. 585/UN26.21/KU/2017).

References

[1] Assis C, Oliveira L S and Sena M M 2017 Food Anal. Methods. In press
[2] Grinshpun H 2014 *J Consum Cult* **14**:343–64
[3] ICO 2016 http://www.ico.org/profiles_e.asp. Accessed 10 October 2017
[4] Ludwig I, Clifford M, Ashihara H and Crozier A 2014 *Food Funct* **29**: 1695–1717.
[5] Barbin D F, Felicio A L S M, Sun D W, Nixdorf S L and Hirooka E Y 2014 *Food Res Int* **61**: 23–32
[6] Damatta FM, Ronchi CP,Maestri M and Barros RS 2007 *Braz J Plant Physiol* **19**: 485–510
[7] Marcone MF 2004 *Food Res. Int.* **37**: 901–12
[8] Suhand D and Yulia M 2017 *Int. J Food Sci*
[9] Jumhawan U, Putri SP, Yusianto BT and Fukusaki E 2016 *J Biosci Bioeng* **122**: 79–84
[10] Suhandy D and Yulia M 2017 *Proc. of the 1st Int. Conf. on the Role of Agricultural Engineering for Sustainable Agriculture Production* (Bogor: Institut Pertanian Bogor) pp 6–10
[11] Naes T, Isaksson T, Fearn T and Davies T 2002 *A User-Friendly Guide to Multivariate Calibration and Classification* (Chichester: NIR Publications)
[12] Saeys W, Mouazen A M and Ramon H 2005 *Biosys. Eng.* **91**: 393–402
[13] Gomez A H, He Y and Pereira AG 2006 *J Food Eng.* **77**: 313–9
[14] Brereton RG 2000 *Analyst* **125**: 2125–54
[15] Sunoj S, Igathinathane C and Visvanathan R 2016 *Comput. Electron. Agric.* **124**: 234–42.
[16] Zhan H, Fang J, Wu HW, Yang HJ, Li H, Wang Z J, Yang B, Tang LY and Fu M H 2017 *Chinese Herbal Medicines* **9**: 250–7
[17] J Workman and Weyer L 2008 Practical Guide to Interpretive Near-Infrared Spectroscopy, (CRC Press: Taylor & Francis Group)
[18] Kasemsumran S, Thanapase W and Kiatsoonthon A 2007 *Anal. Sci.* **23**: 907–10
[19] Alessandrini L, Romani S, Pinnavaia G and Rosa MD 2008 *Anal. Chim. Acta* **625**: 95–102
[20] Suhandy D and Yulia M 2017 *Int. J. Food Prop.* **20**: S331–9
[21] Shrivastava A and Gupta VB 2011 *Chron. Young Sci.* **2**: 21–5

Optimization of Steamed Meals Based on Composite Flour (Taro, Banana, Green Bean) and Its Predicted Shelf Life

Sunarmani[1], Setyadjit[1] and Ermi S[1]

[1] Indonesian Center for Agriculture Postharvest Research and Development, Indonesian Agency for Agriculture Research and Development, Ministry of Agriculture

E-mail: pascapanen@yahoo.com

Abstract. Ongol-ongol is for food diversification by mixing composite flour of taro, banana and mung bean, then was steamed by hot air. The purpose of this study was to find out the optimum way to produce 'ongol-ongol' from composite flour and to know the storage life by prediction method. The research consisted of two stages, namely the determination of the optimum formula of 'ongol-ongol' with Design Expert DX 8.1.6 software and the estimation of product shelf life of the optimum formula by ASLT (Accelerated Shelf Life Test) method. The optimum formula of the steamed meal was produced from composite flour and arenga flour with ratio of 50: 50 and flour to water ratio of 1: 1. The proximate content of steamed meal of optimum formula is 36.53% moisture content, ash content of 1,36%, fat content of 14.48%, protein level of 28.5%, and carbohydrate of 44.77% (w/w). Energy Value obtained from 100 g of 'ongol-ongol' was 320.8 Kcal. Recommended for steamed meal storage life is 12.54 days at ambient temperature.

1. Introduction

Food demand in Indonesia is increasing as the population grows. Fulfillment of food demand can be done by optimizing local food sources. One of the local food resources that can be an alternative for food diversification is taro (*Colocasia esculenta*).

Taro, which is a source of carbohydrates, has potential as a substitution of rice, as a raw material for crisp industry and so forth[1]. Taro contains 25.0% carbohydrate, 1.4% protein, 0.4% fat some minerals and vitamins [2]. Taro tubers are low fat, gluten-free and easy to digest [3]. Taro as a carbohydrate source is usually used as a breakfast meal. To increase the nutritional content of breakfast meal, banana and mung beans can be added.

Taro contains low protein, lysine and amino acids, this can be overcome by the addition of green beans. Mung beans have a protein content of 22.9%, and lysine of 1650 mg / 100 g. Banana kepok is rich in several vitamins including carotene (provitamin A) of 192 RE and vitamin B1 0.1 mg / 100 g [4].Both mung beans and banana 'kepok' are available within a year.

One way to integrate these three raw materials is by making composite flour. Research on flour-based composite taro, banana kepok, and mung beans has been done in the manufacture of breakfast cereal ready to eat, such as composite cookies, brownies, pao [5], and flake [6]. Ongol-ongol in which contain composite flour (taro, banana, mung bean) is an Indonesia traditional food. In this study the product to be made is steamed cake, composite flour ratio used refers to previous research. To form the chewy texture of steamed meal based composite flour, it can be added arenga starch [7]. Definition of composite flour is a mixture of different types of flour to make bread with yeast, bread products without yeast, porridge, pasta, and snacks. Another definition of composite flour is mixture of non-wheat flour [8]

The study aimed to (1) find out the optimum formula of composite flour-based steamed cake, (2) to know the carbohydrate, protein, fat, water, ash, dietary and energy content of the optimum composite starch-base steamed cake and (3) the shelf life of composite flour-based steamed cake products.

2. Materials and methods

The main raw materials used in the study were taro flour, banana flour, green bean flour, arenga flour, and water. Additional ingredients used were sugar (sucrose), salt, coconut milk, milk powder, vanilla extract, sodium benzoate, potassium sorbate. The research was conducted at Indonesian Centre for Agriculture Postharvest Research Development, Bogor.

2.1. Methods

2.1.1. First stage. The first stage begins with the preparation of composite flour that was taro flour, banana, and green bean flour. Furthermore, taro, banana, and green bean flour were mixed with a ratio of 50: 30: 20, using a mixer to produce composite flour.

The formulation of Ongol-ongol consists of composite flour and arenga flour,. The addition of sugar, milk powder, salt, coconut milk, and vanilla, was made the same in each formula, the percentage of addition based on total weight of flour. The percentage of ingredients added was 46% sugar, 2% milk powder, 1.46% salt, 51% coconut milk, and vanilla extract 0.8%. The formulation of steamed cake based on composite flour is presented in table 1. The composite flour, arenga flour, and other ingredients was mixed using mixer for 5 min. The dough was then placed in the tray and was steamed at 90^0C for 60 min. The cake was put in the vacuum-polypropylene bag and was analyzed for sensory evaluation and physicochemical properties. The procedure of making steamed cake is shown in figure 1.

Table 1. Steamed meals formulation based on taro composite flour

Formulation	Composite flour: Arenga starch	Flour: Water
F1	90:10	1:1
F2	90:10	1:3
F3	90:10	1:5
F4	70:30	1:1
F5	70:30	1:3
F6	70:30	1:5
F7	50:50	1:1
F8	50:50	1:3
F9	50:50	1:5
F10	30:70	1:1
F11	30:70	1:3
F12	30:70	1:5
F13	10:90	1:1
F14	10:90	1:3
F15	10:90	1:5

*Fifteen samples of Ongol-*ongol *were tested by 15 panelists [9], for properties of color, flavor, taste, texture* and *appearances. All responses are used to determine the optimum formula* of *a composite flour-based steamed meal.*

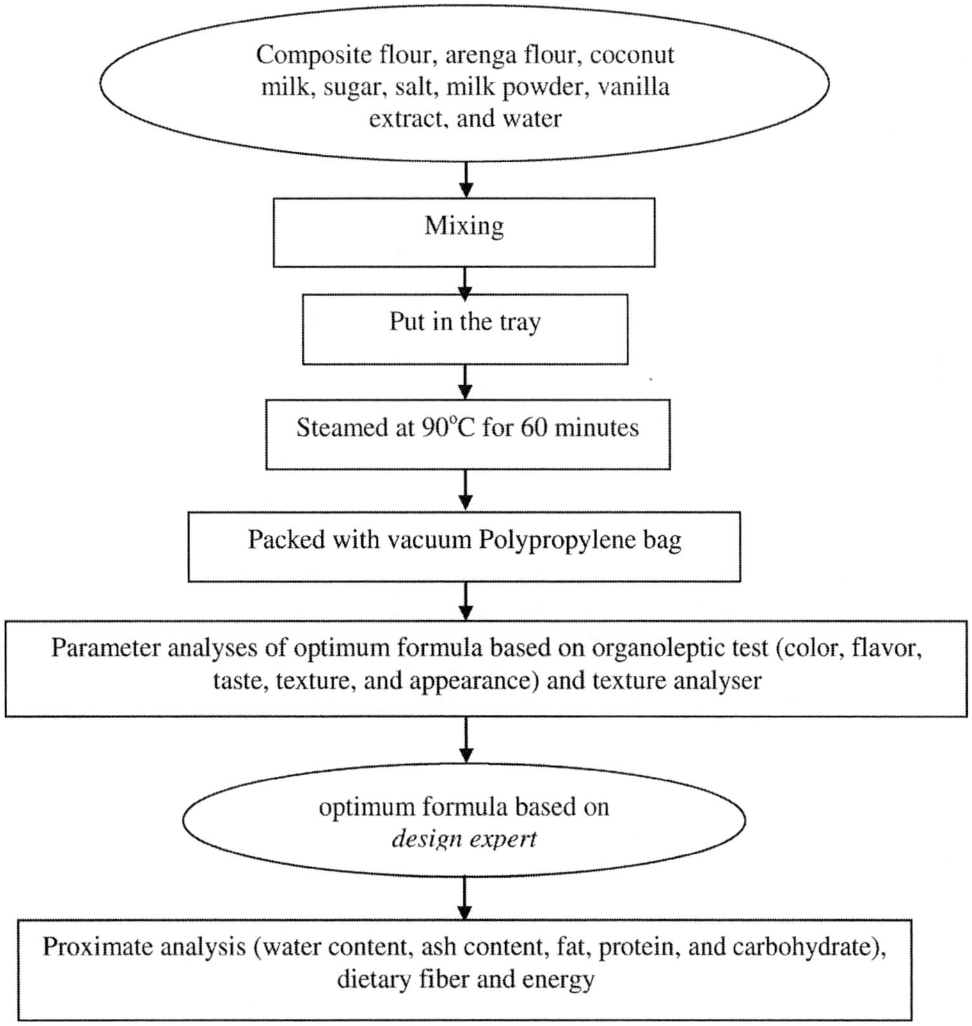

Figure 1. Flowchart of processing steamed meal from taro composite flour

2.1.2. Second stage. This study was to determine the shelf life of steamed cake base on composite flour using ASLT (Accelerated Shelf Life Test)[11] method. To determinate accelerated shelf life is by storing food products in various environments. Storage temperatures used to determine shelf life are 10°C, 20°C, and 30°C with parameters of water content, pH, Aw, and free fatty acid as critical factors in the determination of shelf life [10]. The test was carried out at 10, 20, and 30 °C. pH, water activity, moisture content, and FFA were measured every 4 days within a month

2.2 Determination of steamed meals shelf life prediction

2.2.1. Critical parameters. Storage of food in certain environmental conditions causes a change in the foodstuff, which is the result of adjustments to environmental conditions. ASLT (Accelerated Shelf Life Testing) method is one to determine the shelf life of food product. Achour et al [16] used to ASLT analyzed the food product. One of the environmental conditions that affect the changes in the food is temperature [17]. In the second stage of research, the prediction of shelf life of steamed cake based on composite flour were determined which 4 critical parameters namely water content, pH, aw, and free fatty acid content. Selection of these four parameters is because the four parameters, affect the shelf life of the product.

2.2.2. Acceleration. The slope value (k) represents the amount of quality degradation during storage. If the slope value in ln (natural log) is plotted and plotted with 1 / T (1 / temperature in Kelvin) Arrhenius equation will be obtained. Table 5 shows the Arrhenius equation for water, aw, pH, and free fatty acid parameters. Broadly speaking the value of the constant rate of quality degradation reaction (K) at each storage temperature can be estimated using the obtained Arrhenius equation.

3. Results and discussion

3.1. Formulation of steamed based on composite flour

Determination of optimum formula of this product was based on organoleptic test and texture analysis. The descriptive organoleptic test was using the sensory attributes of a product are identified, described, and quantified using a trained panelist for this purpose. The intensity or aspect of the descriptive analysis indicates the certain level of each character. The levels were depicted using a measurement scale. Scale selection can use the scale line [9]. Line scale used 10 cm long. Scale number 0 indicates the lowest intensity, while the scale is closer to the number 10, shows highest intensity.

The sensory analysis of Ongol-ongol samples was given as means of all scores given by the panelist (table 2). The lowest color value was found at 90:90 (1:5), while at 70:30 (1:1) was recorded as the highest value. In term of flavor, at 90:10 (1:5) panelist judged 3.34, while at 50:50 (1:1) they gave 6.85. As observed in color and flavor, the taste was recorded as the lowest acceptance at 90:10 (1:5), on the other hand at 50:50 (1:1) was the highest one. The texture was also found to be lower at 90:10 (1:5) and the highest at 50:50 (1:1). The appearance of samples was recorded as the likeness of all sensory properties. From table 2, the panelist judged 50:50 (1:1) was the optimum formula in producing of Ongol-ongol cake.

Table 2. Value of organoleptic and texture response

Composite flour: Arenga flour	Flour total : water	Organoleptic test					Texture		*Desirability*
		Color	Flavor	Taste	Texture	Appearance	*Total Work Cycle* (mJ)	*Resilience*	
90:10	1:1	6.48	5.29	4.1	4.77	5.54	19.67	0.18	0.509
90:10	1:3	3.7	5.43	4.21	3.9	3.91	6.11	0.53	0.391
90:10	1:5	1.76	3.34	2.55	1.09	1.51	2.03	0.5	
70:30	1:1	7.36	5.44	5.52	5.49	5.81	3.36	0.15	0.461
70:30	1:3	3.97	5.34	4.91	4.76	4.06	8.67	0.63	0.487
70:30	1:5	1.85	4.41	2.86	2.39	3.15	4.76	0.44	0.147
50:50	**1:1**	**6.72**	**6.85**	**6.55**	**7.09**	**7.44**	**39.51**	**0.43**	**0.835**
50:50	1:3	3.46	5.07	3.99	3.87	4.73	7.08	0.52	0.353
50:50	1:5	1.85	3.75	3.3	1.46	2.32	2.93	0.52	0.089
30:70	1:1	6.75	5.82	5.87	6.28	6.42	38.65	0.6	0.809
30:70	1:3	2.84	5.19	4.55	5.17	4.21	2.1	0.27	0.248
30:70	1:5	2.23	4.2	3.63	3.62	3.11	6.54	0.49	0.235
10:90	1:1	7.44	6.53	7.05	7.37	7.55	18.91	0.62	0.806
10:90	1:3	2.15	4.25	3.6	3.51	3.22	12.24	0.52	0.246

10:90	1:5	1.96	4.25	3.25	2.93	2.81	3.17	0.43	0.144

Parameter on texture analysis was Total Work Cycle (TWC) and Resilience. TWC is expressed in units of milliJoule (mJ). TWC is a combination of Hardness Work Cycle and Recoverable Work Cycle. Hardness Work Cycle is defined as the work required to resist the internal strength of the bond in the food material. Meanwhile, Recoverable Work Cycle is defined as work done by food in opposing forces that suppresses when the force is released. TWC test of samples showed that at 50:50 (1:1) obtain 39.5, which consider as elastic. Meanwhile, at 30:70 (1:3) TWC was recorded 2.1, which consider as soft. *Resilience* is value between *Recoverable Work Done* and *Hardness Work Done*. Value of *Resilience* was 0.1. *Resilience* as value where food come back to their form as a result of work of speed and strength. Value of *Resilience* was 0.15-0.63.

Table 2 shows the results of numerical optimization, in which 15 formulas are considered optimal because they have desirability values. Based on table 2, it was found that the formula with composite flour ratio composition: 50:50 arenga starch and water ratio of 50:50 (1: 1) has the highest desirability value of 0.835, so the formula is the optimum formula.

3.2. Formula optimum validation

Validation is done on the optimum formula to see the predicted characteristics by DX7 software with the results obtained. The value of the obtained characteristics is between the minimum and maximum predictions. Maximum and minimum prediction values are set by 95% confidence interval. Table 3 shows the actual value obtained, compared to Design Expert DX 8.1.6 (trial version) predictive value.

In the organoleptic test on color the prediction value ranges from 2.59-6.72, the actual value obtained is 6.72. On the flavor of the predicted value and actual value is the same of 6.85. At the minimum predictive value taste parameter of 2.56 and a maximum prediction value of 6.55. In panelist response to texture, the predicted value and the actual value are the same *i.e* 7.09. The panelist response to the appearance based on the actual value and the predicted value of 7.44.

In texture test response with texture analyzer, it is found that the parameter of TWC is 39.51, while the prediction value is 15.62-63.92 mJ. In the Resilience parameter, the prediction value ranges from 0.31-0.62, while the actual value obtained is 0.43. The value of the organoleptic test and texture test still falls in between the minimum and maximum prediction, so it is still acceptable at 95% confidence interval.

Table 3. Result of validation of optimum formula

Parameters		Actual Value	Prediction value at 95% confidence interval	
			Minimum	Maximum
Organoleptic	Color	6.72	2.59	6.72
	Flavor	6.85	6.85	6.85
	Taste	6.55	2.56	6.55
	Texture	7.09	7.09	7.09
	Appearance	7.44	7.44	7.44
Texture	*Total Work Cycle* (mJ)	39.51	15.62	63.92
	Resilience	0.43	0.31	0.62

3.3. Chemical properties of steamed meals

The optimum composite flour-based steamed cake is obtained with composite flour ratio composition: 50:50 arenga starch and total flour ratio: 1: 1 water. Table 4 shows the results of Proximate analysis and value.

Table 4. Proximate analysis and energy value of steamed meals of optimum formula

Characteristics	Values
Water content (%)	36.53
Ash content (%)	1.36
Fat (%)	14.48
Protein (%)	2.85
Carbohydrate (%)	44.77
Energy (Kcal)	320.8
Dietary fiber (%)	11.38

The water content of steamed meal based on composite flour is 36.53%. The presence of water in the product can affect the deterioration of chemical and microbiological quality. The steamed meal sample had 36.53 %. This sample is in the range of food with medium moisture content (20-40%) [12]. Ash content is determined by combustion organic compound [12]. The ash content of steamed meal is 1.36%. The fat content of steamed meal is 14.48%, thus producing energy value of 179.08 Kcal. Each gram of fat can produce energy value as much as 9 Kcal [2]. The high content of fat obtained from raw materials of steamed meal is from milk powder. Protein content in steamed meal based is 2.85%. Protein content obtained from green bean powder and milk powder. The energy value generated from the protein is 11.4 Kcal because each gram of protein produces an energy value of 4 Kcal [2].

Carbohydrates are total carbohydrates, so they include fiber content. One gram of carbohydrate produces energy value of 4 Kcal [2]. The content of carbohydrates in the product is 44.7%, resulting in energy value of 179.08 Kcal. The total energy value that can be obtained in 100 of steamed meals is 320.8 Kcal. The value is sum of carbohydrates, proteins, and fats. In this study, the manufacture of steamed meals is expected to be a breakfast substitute food, in one slice of steamed meal for 25 g weight can be produced 80.2 Kcal. In order to meet the needs of calories in the breakfast that is equal to 250 calories, then the steamed cake should be consumed is three pieces of about 75 g.

The dietary fiber content in the product is of 11.38%. Fiber is a component of plant material that cannot be digested [14]. Dietary fiber cannot be digested enzymatically (enzymes produced by humans), so it is not a source of nutrients [13] but contributes to our digestive health [14]. Cellulose, hemicellulose, pectin, lignin, and other substances in indigestible plants are included in dietary fiber [14] The role of dietary fiber for health, prevent coronary heart disease, colon cancer, diabetes, diverticular disease, and obesity [15].

3.4. Shelf life determination

The result shows that the initial quality value of moisture content is 38.27% and the critical limit is 36.39%. Studying water in foods should start with an analytical determination of water content for commercial and legal reasons which are evident [18]. The water content has significant importance for a number of reasons. The determination of water content is therefore the most frequent general analysis performed on foodstuffs [19].

Table 5. Arrhenius equations of quality for water content, aw, pH, and free fatty acid of steamed meals based on composite flour

Parameters	Temperature		1/T (K)	ln K	Linear equation	Arrhenius equation
	°C	K			Ln K vs 1/T(K)	Ln K=Ln Ko-Ea/R (1/T)
Air	10	283	0.0003	0.0024	y = 0.7449x - 8.4364	Ln K=- 8.4364 + 0.7449(1/T)
	20	293	0.0021	0.2536	R^2 = 0.5626	
	30	303	0.0014	0.1466		
a_w	10	283	-0.0001	0.0106	y = 0.1414x - 9.2432	Ln K= -9.2432 + 0.1414(1/T)
	20	293	-0.0001	0.1671	R^2 = 0.8913	
	30	303	-0.0001	0.2852		
pH	10	283	-0.0080	0.4376	y = 0.6967x - 5.3	Ln K= - 5.3 + 0.6967(1/T)
	20	293	-0.0314	0.8987	R^2 = 0.7658	
	30	303	-0.0323	0.8318		
Free fatty acid	10	283	0.0479	0.4661	y = -0.0392x - 3.0037	Ln K= - 3.0037 -0.0392(1/T)
	20	293	0.0454	0.6293	R^2 = 0.9532	
	30	303	0.0443	0.3133		

The initial quality value of aw is 1.000, while the critical limit is 0.997. The effect of water activity on chemical reactions which are important to food stability has been studied extensively [11].The initial quality value of pH is 5.66 and the critical limit is 5.52. The initial quality value of free fatty acids is 0.07, while the critical limit is 0.603. Basic consideration of the critical limit is when finding steamed meal based on composite flour that has been covered with mold on the surface. The product is a sample stored for three days at room temperature (30°C).

Steamed meals based on composite flour on water parameter have the quality decreasing equation Ln K = - 8.4364 + 0.7449 (1 / T), then to predict shelf life when steamed meal is stored at 10°C or 283°K, will yield Ln K = -8.43377 or k = 0.00022, meaning that there will be an increase in water content of 0.00022 per day, so the total quality unit to expire is 2.23.

Table 5 presents estimates of shelf life of steamed meal based on parameters of water, aw, pH, and free fatty acid content at various storage temperatures. The total quality unit is calculated based on the initial quality difference with the critical limit. In the parameters of moisture content and free fatty acid content, the total quality unit is distributed, because in both parameters the quality decrease follows the reaction of the first order. In the parameter aw and the pH of the quality decrease follow the zero order.

In the estimation of shelf life at the various temperatures of 15°C, 25°C, and 35°C, each temperature is converted into Kelvin (K), then included in the Arrhenius equation, to obtain K value (quality degradation constant). Assumption the shelf life on the parameters of water content, aw, pH, and free fatty acid content varies. Estimated shelf life at water content parameters ranged from 24.10-29.09 days. In the aw parameter predicted shelf life is ranged from 29.78-30.96 days. At the pH parameter of presumed shelf life is between 15.35-28.65 days. The presumption of shelf life on the parameters of free fatty acid content ranged from 12.54-12.83 days.

There is a variation in the estimation of shelf life, in which each critical quality parameter produces a different prediction of shelf life. The decision to recommend predicted shelf life data based on the food safety of the product, the data used are short shelf life data [20]. The composite flour-based steamed cake is made with the manual process, and is presumably still lacking in sterility especially in the process of mixing and packaging.

4. Conclusions

The optimum formula of steamed meal are made from ratio composite flour: arenga starch 50:50 and ratio for total composite flour:water 1:1. The prediction of shelf life of the steamed meal is 12.54 day.

References

[1] Deptan 2001 *Talas* http://tanamanpangan.deptan.go.id/doc_upload/Talas.pdf. Accessed: 28 Januari 2012.
[2] Widarso T D 2009 Development of curry taro as ready to eat product (in Indonesian) (Bogor: Bogor Agricultural University)
[3] Apriani RRN 2011 Composit flour formulation from taro, green beans, and banana flour for roasted brownies (in Indonesian) (Bogor: Bogor Agricultural University)
[4] Richana N 2012 *Cassava, sweet potato, botany, process tecjnology, cultivation, postharvest* (in Indonesian) (Bandung: Nuansa) p 95
[5] Yuliani S 2009 *Development of taro flour production process with low oksalat concentration (max 100 ppm) with three processed products* (in Indonesian). (Bogor: Balai Besar Penelitian dan Pengembangan Pascapanen Pertanian, Bogor)
[6] Setyadjit 2009 *Composit flour formulation from taro as ingrediant of high energy ready to eat breakfast meal* (in Indonesian) (Bogor: Balai Besar Penelitian dan Pengembangan Pascapanen Pertanian Bogor)
[7] Fennema O R 1996 *Food Chemistry Third Edition* (New York: Marcel Dekker, Inc,)
[8] Vendy D A V 1993 *Review of Composite Flour Technology in the Context of Tanzania* (Arusha: Tanzania)
[9] Setyaningsih D, Apriyantono A and Sari M P 2010 *Sendori analysis for food and agro industry* (in Indonesian). (Bogor: IPB Press)
[10] Apriyantono A, Fardiaz , Puspitasari N L, Sedarnawati Y and Budiyanto S 1989 *Food Analysis* (in Indonesian) (Bogor: Pusat Antar Universitas Pangan dan Gizi)
[11] Nelson K A and Labuza T P 1994 *Food Sci. and Nutrition* **22**, 1–4: 271-289
[12] deMan J M 1997 *Food Chemical* (in Indonesian). (Bandung: ITB Press)
[13] Linder M C 1992 *Biochemistry of Nutrition and Metabolism* (in Indonesian). (Jakarta: UI Press)
[14] Potter NN 1995 *Food Science* (USA: Chapman and Hall)
[15] Muchtadi D, Palupi N S and Astawan M 1992 *Chemical biochemical and biological method in processed food nutritient evaluation* (in Indonesian) (Bogor: Dikti)
[16] Achour M, Mtimer N, Cornelius C, Zgouli S, Mahjoub A, Ph Thonart and Hamdi M J 2001 *Chem Technology and Biotechnol* **76**(6): 624-28
[17] Sukasih E, Sunarmani and Budiyanto J P 2007 *Postharvest Pert* **4**(2): 72-82
[18] Mathlouthi M 2001 *Food Control* **12**(7): 409-417
[19] Isengard H D 2001 *Food Control* **12**(7): 395-400
[20] Koswara S and Kusnandar F 2004 *Case study of food product shelf life estimation* (in Indonesian) (Bogor: IPB)

Determination of the Characteristics and Classification of Near-Infrared Spectra of Patchouli Oil (*Pogostemon Cablin Benth.*) from Different Origin

Diego M C R[1], Y A Purwanto[1], Sutrisno[1] and I W Budiastra[1]

[1] Department of Mechanical and Biosystem Engineering, Bogor Agricultural University, Indonesia

E-mail: diegok91@gmail.com

Abstract. Research related to the non-destructive method of near-infrared (NIR) spectroscopy in aromatic oil is still in development in Indonesia. The objectives of the study were to determine the characteristics of the near-infrared spectra of patchouli oil and classify it based on its origin. The samples were selected from seven different places in Indonesia (Bogor and Garut from West Java, Aceh, and Jambi from Sumatra and Konawe, Masamba and Kolaka from Sulawesi Island). The spectral data of patchouli oil was obtained by FT-NIR spectrometer at the wavelength of 1000-2500 nm, and after that, the samples were subjected to composition analysis using Gas Chromatography-Mass Spectrometry. The transmittance and absorbance spectra were analyzed and then principal component analysis (PCA) was carried out. Discriminant analysis (DA) of the principal component was developed to classify patchouli oil based on its origin. The result shows that the data of both spectra (transmittance and absorbance spectra) by the PC analysis give a similar result for discriminating the seven types of patchouli oil due to their distribution and behavior. The DA of the three principal component in both data processed spectra could classify patchouli oil accurately. This result exposed that NIR spectroscopy can be successfully used as a correct method to classify patchouli oil based on its origin.

1. Introduction

Indonesia is a rich country of aromatic plant species with around 40 kinds produced. They are easy to plant, maintain and harvesting, easy to process and transport making them preferable for small farmers and remote communities.

Patchouli (*Pogostemon cablin* Benth.) is a plant from *Lamiaceae* family that is well known for its medicinal and aromatic properties. Patchouli is cultivated mainly for its essential oil and especially notable how this extracted oil is internationally important and valuable, principally for the aromatherapy, perfumery, cosmetics, incense stick production and food flavoring industries [1].
Indonesia shares around 80% of all the production of Patchouli plant in the world market, meanwhile the other 20% comes from countries like Malaysia, Philippine, China, India, and Brazil. Nevertheless, despite its reputation in the industry, continually it suffers irregularities due to the sources of its productions [2]; Doing necessary the study of the origin of the plant for the classification and assessment of the quality of patchouli production.

Several destructive methods have been applied for determining the quality of patchouli oil such as gas chromatography [3] [4] [5] and gas chromatography-sniffing [6] [3]. The problem with destructive methods like gas chromatography is that they are time-consuming and complex. Given the current issues associated with aromatic plants, it is necessary to develop an appropriate tool to assess the quality of the patchouli production. Despite some non-destructive methods as a Raman spectroscopy has been used to analyze the quality of several essential oils [7] [8], due to the characteristics of its light dispersion analysis, the assessment of a result in short time with high accuracy still is complicated. In this context, vibrational spectroscopic methods such as NIRS (Near Infra-Red Spectroscopy) together with chemometrics methods can be successfully introduced as a rapid analysis for a non-destructive determination of metabolites occurring in the aromatic plants including patchouli.

The purpose of this study is to determine the characteristics and classifying the NIR spectra samples of patchouli oil from different origin. The research tries to expose that there is a significant difference in the NIRS spectra patterns between each place, despite it is complicated to determine visually and explaining how this makes an important incidence in the quality presentation of the patchouli oil and its assessment process.

2. Material and method

2.1. Samples and device

In this study were selected 84 oil samples of Patchouli aromatic plant from seven different places around Indonesia; Konawe, Kolaka and Masamba from Sulawesi island, Bogor, and Garut from west java and Aceh and Jambi from Sumatra. In total it was around 50 ml per samples in recipient based on every field production.

The instrument used for transmittance measurement was an FT-NIR Spectrometer (NIRFlex N-500, manufactured by BÜCHI Labortechnik AG, Switzerland). The methodology applied for taking the spectrum of the samples is known as a transflectance [9], being this innovative procedure. The device was adapted to measuring liquid samples taking advantage of its characteristics. According to this, the signal processed was in function of the transmittance and reflectance of the spectrum before the respective data analysis.

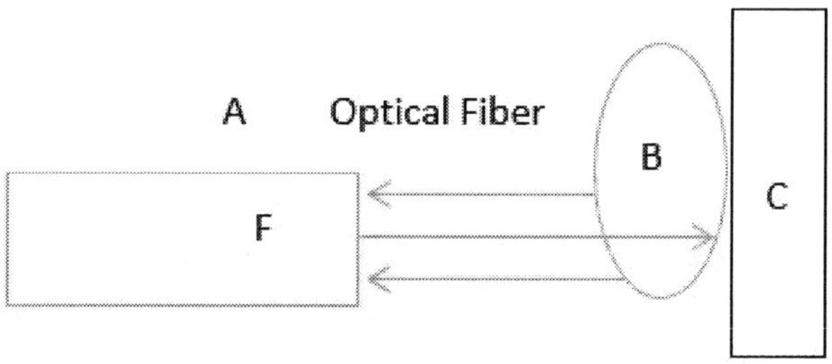

Figure 1. The figure of the explanation of the way of NIRS machine use in the laboratory for taking the transflectance data. A: Gun with the NIR signal and spectrometer detector, B: Oil sample, C: White background, F: Transmittance Sensor. The NIR radiation coming from the optical fiber (gun) goes through the liquid sample, it reflects in the white background surface making contact with the sensor. Then, it backs through the liquid sample, heading to the spectrometer detector.

This methodology has not been well documented because normally the NIRS investigation results focus more on the analysis and calibration of the model and only special name arrangements in the machine if those are required.

2.2. Patchouli spectra acquisition

The transmittance of samples was measured by scanning the samples 3 times at 3 different points, taking the data in there and setting the gun inside the recipient of the oil during the process. After that, the spectra was transformed to absorbance for the subsequent analysis. The wavelength interval used was 1000-2500 nm.

2.3. Chemical analysis

The chemical analysis was assessed to get the chemical content of patchouli oil. The proximate analysis was estimated according to INS (Indonesian National Standard). The main chemical compound of patchouli oil (Patchouli-Alcohol - $C15H26O$) was measured by a liquid chromatography-mass spectrometry (LC-MS) method in every sample studied [10] [11]. The results were assorted based on the percentage of concentration of the patchouli alcohol per sample.

2.4. Spectra processing

The transmittance and absorbance spectra were processed with several spectra data pre-treatments such a Smoothing Savitzky-Golay [12] and first and second derivatives (Polynomial order of 2, 3 smoothing points, left and right sight point of 1) [13] [14]. That method was applied to the standard spectrum with the aim of sharpening their profile and eliminating the interference which is caused by baseline shifts and noise disturbances. Due to the liquid characteristics of the samples, there were not uneven surface effects, multiplication effects like scattering, particle size and multi-colinearity changes, which can cause large variations in the transmittance spectrum [15]. Spectrum data processing was carried out using Unscrambler software X 10.5.

Principal Component Analysis (PCA) was used to cluster the different kinds of patchouli oil based on principal components of processed spectra data. Display of the plot spectra scores of various peaks at the specific wavelengths of spectrum analysis results by PCA was plotted and could become a characteristic or fingerprint of the oil generating a basic in grouping the patchouli cultivars [16] [17]. Discriminant analysis of the result of qualitative analysis by PCA was constructed and applied for classification of the seven kinds of patchouli oil. Discriminant analysis creates functions based on a linear combination of predictor variables to produce the best discrimination between groups. The discriminant function is generated from the sample where the group identity of the subject has been known. The input data were the results of PC1, PC2, and PC3.

3. Results and discussion

3.1. The original spectra of patchouli oil

Some differences are exposed by results of the chemical analysis (Table 1), which indicates the valor of major chemical component (Patchouli-Alcohol - $C15H26O$ among the patchouli oil types). It is important to notice that there are other chemical components like Seychellene and Alpha Bulnesene which are mainly in the patchouli oil. The idea of showing this analysis is to expose a descriptive different between the seven kinds of oil and its behavior based on its constitution.

Figure 2 shows the transmittance spectra as an example, of the different patchouli oil with several peaks and valleys in a specific wavelength for each origin sample. The valley existed related to electromagnetic wave and its responses to the molecular bonds of O-H, C-H, C-O, and C-C (figure 3). The result is subjected to vibrational energy, both of stretch and bent vibration [18].

Despite the spectra are likely similar, if it is possible to review in detail that some interesting data are shown. The patchouli alcohol content in every type of oil has a significant difference. Jambi has the highest content (32.34%) followed by Aceh (30.91%), and Garut (30.98%) and the lowest are Bogor together with Masamba, which is 24.18% and 24.72 % respectively.

Table 1. Chemical characteristic of the patchouli oils based on the main chemical component (Patchouli-Alcohol $C_{15}H_{26}O$). *sd: Standard deviation

Patchouli oil	
Patchouli alcohol $C_{15}H_{26}O$	
Sample	% of the component
Jambi	32.34
Kolaka	26.45
Aceh	30.91
Konawe	29.81
Masamba	24.72
Garut	30.98
Bogor	24.18
Average	28.48
sd*	3.31

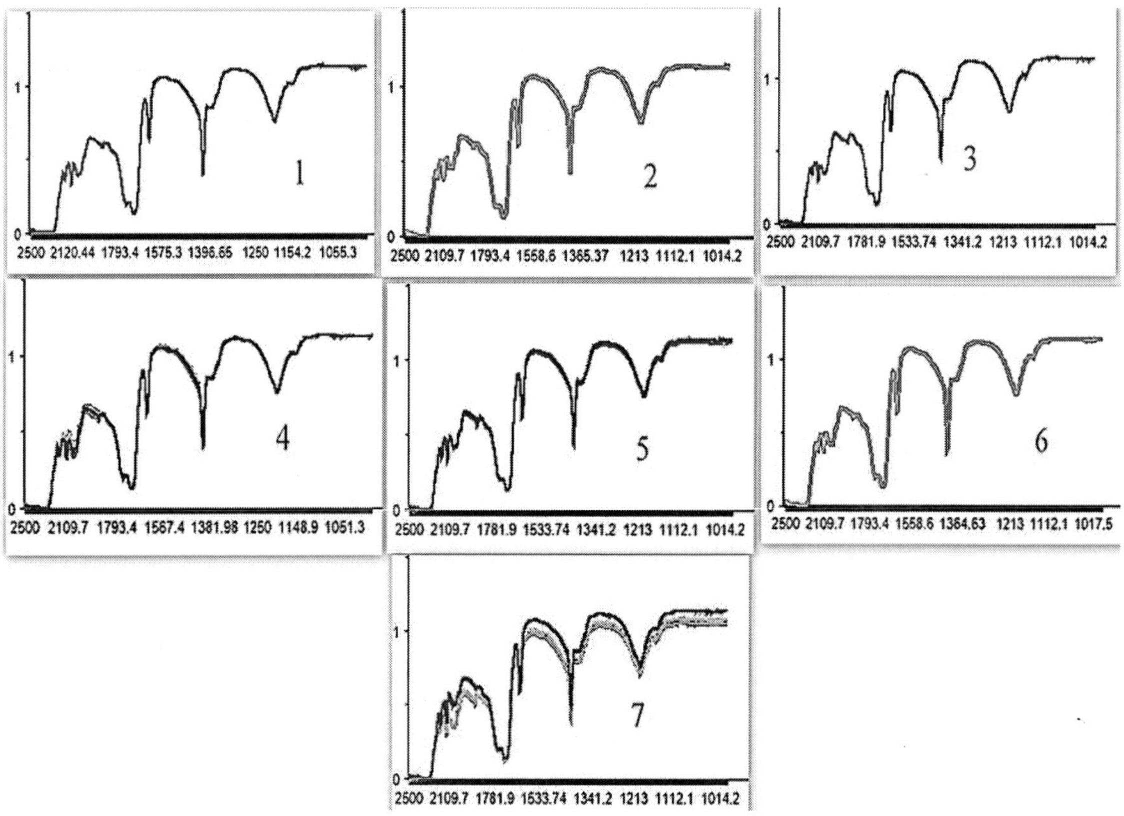

Figure 2. Transmittance spectra of different types of patchouli oil tested. Axis X: Wavelength (nm), Axis Y: Transmittance. 1: Aceh, 2: Bogor, 3: Garut, 4: Jambi, 5: Kolaka, 6: Konawe and 7: Masamba.

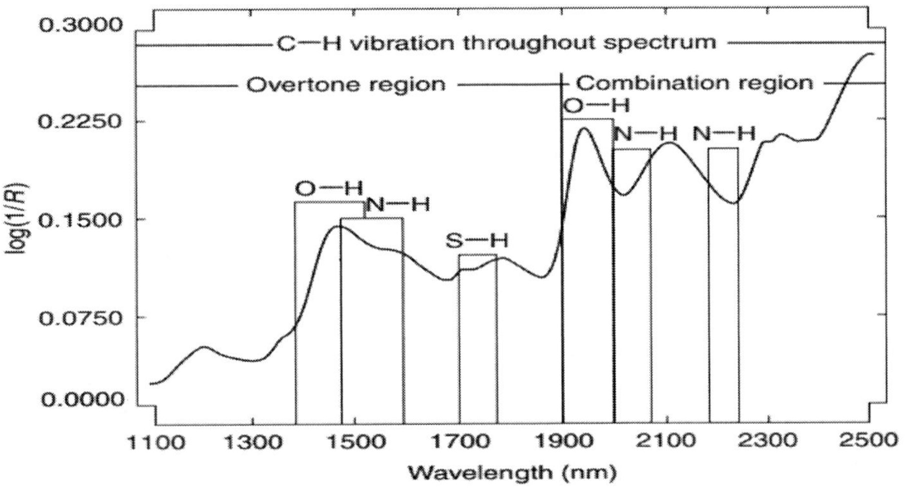

Figure 3. General representation of the valley related to electromagnetic wave and its responses to the molecular bonds.

If it is analyzed the absorbance spectra graphic of the patchouli oil, which contains the spectra of the seven origin places (figure 4), there are several peaks and valleys that exist in there. Based on [19], the wavelength of 1680 and 2230 nm are reference values, 1940 nm to water, 2270 nm to lignin, 2336 nm to cellulose, 2180 nm to protein, 2100 nm to carbohydrate and finally 2310 nm to oil in general, in the case of the combination region of the vibration throughout the spectrum. In the case of the oil, the literature exposed that it is possible to find its main characteristics in the combination region and overtone region of NIR spectrum [20] [21].

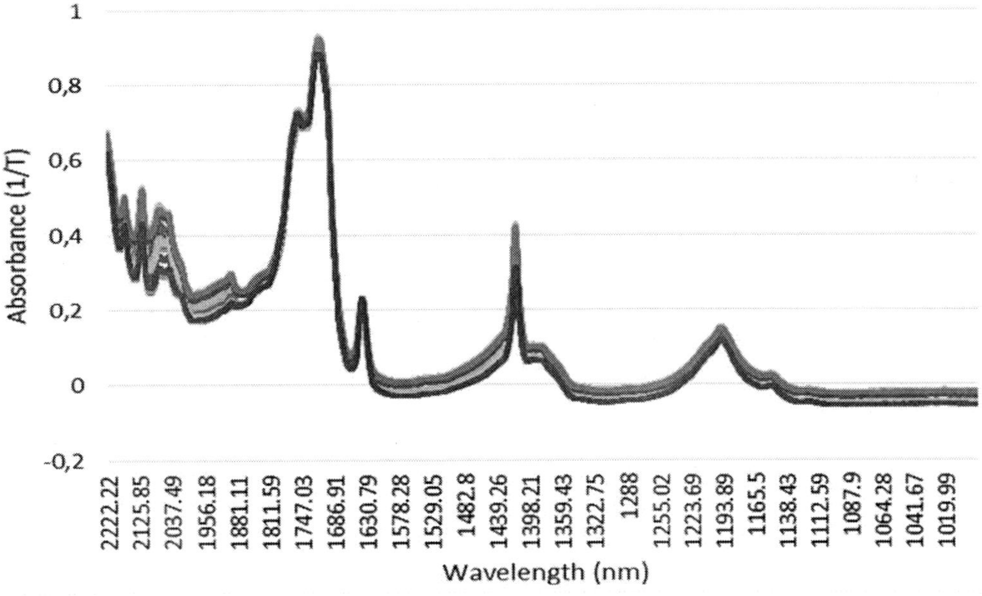

Figure 4. Absorbance spectrum between 2222 – 1019.99 nm, which contain all the spectra of the seven types of oil. Axis X: Wavelength (nm), axis Y: Absorbance (log 1/T). There is representative peak in the range between 1800 to 1600 nm, which can provide information about the chemical composition of the samples.

IOP Conf. Series: Earth and Environmental Science **147** (2018) 012013 doi:10.1088/1755-1315/147/1/012013

3.2. Clustering analysis using PCA of spectra data pretreatment

Figure 5 shows the loading transmittance spectra of first and second principal component. The first and second principal components (PC1 & PC2) represent most variation in original spectra. Furthermore, the principal component was emphasized in the spectra signal that was centered on the wavelength of 1780-1680 nm, which allegedly shows the aromatic characteristics of patchouli oil and its components according to [19]. For example, the Patchouli alcohol ($C_{15}H_{26}O$) tested before, is considered the main chemical compound because this gives the basic aromatic properties of patchouli oil [22] [23].

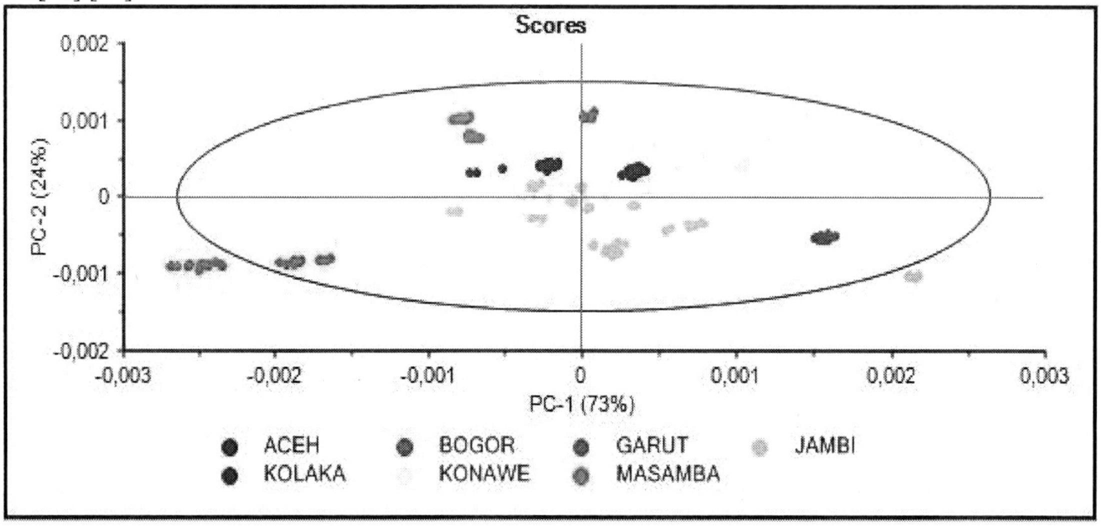

Figure 5. Cluster score graphic of the principal component analysis of the transmittance spectra based on the wavelength of 1780-1680 nm for the seven kinds of patchouli oil tested.

The result of clustering analysis using PCA from the transmittance spectra data after applying the respective pretreatment (first and second derivative Savitzky-Golay due to the region where the PCA was applied) are exposed in figure 5. It is possible to conclude that there is a clear grouping of the data and moreover, in many of them there is no a high correlation, which means that by this analysis it is possible to determine how is different between the spectra of the one place to others. In the general, the variance of all the samples is very significant. The results of the PCA (PC1 and PC2) explain around 97% of the variance spectrum analyzed with 1% of the PC3. There are some samples in data spectra of Jambi and Masamba that are outside of the variance explained, also it can be exposed in the residual plot where around 7 samples of Masamba and 5 of Jambi are identified as outliers (figure 7). it could be related to the range of the chemical data manipulated (highest and lowest values), which are reproduced in the spectra taken it.

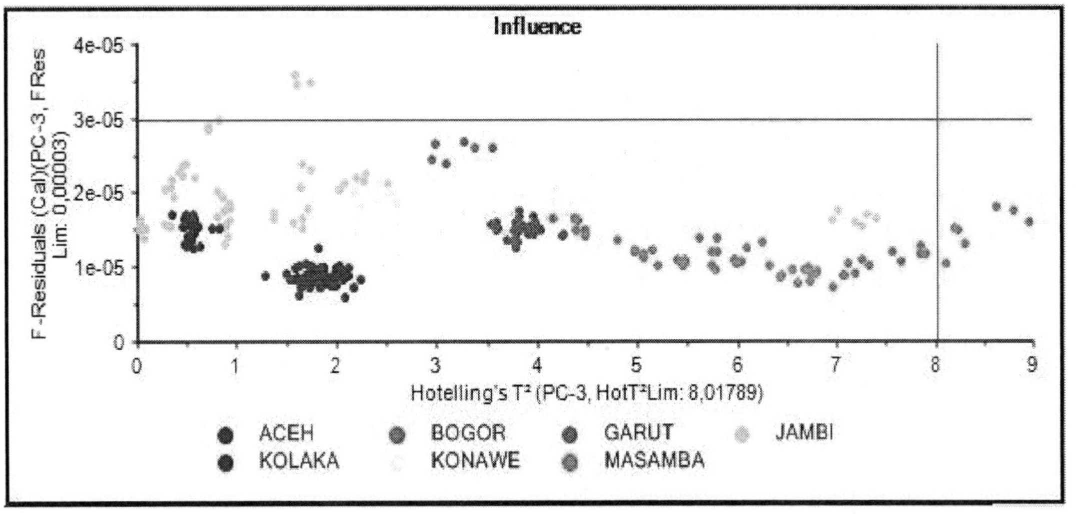

Figure 6. Residual plot of the PCA analysis of the samples for the transmittance spectra

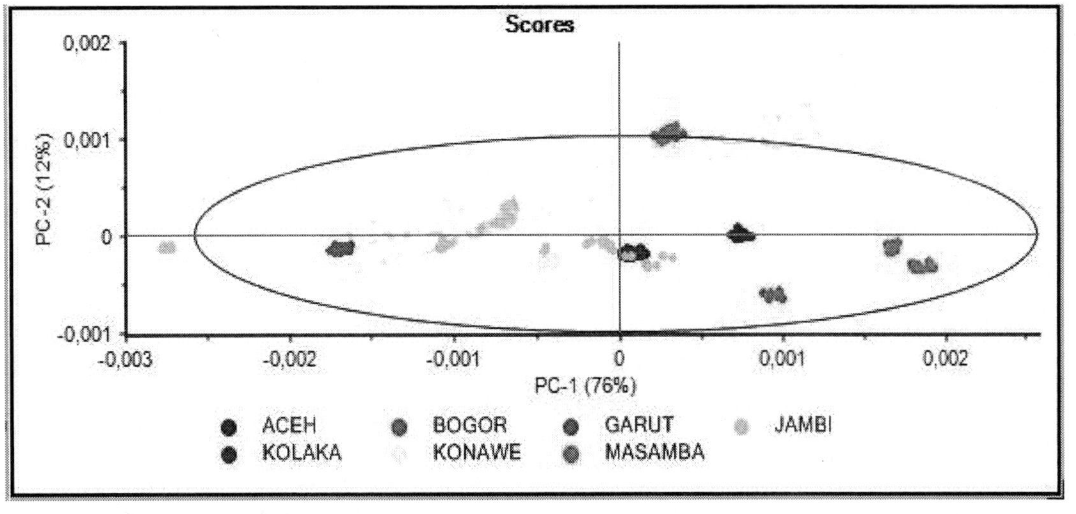

Figure 7. Cluster score graphic of the principal component analysis of the absorbance spectra based on the wavelength of 1780-1680 nm for the seven kinds of patchouli oil tested.

In this case, the result of clustering of PCA also exposed that there is a clear grouping of the data and in the same way of the transmittance, many of them are not highly correlated, which means that by this analysis it is also possible to determine how is the difference between each the spectra. The results of the PCA (PC1 and PC2) explain around 86% of the variance spectrum analyzed with 7% of the PC3.

By this graphic is show that the covariance of the date is a little bit higher compared with the transmittance results and at the same time the correlation of some data increase, which is the case of Aceh, Masamba and Konawe samples. The stated purpose before, open the possibility to determine that transmittance spectra in some way can be a better way to classify the spectrum of the patchouli oil nevertheless. There are some samples outliers in the absorbance spectra, in this case only for Jambi with similar behavior to the PCA results of the transmittance spectra.

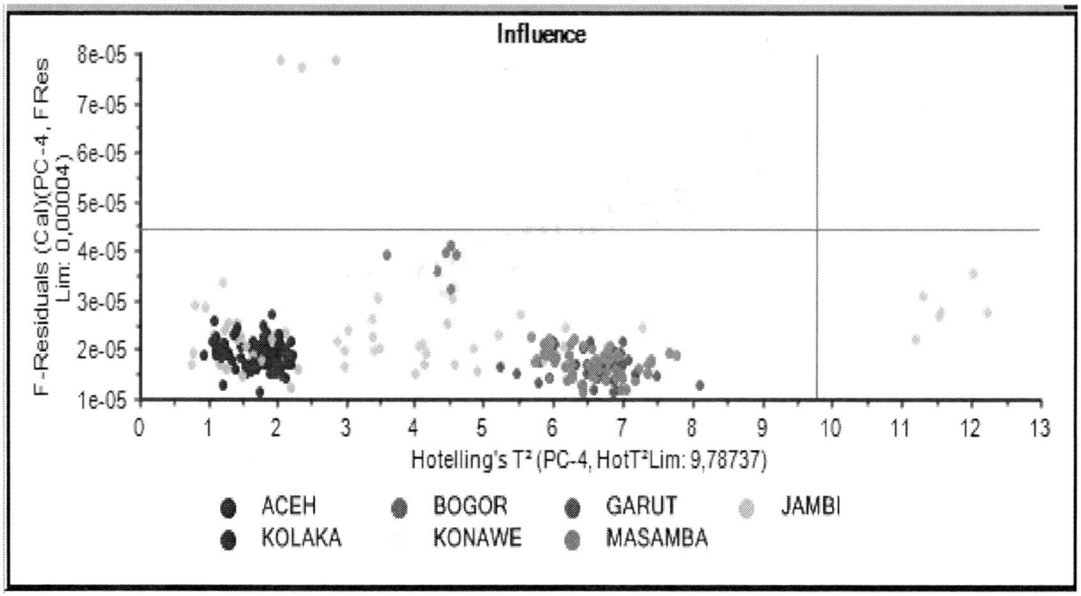

Figure 8. Residual plot of the PCA analysis of the samples for the absorbance spectra

3.3. Discriminant analysis

The discriminant analysis was applied focus in the wavelength close to the peak of 1780-1680 nm in where was carried out the PCA analysis, according to the information of [19]. The result of classification using discriminant analysis for the transmittance spectra showed that the overall prediction values were the highest, which a probability of classification of 100% for all the spectra sample. The discriminant analysis was applied to the PC data. The results exposed that it is possible to classify the spectra of the oil based on its origin due to the variance of the data exposed in the clustering beforehand.

The percentage of the prediction shown in table 3 explains that all the spectra samples can be discriminate in function of the others with the highest probability of success. The samples of Jambi, Kolaka, Garut, Aceh, Masamba, Bogor and Konawe have all perfect predicted values, which means that its data spectrum can be identified as a its specific place of origin.

In the absorbance case, the results are similar to the transmittance case. However, it is not possible to say which spectroscopy is better, because despite transmittance spectra has better PCA results and lower colinearity in the clustering, the absorbance spectra expose a better-explained variance with less outliers and moreover, the results of the discriminant analysis are similar for both of them.

Table 2. Table of classification of each sample for transmittance spectra based on its origin applied to the data spectra in the range wavelength of 1780-1680 nm.

	Aceh	Bogor	Garut	Jambi	Kolaka	Konawe	Masamba	
Aceh	60	0	0	0	0	0	0	60
Bogor	0	6	0	0	0	0	0	6
Garut	0	0	30	0	0	0	0	30
Jambi	0	0	0	60	0	0	0	60
Kolaka	0	0	0	0	30	0	0	30
Konawe	0	0	0	0	0	6	0	6
Masamba	0	0	0	0	0	0	60	60
	60	6	30	60	30	6	60	252

Table 3. Summary table with the percentage of discrimination between each sample for transmittance spectra

	Correct	Total	% correct	% missed
Aceh	60	60	100	0
Bogor	6	6	100	0
Garut	30	30	100	0
Jambi	60	60	100	0
Kolaka	30	30	100	0
Konawe	6	6	100	0
Masamba	60	60	100	0
	252	252	100	0

Table 4. Table of classification of each sample for absorbance spectra based on its origin applied to the data spectra in the range wavelength of 1780-1680 nm.

	Aceh	Bogor	Garut	Jambi	Kolaka	Konawe	Masamba	
Aceh	60	0	0	0	0	0	0	60
Bogor	0	6	0	0	0	0	0	6
Garut	0	0	30	0	0	0	0	30
Jambi	0	0	0	60	0	0	0	60
Kolaka	0	0	0	0	30	0	0	30
Konawe	0	0	0	0	0	6	0	6
Masamba	0	0	0	0	0	0	60	60
	60	6	30	60	30	6	60	252

Table 5. Summary table with the percentage of discrimination between each sample for absorbance spectra

	Correct	Total	% correct	% missed
Aceh	60	60	100	0
Bogor	6	6	100	0
Garut	30	30	100	0
Jambi	60	60	100	0
Kolaka	30	30	100	0
Konawe	6	6	100	0
Masamba	60	60	100	0
	252	252	100	0

4. Conclusions

The final results showed that it is possible to classify the oil samples based on its origin which a high accuracy using NIR spectroscopy. The results of the PCA and the discriminant analysis suggest that there is different between each grouping sample which is related to the existence of a significant difference between the data of the chemical composition.

Some of the results got it in this research could be due to the liquid characteristics of the samples. There were no effects of scattering, particle size and multi-colinearity, which could change in each sample, this produces a spectrum very similar in every case. Therefore, it makes a huge influence in the moment of analyzing the data.

Transmittance and absorbance spectra give a similar response in the moment of classifying the data; both of them give the same values in the discriminate analysis. This behavior does not permit to make a clear preference in the moment of selecting which one have a better accuracy for the objective purpose in this experience.

The variance was decent, therefore, the result achieved in the PCA which many data not highly correlated. On the other hand, the data pre-treatment carry out here were the best that the literature suggests.

References

[1] Ramya H, Palanimuthu V and Singla R 2013 An introduction to patchouli (Pogostemon cablin Benth.) – A medicinal and aromatic plant: It is importance to mankind *J. Agric. Eng. Int* **15** 243.

[2] Orellana A 2009 *Agrotechnology for the cultivation of patchouli* (San José de Guatemala: Institut of Science and Technology (ICTA))

[3] Cserháti T, Forgács E, Deyl Z and Miksik I 2005 Chromatography in authenticity and traceability tests of vegetable oils and dairy products: a review *J. Bio. Chr* **19** 183

[4] Daferera D J, Tarantilis P A and Polissiou M G 2002 Characterization of essential oils from Lamiaceae Species by Fourier transform Raman spectroscopy *J. Agric. Food. Chem* **50** 5503

[5] Nikoli´c M, Jovanovi´c K, Markovi´c T, Markovi´c D, Gligorijevi´c N, Radulovi´c S and Sokovi´c M 2014 Chemical composition, antimicrobial, and cytotoxicproperties of five Lamiaceae essential oils *J. Ind. Crops. Prod* **61** 225

[6] Chin S and Marriott P J 2015 Review of the role and methodology of high resolution approaches in aroma analysis *J. Chim. Acta* **854** 1

[7] Strzemski M, Wojciak- Kosior M, Sowa I, Agacka-Moldoch M, Draczkowski P, Matosiuk D, Kurach L, Kocjan R and Dresler S 2017 Application of Raman spectroscopy for direct analysis of *Carolina acanthifolia* subsp. Utzka root essential oil *J. Tal* **174** 633

[8] Asif M, Nawaz H, Naz S, Mukhtar R, Rashid N, Ahmad I and Saleem M 2017 Raman spectroscopy for the characterization of different fractions of hemp essential oil extracted at 130°C using distillation method *J. Spect. Acta* **182** 168

[9] Cayuela J and Garcia J 2016 Sorting Olive oil based on alpha-tocopherol and total tocopherol content using near infra-red spectroscopy (NIRS) analysis *J food and Eng* **202** 79

[10] Gokulakrishnan J, Kuppusamy E, Shanmugam D, Appavu A and Kaliyamoorthi K 2013 Pupicidal and repellent activities of Pogostemon cablin essential oil chemical compounds against medically important human vector mosquitoes *J. Asi. Pac. Trop. Dis* **3** 26

[11] Murugan R and Mallavarapu G R 2013 Bisabolol, the main constituent of the essential oil of Pogostemon speciosus *J. Indus. Crop. Prod* **49** 237.

[12] Savitzky A and Golay M J 1964 Smoothing and differentiation of data by simplified least squares procedure *J. Anal. Chem* **36** 1627.

[13] Williams P and Karl N 1990 *Near-Infrared Technology in the Agricultural and Food Industries* (Minnesota: American Association of Cereal Chemists. Inc).

[14] Lee M S, Hwang Y S, Lee J and Choung M G 2014 The characterization of caffeine and nine individual catechins in the leaves of green tea (Camellia Sinensis L.) by near-infrared reflectance spectroscopy *J. Food Chem* **158** 351.

[15] Martens H and Naes T 1987 *Multivaraiate calibration by data compression*, In: *Near-Infrared Technology in the the Agricultural and Food Industries* (Minnesota: American Association of Cereal Chemists. Inc).

[16] Cozzolino D, Cynkar W U, Shah N and Smith P 2011 Multivariate data analysis applied to spectroscopy: potential application to juice and fruit quality *J. Food Res. Int* **44** 1888.

[17] Kumar N, Bansal A, Sarma G S, and Rawal R K 2014 Chemometrics tools used in analytical chemistry: an overview *J. Talanta* **123** 186.
[18] Cen H and He Y 2007 Theory and application of near infrared reflectance spectroscopy in determination of food quality *J. Tren. Food Sci.Tech* **18** 72.
[19] Burns D A and Ciurzak E W 2008 *Handbook of Near-Infrared Analysis.* (New York: 3rd Edition).
[20] Kuriakoe S and Joe H 2013 Feasibility of using near-infrared spectroscopy to detect and quantify an adulterant in high quality sandalwood oil *J. Spec. Acta Part A: Mol. Bio. Spect* **115** 568.
[21] Dupuy N, Gaydou V and Kister J 2014. Quantitative Analysis of Lavender (Lavandula angustifolia) Essential Oil Using Multiblock Data from Infrared Spectroscopy *J. Ame. Anal. Chem* **5** 633.
[22] Yougen W, Chungong L, Xianchao L, Meng Y and Xinwen H 2013 Comparison of the Essential Oil Compositions between *Pogostemon cablin* and *Agatache rugosa* Used as Herbs *J. Essen. Oil Bear. Plants* **6** 705.
[23] Nguyên Xuân D, Piet A, Tran Huy T and La Dinh M 1989 Chemical Compositionof Patchouli Oil from Vietnam *J. Essen. Oil Res* **2** 99.

Vaname (*Litopenaeus vannamei*) Shrimp Fry Counting Based on Image Processing Method

M Solahudin[1], Slamet W[1] and Dwi A S[1]

[1] Department of Mechanical and Biosystem Engineering, Bogor Agricultural University, Indonesia

E-mail: msoul9@yahoo.com

Abstract. Manually counting for vaname shrimp fry has some weaknesses, such as subjectivity in counting, time payload, exhausted in counting and inadequate accuracy especially in calculate large quantities of shrimp fry. Considering that condition, therefore a fast and good accuracy method in shrimp fry counting is needed. The purposes of this study are to develop a method in vaname shrimp fry counting through image processing by using, also determine the value of parameters which is needed in creating a machine to count shrimp fry automatically based on CCD camera. Images taken using CCD camera with backlight system. Image processing processes are consist of image cutting, thresholding, dilation and labelling. Accuracy value of shrimp fry counting using image processing program is 98.49%. The average time of shrimp fry counting using image processing program is 1.70 second/image with standard deviation is 0.15 seconds.

1. Introduction

Vaname shrimp and tiger shrimp are the two main commodities for the export of Indonesian shrimp commodities. Currently, shrimp farmers prefer vaname shrimp than tiger prawns. Vaname shrimp is a commodity imported from South America to cope with the constraints of tiger shrimp cultivation that many attacked by White Spot Virus (WSV). Vaname shrimp have several advantages, such as: domesticated mothers (not dependent on nature), more resistant to virus attacks, and can be maintained with high density [1].

Production of vaname shrimp in Indonesia over the past five years has continued to increase. Increased production of shrimp can be seen in table 1.

Table 1. Vaname shrimp production (2010-2014)

Year	Production (ton)
2010	206578
2011	246420
2012	251763
2013	390278
2014	411729

Source : Directorate General of Aquaculture [3]

The cultivation of vaname shrimp is broadly divided into two main activities, namely hatchery and cultivation. Hatchery activities conducted by hatcheries or known as hatchery or by household-scale pembenih. In the hatchery activities there is the process of calculating shrimp fry at the time will be sold to consumers. Until now the calculation process is still done manually. Calculation of fish or shrimp fry conducted manually has many weaknesses, among others: subjectivity calculation, slow time, fatigue in calculation, and inaccurate accuracy, especially to calculate the fry of fish or shrimp in large numbers [2]. Manual calculation of fry loss is a problem in itself, especially wasted time to do the activity.

Currently the utilization of machine vision technology has been widely used to replace the function of vision in humans. The use of camera CCD camera or digital camera in various fields proved able to replace the function of the sense of human vision. The advantages of such technology is able to do the job quickly and consistently. The technology is a great opportunity to develop because it is cheap, available and easy to make.

The development of image-based technology with camera as image capture is very possible developed in the process of calculating shrimp fry. Image processing technology is expected to help the calculation of shrimp fry quickly, accurately and cheaply. Shrimp fry can be counted using an image processing program from shrimp or photo shrimp. This study aims to develop the method of calculating vaname shrimp fry with image processing using C # (C-Sharp) programming language, and to determine the value of the parameters required in CCD-based shrimp fry counter machine design.

2. Materials and methods

2.1. *Vaname Shrimp (Litopenaeus vannamei)*

Vaname Shrimp (Litopenaeus *vannamei*) is a native shrimp of South America. The inauguration of vaname shrimp in Indonesia by the government (Ministry of Marine Affairs and Fisheries) was carried out in 2002 [1].

The shrimp fry used in this study have a length of 6 mm - 8.5 mm. Transparent glass tray is prepared as a place for shrimp fry when shooting the image. The water level in the tray is set as high as 0.5 cm so that the shrimp fry do not overlap each other. The glass tray is placed on top of a backlight system with LEDs as a light and the background is white. The CCD camera is placed perpendicular to the tray.

Image capture is done using CCD camera. Image captured images from CCD cameras are displayed on a computer monitor and stored in JPG format. The image that has been taken is inserted into the pre-made image processing program. Some processes are done at the image processing stage. The process used is image cutting, binary (thresholding) and dilation three times. The next process is enumeration of the image of shrimp fry by labeling method, which serves to calculate the number of shrimp fry.

2.2. *Experimental Setup*

The digital image is obtained by a digital image device system that performs an image exploration to form a matrix in which its elements express the value of the light intensity in a discrete set from the point. According to Prasetyo [4] Image origin is defined at coordinates (0,0). The image coordinate system can be represented by a two dimensional matrix f (x, y). Image processing is needed as a process of processing pixels in digital images to improve the quality of the image where the resulting image can display information clearly, and the second extracts the prominent feature information in an image, where humans get characteristic information from the image numerically. With other words the computer to interpret the existing information on the image through the amount of data that can be clearly distinguished [5].

Image thresholding is applied as a method to separate objects and backgrounds. Thresholding is a simple and effective technique for image segmentation. Image segmentation is the process of separating images into several homogeneous parts and extracting them into several objects to be observed [7]. Ahmad [6] also explains that the process of thresholding is to convert the pixels of an object to a color image into pixels of maximum intensity (255) in binary imagery and to change background pixels on the color image into pixels of minimum intensity (0) on a binary image, or vice versa (an object with an intensity value of 0 and a background with an intensity value of 255 on the resulting binary image).

Morphological operations in image processing is needed to change the shape of the object structure contained in the image. Morphological operations are usually applied to binary imagery. According to Tambe et al. [8] there are four basic operations of morphological processing ie dilation, erosion, opening and closing. The dilation operation is to adjust the object in an image where the expansion is based on the structuring element. The erosion operation is shrinking the object in an image where the amount of depreciation is based on the structuring element [9]. According to Raid et al. [10] opening is an erosion operation followed by a dilation operation using the same structuring element and closing is a dilation operation followed by erosion operation using the same structuring element.

3. Result and discussion

3.1. Setup of image capture devices

The camera height or distance of the shoot affects the length and width of the image capture. The camera height should be adjusted so that the resulting image capture represents the entire desired image area. The camera image capture format in this study is 720 x 576 pixels or with 5 : 4 image ratios. The camera height (T) relationship to the length (P) and width (L) of the image follows the following equation:

$$P = 1.0282 \times T + 2.0537 \tag{1}$$

$$L = 0.7464 \times T + 1.4907 \tag{2}$$

The above equation is obtained from the measurement test of the distance of the shooting distance to the length and width of the image. The minimum distance of shooting used in the experiment is 10 cm. Figure 1 below shows a graph of the relationship between the distance of the shoot by the length and width of the image. Backlight system also has other advantages that there is no reflected light from the water because the light is given from below. The backlight system was finally selected in this study. Image acquisition system in this study can be seen in figure 2.

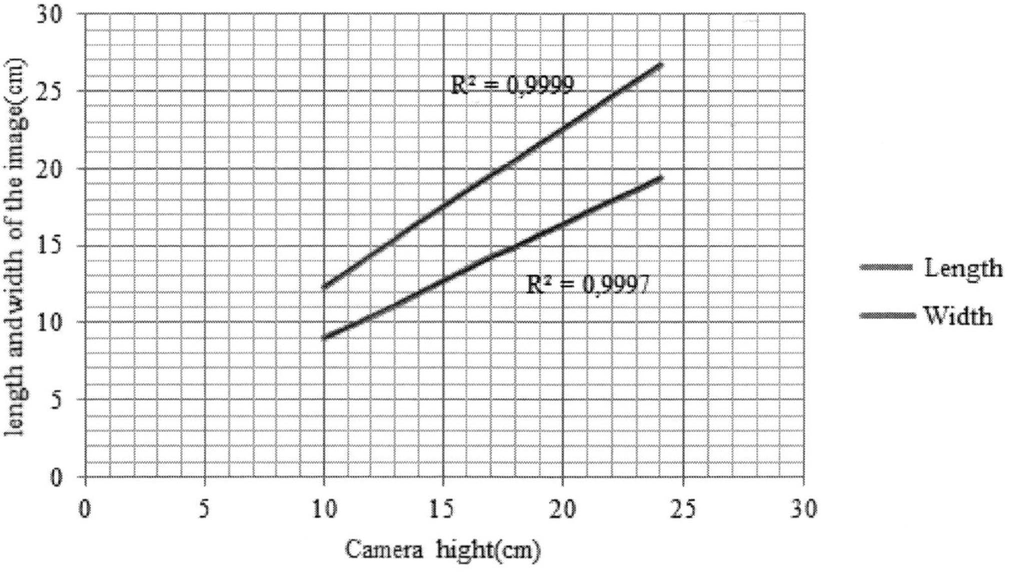

Figure 1. The relationship between the distance of the shoot by the length and width of the image

The image acquisition system in figure 2 used the LED as a light sources. White color selected to facilitate the separation of objects against the background in the thresholding process. Glass tray containing shrimp fry placed on a transparent plate made of acrylic and the camera placed perpendicular to it.

3.2. Thresholding Determination.
The thresholding process is used to separate the object (shrimp fry) with the background. The thresholding process converts the background pixels to black (R,G,B : 0,0,0) and the shrimp fry pixels to white (R,G,B : 255, 255, 255). The greyscale value is used as the thresholding limit in this study.

Determination of thresholding limit is done by adjusting (tuning) the greyscale value until the image obtained where the object and the background look real different. The *ImageJ* app is used for tuning greyscale values. The value of greyscale 70 is chosen as the thresholding limit of the tuning result because it is able to separate the object and the background well. Figure 3 shows the tuning process of thresholding boundary determination.

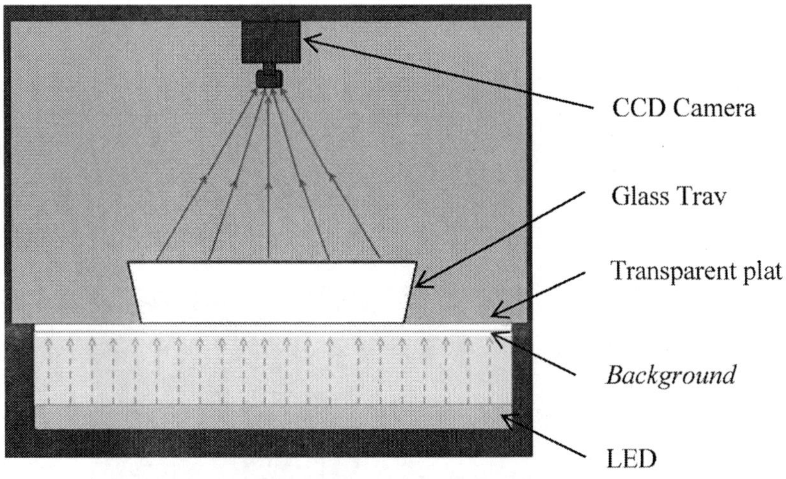

Figure 2. Image acquisition design

Pixels on images that have a greyscale value above 70 are considered as backgrounds and changed to black. Pixels that have a greyscale value below 70 are considered as objects and changed to white. The result of thresholding shows that the greyscale 70 delimiter value can separate object and background well.

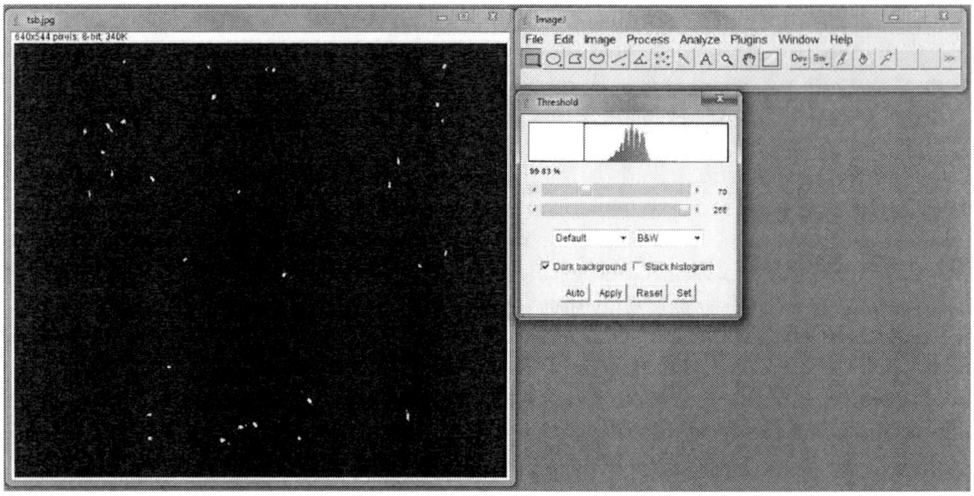

Figure 3. Interface of tuning process for thresholding delimiter value in Image-J application

3.3. Result of Image Processing.

Processing of shrimp fry image begins with thresholding the original image (figure 4) where the background is changed to black and the object is changed to white. The next stage is to perform a dilation operation on the binary image. Dilation operation is the process of adjusting the object, which in this study aims to combine the pixels of a separate object (shrimp fry). Dilation process is done three times for each image. The result of the dilation process of the threshold image is shown in figure 5.

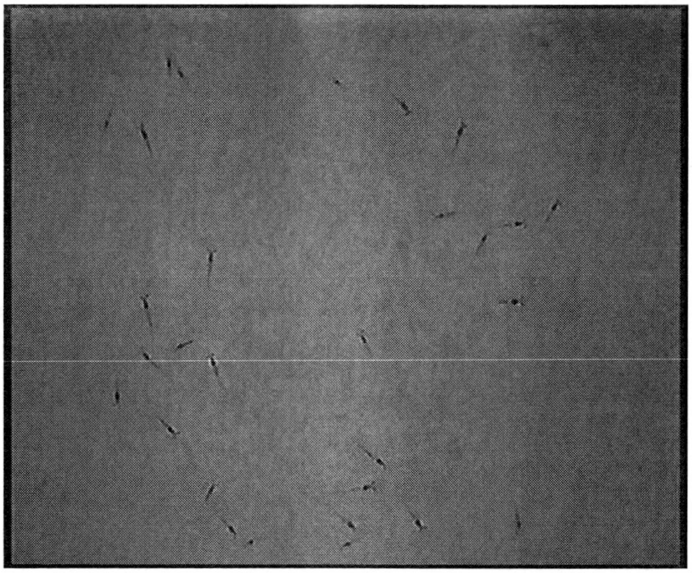

Figure 4. Original image

IOP Conf. Series: Earth and Environmental Science **147** (2018) 012014 doi:10.1088/1755-1315/147/1/012014

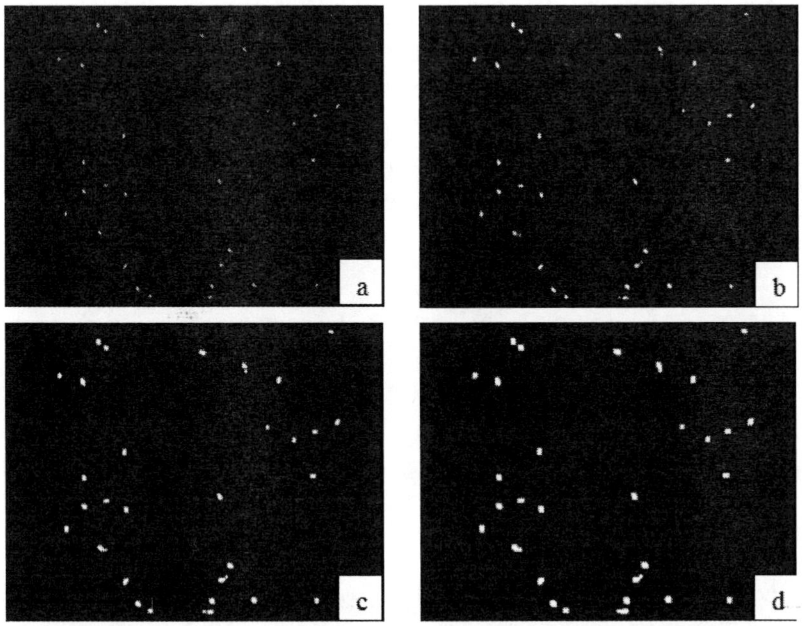

Figure 5. (a) The thresholding result (b) results once a dilation (c) the result of two dilations (d) the result of three dilations

The operation is dilationed three times after the thresholding process is able to unify separate object pixels, so counting errors can be minimized. The merging process of separate object pixels (one shrimp fry) can be seen in figure 6. Testing the program conducted to determine the level of accuracy and time consume of shrimp fry calculation. Testing the program is done by processing the image of shrimp fry that have been taken earlier into the image processing program. The processed image has a variety of shrimp fry. In the container used, the number of fry is limited to 66. The number of fry above 66 is ineffective, as more and more shrimp fry are overlapped. The average accuracy of program testing to calculate shrimp fry in images with different number of fry is shown in table 2.

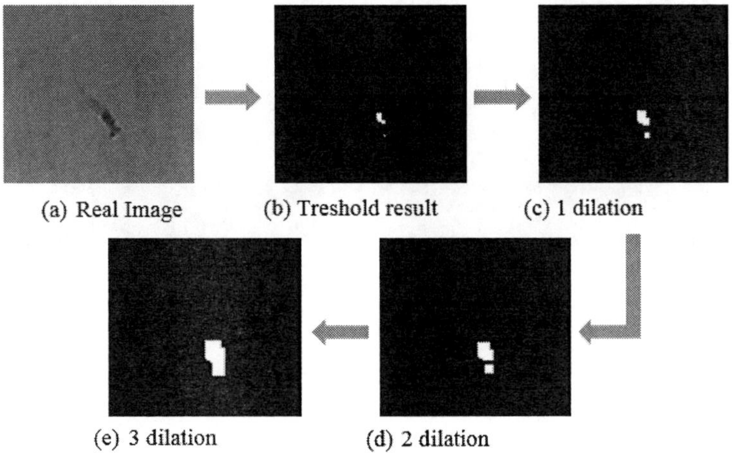

Figure 6. Pixel merging process of one separate shrimp fry

Table 2. Average accuracy of shrimp fry counter program

No	Number of Shrimp fry	Average accuracy (%)
1	5	100.00
2	10	100.00
3	15	100.00
4	20	100.00
5	25	100.00
6	30	100.00
7	35	100.00
8	40	99.00
9	46	97.39
10	56	96.07
11	61	96.72
12	66	92.73
	Average	98.49

The results of the test program showed a fairly high average accuracy of 98.49%. Lower accuracy occurs in higher number of shrimp fry. The calculation result is less than the actual amount caused by two conditions. The first condition is the presence of shrimp fry (more than one) that are interconnected or overlap , as shown in figure 7. The shrimp fry are mutually counted as one by the program. The second condition is that shrimp fry are detected as background because the black spots on the head are not visible, so it has a greyscale value above the threshold limit. This condition can be seen in figure 8. The calculation result is more than the actual amount caused by pixels from one shrimp fry that is still separated (over segmentation) even though it has been through dilation process. This condition causes one shrimp fry can be counted as more than one fry, as shown in figure 9.

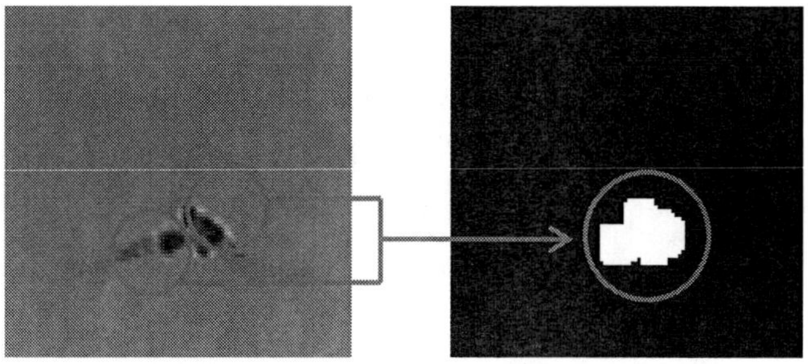

Figure 7. Two tangled shrimp fry are counted as one

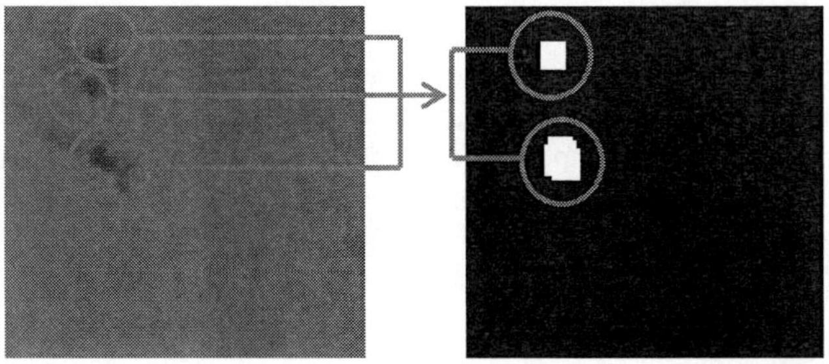

Figure 8. One shrimp fry missing as detected as background

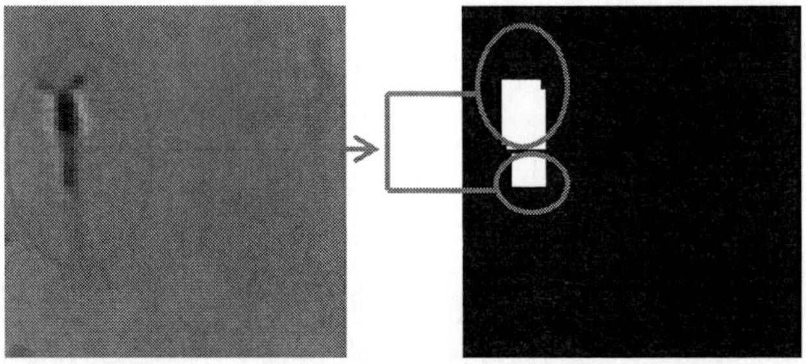

Figure 9. One shrimp fry count is counted as two

Table 3 shows the average calculation time using a program for twelve different densities, where the water level in the tray is 0.5 cm. Image processing time (calculation time) uses an average program of 1.70 seconds with standard deviation of 0.15 seconds. The amount of 61 shrimp is the optimal number of shrimp where the count errors are still below 5%. The calculation time without loading time to bring the fries to the tray and reload back to the original container per shrimp at 61 shrimp is 0.027 seconds, while the conventional counting time (manual) per shrimp is 1.04 seconds.

Table 3 Calculation time using shrimp fry counter program

No	Number of Shrimp Fries	Density (Shrimp/cm^3)	Calculation process (second)
1	5	0.07	1.66
2	10	0.15	1.75
3	15	0.22	1.70
4	20	0.30	1.62
5	25	0.37	1.54
6	30	0.45	1.60
7	35	0.52	1.82
8	40	0.60	1.76
9	46	0.69	1.76
10	56	0.84	1.70
11	61	0.92	1.72
12	66	0.99	1.74
	Average		1.70

4. Conclusions

Thresholding operation to process the image of shrimp fry using greyscale value 70. Calculation of shrimp fry using image processing program that has been built has 98.49% accuracy. Average shrimp fry counting time using image processing program of 1.70 is seconds with standard deviation of 0.152 seconds. The counting time without loading time to bring the fries to the tray and reload back to the original container per shrimp fry in an image with 61 fry is 0.027 seconds. The weakness of the image processing program that has been built that is not able to separate the shrimp fry are mutually intersect, have not been able to detect the shrimp fry that are not visible black spots and there are still images of shrimp fry over segmentation.

References

[1] Widigdo B 2013 *Shrimp Cultivation with Biocrete Technology* (in Indonesian) (Jakarta: Kompas Media Nusantara)

[2] Seminar K B 2000 Design of Baby Fish Counter with Parallel Sensors (in Indonesian) *Jurnal Keteknikan Pertanian* **14**

[3] Directorate General of Aquaculture 2016 Shrimp Production in 2010-2014 (Jakarta: Ministry of Marine and Fisheries)

[4] Prasetyo E 2011 *Digital Image Processing and Its Application in Matlab* (in Indonesian) (Yogyakarta: Andi Publishing)

[5] Noercholis A, Muslim M A and Maftuch M 2013 Extraction of Roundness Feature for Counting the Number of Leukocytes in Fish Blood Cell Image (in Bahasa) *Jurnal EECCIS* **7** 35-40

[6] Ahmad U 2005 *Digital Image Processing and Its Programming Technique* (in Indonesian) (Yogyakarta: Graha Ilmu)

[7] Gonzalez R C, Woods R E 2002 *Digital Image Processing* (New Jersey: Prentice Hall, Inc)

[8] Nilasari N K N, Suprapto, and A A Supianto 2014. Identification of number of colonies in bacterial image with improved counting morphology method (in Indonesian) *Jurnal Mahasiswa PTIIK*

[9] Amalorpavam G, Naik H T, Kumari J and Suresha M 2013 Analysis of Digital Images Using Morphlogical Operations *International Journal of Computer Science & Information Technology* **5** 145

[10] Raid A, Khedr W, El-Dosuky M and Aoud M 2014 Image restoration based on morphological operations *Int. J. Comput. Sci. Eng. Inf. Technol* **4**

Application of Cold Storage for Raja Sere Banana (*Musa acuminata colla*)

Crismas S R S[1], Y A Purwanto[1,2] and Sutrisno[1]

[1] Department of Mechanical and Biosystem Engineering, Bogor Agricultural University, Indonesia
[2] Center for Tropical Horticulture Studies, Bogor Agricultural University, Indonesia

E-mail: arispurwanto@ipb.ac.id

Abstract. Raja Sere is one of the indigenous banana cultivars in Indonesia. This cultivar has a yellow color when ripen, small size and sweet taste. Traditionally, the growers market this banana cultivar to the market without any treatment to delay the ripening process. Banana fruits are commonly being harvested at the condition of hard green mature. At this condition of hard green mature, banana fruits can be stored for a long-term period. The objective of this study was to examine the effect of cold storage on the quality of raja sere banana that stored at 13^{0}C. Banana fruits cultivar Raja Sere were harvested from local farmer field at the condition of hard green mature (about 14 weeks age after the flower bloom). Fifteen bunches of banana were stored in cold storage with a temperature of 13°C for 0, 3, 6, 9, and 12 days, respectively. For the control, room temperature storage (28^{0}C) was used. At a storage period, samples of banana fruits ripened in the ripening chamber by injecting 100 ppm of ethylene gas at 25^{0}C for 24 hours. The quality parameters namely respiration rate, hardness, total soluble solids (TSS), change in color, and weight loss were measured. For those banana fruits stored at room temperature, the shelf-life of banana was only reached up to 6 days. For those banana fruits stored in cold storage, the condition of banana fruits was reached up to 12 days. After cold storage and ripening, the third day measurement was the optimal time for bananas to be consumed which indicated by the yellow color (lightness value = 68.51, a* = 4.74 and value b* = 62.63), TSS 24.30 °Brix and hardness 0.48 kgf, weight loss about 7.53-16.45% and CO_2 respiration rate of 100.37 mLCO_2/ kg.hr.

1. Introduction

Banana is one of the fruits that has a complete nutrient content, can be consumed by all age levels at any times. Bananas can be used as an alternative source of carbohydrates because they contain high carbohydrates. Also, bananas contain vitamin A, B_1, B_2, B_6, B_{12} and vitamin C. Bananas can be consumed as a fresh fruit or processed into snacks, such as banana chips, fried bananas, boiled bananas and juice [1,2].

The problems often encountered in bananas are susceptible to damage and decay due to the respiration process, and ethylene produced itself, so it cannot be stored for longer and cannot reach a wider marketing area. Postharvest handling of bananas among farmers, traders or exporters still needs to be repaired because there are still many neglected, causing bananas that reached the consumers have low quality and short shelf-life, so the impact on the selling price and consumer acceptance will also decrease. Generally, parameters of banana quality, namely have a perfect shape, same maturity level, bright fruit skin tone, smooth, fruit flesh is not flabby and good aroma and taste. Preferably, banana is harvested in mature green conditions so that distribution can reach the other regions, countries and also the maturation can be controlled.

Banana is climacteric fruits that in maturation stage marked by increase a peak of respiration rate and ethylene production, as well as changes in composition, discoloration, fruit softening and aroma production [3]. In climacteric fruits, the respiration process continues after the fruit is harvested and will continue the maturation process. Bananas are also included in perishable fruit, especially during the transportation and distribution process, so that the good post-harvest handling is required.

Raja Sere banana (Musa acuminata colla var Raja sere) is one of the indigenous banana that has much demand by the consumers because it can be consumed in fresh form and easy to find. Also, Raja Sere banana has a special aroma, good taste, and a small size so easy to consume directly and has a fairly high economic value. However, post-harvest handling of Raja Sere banana still has problems, especially among farmers who harvest not by the age of harvest-time and post-harvest handling is still not good cause banana has a low selling value and short shelf-life relatively due to the low quality of bananas. Banana damage can happen since from the field, during the process of transportation and distribution, and post-harvest handling so that it affects the acceptance of consumers.

Therefore, post-harvest handling is needed to maintain the quality of bananas, forms, taste, texture and nutritional content, to fulfill of consumers demand for longer periods of time and can be reached to a wider area by delayed the ripening process and maintaining the mature green of bananas so not to ripen quickly during storage. One way that can be used to extend banana shelf life is low-temperature storage. [4] stated that the temperature of bananas during storage and transportation ranges from 13-14°C with RH 90-95% because it can inhibit metabolism process, to slow the maturation process, extend the shelf-life of banana and also at that temperature banana will avoid from chilling injury [5].

According to [6], storage at low temperatures can decrease biochemical reactions that occur in fruit, inhibit respiration rate, reduce of ethylene production and it works and inhibits the process of softening the fruit flesh so that it can extend the shelf life of the fruit. [7], stated that based on the simulation model obtained, the tolerance of maturation delay (storage) of Ambon bananas at 15°C and then substituted with the addition of 100 ppm ethylene could reach up to 30 days of shelf life. Therefore, it is also necessary to do a low-temperature storage on the Raja Sere banana to extend the shelf life and overcome the problems that faced by banana growers, traders, and consumers to keep the good quality. The objectives of this study were to examine the effect cold storage for banana fruits cultivar Raja Sere on the changes in the quality after ripening process.

2. Materials and methods

The study was conducted in two stages, preliminary research conducted in September 2016 and the main research was conducted in November 2016 until February 2017 at the Laboratory of Food Processing and Agricultural Products (TPPHP) Department of Machinery and Biosystem Technology, Faculty of Agricultural Technology, Bogor Agricultural Institute. The storage temperature consists of two temperatures, i.e., room temperature (28°C) as control and low-temperature 13°C, the storage period of 0, 3, 6, 9, and 12 days and each treatment was carried out in three replications. The quality parameters measured after the ripening with ethylene, namely respiration rate, total soluble solids, firmness, color, and weight loss.

3. Results and discussion

3.1. *Respiration Rate*

The respiration rate during cold storage is relatively stable compared to banana storage at room temperature [10]. The results of the study in figure 1 show that the respiration rate of Raja Sere banana leaves a pattern of respiration of climacteric fruits. After ripening, measurement of quality parameters is only done until day four because, after that day, the fruit has been threshed.

Banana respiration rate (figure 1) after ripening on 0-day storage has the highest respiration rate on the third day after ripening, which was 102.05 $mLCO_2$/kg.hr. Banana respiration rate on three days storage at room temperature, after ripening has the highest value of CO_2 respiration rate on the third day, that is 120.99 mLCO_2/kg.h, while at 13°C storage occurs on the 4th day, that is 59, 91 mLCO_2/kg.hr). After 6 days storage at room temperature and at 13°C, the highest respiration rate after ripening was on day 3, i.e. 117.74 mLCO_2/kg.h (room temperature) and 89.79 mLCO_2/kg.h (13°C). At the storage temperature of 13°C for 9 days and 12 days, the rate of banana respiration after peak ripening occurred on the 3rd day, which was 115.83 mLCO_2/kg.h (storage 9 days) and 100.37 mLCO_2 / kg.h on second day (12 days storage).

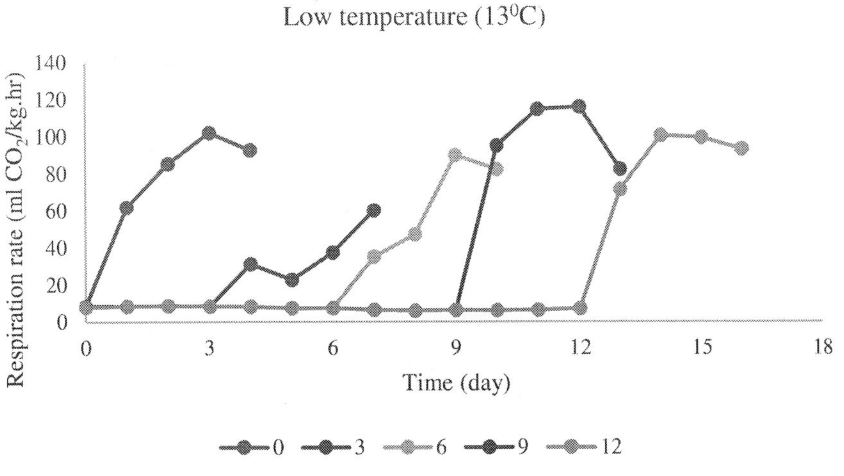

a

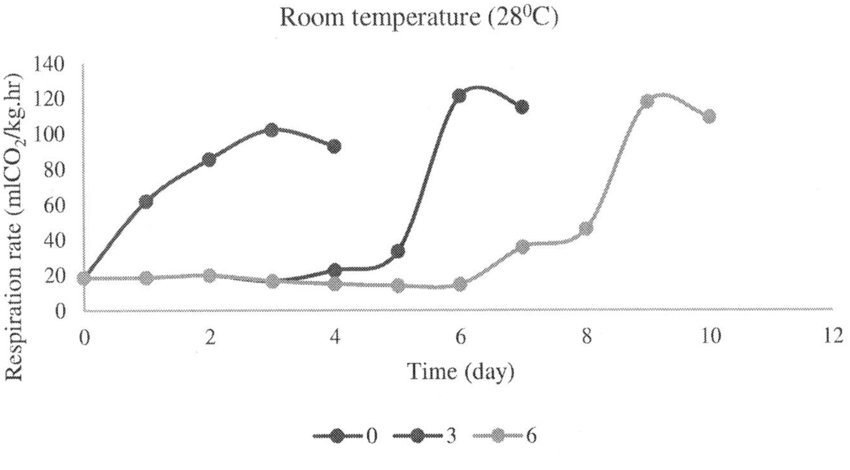

b

Figure 1. Respiration rate (mLCO_2/kg.h) of Raja Sere banana after storage time and ripening for a) 0, 3, 6, 9, and 12 days at 13⁰C and b) 0,3, and 6 days at room temperature

Cold storage can inhibit metabolic processes (transpiration, respiration, ethylene production) in bananas so that the maturation process can be slowed down. Bananas storage at room temperature (28°C) is only able to survive until the sixth day, while cold storage (13°C) can survive until the twelveth day. After ripening with ethylene, it further promotes the internal ethylene production of the banana itself and accelerates the maturation process, both stored at 13°C and 28°C and increasing respiration rates characterized by an increase in the amount of CO_2 concentration [11]. Threshing fruits after ripening showed the fruit has begun to senescence (already past the optimal of eating time).

3.2. Total Soluble Solid

The total soluble solids (TSS) of Raja Sere banana that has been ripened getting higher every day of measurement. This is because, during storage and maturation there will be a change of starch into sugars (sucrose, glucose, and fructose) which is soluble in water [12]. According [10], banana cv. Raja buluh have a total soluble solid that continue to increase from storage to maturation, which ranges from 18-32°Brix. The increase in total soluble solid during ripening may result from an increase in the concentration of organic solutes as a consequence of water loss. The increase may also be possible due to numerous anabolic and catabolic process taking place in the fruit preparing it for senescence [13].

The total soluble solids of Raja Sere banana at room temperature ranged from 13.4 - 24.67 °Brix (on storage 0 days to 6 days storage), whereas at cold temperatures ranged from 13.4 to 24.30 °Brix (on storage 0 days up to 12 days storage) (figure 2). The increase in total soluble solids occurring at room temperature occurs due to high-temperature differences, leading to chemical reactions and faster breakdown of carbohydrates by enzymes [14].

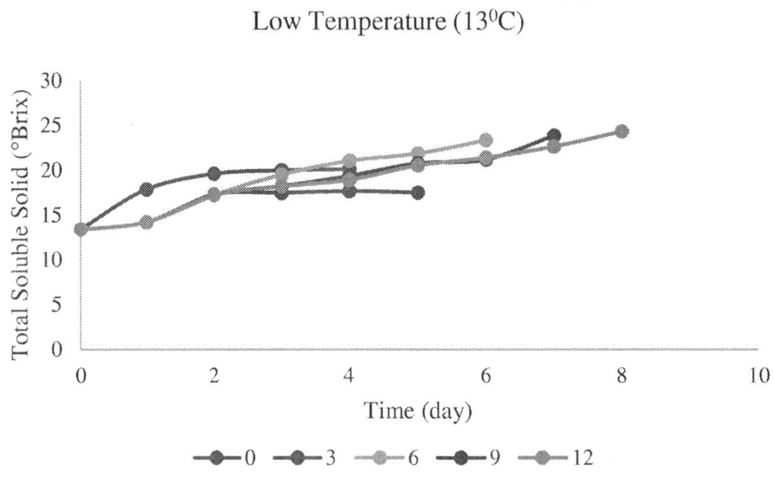

a

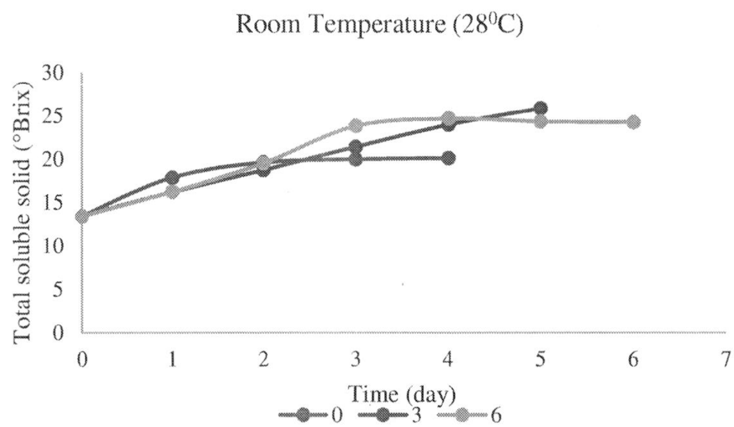

b

Figure 2. Total soluble solids of Raja Sere banana after storage time and ripening for a) 0, 3, 6, 9, and 12 days at 13^{0}C temperature and b) 0, 3, and 6 days at room temperature

According to [15], an ethylene-injected on banana will have a total soluble solid larger than un-injected bananas. Also, bananas stored at higher temperatures will result in higher total soluble solids (22.1 °Brix) compared to bananas stored at low temperatures (21.1 °Brix). This is by the measurement data obtained that the bananas stored at room temperature (28°C) are higher than the total soluble solids rather than at low temperatures (13°C). Increasing of total soluble solid was related with other parameters, such as the increase in weight loss that caused by the hydrolysis of starch into simple sugars, the decreasing hardness value is also caused by the hydrolysis of starch that causes the flesh of banana will soften than before.

3.3. Firmness

Firmness is one of the most crucial factors in determining the post-harvest quality of fruits, as banana. The level of banana firmness indicates the level of banana maturity. The more mature banana, the firmness of the flesh fruit will decrease (more softened) because during the process of maturation occurs the reshuffle of starch into sugar and wall cell reshuffle [15]. According to [13] said that the decrease in firmness,during ripening may be due to breakdown of insoluble protopectin into soluble pectin or by cellular disintegration leading to membrane permeability. The hard or semi-hard banana fruits are not liked by most of the consumers. Therefore, to develop desirable quality, the banana fruits need to be ripened artificially for enhancing consumer acceptability. The value of firmness of Raja Sere banana after the ripening continues to decrease every day of measurement.

The value of firmness at room temperature storage ranged from 0.36 - 6.06 kgf, whereas in cold storage 0.81 - 6.06 kgf. The highest firmness value is in initial storage, which is 6.06 kgf, while the lowest is in 6 days storage (room temperature) and 12 days (temperature 13°C) (figure 3). This occurs because during storage also occurs metabolic processes such as transpiration, respiration, breakdown of cell walls, although it is slow and will be faster after ripening.

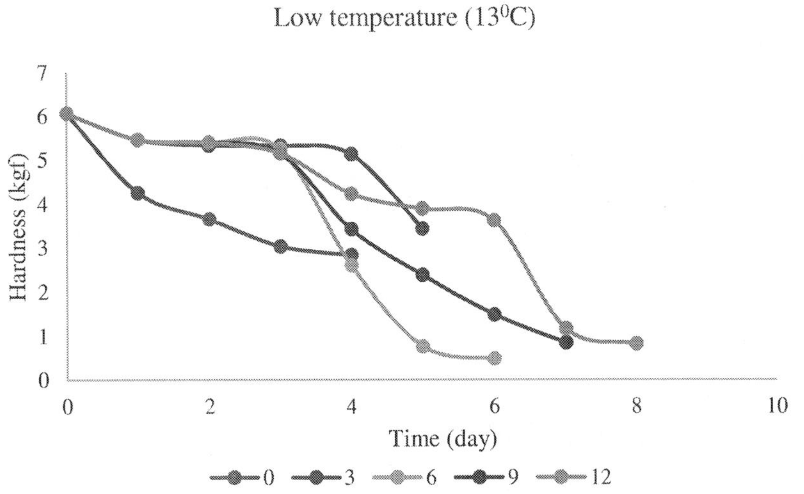

a

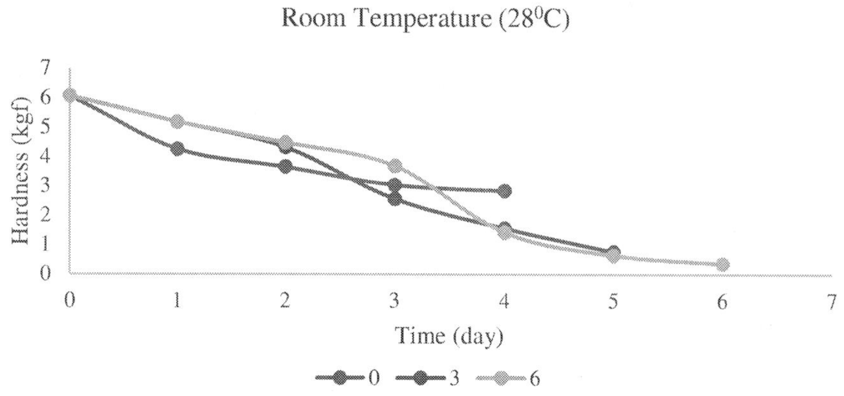

b

Figure 3. The firmness value of Raja Sere banana after storage time and ripening for a) 0, 3, 6, 9, and 12 days at 13^0C temperature and b) 0, 3, and 6 days at room temperature

The injection of ethylene on banana ripening can accelerate the softening of banana flesh compared with bananas without ethylene [11]. The higher the temperature of banana ripening, the lower effect of ethylene being injected against the softening of the fruit flesh because as long as the storage of bananas stored at high temperatures has produced internal ethylene.

According to [16], softening of banana flesh can occur with three processes, including the conversion of starch into sugar, the breakdown of pectin components and the movement of water from the skin of the fruit to the flesh during the maturation process. Bananas stored for 3 days have a higher firmness value than 6, 9, and 12 days storage. This proves that storage duration affects the firmness of bananas after ripening. In addition to the long-term storage, storage temperature also affects the firmness value. The other opinion come from [17], that said that hydrolysis of starch to sugar during ripening causes an increase in osmotic pressure is usually associated with a decrease in turgor pressure which may account for shortening during ripening and would cause the firmness decrease.

According to [15], fruit size also determines the fruit firmness, because of the smaller size of fruit the faster the occurrence of decomposition of sugar and starch content during maturation, thus accelerating the process of banana maturation of smaller size than larger. It is also supported by research conducted by [18] which states that smaller tomatoes will have a higher total soluble solid value than large tomatoes because the smaller size of the fruit the easier it will be to degrade the sugar and starch in the fruit. Decreasing level of banana firmness associated with the increasing value of total soluble solids.

3.4. Color

The color parameter is the crucial parameter of consumer acceptance. During storage, the measurement of color to the value of Lightness (brightness), a value, and b value to determine the level of maturation of bananas (figure 4). The color of bananas at 13°C did not change too much, which is still green with maturity index 1 (green) in accordance with the index of banana maturity [19], while storage at room temperature, bananas began to change color on the 15th day to yellow with a maturity index of 4, ie more yellow than green, so it was decided to store temperature in 13°C temperature for 12 days.

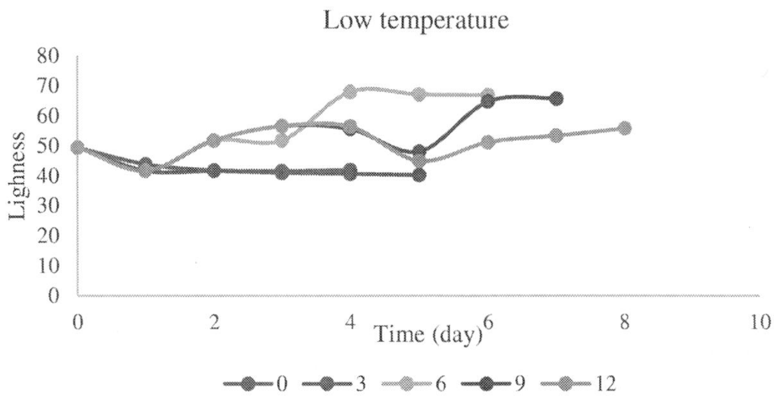
Low temperature
Lighness
Time (day)
0
3
6
9
12

a

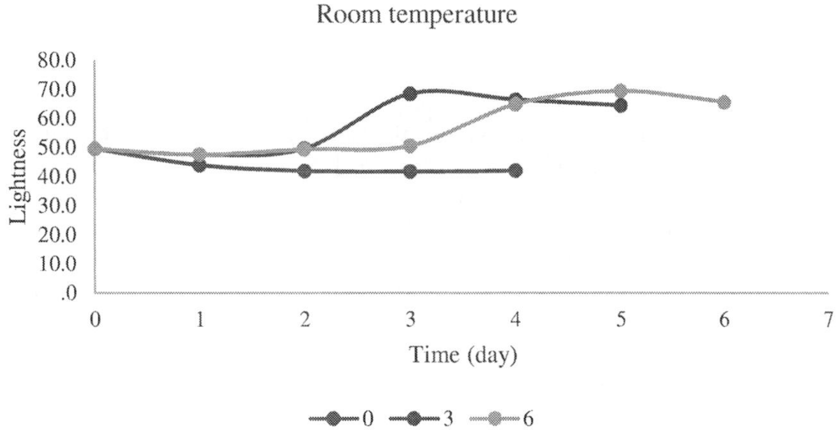
Room temperature
Lightness
Time (day)
0
3
6

b

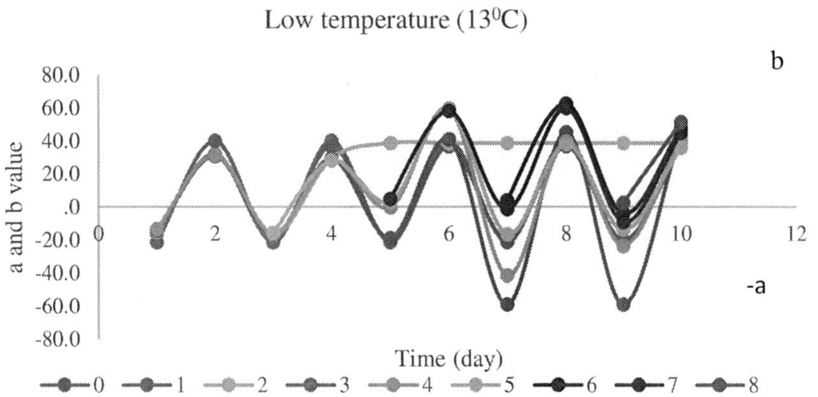
Low temperature (13⁰C)
b
-a
a and b value
Time (day)
0
1
2
3
4
5
6
7
8

c

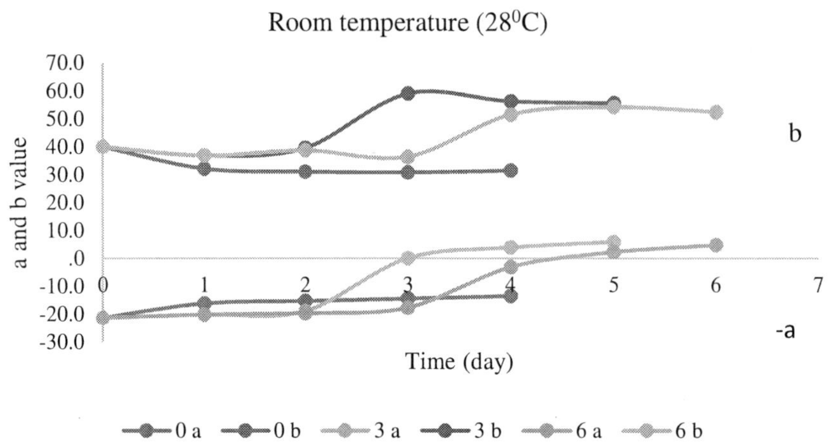

d

Figure 4. The color (Lightness, a and b value) of Raja Sere banana after storage time and ripening for (a and c) 0, 3, 6, 9, and 12 days at 13^{0}C temperature and (b and d) 0, 3, and 6 days at room temperature

The change of banana skin color from green to yellow is due to the degradation of chlorophyll (green pigment) to a carotenoid (yellow pigment). Degradation of chlorophyll occurs immediately near the climacteric peak of fruit [20]. According to Burg and Burg (1965) in [21], states that ethylene has an effect on the degradation of chlorophyll during maturation, that also said by [13] that ethylene gas is known to accelerate the chlorophyll degradation and induce yellowness in green tissue of many fruits. The rate of degradation of banana chlorophyll along with the increase in temperature [12] also as [22 & 23] said that banana fruit has a unique contradiction compared to the most other fruits, which degreens at a faster rate at higher temperature. According to [24], the amount of chlorophyll during the maturation decreases to zero and then during the senescence period, enzyme activity becomes high and then dark color will cover the skin of the fruit (the fruit becomes rotten).

3.5. Weight loss

During storage, bananas are heavily weighted due to a chemical component degradation. Weight loss in 13°C temperature storage is lower (constant) than to room temperature storage. It proves that the higher the temperature storage of the fruit, the greater weight loss activity. Weight loss caused by the metabolism process in banana tissue. Respiration and transpiration play an important role in the shrinkage of banana weight. According to [15], smaller bananas will experience greater transpiration and respiration than larger bananas, causing a high weight loss. Raja Sere banana is a small banana fruit, with a fruit length of 10-15 cm [2].

The value of weight loss at room temperature storage ranged from 6.25 to 10.67%, while at storage temperature13°C 7,53-13,05% after storage for 12 days. During observation 4 days after ripening, the weight loss of banana was smaller after storage at room temperature compared with storage at 13°C. This is because, during storage at room temperature, the weight loss of weight occurs greater than at the storage temperature of 13°C (figure 5).

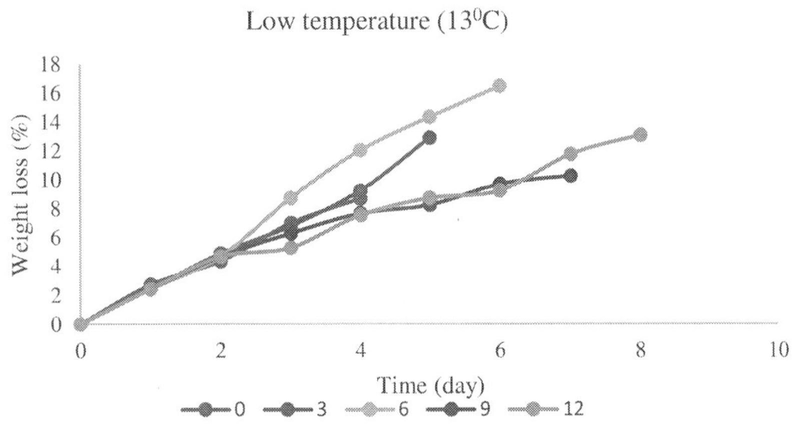

a

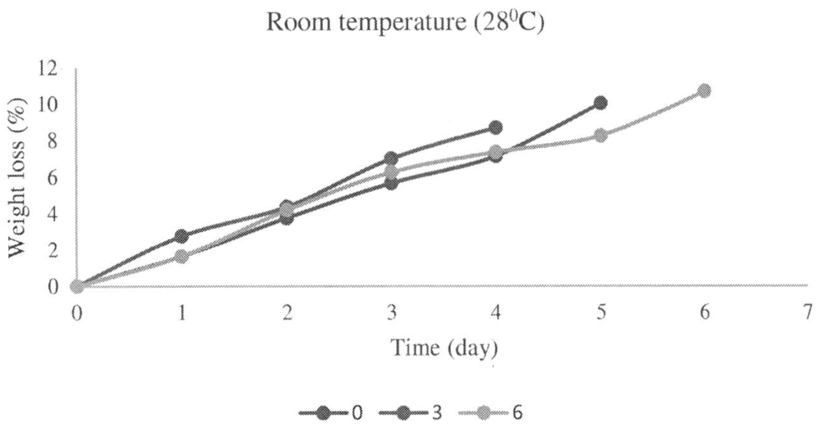

b

Figure 5. The weight loss of Raja Sere banana after storage time and ripening for a) 0, 3, 6, 9, and 12 days at 13^0C temperature and b) 0, 3, and six days at room temperature

According to [25], the softening of fruit texture is a major process of ripening, leading to a loss of firmness, facilitating pathogen infection, increasing postharvest decay, and reducing shelf-life and fruit quality. It also is shown that the weight loss of banana fruit will increase as long as ripening process. The weight loss of banana fruit was increased as long as the storage-time so it also can show that the quality of banana decreased or come into the senescene process. The weight loss parameter is related to the firmness parameter, the firmness will decrease so the weight loss will high.

4. Conclusions

The conclusion from this study:

1. Storage of Raja Sere banana at room temperature can only reached up to 6 days, while at 13°C reached 12 days.
2. The longer the storage time was done, the value of firmness will decrease. Otherwise the total soluble solids and weight loss will higher and will decrease when starting to senescence.
3. The 3rd day after-ripening is the optimal eating time of banana with the total soluble solids of 24.30°Brix, hardness of 0.36 kgf and full yellow color (a * = 4.74 and b * = 52.49).

References

[1] Purwoko B S, Suryana K 2000 Effect of simple temperature and coating for cavendish banana fruit quality (in Indonesian) *Buletin Agron* **28**(3) 77-84

[2] Prabawati S, Suyanti and Setyabudi D A 2008 Postharvest and processing technology for banana fruit (in Indonesian) Balai Besar Penelitian dan Pengembangan Pascapanen Pertanian *Badan Penelitian dan Pengembangan Pertanian* 1-64

[3] Ritenour M A, Mangrich M E, Beaulieu J C, Rab A and Saltveit 1997 Ethanol effects on the ripening of climateric fruit *Post. Biol. Tech.* **12** 35-42

[4] Zhang H, Yang S, Joyce D C, Jiang Y, Qu H and Duan X 2010 Physiology and quality response of harvested banana fruit to cold shock *Post. Biol. Tech.* **55** 154-159

[5] Wills R B H, Harris D R, Spohr L J and Golding J B 2014 Reduction of energy usage during storage and transport of bananas by management of exogenous ethylene levels. *Post. Biol. Tech.* **89** 7-10

[6] Wills R B H, Wc Glasson W B, Graham D, Lee T H and Hall E G 1989 Postharvest an introduction to the physiology and handling of fruits and vegetables *AVI Publ.*

[7] Iswari K 2002 *Study of Storage and Ethylene Utilization for Ambon Banana Artificial Maturation with Stepped Temperature Method* (in Indonesian) Magister thesis (Bogor: Bogor Agricultural University)

[8] Winarno F G 2002 *Physiology of Horticulture Off Harvest* (in Indonesian) (Bogor: M-Brio Press Bogor)

[9] Lamona A 2015 Packaging and low temperature storage influence for fresh red curly chili quality (in Indonesian) *J. Keteknikan Pertanian* **3**(2) 145-152

[10] Werdiningsih W 2008 *Study of Raja Bulu Banana Quality in Storage and Curing* (in Indonesian) Magister thesis (Bogor: Bogor Agriculutral University)

[11] Ahmad S, Thompson A K, Hafiz I A and Asi A A 2001 Effect of temperature on the ripening behavior and quality of banana fruit *Int. J. Agri. Biol.* **03** (2) 224-227

[12] Mitra S K 1997 Postharvest Physiology and Storage of Tropical and Subtropical Fruits CAB International (London)

[13] Mahajan B V C, Kaul T, Gill M I S, Dhaliwal H S, Ghumam B S and Chahil B S 2010 Studies on optimization of ripening techniques for banana *J. Food Sci. Technol (May-June 2010)* **47**(3) 315-319

[14] Broto W, Suyanti, Sjaifullah and Duracham 1994 Papaya (*Carica papaya*, L.) nectar quality analysis (in Indonesian) cv Dampit dan Paris *Jurnal Hortikultura* **4**(1) 34-41

[15] Ahmad S, Nasir M A, Hussain Z, Anwar R and Thompson A K 2007 Effect of fruit size and temperature on the shelf life and quality of ripe banana fruit. *Sarhad J. Agri.* **23**(2007) 45-54

[16] Thompson A K 1996 Effect of temperature on the ripening behavior and quality of banana fruit *Int. J. Agr. Biol.* **3**(2) 224-227

[17] Jaiswal P, Jha S H, Kaur P P, Bhardwaj R, Singh A K and Wadhawan V 2012 Prediction of textural attributes using color values of banana (*Musa sapientum*) during ripening AFSTI *J. Food Sci. Tech.*

[18] Salunkhe D K, Bolin H R and Reddy N R 1991 *Processing and nutritional quality of fruits and vegetables 2nd Ed.* vol 1 (Florida: CRC Press Boca Raton)

[19] Kader A A 2005 *Banana: Recommendations for maintaining postharvest quality* (California: University of California Davis)

[20] Eskrin NAM 1990 *Biochemistry of Foods* (London: Academic Press)

[21] Siriboon N and Propapan B 2000 *A Study on the Ripening of "Namwa" Banana* (Bangkok: University Bangkok Thailand)

[22] Yang X, Zhang Z, Joyce D, Huang X, Xu, L and Pang X 2009 Characterization of chlorophyll degradation in banana and plantain during ripening at high temperature *Food Chemistry* **114** 383-390

[23] Yang X, Song J, Fillmore S, Pang X and Zhang Z 2011 Effect of high temperature on color, chlorophyll fluoresence and biosynthesis in green-ripe banana fruit *Postharvest Biology and Technology* **62** 246-257

[24] Wills R B H, Mc. Glasson W B, Graham D, Lee T H and Hall E G 1981 *Postharvest Introduction to The Physiology and Handling of Fruit and Vegetables* (New York: An Avi Book Van Nostrand Reinhold)

[25] Wei J, Ma F, Shi S, Qi X, Zhu X and Yuan J 2010 Changes and postharvest regulation of activity and gene expression of enzymes related to cell wall degradation in ripening apple fruit *Postharvest Biology and Technology* **56** 147-154

Effect of Agitation in Alkalization Process on the Characteristics of Sodium Carboxymethyl Sago and Cassava Starches

Titi C S[1], R Fachrudin[1], E Ruriani[1] and I Yuliasih[1]

[1] Department of Agroindustrial Technology, Bogor Agricultural University, Bogor, Indonesia

E-mail: titi-cs@apps.ipb.ac.id

Abstract. Sodium carboxymethyl starch (Sodium CMS) is a modified starch prepared by two successive processes, alkalization and etherification. Alkalization will change the activated hydroxyl group of starch to more reactive alkoxide (St-O$^-$), and then carboxymethyl group will substitute the hydroxyl group into sodium CMS. This research investigated the effect of agitation (1000 rpm of stirring and 4000 rpm of homogenization) in alkalization process to the modification of native starch into sodium CMS. Cassava and sago starches were mixed with sodium hydroxide (1.8 and 1.9 moles per mole anhydrous glucose units). The combination of NaOH and homogenizing gave the highest degrees of substitution for cassava (DS 0.73) and sago (DS 0.55) starches. The sodium CMS characteristics (paste clarity, water and oil absorption capacities, solubility, swelling power) were a function of mixing method but not on the amount of NaOH used.

1. Introduction

Starch is one of natural polymers that has been used in food and non-food applications. However, the utilization of native starch is limited in physical and chemical properties, since it is insoluble in cold water, easily degraded in high temperature, has low paste clarity and low stability. A common way to overcome these weaknesses is to modify the molecular structure of starch by physical, chemical, and or a combination of both treatments, which will improve the properties of native starch [1]. Modification of starch can be carried out by cleavage the molecular structure, rearranging the molecular structure, oxidation or substitution of the chemical group on the starch molecule [2], changing the chemical structure through D-glucopyranosyl starch linkage which consists of anhydrous glucose units (AGU). Carboxymethylation is one method of modification by substitution of the starch molecule, which produces starch with low gelatinization temperature, high solubility, and high shelf life. The carboxymethylation process takes place by substituting the native starch hydroxyl group (-OH) with the carboxymethyl group (CH_2COO-) to produce Na-carboxymethyl or carboxymethyl starch (CMS)[3]. The utilization of Na-carboxymethyl starch is used as a disintegrant in the pharmaceutical industry [4] as well as sizing and printing agent in the textile industry [5-6].

Carboxymethylation takes place through two stages, namely alkalization and etherification. The Na-carboxymethyl starch can be produced by the reaction between native starch and sodium monochloro-acetate (SMCA) by induction of NaOH [7]. Alkalization stage uses a strong base which converts an activated starch hydroxyl group and transforms it into a more reactive form of alkoxide (St-O-) [3]. The etherification process is conducted in the second stage by substituting the carboxymethyl group of SMCA with alkoxide and producing Na-carboxymethyl starch.

The formation of Na-carboxymethyl starch can be detected through the degrees of substitution (DS) produced. Factors that affect the production of Na-carboxymethyl starch include starch type, NaOH, and SMCA concentration. The alkalization stage also influences Na-carboxymethyl starch production. Agitation at this stage will determine the effectiveness of alkalization to activate the starch carboxyl group. Hasanudin [8] states that more collisions between molecules of a substance will cause the reaction to take place more quickly. Mixing can also expand the contact area so as to increase the chance of collisions between molecules. Each type of agitation will produce different reaction rates. Studies on the optimum concentrations of NaOH and SMCA have been conducted previously [3]. The type of starch used will influence the different characteristics of Na-carboxymethyl starch [3, 9].

This study aims to investigate the effect of agitation and mixing (stirring and homogenization) in the alkalization stage and the concentration of NaOH used on the characteristics of the Na-carboxymethyl starch produced from sago and cassava starches.

2. Materials and method

2.1. *Production of Na-Carboxymethyl Starch*

Production of Na-carboxymethyl starch is carried out based on Ref. [3] with slight modifications. The process was divided into two successive processes, alkalization and etherification (carboxymethylation).

2.1.1. Alkalization process. Cassava and sago starch were obtained from small industries in Bogor. Starch was dried and sieved to pass 100 mesh, and its moisture content determined for calculating the number of chemicals used in the reaction. About 50 g of starch is suspended in isopropanol with starch ratio: isopropanol 1:14 (w: v). Two types of agitation were used in this experiment, (1) hot magnetic stirrer with 4 cm magnetic bar and agitation speed of 1000 rpm and (2) Ultra Turrax homogenizer with 4 cm of rotor diameter and agitation speed of 4000 rpm. The suspension was then added NaOH at a ratio of 1.8 - 1.9 moles/mole of AGU (anhydroglucose unit) and mixed for 20 minutes at 40°C. Viscosity and density of each mixture was measured for determining the Reynold number of fluids

2.1.2. Carboxymethylation Process. The reactor used consists of a three-neck flask with reflux. Sodium monochloroacetate (SMCA) was added with a ratio of 1.5 moles/mole AGU. Continous stirring at 250 rpm was continued for 3 hours after the addition of SMCA with the reaction temperature kept constant at 40°C. After the mixing process, the pH value of the medium was set in the range pH 5.5 - 6.5 by adding 50% HCl solution to stop the substitution reaction process. The formed Na-carboxymethyl starch was washed with 85% ethanol and filtered until the filtrate gave a negative response to the silver nitrate solution indicative of the absence of SMCA. The Na-carboxymethyl starch is then dried with oven at 60°C.

2.2. *Characterization of Na-Carboxymethyl Starch*

Etherification parameters; the degree of substitution (DS) and reaction efficiency (RE) were determined according to ISO 11216-1998 method [10]. Characterization of Na-carboxymethyl starch solubility and swelling power, water and oil absorption capacities, and 1 % paste clarity at 650 nm were conducted with slight modification [11].

3. Results and discussion

The starch used in this study has amylose content of 25.12% for sago and 19.15% for cassava. Ref. [14] states that sago starch contains 27% amylose and 73% amylopectin, whereas according to Ref. [15], cassava amylose component was 17% and 83% amylopectin, which is in reasonable agreement with our findings. The amylose and amylopectin ratio will affect the characteristics of the Na-carboxymethyl starch formed. Amylose tends to be easier to swell when compared with amylopectin and this helps the distribution of chemical compounds into starch granules.

3.1. *Agitation and flow pattern in alkalization process*

The first step is alkalizing with NaOH as a promoter; therewith producing alkaline starch. Isopropanol serves as a reaction medium and will dissolve minor components such as fiber, ash, fat, and protein. The alkalization step is an opening step for activating the starch hydroxyl group (St-OH) into a negatively charged alkoxide group [3]. Repulsive force of the negative charge will make the starch granules swell [16] thus facilitating diffusion of chemicals into the starch granules [3].

Swelling is rarely a spontaneous process, and requires the input of energy, usually supplied by mechanical shear provided by various types of mixers; in this research, a magnetic stirrer and rotor-stator homogenizer are used. It is important to match the agitator and agitation condition to the characteristics of product; therefore, dimensionless analysis is used, as described in table 1. Cassava starch produced higher viscosity and swelling capability compared to sago starch, as well as low density of mixture. It is expected that these properties will be beneficial in the etherification process [17].

Table 1. Effects of agitation types and starch mixture to the Reynolds Number (N_{Re})

Type of agitation	Viscosity (Pa.s)	Density (kg/m^3)	Ø impeller (m)	Speed (rpm)	N_{Re}
Cassava Starch					
Stirring	0.12	857.35	0.04	1 000	11 431
Homogenizing	0.12	857.35	0.04	4 000	45 725
Sago Starch					
Stirring	0.02	876.05	0.04	1 000	70 084
Homogenizing	0.02	876.05	0.04	4 000	280 336

The results of N_{Re} calculation (table 1) for each type of stirring or agitation using both cassava and sago starch indicate a Reynolds number over 4000, which is indicative of turbulent flow [18] that is stronger in the homogenizer than in the magnetic stirrer.

3.2. *Characteristics of Na-Carboxymethyl Starch*

The alkalization stage weakens the alpha-helix in starch and breaks down the crystalline structure [16], which can facilitate solvent and SMCA entering the starch granules and substituting the alkoxide group with the carboxymethyl group from SMCA [17].

Table 2. Substitution parameters for carboxymethyl starch production

Parameter	Magnetic Stirring		Homogenizing	
	Moles of NaOH per mole AGU			
	1.8	1.9	1.8	1.9
Cassava Starch				
Degree of substitution	0.50±0.02^a	0.53±0.01^a	0.79±0.02^b	0.82±0.03^b
Reaction Efficiency (%)	33.27±1.39^a	35.53±0.68^a	52.56±1.55^b	54.70±1.94^b
Sago Starch				
Degree of substitution	0.54±0.02^a	0.55±0.02^a	0.70±0.01^b	0.73±0.02^b
Reaction Efficiency (%)	35.90±1.63^a	36.93±1.60^a	47.18±0.97^b	48.60±1.06^b

*Means within a row related to particular parameter with the same superscript letter are not significantly different as α=0.05 confidence level.

3.2.1. The degree of Substitution and Reaction Efficiency. The degree of substitution (DS) is the average number of substituents per anhydroglucose unit (AGU). Each AGU contains three hydroxyl groups, therefore the maximum DS value obtainable is 3. The degree of substitution in carboxymethylation of starch is defined as the average number of starchy hydroxyl groups substituted by the carboxymethyl group [19]. The factors that affect DS and are investigated here are starch type and amount of alkaline used; other factors can be substituent concentration, reaction temperature, and time, type of solvent and water content (see table 2).

The results of DS measurements (table 2) showed a medium degree of substitution (DS 0.50-0.82) on both sago and cassava starches, and the amount of NaOH concentration added does not affect the effectiveness of the alkalization reaction. This fact is related to the statement of Ref. [3], that the increasing of NaOH concentration in the reaction will increase the DS value. The higher NaOH concentration used will make the etherification reaction better because of the increased reactivity of the alkali starch to SMCA used in the reaction. The DS value for samples prepared with the homogenizer are higher than when mixed with a magnetic stirrer. During homogenization smaller droplets are made that are more reactive, leading to more reactive alkali starch groups so that etherification reaction can run more optimally.

3.2.2. Paste Clarity. The clarity level of starch paste is positively correlated with the swelling of starch granules (table 3); high swellability increasing the clarity of pastes [20]. Table 3 shows that clarity of Na-carboxymethyl starch pastes is higher than for native starch. The starch produced using a magnetic stirrer tends to have higher clarity than those produced using a homogenizer; the differences between cassava and sago are very small. Homogenization will cause the expansion of the amorphous regions resulting from the alkalization step which is positively correlated with water and oil absorption capacity.

Table 3. Physico-chemical characteristics of Na-carboxymethyl starch

Parameter	Native Starch	Magnetic Stirring		Homogenizing	
		Moles of NaOH per mole AGU			
		1.8	1.9	1.8	1.9
Cassava Starch					
1% Paste Clarity (%T)	81.03 ± 0.21^{a}	98.05 ± 0.48^{c}	97.58 ± 0.49^{c}	90.15 ± 0.27^{b}	89.80 ± 0.36^{b}
Water absorption capacity (%)	172.00 ± 0.01^{a}	986.44 ± 0.54^{b}	987.00 ± 0.56^{b}	989.56 ± 0.81^{c}	990.78 ± 1.36^{c}
Oil absorption capacity (%)	131.00 ± 0.33^{a}	200.89 ± 0.54^{b}	201.56 ± 1.56^{b}	204.67 ± 0.94^{c}	205.89 ± 0.98^{c}
Solubility at 70 °C	14.00 ± 0.01^{a}	44.00 ± 2.53^{b}	46.67 ± 2.07^{b}	68.00 ± 1.89^{c}	70.00 ± 0.94^{c}
Swelling power at 70 °C	9.07 ± 0.01^{a}	13.57 ± 0.64^{b}	14.01 ± 0.40^{b}	15.32 ± 0.18^{c}	15.97 ± 0.72^{c}
Sago Starch					
1% Paste Clarity (%T)	76.47 ± 0.31^{a}	96.50 ± 0.35^{c}	96.28 ± 0.68^{c}	89.85 ± 0.63^{b}	89.82 ± 0.57^{b}
Water absorption capacity (%)	110.67 ± 0.94^{a}	836.11 ± 1.86^{b}	838.67 ± 3.10^{b}	982.44 ± 0.81^{c}	982.89 ± 1.96^{c}
Oil absorption capacity (%)	120.67 ± 0.67^{a}	182.11 ± 0.89^{b}	184.11 ± 0.50^{c}	201.00 ± 0.81^{d}	202.00 ± 1.84^{d}
Solubility at 70 °C	20.00 ± 0.01^{a}	60.00 ± 2.53^{b}	62.67 ± 2.07^{b}	85.33 ± 2.07^{c}	87.33 ± 1.63^{c}
Swelling power at 70 °C	9.30 ± 0.01^{a}	12.73 ± 0.33^{b}	12.80 ± 0.30^{b}	14.11 ± 0.29^{c}	14.32 ± 0.73^{c}

*Means within a row related to particular parameter with the same superscript letter are not significantly different as α=0.05 confidence level.

3.2.3. Water and Oil Absorption Capacity. Table 3 shows that water absorption of starch increased after modification for both cassava and sago starch; which is in line with [23]. As mentioned, this increase occurs due to the loss of the crystalline structure [24], which weakens the structure of starch granules so water can easily enter. The ability of starch to absorb oils shows that starch also has a lipophilic portion. Oil absorption capacity is used to measure the product hydrophobicity; the increase in oil absorption is caused by the introduction of the Na-carboxymethyl group (CH_2COONa). The Na-carboxymethyl starch will absorb the oil in the form of a fatty acid (R-COOH). The tendency to bind between fatty acids and Na-carboxymethyl starch is related to the same molecular polarity. Natural starch (R-OH) will have more difficulty to bind to fatty acid (R-COOH) due to differences in polarity.
3.2.4. Swelling and solubility. Sodium carboxymethyl starch is cold water soluble starch, has high dispersed distribution, and excellent water absorption capacities compared to native starch. Carboxymethylation can improve the swelling and solubility of native cassava and sago starches (table 3). Sago starch revealed slightly higher solubility compared to cassava starch, since sago and cassava starches contained different amounts of amylose 25.12% and 19.15%, respectively and as a result different crystallinity. During the modification process, the molecular weight decreases and the degree of polymerization decreases [25], leading to an increase in the hydrophilic nature of starch [26]. The crystalline structure weakens and makes the starch more easily hydrated by water [16].

4. Conclusions
The production of Na-carboxymethyl starch can be carried out by carboxymethylation process using NaOH as an alkalizing agent and SMCA as etherification agent. The alkalization step becomes the initial stage that will affect to the etherification stage. The stirring type of the alkalization stage affects the efficiency of the reaction. The type of starch affects the characteristics of Na-carboxymethyl starch which is related to amylose and amylopectin compositions, but not too much. The highest DS values were found for sago starch treated with NaOH using a homogenizer that also has positive effect on paste clarity, solubility, swelling power, water absorption, oil absorption.

Acknowledgement
This publication is part of the research funded by The Indonesian Ministry of Research, Technology, and Higher Education through National Strategic Research Program No. 079/SP2H/LT/DRPM/II/2016 to Titi Candra Sunarti, Bogor Agricultural University.

References
[1] Liu Q 2005 *Food Carbohydrates: Chemistry, Physical Properties and Applications* (Boca Raton: Taylor & Francis)
[2] Wurzburg O B 1986 *Modified Starches: Properties and Uses* (Florida: CRC Pr.)
[3] Sangseethong K, Ketship S and Sriroth K 2005 *Starch* **57** 84
[4] Shah U and Augsburger L 2002 *Pharm Develop Technol* **7** 345–359
[5] Ragheb A A, El-Sayiad H S and Hebeish A 1997 *Starch* **49** 238–245
[6] Tatongjai J and Lumdubwong N 2010 *Carbohyd Polym* **81** 377–384
[7] Lawal O S, Lechner M D and Kulicke W M 2008 *Int J Biol Macromol* **42** 429–435
[8] Hasannudin M 2015 Factors affecting reaction rate.(in Indonesian). [accessed at 31 Januari 2017]. http://kimiadasar.com/faktor-faktor-yang-mempengaruhi-laju-reaksi/
[9] Yaacob B, Amin M, Hashim K and Bakar B *Iranian Polym.* **20**(3): 195–204
[10] [ISO] International Organization for Standardization 1998 (Geneva: Intenational Organization for Standardization)
[11] Li Y, Liu C, Tan Y, Xun K, Lu C and Wang P 2014 *Carbohydr. Polym.* **110** 87
[12] Hu X 2013 *Cereal Chem.* **90** 24
[13] Ester R F, John K, Xin Q 2004 *J Cereal Sci.* **39** 151–165
[14] Flach M 1983 *Thes Sago Palm. Food and Agriculture Organization of United Nation* (Rome: FAO)

[15] Rickard J E, Blanshard J M V and Asaoka M 1992 *J Sci Food Agric.* **59** 53–58
[16] Chen J and Jane J 1994 *Cereal Chem.* **71** 623
[17] Kooijmana L M, Ganzeveld K J, Manurung R M and Heeres H J 2003 *Starch* **55** 495
[18] Earle R L 1983 *Unit operations in food processing 2nd Edition* (Oxford: Pergamon Press)
[19] Zhou X and Yang Q G 2007 *J Mater Process Technol* **183** 407–411
[20] Suriani A R 2008 Study of Effects of Heating and Repeated Cooling on Physical and Functional Characteristics of Garut Modified Starch (Marantha arundinacea) (In Indonesian) (Bogor: Bogor Agricultural University)
[21] Thitipraphunkul K, Uttapap D, Piyachomkwan K and Takeda Y 2003 *Carbohyd Polym.* **53** 317–324
[22] Zayas J F 1997 *Functional of Proteins in Food* (Jerman: Springer-Verlag)
[23] Collado L S and Corke H 1999 *Food Chem.* **65** 339–346
[24] Spychaj T, Zdanowicz M, Kujawa J and Schmidt B 2013 *Starch* **65** 22
[25] Jiao L V, Tan S, Peng L, Chen H and Liu X 2013 *Mater Appl.* **2**(2): 1–11
[26] Cardoso M B, Putaux J L, Samios D and da Silveira N P 2007 *Carbohydr Polym* **70** 160–165

Measurement of Moisture Sorption Isotherm by DVS Hydrosorb

Y R Kurniawan[1,2], Y A Purwanto[2], N Purwanti[2] and S Budijanto[3]

[1] Center for Appropriate Technology Development, Indonesian Institute of Sciences, Indonesia
[2] Department of Mechanical and Biosystem Engineering, Bogor Agricultural University, Indonesia
[3] Department of Food Science and Technology, Bogor Agricultural University, Indonesia

E-mail: yorizk1@gmail.com

Abstract. Artificial rice made from corn flour, sago, glycerol monostearate, vegetable oil, water and jelly powder was developed by extrusion method through the process stages of material mixing, extrusion, drying, packaging and storage. Sorption isotherm pattern information on food ingredients used to design and optimize the drying process, packaging, storage. Sorption isotherm of water of artificial rice was measured using humidity generating method with Dynamic Vapor Sorption device that has an advantage of equilibration time is about 10 to 100 times faster than saturated salt slurry method. Relative humidity modification technique are controlled automatically by adjusting the proportion of mixture of dry air and water saturated air. This paper aims to develop moisture sorption isotherm using the Hydrosorb 1000 Water Vapor Sorption Analyzer. Sample preparation was conducted by degassing sample in a heating mantle of 65°C. Analysis parameters need to be fulfilled were determination of P_o, sample data, selection of water activity points, and equilibrium conditions. The selected analytical temperatures were 30°C and 45°C. Analysis lasted for 45 hours and curves of adsorption and desorption were obtained. Selected bottom point of water activity 0.05 at 30°C and 45°C yielded adsorbed mass of 0.1466 mg/g and 0.3455 mg/g, respectively, whereas selected top water activity point 0.95 at 30°C and 45°C yielded adsorbed mass of 190.8734 mg/g and 242.4161mg/g, respectively. Moisture sorption isotherm measurements of articial rice made from corn flour at temperature of 30°C and 45°C using Hydrosorb showed that the moisture sorption curve approximates sigmoid-shaped type II curve commonly found in corn-based foodstuffs (high-carbohydrate).

1. Introduction

Artificial rice, also called rice analogues, is made from local carbohydrate sources besides rice and wheat flour. This rice was created in response to the problem of the availability of rice as the staple foods. The high consumption of rice that is not offset by increase in rice productivity has the potential to flooding of imported rice in market. In addition, increasing consumption of food made from wheat flour that in fact is made from imported wheat makes out the local carbohydrate sources such as cassava, corn, sweet potatoes and so on [1].

IPB as a leading agricultural education institution has been researching and developing artificial rice. Artificial rice produced mini-plant scale and has been spread on the market. The artificial rice is produced using extrusion technology and is made from the main ingredients of corn flour and sago as well as some other additives. The fabrication process of artificial rice using extrusion process passes several stages commonly used by other extruded products i.e. formulation, extrusion, and drying. After process of drying, artifial rice is packed and stored before it is distributed.

One of the most influential food properties in production process of artificial rice is moisture sorption isotherm. Sorption isotherm is relationship between water activity with moisture content of food materials at certain temperature and pressure (equilibrium moisture content). The pattern of water sorption isotherm shows interaction between food with water molecules contained therein and the water molecules in surrounding air. Equilibrium moisture content because material lost water (desorption) can be useful for drying process analysis, and equilibrium moisture content that occurs due to material adsorb water (adsorption) to determine storage method.

Study of measurement sorption isotherm generally still using the traditional method of saturated salt slurry method, no exception for Indonesia. This method has disadvantages of long period of measurement time and periodic handling of samples. Nowadays, there is new method, well-developed method, that is humidity generating method or better known as Dynamic Vapor Sorption (DVS) method. DVS technique is performed using instrument that utilize divided flow technology. This technology allows the instrument to generate sorption isotherm in relatively short period of time. User selects the desired relative humidity or relative humidity series using the program computer. Relative humidity is automatically controlled by mass flow controller, dry air flow and wet air flow, by mixing them in right ratio will yield selected relative humidity. Sample is conditioned in desired relative humidity, then mass or volume of sorption is measured. The instrument is automatically programmed to cycle between humidity levels, and keeps the humidity constant in each level until equilibrium is reached [2][3].

There are already some commercially available brands as mentioned by Mermelstein [3]: Dynamic Vapor Sorption (UK), IGA-Sorp (England), VTI and Q5000SA (USA), Cisorp Water Sorption Analyzer (UK), SPS Moisture Sorption Analyzers (Germany), Hydrosorb™ 1000 Water Vapor Sorption Analyzer (USA), AquaSorp Isotherm Generator (USA). This DVS method has a number of advantages compared to the saturated salt slurry method. The equilibrium time is around 10 to 100 times faster than the standard method. The DVS instruments are automated and use computer interface thereby reducing need for supervision work ofthe slurry methods and maintaining relative humidity over the duration of the experiment.

Most of the above mentioned DVS instruments work gravimetrically by measuring mass changes per unit of time, and partly volumetrically by measuring pressure changes. Hydrosorb™ 1000 analyses by vacuum-volumetric system and figure 1 shows the instrument. Hydrosorb™ 1000 is better known as apparatus for measuring surface area and porosity. This paper aims to develop moisture sorption isotherm using the Hydrosorb™ 1000 Water Vapor Sorption Analyzer.

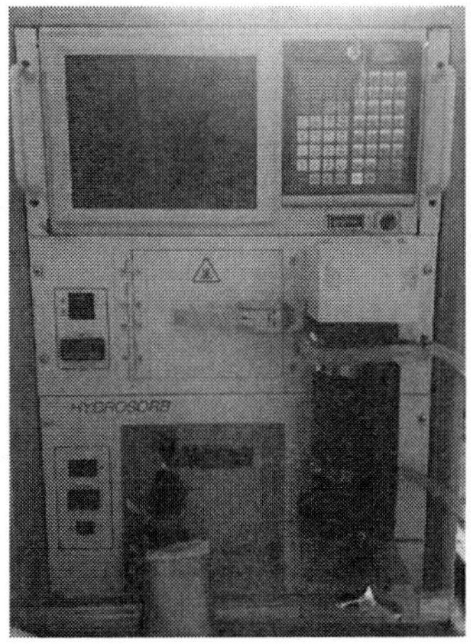

Figure 1. Pictorial view of Hydrosorb 1000 Water Vapor Sorption Analyzer.

2. Materials and methods

2.1. Materias and equipments
Artificial rice samples were rice analogues manufactured by FITS Mandiri, Taman Kencana Science Park, Bogor, Indonesia. Artificial rice made from corn flour, sago, glycerol monostearate, vegetable oil, water and jelly powder was developed by extrusion method. The artificial rice has been sold commercially. Equipment were used in this study :*Differential Scanning Calorimeters* DSC 60 Shimadzu, *Water Vapor Sorption Analyzer* Quantachrome Hydrosorb^{TM} 1000, desk computer, drying oven, 10 mL measuring cylinder, desiccator, and analytical balance.

2.1.1. Moisture content and bulk density Clean empty cup is dried in an oven at temperature of 100±5°C until it weighs constant (1 hour). The cup is cooled in desiccator for 30 minutes and weigh (W_o). Two to three grams of sample are weighed in the cup (W_1). The cup containing the sample is dried in a drying oven at temperature of 103-105°C to constant weight. The cup containing the dried sample is cooled in the desiccator for 30 minutes and weigh (W_2). Weight-loss is calculated as moisture content (m) using equation (2.1).

$$m(\%) = \left(\frac{W_1 - W_2}{W_1 - W_o} \right) \times 100\% \quad (2.1)$$

Bulk density is total granules mass occupying a particular volume unit. Rice analogues sample was poured into a certain known volume from a fixed height and mass of samples occupying the volume as determined. Ratio was calculated as g/mL [4].

2.1.2. DSC Analysis. Thermal characteristics of artificial rice were analyzed by using DSC 60 Shimadzu, Tokyo, Japan. The artificial rice were ground into flour and about 3 mg was weighed in a special sample container (pan cell no. 20153090). Distilled water 10 μL was added using a micropipette and then the pan cell was sealed tightly. The pan cell containing sample and the distilled water wererested for 2 hours at room temperature to achieve equilibrium. The pan was then heated from 40°C to 120°C with heating rate of 5°C/min[5].

2.2. Apparatus Hydrosorb 1000

Hydrosorb is innovation in moisture sorption characterization and works automatically to rapidly measure adsorption and desorption isotherms of water, including BET surface area calculations and heats of adsorption. Adsorbent temperature is thermostatted in the range 12-47°C or 12-85°C using thermal jacket controlled by temperature circulator bath with accurate water vapor dosing performed by special, heated (100°C) manifold design. Hydrosorb-1000 is the constant-volume variable-pressure (cvvp) apparatus and volumes of charging and dosing were calibrated by charging nitrogen gas (N2) with purity of 99.99%. The standard volume had been calibrated by liquid filling method using distilled water[6]. The integrated control system provides real-time status, graphical data review and reporting functions via built-in color monitor and printer port.

Control system of the Hydrosorb among others :

1. Computer
   The main parts of computer include computer controls, TFT display monitor and keypad. In the computer controls there is a button to activate Hydrosorb computer and set up the display monitor. TFT monitors displays the status of Hydrosorb operation, and to the right of the monitor is a keypad. Alternatively, standard keyboard can be connected to the computer via connnection jack on the lower right of the keypad.
2. Panel and port
   There are main power switch and serial port in rear panel. Main power switch controls power across instrument including oven manifold and computer. When the button is ON, the power is supplied to Hydrosorb, although the computer and the oven might be switched off. RS-2232 serial port serves to relate Hydrosorb to an external PC for control, enhanced data reduction and reporting capabilities using HydroWin software.Gas connection connects high-purity nitrogen supply. Gas regulator must be set to deliver 10 psig (70 kPa). Vacuum pump is connected to back of Hydrosorb via vacuum fitting using vacuum hose.
3. Degasser
   Degassing of sample is performed insitu using built-in heating mantle. Main components are heating mantle and temperature control. Temperature control regulates the temperature of the heating mantle. Degasser is also equipped with thermocouple, LED temperature display, and power switch.
4. Oven
   Water for analysis is stored in the water reservoir in a vial glass installed inside the oven. The oven functions to heat the water vapor dosing system and has been equipped with temperature controller and oven power switch. The appropriate operating temperature has already been set to 100°C. Adjacent to the water reservoir,there are needle valve settings for N_2 inlet, vapor source inlet, and fine vacuum inle thathave been pre-set at factory and need to be adjusted periodically.
5. Analysis station
   Sample cell is mounted on sample station secured by bulkhead fiting. Sample cells are available with 3 diameters choice conforming sample size, outside diameters of stem 6, 9, and 12 mm. Bulkhead fitting equipped with Viton O-ring, adaptersleeve and retainer ring, serves to keep the sample cell with a vacuum firm seal to the bulkhead on the analysis port. Dewar is used in conjunction with external water bath circulator (figure 2) and functions to thermostat the sample cell at the desired isothermal temperature. The bath circulator allows analysis at different temperature.

Figure 2. Dewar and steps to use [6][7].

Significant components in the Hydrosorb system are manifold and three-way valve[8]. Working principle of the system is preceded by pressure conditioning (using vacuum pump) in the Hydrosorb and sample cell. After the pressure has stabilized, water vapor is supplied to the manifold until the desired relative pressure is reached. Then, the water vapor is channeled into the sample cell so that the vaporoccupy the sample cell. As water molecules adsorb on the sample, the pressure in the sample cell will drop until the relative pressure reaches equilibrium. Pressure change that occur at equilibrium will be calculated to obtain the volume/mass of adsorbed water vapor. Three-way valve plays key role of monitoring the pressure in the manifold and in the sample cell alternately.

3. Results and discussion

Determination of water sorption curve was carried out at 30°C and 45°C. At temperature 30°C, the measurement was carried out by adsorption at water activity of 0.05-0.95 by an increase of 0.05 steps, and at temperature of 45°Cwas performed by adsorption and desorption at water activity of 0.05, 0.1, 0.15, 0.2 , 0.25, 0.3, 0.4, 0.5, 0.6, 0.7, 0.8, 0.85, 0.9, and 0.95. Measurement using DVS Hydrosorb can only be done for one sample, thus for analysis with different sample must be resetting.

3.1. Material characterization

Preliminary measurement should include moisture content, density, and glass transition point. The density value will be input in filling the parameters of sample when initiating the analysis, whereas from the measurement of moisture content using oven method, it can be estimated the moisture content after sample preparation stage or before the start of the analysis. The glass transition point becomes degassing temperature limit. The degassing temperature, as a general guideline, must be kept below glass transition point to prevent undesirable changes in the structure of the material. Valuable glass transition point information can be collected from thermal analysis, particularly DSC (differential scanning calorimetry) and TGA (thermogravimetric) analysis.

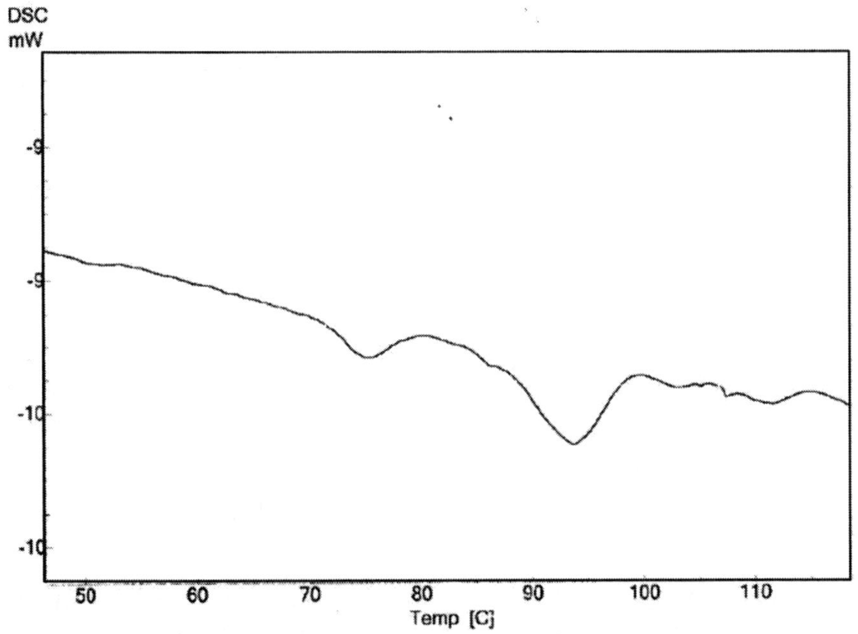

Figure 3. DSC thermogram of artificial rice.

The bulk density was 0.58 g/mL, while the sample moisture content of measurement at the sorption temperatures of 30°C and 45°C were 14.74% db and 14.39% db, respectively. The result of the thermal characteristic test in DSC was thermogram presented in figure 3. The thermogram shows the presence of two melting curves at temperatures around 75°C and 93°C. It means the gelatinization temperature is at the top of the first melting curve at 75.37°C and has curve peak point/melting point of 93.74°C. There is no shift that can indicate glass transition point. Herawati *et al.* [9]stated that determination of glass transition point using conventional linear heating in DSC was very difficult. Glass transition temperature of instant artificial rice was determined by isothermal relaxation method in DSC. The result of analysis using Gordon-Taylor formula approach predicted glass transition temperature 75.5°C.

3.2. Sample regeneration

Each sample should be degassed before performing an analysis to remove unwanted contaminants or vapors in the sample. Degassing is executed in-situ by placing approximately 0.5 g of sample in the sample cell and then dryed in the heating mantle. The degassing variables are temperature and time. The proper temperature of degassing varies depending on the chemical structure of the material and the desired dryness state. The degassing time varies according to the initial moisture level, porosity and temperature used. To improve sample preparation capability, additional instruments can be used for flow outgassing and vacuum outgassing.

The temperature used for artificial rice degassing was 65°C. Determination of degassing temperature was based on Herawati *et al.*[9] which states that the glass transition temperature of instant artificial rice made from corn flour is 75.5°C. The degassing temperatures used are still relevant to the conditioning of samples performed on the research of sorption isotherm by Wulandari and Soekarto [10] which dried material up to 2% moisture content by heating the sample at oven temperature of 70°C. During the degassing process, water spots stick to the sample cell wall indicating the water contained in the material is gradually released. Degassing can be stopped by disappearance of water spots on the sample cell wall.

3.3. Sample analysis

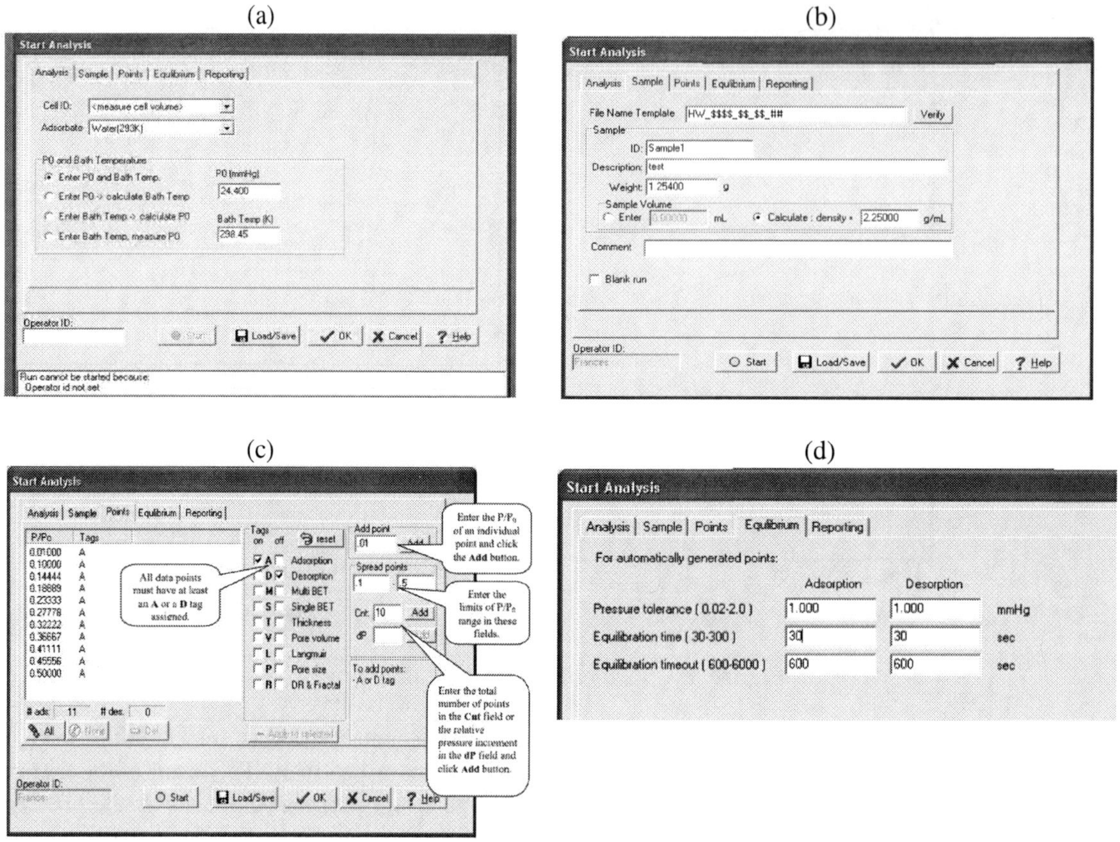

Figure 4. Window of “Start Analysis” [11].

After completing all parameters in the “Start Analysis” window (figure 4), the user can click “Start” box to commence analysis. Sample analysis progress can be monitored in continuation window. The total analysis time might be hours or days depending on the quantity and type (hydrophobic, hydrophilic, porous or non-porous) of sample. The results of water sorption isotherm analysis of the artifical rice at temperature of 30°C and 45°C are presented in figure 5. The sorption isotherm at 30°C was measured at 19 points of adsorption taking place for approximately 25 hours, whereas at temperature of 45°C was analyzed at 26 points of adsorption-desorption for about 36 hours.

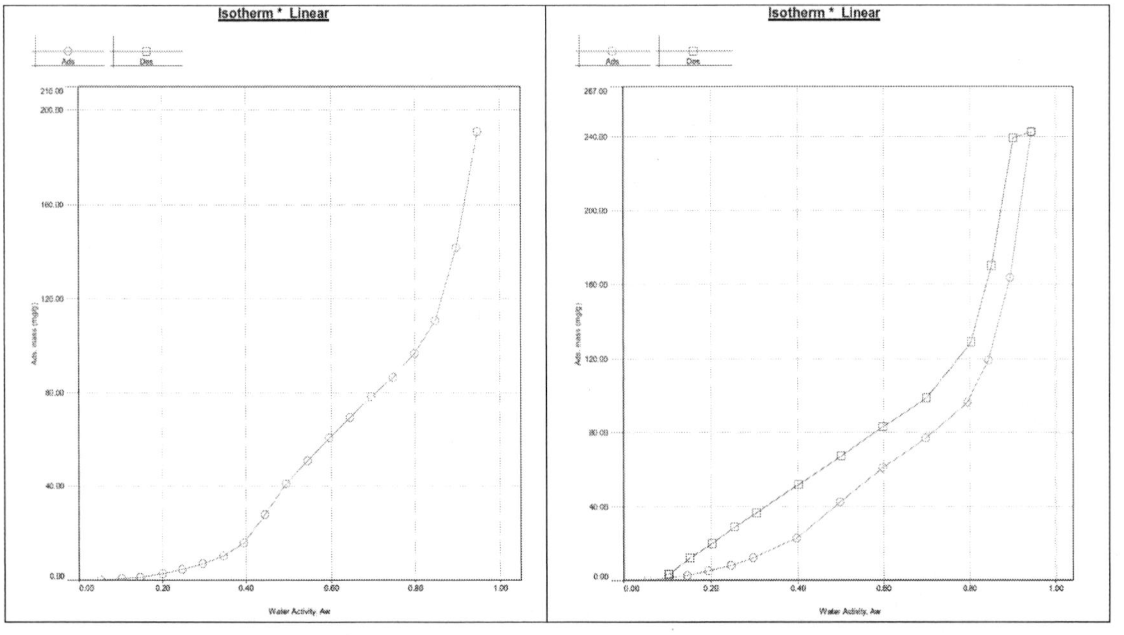

Figure 5. Water sorption isotherms of artificial rice at 30°C and 45°C.

Water sorption curve expressed in water activity (a_w) in abscissa and adsorbed mass (mg/g) in ordinate. Selected bottom point of water activity 0.05 at 30°C and 45°C yielded adsorbed mass of 0.1466 mg/g and 0.3455 mg/g, respectively, whereas selected top water activity point 0.95 at 30°C and 45°C yielded adsorbed mass of 190.8734 mg/g and 242.4161 mg/g, respectively. By converting adsorbed mass unit from mg/g to g/100g, it will be retrieved curves illustrating the relationship between water activity (a_w) with the equilibrium moisture content (% dry basis) which are more commonly encountered in water sorption isotherm curve. At the lowest water activity (0.05), samples of artificial rice show different levels of moisture content, which at the analysis temperature of 30°C by 0.02% and at temperature of 45°C by 0.04%. At the highest water activity (0.95), the equilibrium moisture content at 30°C and 45°C were 19.1% and 24.2%, respectively.

The characteristic of moisture sorption isotherms showed the increase in equilibrium moisture content is positively correlated with the water activity at fixed temperature [12]. The higher the relative humidity level the more water is adsorbed by the artificial rice. Initially the rapid increase in equilibrium moisture content occurs in the first layer (aw <0.5), known as monolayer moisture. In this layer, single water adsorption occurs on the material. The second layer occurs above the monolayer until activity water of about 0.7-0.8, known as multilayer moisture showing the water adsorption transition from free state to chemically bonded state. The third layer is an area with rich of free water content [13][14].

The research of moisture sorption isotherm of instant artificial rice and commercial rice by Herawati [15] showed that the pattern of mositure content graphic of commercial rice is higher than instant artificial rice. Measurement of equilibrium moisture content using saturated salt solution resulted in moisture content of 28.8% db and 28.1% db for paddy rice and instant artificial rice at water activity 0.97, respectively. At the lowest water activity 0.08, equilibrium moisture content were 5.3% db and 4.6% db for paddy rice and instant artificial rice, respectively. The experimental artificial rice used glycerol mono-stearate and vegetable oil which are hydrophobic component, while the instant artificial rice used glycerol mono-stearate. The hydrophobic components are thought to affect the material's binding capacity to water. Hydrophobic component factor and addition of alginate hydrocolloid of instant artificial rice are suspected to be the cause of higher equilibrium moisture content than experimental artifical rice at various levels of water activity.

Curves of moisture sorption isotherm of experimental artificial rice close to type II, resembling sigmoid shape or like the letter S. Labuza [16] classifies moisture sorption isotherm curves into 3 types, namely type I is Langmuir type, type II is sigmoid or S-shaped curve and type III (Flory-Huggins) which is shaped like the letter J. The moisture sorption isotherm curve has a sigmoidal shape in many foods, although for foods with high sugar content or low soluble molecules having sorption curve in the form of J. J-shaped curves are typical of products holding small quantities of water at lower a_w levels and higher quantities of water at high a_w values. Dry food ingredients generally include sorption isotherm of type II. Corn and its derivative products such as corn flour, corn starch, corn snack, tortilla, instant artificial rice belong to the group [17][18][19][20][21][22]. The type II sorption isotherm curve is due to the accumulative effect of hydrogen bonding, Raoult law, capillaries, and the interaction between the surface of the material and the water molecule. The type II sorption curves are common pattern in amorphous food systems. The carbohydrates and starches used are mostly amorphous materials.

Moisture sorption isotherm curve at 45°C is presented by adsorption and desorption so that it appears hysteresis. In many cases, the adsorption and desorption curves do not coincide so that hysteresis phenomena are discovered. In this state, the adsorption process produces a lower equilibrium moisture content than the desorption process. Large hysteresis size occurs in multilayer areas. Material characteristics affect the size of hysteresis [10].

4. Conclusions

Thermal characterization of test material needs to be done to determine the degassing temperature. The degassing temperature used for experimental artificial rice was 65°C. Degassing can be completed after water spots on the sample cell wall have disappeared. The results of moisture sorption isotherm of articial rice made from corn flour at temperature of 30°C and 45°C using Hydrosorb showed that the moisture sorption curve approximates sigmoid-shaped type II curve commonly found in corn-based foodstuffs (high-carbohydrate). The content of hydrophobic material and hydrocolloid material in the artificial rice composition are thought to affect the water holding capacity so that it affects the moisture sorption isotherm curve.

References

[1] Budijanto S and Yuliyanti 2012 Study of preparation sorghum flour and application for analogues rice production *J. Teknol. Pertan.* **13** 177-86

[2] Yu X 2007 *Investigation of Moisture Sorption Properties of Food Materials Using Saturated Salt Solution and Humidity Generating Techniques* PhD thesis(Urbana: University of Illinois at Urbana-Champaign) p73

[3] Penner E A 2013 *Comparison of the New Vapor Sorption Analyzer to the Traditional Saturated Salt Slurry Method and the Dynamic Vapor Sorption Instrument* thesis (Urbana: University of Illinois at Urbana-Champaign) pp 15-6

[4] Singh N, Kaur L, Sodhi N S and Sekhon K S 2005 Physicochemical, cooking and textural properties of milled rice from different Indian rice cultivars *Food Chem.* **89** 253-59

[5] Gelders G G, Vanderstukken T C, Goesaert H and Delcour J A 2004 Amylose–lipid complexation: a new fractionation method *Carbohydr. Polym.* **56** 447–58

[6] Cevallos O R F 2012 *Adsorption Characteristics of Water and Silica Gel System for Desalination Cycle* thesis (Thuwal: King Abdullah University of Science and Technology)

[7] Hydrosorb Manual 2007 *Hydrosorb 1000 Operating Manual*(Boynton Beach: Quantachrome Corporation) p 23

[8] Seymour L 2006 *Apparatus and Method for Water Sorption Measurement* US patent 2006/0027014A1 pp 3-4

[9] Herawati H, Kusnandar F, Adawiyah D R, Budijanto S and Rahman M S 2014 Thermal characteristics and state diagram of extruded instant artificial rice *Thermochimica Acta* **593** 50-7

[10] Wulandari N and Soekarto S T 2003 Hysteresis phenomena of moisture sorption isotherm in amylose, amylopectin, protein and cellulose *J.Teknol.dan Ind.Pangan* **14** 21-8
[11] HydroWin Manual 2007 *HydroWin 5.0+ Operation Manual* (Boynton Beach: Quantachrome Corporation)
[12] Oliveira E G, Rosa G S, Moraes M A and Pinto L A A 2009 Moisture sorption characteristics of microalgae *Spirulina platensis Brazilian J. Chem. Eng.* **26** 189-97
[13] Widowati S, Herawati H, Syarief R, Suyatma N E and Prasetia H A 2010 Effect of moisture sorption isotherm to stability of "sweet potato rice" *J. Teknol. dan Ind. Pangan* **21** 123-8
[14] Menkov N D and Durakova A G 2007 Moisture sorption isotherms of sesame flour at several temperatures *Food Technol. Biotechnol.* **45** 96–100
[15] Herawati H 2015 *Process Optimization, Moisture Sorption Isotherm Profile and Thermal Analysis of Instant Artificial Rice* PhD thesis (Bogor : Bogor Agricultural University) pp 76-83
[16] Labuza T P 1984 *Moisture Sorption: Practical Aspects of Isotherm Measurements in Use* (St Paul : American Association of Cereal Chemist)
[17] Sukainah A, Tawali A B, Salengke and Laga A 2013 The effect of fermentation on adsorption isotherm corn flour and corn crackers *Int. J. of Sci. and Tech. Research* **2** 263-7
[18] Budijanto S, Sitanggang A B and Kartika Y D 2010 Shelf life study of tortilla using accelerated shelf life testing method and its mathematical modelling of moisture sorption isotherms *J. Teknol. dan Ind. Pangan* **21** 165-70
[19] Aini N, Prihananto V and Wijonarko G 2014 Moisture sorption isotherm of instan corn flour from four variety of corn *Agritech* **34** 50-5
[20] Palou E, Lopez-Malo A and Argaiz A 1997 Effect of temperature on the moisture sorption isotherms of some cookies and corn snacks *J. of Food Eng.* **31** 85-93
[21] Peng G, Chen X, Wu W and Jiang X 2007 Modeling of water sorption isotherm for corn starch *J. of Food Eng.* **80** 562-7
[22] Oyelade O J, Tunde-Akintunde T Y, Igbeka J C, Oke M O and Raji O Y 2008 Modelling moisture sorption isotherms for maize flour *J. of Stored Prod. Research* **44** 179-85

Liquid Fertilizer Spraying Performance Using A Knapsack Power Sprayer On Soybean Field

Gatot P[1] and Anang R[1]

[1] Department of Mechanical and Biosystem Engineering, Bogor Agricultural University, Indonesia

E-mail: gph3025818@gmail.com

Abstract. An effort for increasing soybean production can be conducted by applying liquid fertilizer on soybean cultivation field. The objective of this research was to determine liquid fertilizer spraying performance using knapsack power sprayer TASCO TF-900 on a soybean cultivation field. Performances test were conducted in the Laboratory of Spraying Test and on a soybean cultivation field to determine (1) effective spraying width, (2) droplets diameter, (3) droplets density, (4) effective spraying discharge rate, and (5) effective field capacity of spraying. The research was conducted using 2 methods: (1) one-nozzle spraying, and (2) four-nozzles spraying. Results of the research showed that at a constant pressure of 900 kPa effective spraying width using one-nozzle spraying and four-nozzles spraying were 0.62 m and 1.10 m. A bigger effective spraying width was resulted in a bigger average effective spraying discharge rate and average effective spraying field capacity of 4.52 l/min and 83.92 m^2/min on forward walking speed range of 0.94 m/s up to 1.77 m/s. On the contrary, bigger effective spraying width was result in bigger droplets diameter of 502.73 μm and a smaller droplets density of 98.39 droplets/cm^2, whereas smaller effective spraying width was resulted in a smaller droplets diameter of 367.09 μm and a bigger droplets density of 350.53 droplets/cm^2. One-nozzle spraying method produced a better spraying quality than four-nozzles spraying method, although four-nozzles spraying was resulted in a bigger effective field capacity of spraying.

1. Introduction

Soybean (*Glycine max* (L.) Merril) was one of foodstuff that contains protein and it can be supplied to industry and livestock. An effort for increasing soybean production can be conducted by applying liquid fertilizer on soybean cultivation field (Suhaeni 2007). Soybean plant needed effective maintenance because mainly for facing competition with weeds that can decrease soybean production.

Liquid fertilizing can be conducted by applying sprayers to spray it onto soybean leaves. Spraying quality was influenced by some factors i.e. sprayer type, nozzle number, nozzle type, forward speed, spraying direction, and spraying pressure.

Soybean plants were planted on rows with certain distance for each row. A spraying method for the rows was one-way spraying with constant spraying pressure utilized a knapsack power sprayer. A research can be conducted to analyze spraying effectiveness and efficiency by applying some treatments like sum of nozzles and forward speed of spraying. This research can be used to determine liquid fertilizer spraying performance utilized knapsack power sprayer on a soybean cultivation field. The objective of this research is to determine liquid fertilizer spraying performance using a knapsack power sprayer on soybean field.

2. Materials and methods

The research was conducted from February 2017 to June 2017 in the Laboratory of Spraying Test and on a soybean cultivation field to determine (1) effective spraying width, (2) droplets diameter, (3) droplets density, (4) effective spraying discharge, and (5) effective field capacity of spraying as seen in figure 1.

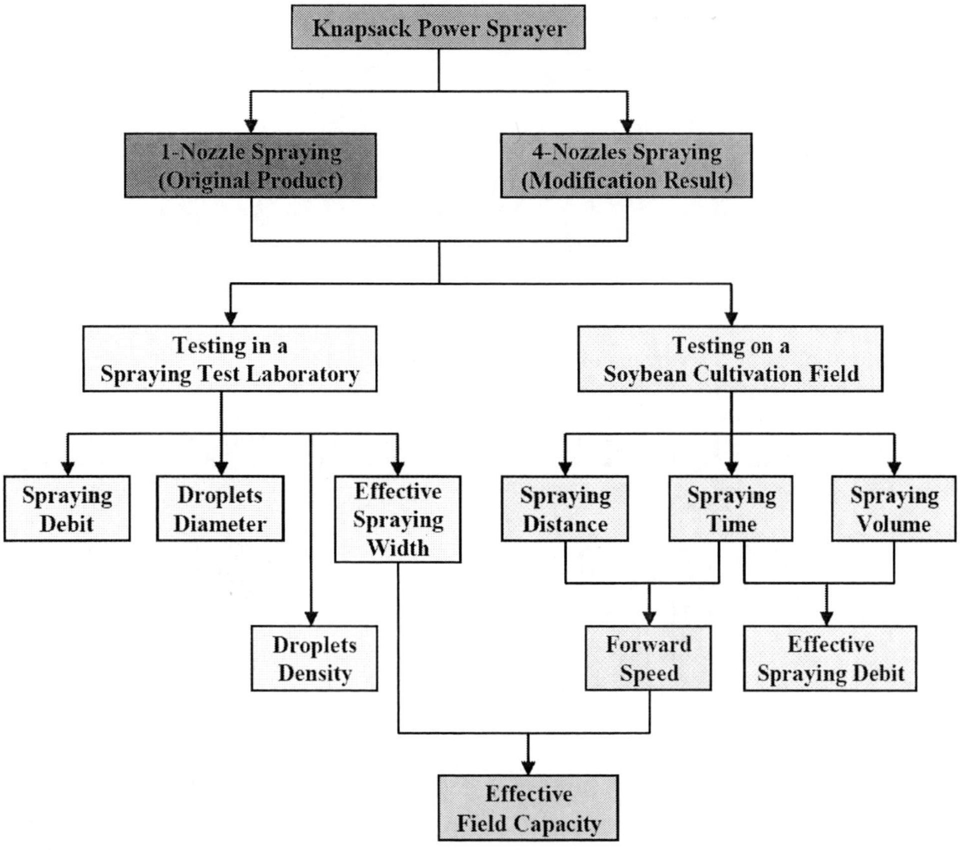

Figure 1. Research method to determine liquid fertilizer spraying performance utilized knapsack power sprayer

The tools which were used in this research are knapsack power sprayer TASCO TF-900 and some spray performance measuring instruments such as patternator, timer, scales, measuring cup, and tape meter. The materials used were liquid fertilizers, pipes, clean water, ink, concord paper, and glue.

The research was conducted using 2 methods: (1) one-nozzle spraying, and (2) four-nozzles spraying. The one-nozzle spraying was given from original product, whereas the four-nozzles spraying was made from modification result. All nozzles which are used in this research are solid cone nozzles with funnel as seen in figure 2.

Figure 2. An example of a solid cone nozzle with funnel

Determination of spraying performance parameters in the Laboratory of Spraying Test is important to do before it is applied on a soybean cultivation field as seen on figure 3. This research is mainly used for effective spraying width (ESW) determining.

Figure 3. One-nozzle (left) and four-nozzles (right) spraying on patternator in the Laboratory of Spraying Test

Liquid fertilizer spraying will be more effective with finer droplets size (droplets diameter) and higher droplets density. Smaller droplets diameter and higher droplets density make this condition met so that the effectiveness of liquid fertilizer spraying by using a knapsack power sprayer will be achieved.

Efficient liquid fertilizer spraying is also important in order to save time and money. This can be achieved with quick and cheap liquid fertilizer application. In order to achieve it, no repeated application is advised, droplets distribution during the application time is done once only, and the volume of droplets applied is not excessive. This condition is met if the effective spraying width (ESW) and forward speed of spraying (FFS) are high and effective spraying discharge rate (ESD) is low (economical). In other words, liquid fertilizer spraying will be more efficient (short application time and economical application cost) if ESW and FFS are high and ESD is low. It is noted when ESW and FFS are high, so it will result in high effective field capacity of spraying (EFC).

Optimization of liquid fertilizer spraying needs to be done in so that it can be done optimally (effective and efficient). Optimal spraying is achieved with if liquid fertilizer has the following requirements: (1) minimum droplets diameter, (2) maximum droplets density, (3) maximum ESW, (4) minimum ESD, and (5) maximum EFC. These five parameters of effectiveness and efficiency of liquid fertilizer spraying can be measured in the Laboratory of Spraying Test and on a soybean cultivation field as seen on figure 4.

Figure 4. One-nozzle (left) and four-nozzles (right) sprayings on a soybean field

3. Results and discussion

Results of the research showed that at a constant pressure of 900 kPa effective spraying width using 1-nozzle spraying and 4-nozzles spraying were 0.62 m and 1.10 m, respectively. This result described that the increment of sum of spraying nozzles would result in wider effective spraying width. An example of spraying pattern result to determine effective spraying width can be seen on figure 5.

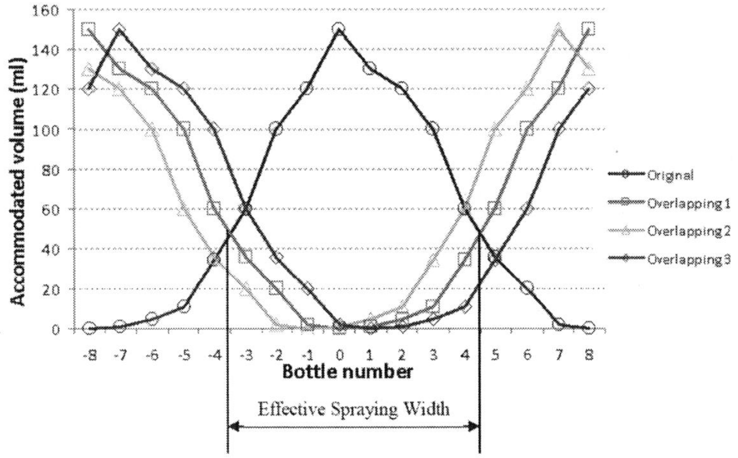

Figure 5. An example of knapsack power spraying pattern result

Effective spraying discharge rate on soybean field was vary on various forward walking speed. Effective spraying discharge rate decreased by increasing forward walking speed. Average effective spraying discharge rate of 4-nozzles spraying method was bigger than average effective spraying discharge rate of 1-nozzle method (figure 6). Effective spraying width of 4-nozzles spraying was bigger than 1-nozzle spraying, so it caused a bigger spraying volume that result in increasing effective spraying discharge rate.

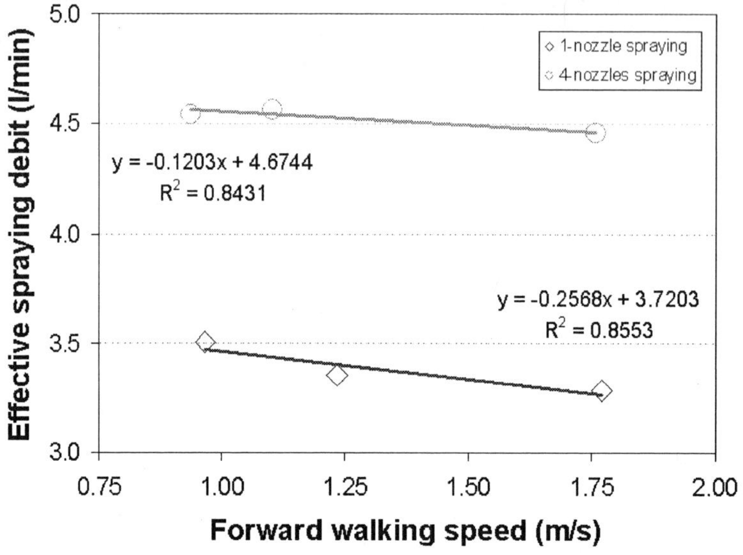

Figure 6. Effective spraying discharge rate on various forward walking speed

Effective spraying field capacity increased by increasing forward walking speed on soybean field and it also increased by increasing the number of nozzle. It can be seen on figure 7.

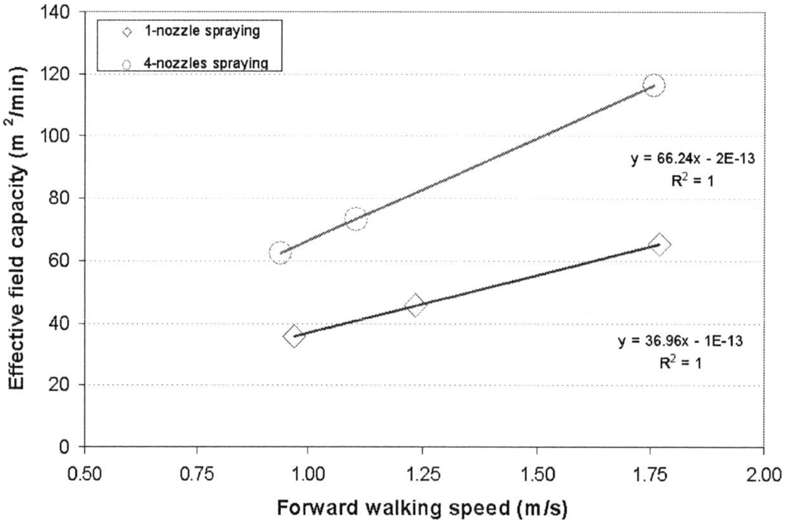

Figure 7. Spraying effective field capacity on various forward walking speed

Different number of spraying nozzle produced different droplets diameter and droplets density. One-nozzle spraying produced smaller droplets diameter than four-nozzles spraying. It also produced bigger droplets density than four-nozzles spraying. Visualisation and value of droplets diameter and droplets density can be seen on figure 8 and figure 9.

(a) (b)

Figure 8. Visualisation of droplets diameter and droplets density of (a) one-nozzle spraying result and (b) four-nozzles spraying result

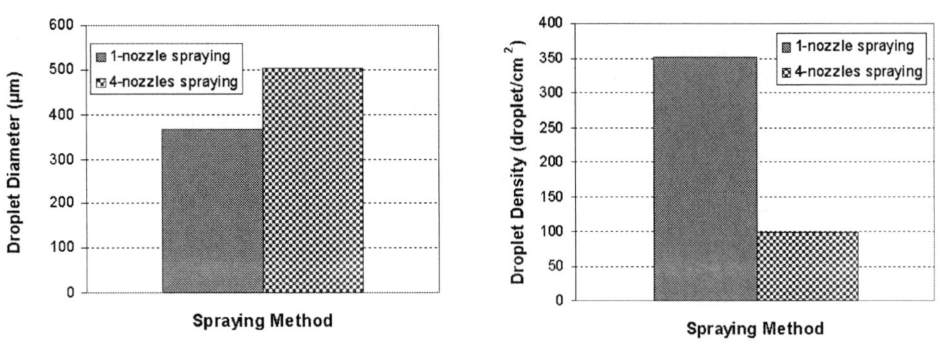

(a) (b)

Figure 9. Value of droplets diameter and droplets density of one-nozzle spraying (a) and four-nozzles spraying (b) results

A bigger effective spraying width of 1.10 m was result in a bigger average effective spraying discharge rate of 4.52 l/min and an average effective spraying field capacity of 83.92 m^2/min on forward walking speed range of 0.94 m/s up to 1.77 m/s. On the contrary, bigger effective spraying width was result in a bigger droplets diameter of 502.73 µm and smaller droplets density of 98.39 droplets/cm^2, whereas smaller effective spraying width was result in smaller droplets diameter of 367.09 µm and bigger droplets density of 350.53 droplets/cm^2. Based on the ASABE S572.1 droplets size classification (table 1), it can be said that droplets diameter of 502.73 µm was classified as very coarse spray quality, whereas droplets diameter of 367.09 µm was classified as coarse spray quality.

Droplet diameter and droplets density determined quality or effectiveness of spraying, whereas effective spraying field capacity determined spraying efficiency. One-nozzle spraying method showed better spraying quality than four-nozzles spraying because it had smaller droplets diameter and had bigger droplets density than four-nozzles spraying, although four-nozzles spraying was result in bigger effective spraying field capacity.

Table 1. ASABE S572.1 Droplet Size Classification (PES 2017)

Spray Quality*	Size of Droplets	VMD Range (Microns**)	Color Code	Retention on Difficult to Wet Leaves	Used for	Drift Potential
Extremely Fine	Small	<60	Purple	Excellent	Exceptions	High
Very Fine		61-105	Red	Excellent	Exceptions	
Fine		106-235	Orange	Very Good	Good Cover	
Medium		236-340	Yellow	Good	Most Products	
Coarse		341-403	Blue	Moderate	Systemic Herbicides	
Very Coarse		404-502	Green	Poor	Soil Herbicides	
Extremely Coarse		503-665	White	Very Poor	Liquid Fertilizer	
Ultra Coarse	Large	>665	Black	Very Poor	Liquid Fertilizer	Low

*Always read the pesticide label to determine which spray quality is required.
**Estimated from sample reference graph in ASABE/ANSI/ASAE Standard S572.1

The ASABE S572.1

4. Conclusions

It was concluded that spraying utilized knapsack power sprayer with one-nozzle produced effective spraying width of 0.62 m, average effective spraying discharge rate of 3.38 l/min, and average effective spraying field capacity of 48.97 m^2/min. Spraying utilized knapsack power sprayer with four-nozzles produced effective spraying width of 1.10 m, average effective spraying discharge rate of 4.52 l/min, and average effective spraying field capacity of 83.92 m^2/min.

A bigger effective spraying width was result in bigger droplets diameter of 502.73 μm (very coarse spray quality) and smaller droplets density of 98.39 droplets/cm^2, whereas smaller effective spraying width was result in smaller droplets diameter of 367.09 μm (coarse spray quality) and bigger droplets density of 350.53 droplets/cm^2. One-nozzle spraying method produced better spraying quality than four-nozzles spraying method, although four-nozzles spraying was result in bigger effective spraying field capacity.

References

[1] Kurniawan T 2014 Modification number nozzle of motorized knapsack sprayer and nozzle performance test at various type and pressure (in Indonesian) (Bogor: Bogor Agricultural University)

[2] PES 2017 ASABE S572.1 Droplet Size Classification. Pesticide Environmental Stewardship. https://pesticidesstewardship.org/pesticide-drift/understanding-droplets-size/ [accessed on 01 January 2018]

[3] Suhaeni N 2007 *Practical Guide to Planting Soybean* (in Indonesian) (Bandung: Nuansa)

[4] Yuwana NAJ. 2014. Design and Construction of Grid Patternator for Sprayer Spray Performance Test (in Indonesian) (Bogor: Bogor Agricultural University)

IOP Conf. Series: Earth and Environmental Science **147** (2018) 012019 doi:10.1088/1755-1315/147/1/012019

Conceptual Design of Fertilizer Applicator for Oil Palm on Terrace Cultivation

W Hermawan[1]

[1] Department of Mechanical and Biosystem Engineering, Faculty of Agricultural Engineering and Technology, Bogor Agricultural University, Indonesia

E-mail: wawanfateta@yahoo.co.id

Abstract. The mechanical application of fertilizer for oil palm planted on terraces is still constrained by the narrow path which is difficult to pass by a power spreader. The objective of this research was to develop a conceptual design of fertilizer applicator for oil palm planted on terraces. The design requirements were developed based on a) terrace and track conditions, b) fertilizers and fertilization conditions, c) available prime movers, and d) user needs. Five design concepts were obtained: 1) an applicator with left and right arms to distribute the fertilizer, 2) an all-terrain vehicle equipped with a manually operated fertilizer injector, 3) an applicator equipped with a hole digger, 4) an applicator equipped with a shovel, and 5) an applicator equipped with a rotary tiller. The concepts were evaluated and compared with the current power spreader. The evaluation results showed that the applicator equipped with a rotary tiller had the most advantages on the expected criteria. The final design concept uses a 110 cm wide mini crawler tractor as the prime mover and is equipped with a hopper and a spinner disk for metering and conveying the fertilizer, and a 20 cm wide rotary tiller in the front of the machine.

1. Introduction

Fertilization work is one of the most important activities in the cultivation of oil palm crops [1]. However, the rapid development of oil palm plantations in Indonesia raises one of the problems of labor constraints for oil palm cultivation activities, especially for fertilizer application. Therefore, large oil palm plantations applied mechanization on the fertilizer application. Fertilization is mechanically carried out mostly on flat palm plantations, using a mechanical fertilizer spreader operated by a four-wheeled tractor [or attached to a small crawler tractor [2-4]. The power spreader operates on the track between the rows of oil palms and spreads the granular fertilizer onto the soil surface near the row of oil palm trees. Mechanized fertilizer application may result in more efficient nutrient use and labor savings.

Some of the oil palm plantations are wavy, sloping and hilly. In sloping areas, the oil palms are planted following contour lines with terracing system. The sloping land is formed terraced and the oil palms are planted along the path of the terraces. Terraces must be installed on sloping land to provide adequate access for field workers and harvesters and to reduce soil erosion and surface run-off. The width of the terrace depends on the steepness of the land and ranges from 2-3 m. On the terrace, oil palms are planted somewhat closer to the side of the cliff terrace. In the palm oil terracing system, fertilization is still done by manual, and cannot be done mechanically.

There are problems faced in providing fertilizer for oil palm planted on the slopes, namely: 1) difficult to be crossed by a power spreader for mechanical fertilization, 2) manually fertilization is less effective and cannot be guaranteed the application rate, 3) fertilization by sowing above the soil surface at risk of rainwater runoff, so fertilization efficiency is low. It is very important to apply the fertilizers in the best manner over the areas of the estate that are likely to result in the most efficient uptake of nutrients, and to minimize negative environmental impacts related to over-fertilization, land degradation, and pollution from heavy metals such as cobalt and eutrophication by P application [1].

In such conditions, fertilization needs to be done more effectively, with the help of an appropriate machine. Current equipment and machinery cannot be used in the area. Therefore it is necessary to develop or design a fertilizer application machine for oil palm planted in slope area. For the purpose, some conceptual design should be developed and evaluated. Conceptual design is the key to achieve innovation, for it is at this stage when product concepts are assessed and selected before they are developed into final products sold in the market [6].

The objective of this research was to develop a conceptual design of fertilizer applicator for oil palm planted on terraces.

2. Materials and methods

2.1. Research activities

For developing the oil palm fertilizer applicator, the following activities have been conducted.

1. Identification of fertilizer characteristics and fertilization system used (in the plantation)
2. Measurement of land and soil conditions on terraces of sloping plantation.
3. Review of tractors available on site and can be used to drive the fertilizer applicator.
4. Identification of machine specifications.
5. Development and evaluation of fertilizer machine design concepts.
6. Design analysis.
7. Engineering drawing and 3D modeling.

The study was conducted in two oil palm plantations owned by Astra Agro Lestari Company located in Riau Province. The intensity of application of fertilizers and application rate of oil palm fertilization conducted by the company was recorded and documented in accordance with the conditions in the field. Sample of fertilizer used by the company was taken and its characteristics are measured, including: water content, bulk density, angle of repose, and grain size distribution.

Physical and mechanical properties of the soil on the terraces were measured, including soil texture, water content, soil bulk density, penetration resistance, and shear resistance. Penetration resistance was measured around the palm, at 0.75-1.5 m from the palm, using a penetrometer with the cone size of 2 cm^2. The measurement was conducted at 0-20 cm depth. The shear resistance of the soil was measured using a penetrometer equipped with a shear ring. The measurement was conducted at the soil surface and a depth of 10 cm. In addition, there were also measurements of the slope of the land, the width of the available track on the terrace, and the length of the planting rows.

Available tractors on site that can be used to drive the fertilizer applicator were identified and studied. Then, discussion with the plantation manager and the technical staffs was conducted to identify the user needs in developing the design specifications.

2.2. Overview of the oil palm planting conditions on the sloping land

There are mainly three conditions of oil palm cultivation on the sloping land found in the field. The first, the oil palms are planted on the sloping land as shown in figure 1(a). Secondly, the sloping land that has been planted with oil palm is made of terraces so that there is a machine-passable track (for fruit evacuation or a crawler type tractor) as shown in figure 1(b). In this condition, the palm is located at the upper terrace about 0.5-1 m above the passable terrace. The third is the sloped land which is terraced and then planted with oil palm on the terrace as shown in figure 1(c). In the last condition, palms are located on the same terrace with a track for fruit evacuation as wide as 1-1.5 m which can be traversed on the contour slope. The track condition is relatively flat and the slope is up to 10°.

(a) (b) (c)

Figure 1. Oil palm planting conditions on sloping land.

The soil condition around the oil palm tree (palm circle) was quite dense. The rooting area was at a depth of 10 to 40 cm. The soil density conditions around the oil palm are shown in table 1. These conditions were quite dense and could be easily crossed by crawler-type vehicles or large tractor wheels. The bulk density of the soil ranges from 0.85 - 1.22 g/cm^3. The average bulk density of the soil around the palm was 1.16 g/cm^3, whereas in the track the average bulk density was 0.94 g/cm^3. The deeper, the soil becomes lighter, especially after passing through the dense layer of oil palm root zone. The shear strength of the soil around the palm is higher, compared to that of the track.

Table 1. Soil penetration resistance around the area of 0.75-1.5 m from the palm and on the track.

Depth	Penetration resistance (kPa)	
(cm)	Near the palm	On the track
0-5	10.2	4.6
5-10	14.7	6.4
10-15	20.6	6.8
15-20	24.2	6.8

2.3. Overview of fertilizer and fertilization system

Most plantations were fertilized with NPK fertilizer (and other fertilizers), which were spread manually, and spread using a power spreader. Based on the survey results in the oil palm plantation and discussion with the plantation manager, the condition of fertilization and fertilization system that are recently implemented are as follows.

1. The fertilizer (NPK, urea, etc.) are sowing on the ground, near the oil palm trees.
2. Application rate of fertilizer according to the needs of plants, in the range of 2-3 kg NPK per palm.
3. In the flat area, fertilization was done mechanically using a power spreader attached to a crawler tractor or a four-wheeled tractor (figure 2).
5. Field working performance of the mechanical fertilization was 4 ha/h (using a crawler type tractor (Canycom)); or 5 - 8 ha/hour (using a four-wheeled tractor).
6. The machine operating costs was in the range of IDR 11,500- IDR 15,500 per ha.
7. The fertilizer characteristics are shown in table 2.

Table 2. Physical characteristics of NPK fertilizer used.

Bulk density	Water content (db)	Angle of repose	Granules size		
			average	maximum	minimum
1.2 g/cm^3	7.05 %	40°	1.4 mm	3.3 mm	0.7 mm

(a)

(b)

Figure 2. (a) Fertilizer application using a power spreader attached on a crawler type tractor "Canycom", (b) a fertilizer spreader attached to a four-wheeled tractor.

2.4. Design Specification and Criteria

On the occasion of this research, oil palm fertilizer applicator that will be developed is for the cultivation of oil palm planted in the contour line of sloping land, with a narrow track. From the discussion with the company technical team, and taking into account the conditions of the land and oil palm plantations to be applied, and by including the results of the discussion with the user, the machine to be developed has the characteristics as listed in table 3.

Table 3. Characteristics of fertilizer applicator developed and the design criteria

No.	Component/ conditions	Characteristics
1	Fertilizer placement	Fertilization beneath the soil with a depth of 10-15 cm, in the root zone (according to the condition of the plant) at a certain distance from the oil palm tree. Placement of fertilizer in the disconnected furrow.
2	Fertilizer loading and refilling	• Able to load the required fertilizer (on the fertilizer hopper) the required amount (minimum for two-row crops, for refilling on one side only). • Types of granular fertilizers commonly used on site
3	Machine operation	• Operated across the pathways of oil palm planting terrace, which was provided and can be safely passed by a four-wheeled tractor or mini crawler tractor. • Implementation of fertilizer applicator is not a separate unit that is pulled (trailing)
4	Main construction	A fertilizer hopper; a metering system; a spinner and fertilizer distributors; power transmission systems, frames and supporting parts
5	Prime mover (tractor)	• A four-wheel tractor or small crawler tractor available at the site, or can be provided by the company. • The tractor needs to have a power take off (PTO) or a hydraulic power supply to drive the metering device and drive the fertilizer applicator mechanism in the soil. • The tractor has a three-point linkage or hydraulic system for lifting and lowering the fertilizer applicator unit.

2.5. Design concept selection method

The design requirements were developed from the results of analysis based on a) terrace and track conditions, b) fertilizers and fertilization conditions, c) available prime movers, and d) user needs. Then design concepts were developed by considering the design requirements. The design concepts were then evaluated, and the best concept was selected. Evaluation criteria were extracted from stakeholder requirements. The key requirements then were identified through prioritization as primary, secondary and tertiary.

The evaluation of the design concepts was conducted using the following evaluation procedure:

Step 1. Screening

Step 2. Comparison

Step 3. Decision-making.

The purpose of screening was to eliminate any concepts that were not feasible or acceptable. Screening involved comparing concepts to a series of absolute criteria [5]. The following two questions were used for screening concepts:

Question 1. Is the concept currently feasible? Alternative questions: Will it work? Will it require additional technological research and development? Unless strong evidence suggested that a concept was not feasible, then proceeded to Question 2.

Question 2. Will the concept meet each of the customer requirements? Unless strong evidence suggested that a concept will not meet one or more customer or design requirements, the concept was saved for Step 2 and continued screening other concepts. After screening every concept, then proceeded to Step 2.

The purpose of comparison was to evaluate the relative advantages and disadvantages of each concept. For evaluating the relative advantages and disadvantages of each concept, the Pugh method was used. The first seven steps for using Pugh's method during concept selection were:

1. Choose the criteria by which the concepts will be evaluated.
2. Formulate the decision matrix.
3. Clarify the design concepts.
4. Choose the initial datum concept.
5. Run the matrix. During this step, each concept was compared with the datum for each criterion.
6. Evaluate the ratings.
7. Establish a new datum and rerun the matrix.

The power spreader that was recently used by the company was used as the datum concept, in the selection process. The purpose of decision-making was to choose the concept(s) to develop further. When one design concept was clearly superior to the rest, the decision-making step was straightforward in choosing the superior concept. If the first comparison step resulted more than one better concepts with the same score, then the remaining concepts were evaluated using Decision Matrix Analysis [5, 6]. New evaluation criteria were specified form the main factor, and relative-weighted using tree diagram method.

For the evaluation, following seven criteria were used.

1) Able to apply the fertilizer beneath the soil with a depth of 10-15 cm, in the root zone at a certain distance from the oil palm tree.
2) Able to load the required fertilizer (on the fertilizer hopper) the required amount (minimum for two row crops, for refilling on one side only).
3) Can be operated across the pathways of oil palm planting terrace.
4) The metering device and spreading system are suitable for the type of granular fertilizer
5) Can be operated across the pathways of oil palm planting terrace.
6) The fertilizer applicator unit is not a separate unit that is pulled (trailing) by the prime mover
7) Using a four-wheel tractor or a small crawler tractor available at the site, or can be provided by the company.
8) Has a high working capacity.
9) Low manufacturing cost and operation cost.

IOP Conf. Series: Earth and Environmental Science **147** (2018) 012019 doi:10.1088/1755-1315/147/1/012019

3. Results and discussion

3.1. Design Concepts

Based on the design criteria, several design concepts were studied and developed. Some principle mechanisms and fertilizer applicator design were studied, in generating the concepts. As a result five design concepts were proposed, including:

1. An applicator with left and right arms to distribute the fertilizer to the upper and lower terraces as shown in figure 3 (Concept-1)
2. An all-terrain vehicle equipped with a manually operated fertilizer injector (Concept-2)
3. An applicator equipped with a hole digger to place the fertilizer into the soil (Concept-3)
4. An applicator equipped with a shovel or a disk for making fertilizer furrows (Concept-4)
5. An applicator equipped with a rotary tiller to place and mix the fertilizer beneath the soil surface (Concept-5).

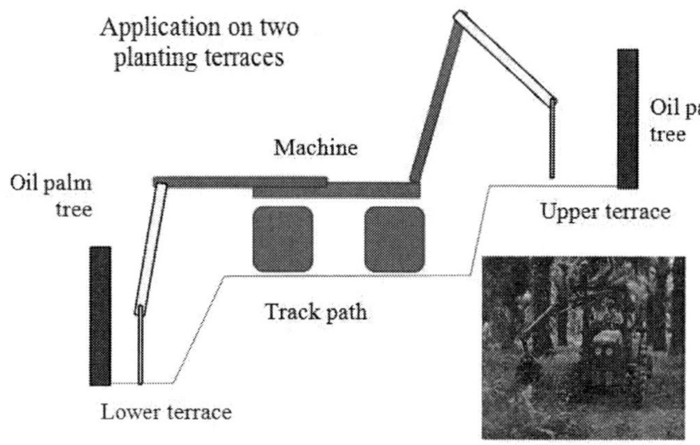

Figure 3. Fertilizer application concept on two planting terraces (Concept-1).

The first concept of fertilizer application system on sloping plantation as shown in figure 4, can apply fertilizer on the terraces using a two-channel system with a machine operated in the track of a terrace. The applicator's arms spread fertilizer on both sides (upper and lower terraces). This concept can answer the steep slopes conditions, but using a complex mechanism, big in size, and difficult to operate the machine in young oil palm planting areas. The track used should be wide enough at least 2.5 m wide so that the machine can be stable running.

The second concept of fertilizer application method is using an injector mechanism that is manually positioned by the operator and using an all-terrain vehicle (ATV) or a motorcycle. The ATV is equipped with a trailer on the left side for carrying the fertilizer and the pneumatic mechanism unit. The third concept uses a digger mechanism for making holes on soil so that the fertilizer can be placed in the holes. Preparation of fertilizer holes can be carried out using a sharply-tapered wheel (can be a gear or the like). The hole depth is about 5 cm, the width of the digger point is about 8 cm, the number of holes is 20-30 holes per palm (considering the fertilizer application rate 2-3 kg/palm). Two concepts of fertilizer placement on the holes around the oil palm are shown in figure 4. An auger-type metering device is used for metering the fertilizer and a pneumatic system equipped with a blower is used for spreading and conveying the fertilizer to the holes.

(a) (b)

Figure 4. Concepts of fertilizer placement on the holes around the oil palm of Concept-3: (a) straight pattern and (b) half circle curved pattern.

The fourth concept is similar to the third concept, except the fertilizer placement method. In this concept, the fertilizer is placed in a furrow cut by a shovel or a disc as shown in figure 5(a). In the fifth concept, the applicator is equipped with a rotary tiller to place and mix the fertilizer beneath the soil surface at a depth of 10 cm, as shown in figure 5(b).

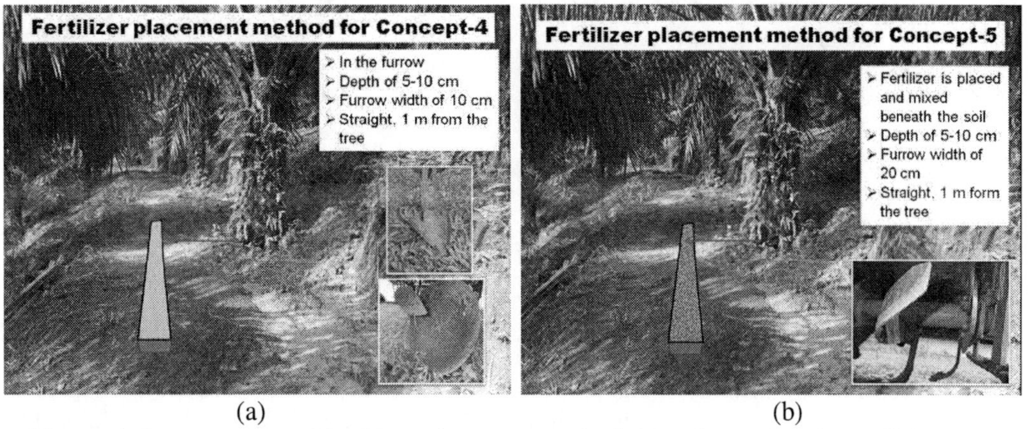

(a) (b)

Figure 5. Two concepts of fertilizer placement method: (a) using a shovel or a disk of Concept-4, (b) using a rotary tiller of Concept-5.

3.2. Design evaluation result

In the screening step, all the five design concepts were evaluated using two questions as described above. Base on the question 1, the concept-1 was decided to be no feasible. The machine construction will be complex and the operation will be difficult especially for the young oil palm. The arms of the machine will disturb the palm frond. Discussion with the user resulted that the concept should be eliminated in the first step. However, four other concepts passed in this evaluation step.

To compare the remaining concepts, using the Pugh method, the design criteria was used as the evaluation criteria on the user requirements mentioned previously and selected. The power spreader that recently used in the company, was selected to be the initial datum. The Pugh concept selection matrix shown in table 4 resulted from the first analysis. As shown in table 4, the four concepts were superior to the datum. However, in the first run concept-4 and concept-5 have the same level of advantages.

Table 4. Pugh matrix for fertilizer applicator concepts (run #1).

Article o.	Article II. Criterion	The power spreader	Concept-2	Concept-3	Concept-4	Concept-5
1.	Able to apply the fertilizer beneath the soil with a depth of 10-15 cm, in the root zone at a certain distance from the oil palm	D	+	+	+	+
2.	Able to load the required fertilizer		S	S	S	S
3.	Can be operated across the pathways of oil palm planting terrace	A	+	+	+	+
4.	The metering device and spreading system are suitable for the type of granular fertilizer	T	S	S	S	S
5.	Can be operated across the pathways of oil palm planting terrace	U	+	+	+	+
6.	The fertilizer applicator unit is not a separate unit that is pulled by the prime mover	M	-	+	+	+
7.	Using a four-wheel tractor or a small crawler tractor available at the site		-	+	+	+
8.	Has a high working capacity		-	-	S	S
9.	Low manufacturing cost and operation cost		+	S	S	S
	+		4	5	5	5
	S		2	3	4	4
	-		3	1	0	0

Two concepts have a different method of the fertilizer application. In the concept-4 the fertilizer is placed in a row formed by a disk or a shovel, however, in the concept-5 the fertilizer is placed and mixed with the soil using a rotary tiller. The most related criterion with this condition is the first criterion, i.e., the ability to apply the fertilizer beneath the soil with a depth of 10-15 cm, in the root zone at a certain distance from the palm. The criterion can be specified to be new criteria, including a) effectively mixed the fertilizer with the soil, b) fertilizer well covered by the soil, and c) not leaving the open furrow. The two concepts use a different mechanism for placing the fertilizer beneath the soil so that the manufacturing cost for the mechanism should be considered in the evaluation. According to these reasons, new following evaluation criteria then were used in the decision-making step:

a) effectively mixed the fertilizer with the soil (O_{31}),
b) fertilizer well covered by the soil (O_{32}),
c) not leaving the open furrow (O_{33}), and
d) low manufacturing cost for mechanism of fertilizer placement (O_{22}).

The weighted factors were decided using weighted tree diagram, as shown in figure 6. The evaluation criteria then were relatively scored for each concept, using 0-10 scale. To select the best concept among the two concepts, the *weighted decision matrix* was used as shown in table 5.

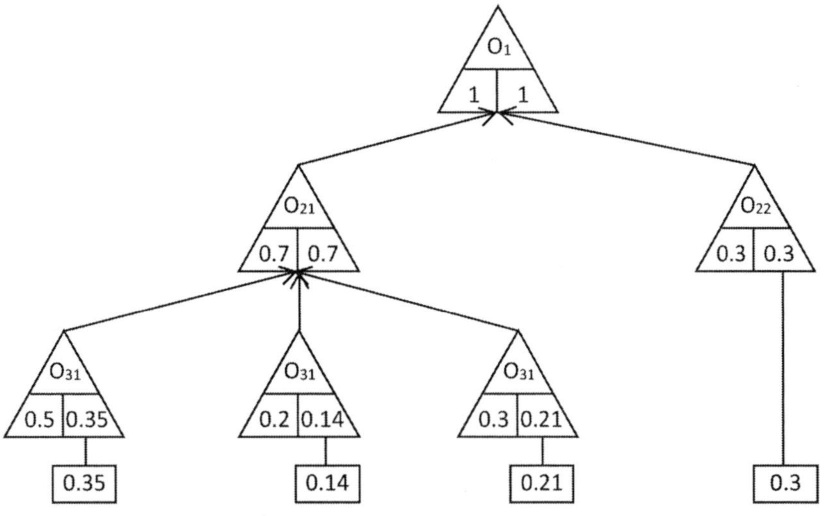

Figure 6. Weighted tree diagram for each evaluation criteria (objectives).

Table 5. Weighted *decision matrix results.*

Criteria	a (O_{31})	b (O_{32})	c (O_{33})	d (O_{22})	Total
	0.35	0.14	0.21	0.3	
Concept-4	3	4	4	10	5.45
Concept-5	10	6	10	3	**7.34**

As shown in table 5, the concept-5 is superior to the concept-4. Then, the concept-5 was decided to be the best design concept that will be developed further.

3.3. Design analysis

In accordance with the specifications and design limits that have been studied, the main function of the developed machine was: "applying granular NPK fertilizer to oil palms planted on terraces effectively". From the main function then sub-functions were developed. Alternative mechanisms and components for each function were proposed. The result of the functional design analysis is presented in table 6. Underlined mechanisms or components were selected to be implemented in the design of the fertilizer applicator.

To get the optimum design, each part and mechanism has been analyzed using proper design analysis methods, including:

1. Material strength analysis for each external loaded component, such as fertilizer hopper, framework of fertilizer hopper, shafts, rotary supporting frame, rotary lifting frame etc [7].
2. Power requirement analysis, to rotate the rotary blades, to lift the rotary frame, and to rotate the spinner of the fertilizer metering device.
3. The kinematic analysis of the rotary lifting frame of the rotary unit, which is driven by the hydraulic cylinders, and the kinematic analysis of the movement of the levers as well as the set of drive linkages for shifting the fertilizer metering disk.
4. Analysis of the size of fertilizer hopper, size of the spinner space, size of the fertilizer discharge gate at the bottom of the hopper.
5. Analysis of machine stability when operated on sloped land or track.
6. Machine working performance analysis and analysis of machine operating costs.

Table 6. Sub-functions and alternative mechanisms of the fertilizer applicator

No.	Function/sub-function	Alternative mechanisms, components
1	Store granular fertilizer	A hopper ←*selected*
2	Convey the fertilizer to the metering system	• A tube • An open-closed gate ←*selected*
3	Meter the fertilizer according to the application rate and convey it to the targeted area	• A grooved rotor type metering device • A centrifugal type metering device and a spinner ←*selected*
4	Hold the hopper and metering device unit	A frame ←*selected*
5	Drive and rotate the metering device	• A power transmission system from tractor engine • A hydraulic motor ←*selected* • A gasoline engine • An electric motor
6	Open furrows for placing the fertilizer, or loosening the soil and mix the fertilizer in the soil	• A shovel • A toothed wheel • A soil drill • A rotary tiller ←*selected* • A disk type furrow opener
7	Drive and rotate the mechanism for mixing the fertilizer in the soil	• A power transmission system from tractor PTO • A hydraulic motor ←*selected* • A gasoline engine • An electric motor
8	Lifting and lowering the mechanism for mixing the fertilizer in the soil	• A lifting frame using hydraulic cylinders ←*selected* • The three-point linkage of the tractor
9	Providing mechanical power for the prime mover	Tractor engine and a hydraulic power system ←*selected*
10	Operate the gate of the fertilizer metering device	A handle and its linkages ←*selected*
11	Support the operator when pouring fertilizer into the hopper	A support frame and a stepper ←*selected*
12	Support the load of all components of the fertilizer applicator	A frame ←*selected*
13	Drive and move the fertilizer applicator unit	• A small four-wheeled tractor • A mini crawler tractor ←*selected* • An all-terrain vehicle (ATV)

3.4. Construction design of the fertilizer applicator

As a result of oil palm fertilizer applicator design, 3D model has been created, as well as engineering drawings. The 3D model image of the proposed machine is shown in figure 7. The construction and drawings are shown in figure 8. The general specifications of the machine are presented in table 7. The final design concept of the fertilizer applicator uses a 110 cm wide mini crawler type tractor (Canycom S 160) as the prime mover. The machine is equipped with a hopper and a spinner disk for metering and conveying the fertilizer. The hopper capacity is 475 l that can store about 520 kg of NPK fertilizer. The spinner is rotated by a hydraulic motor (Rexroth MKM-32). A 20 cm wide rotary tiller is equipped in the front of the machine, to pulverize the soil and mix the fertilizer in the soil at a depth of 5-10 cm. Four rotary blades are installed on the rotary shaft which is rotated by a hydraulic motor (Rexroth MKM-32).

Table 7. General specifications of the fertilizer applicator.

Specifications	Unit	Value
Weight of the machine (without 'Canycom')	kg	187
Weight of the machine with the 'Canycom'	kg	1300
Length x Width x Height	cm	291 X 112 X 238
Hopper capacity	l (kg)	475 (520)
Prime mover		Canycom S 160
Engine Type		20.5 HP Diesel Engine
Hydraulic motor for rotary tiller and spinner		Rexroth MKM-32
Operation speed	km/h	0 - 7.5
Working capacity	ha/h	2-2.5

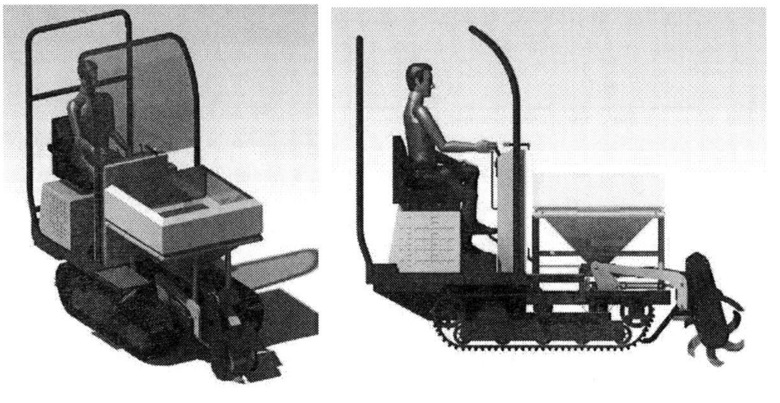

Isometric view Side view

Figure 7. Three-D model of the fertilizer applicator for oil palm on terrace cultivation.

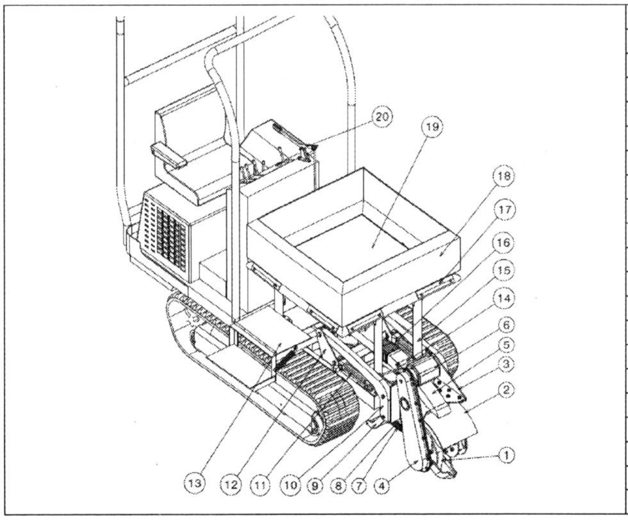

No.	Part name
1	Rotary
2	Upper cover of the rotary
3	Frame of the rotary cover
4	Cover of chain and sprockets
5	Fertilizer conveying tube
6	Hydraulic motor
7	Spring+Pin
8	Holder of the rear cover of the rotary
9	Rotary rear cover
10	Lifting arm
11	Hydraulic cylinder
12	Lifting arm frame
13	Stepper
14	Frame of fertilizer conveyor
15	Frame of hopper holder
16	Rubber tube of fertilizer conveying tube
17	Hopper frame fasteners
18	Fertilizer hopper
19	Fertilizer screen
20	Controlling handle for metering device

Figure 8. The construction of the fertilizer applicator.

4. Conclusions

The evaluation results showed that the concept of fertilizer applicator equipped with a rotary tiller had the most advantages on the expected design criteria. A 3D model and also the engineering drawing of the concept was developed. The design concept uses a 110 cm wide mini crawler tractor as the prime mover and is equipped with a hopper and a spinner disk for metering and conveying the fertilizer and a 20 cm wide rotary tiller in the front of the machine.

References

[1] Goh K J, Hardter R and Fairhurst T H 2003 *Fertilizing for Maximum Returns* (Singapore: Int. Potash Inst. (IPI)) p 279

[2] Wahid B O, Rahim A S, Tarmizi A M, and Basri M W 2004 Variable-rate fertilizer applicator for oil palm *MPOB Information Series (MPOB TT No. 216)* (Kuala Lumpur: Malaysian Palm Oil Board, Ministry of Plantation Industries and Commodities) p 1

[3] Jahis S, Hitam A, Ali Z M and Nor M M 2002 Mechanical fertilizer spreader for young palm MPOB Information Series (MPOB TT No. 135) (Kuala Lumpur: Malaysian Palm Oil Board, Ministry of Plantation Industries and Commodities) p 1

[4] Ishola T A, Yahya A, Shariff A R M and Aziz SA 2012 Variable rate technology fertilizer applicator for oil palm plantation *Int. Conf. on Agricultural and Food Engineering for Life (Cafei2012)* (Putrajaya) (Malaysia) p 520

[5] Dieter G E and Schmidt L C 2009 *Engineering Design* (New York: McGraw-Hill)

[6] Justel D, Vidal R, Arriaga E, Franco V and Ester Val-Jauregi E 2007 Evaluation methods for selecting innovative product concepts with greater potential marketing success *Int. Conf. on Engineering Design, ICED '07*(Paris) (France) p 28

[7] Chaudhari S, Naeem M, Jigar P and Preyash P 2017 Design and development of fertilizer spreader machine *Int. J. of Engineering Sciences & Research Technology* **6**(4) p 62

System Identification and Steering Control Characteristic of Rice Combine Harvester Model

S P Sutisna[1], R P A Setiawan[1], I D M Subrata[1] and T Mandang[1]

[1] Department of Mechanical and Biosystem Engineering, Bogor Agricultural University, Indonesia

E-mail: setperna@gmail.com

Abstract. This study is a preliminary research of rice combine harvester trajectory. A vehicle model of rice combine used crawler with differential steering. Turning process of differential steering used speed difference of right and left tracks This study aims to learn of rice combine harvester steering control. In real condition, the hydraulic break on each track produced the speed difference. The model used two DC motors with maximum speed 100 rpm for each tracks. A rotary encoder with resolution 600 pulse/rotation was connected to each DC motors shaft to monitor the speed of tracks and connected to the input shaft of a gearbox with ratio 1:46. The motor speed control for each track used pulse width modulation to produce the speed difference. A gyroscope sensor with resolution 0.01° was used to determine the model orientation angle. Like the real rice combine, the tracks can not rotate to the opposite direction at the same time so it makes the model can not perform the pivot turn. The turn radius of the model was 28 cm and the forward maximum speed was 17.8 cm/s. The model trajectory control used PID odometry controller. Parameters input were the speed of each track and the orientation of the vehicle. The straight line test showed the controller can control the rice combine model trajectory with the average error 0.67 cm.

1. Introduction

A combine harvester is a versatile machine designed to harvest a variety of grain crops. The grain crops are harvested from the field to deliver clean grains, collected in the machine tank and discharge periodically for transportation and further processing or storage. The following main crops are harvested using combine harvesters: wheat, rice, barley, soybeans, and corn. The combine harvester combines all technological operations of grain crop harvesting: cutting and gathering of the plants, grain threshing and separating, and grain cleaning and collecting in the combine tank [1].

The rice combine harvester operated in the field requires a higher operator skill and performance level. Operator has to simultaneously control the travel speed and direction of the combine harvester while adjusting the height of the header after considering numerous parameters [2]. Combine harvester operating on paddy field under unfavorable environmental conditions such as the dust particles, noise, and vibration from the machine increase the operator's fatigue. The fatigue reducing operator's focus which can decrease the productivity and efficiency. An automatic system is needed to help operator to control the combine. The autonomous guidance system performs unmanned harvesting and can be used to replace all field operations performed by operator [3]. The autonomous guidance system overcomes the lack skill and performance operator and gains more work stability and productivity. The system allows traveling fast and accurately along the target path.

In recent years, numerous sensors, methodologies, computational algorithms, and other supporting technology evolution cause development of autonomous agricultural machine still interested. The automation methodologies have been proposed and developed for an autonomous guidance system because it depends on the field environment, operation, and supporting technologies [4]. The different method can be different for the sensors, navigation, algorithm, and equipment. All of the methods have been developed the same purpose to get high accuracy and replace all operator function.

To make the combine can run automatically, follow the reference path, it is necessary to know the control system characteristics of the steering in operating the combine. Characteristics that needed to be known, there is how to turn, rotate, maximum speed, orientation change, turning radius, and how to control it along operated [5]. Learning characteristics of combine harvester steering control is difficult using the real combine. On real size is needed more space and energy to do this process. A model is created for reducing those needs which is built smaller than the real combine with the similar steering type. The steering type of combine will be used to developed automatic trajectory system is crawler type with two tracked wheel differential drive. This study is a preliminary research of rice combine harvester trajectory to learn how the rice combine harvester steering control.

2. Materials and methods

This study started from the measurement of combine harvester dimensions, especially on the moving parts. After that, determination of model dimensions had been made based on market availability and space requirements for the placement of electronic components such as microcontrollers, sensors, drivers, batteries, and motors. The finished design model manufactured and installed of electronic components. After the model that been made, then functional test was done to determine the suitability of each component function. In addition, calibration of the sensors was also used. In order to control the model carried out the design of controllers and navigation. Finally, trajectory test was done the model to follow the path of reference.

A vehicle model of combine was built using crawler with a differential drive like the real combine. The model main frame was made of 3 mm thickness aluminum plate. Design of vehicle model can be seen in figure 1.

Figure 1. Design of combine vehicle model.

The model used crawler as wheel drive which used plastic material. Comparison between model and real can be seen on table 1.

Table 1. Comparing parts between model and real.

Parts	Model	Real
Track material	Plastic	Iron steel
Track length (cm)	35	112
Sprocket diameter (cm)	5	16
Length between tracks (cm)	30	94
Track width (cm)	4.5	25

IOP Conf. Series: Earth and Environmental Science **147** (2018) 012020 doi:10.1088/1755-1315/147/1/012020

Most of the model parts used the scale of 1:3 to approach the real condition and adjust availability parts of the model. In real condition, the speed difference on combine crawler is produced by the hydraulic brake on each tracked wheel. The model used two DC motors with maximum speed 100 rpm for each track. Each DC motors speed was controlled using pulse width modulation to produce the speed difference. Like the real rice combine, the tracked wheels of model cannot rotate in the opposite direction at the same time so the model does not do pivot turn.

The model trajectory control used odometer PID controller on equation 1. Parameters input were the speed of each tracked wheel and the orientation of the vehicle.

$$u = k_1 e_n + k_2 \int e_n d\alpha + k_3 \frac{de_n}{d\alpha} + k_4 e_\phi + k_5 \int e_\phi d\alpha + k_6 \frac{de_\phi}{d\alpha} \quad (1)$$

where:

u	: ratio of difference and total speed
k_1, k_2, k_4	: PID constant
e_n	: position error
e_ϕ	: orientation error

Dead reckoning navigation was used to determine the model position. The next position was calculated from the initial position, velocity, orientation, and time. Determination of velocity as well as position on x, y coordinate was calculated using equation 2, 3, and 4.

$$\dot{s} = R\dot{\alpha} = R\frac{(\dot{\theta}_1 + \dot{\theta}_2)}{2} \quad (2)$$

$$\frac{dX}{d\alpha} = R\cos\phi \quad (3)$$

$$\frac{dY}{d\alpha} = R\sin\phi \quad (4)$$

where:

R : Wheel radius

ϕ : Orientation angle

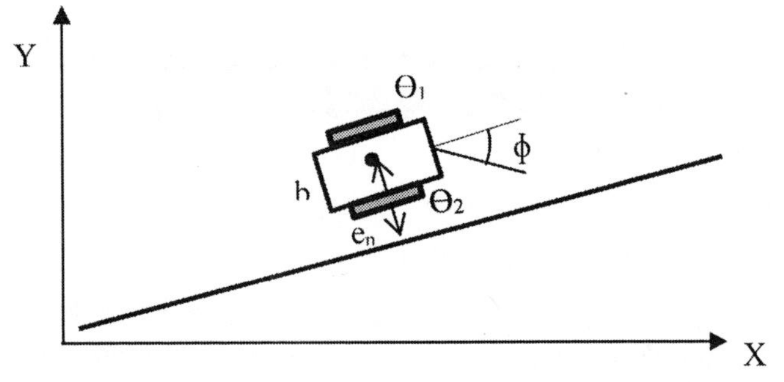

Figure 2. Vehicle relative position, position error, and orientation.

Position error of model from trajectory path is marked e_n and orientation error is marked e_ϕ as shown in figure 2. That errors can be determined using equation 6 and 7.

$$e_n = \sqrt{\Delta X^2 + \Delta Y^2} \quad (5)$$

$$e_\phi = \phi_r - \phi \quad (6)$$

A rotary encoder with resolution 600 pulse/rotation was connected to each DC motors shaft to monitor the speed of crawler and connected to the input shaft gear box with ratio 1:46. Orientation of the vehicle was determined using a gyroscope sensor with resolution 0.01°. A microcontroller was used

IOP Conf. Series: Earth and Environmental Science **147** (2018) 012020 doi:10.1088/1755-1315/147/1/012020

as the main controller. The microcontroller collected the data from sensor, proceed and determined speed of each track. The motor driver receives the pulse modulation signal from the microcontroller to drive the DC motors. The data from rotary encoders, gyroscope, and the speed of each track was recorded using SD card module for data analysis. Control system of the model can be seen in figure 3.

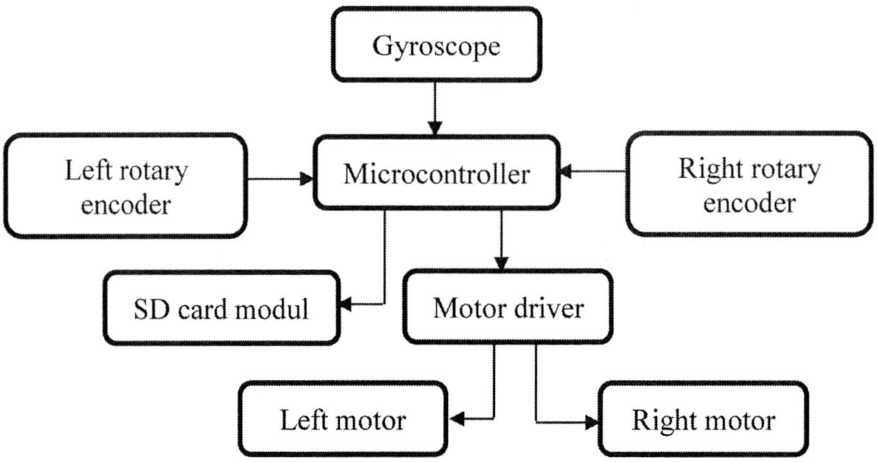

Figure 3. Control system of the combine harvester model.

Steering control characteristic consists of time response each track, time response turn orientation, turn radius, and path following test. Characteristic of time response was used to develope PID equation especially tunning PID process. One of the methods tunning PID use Ziegler-Nichols method which was used in this study. The turn radius and steering control process characteristic as long as the model goes along path trajectory.

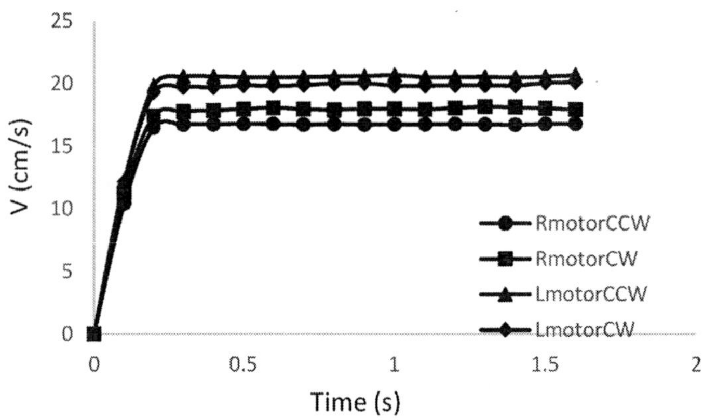

Figure 4. Time response motor.

3. Results and Discussion

Time response test was done for each track motors. The motors were drive to maximum speed from the rest position. It can show the time of each motor to reach maximum speed.in figure 4 shows the result of time response test for each motor. The motors have different response although the same motor in different direction. Right motor have maximum speed 16 cm/s for counterclockwise and 17 cm/s for clockwise direction as well as left motor have maximum speed 20 cm/s for counterclockwise and 19 cm/s for clockwise direction. That difference can be caused by motor manufacturing or track setting.

The speed difference between right motor and left motor cause the model can not trajectory on straight line path without controlled.

In figure 5 shows time response test result for orientation change. One of the motors rotated at maximum speed and the other did not rotate. It was done alternately for right motor and left motor to know the orientation change versus time. The orientation change average for CW direction is 4.1 °/s and CCW direction is 4.2. There is little difference because left motor speed higher than right motor.

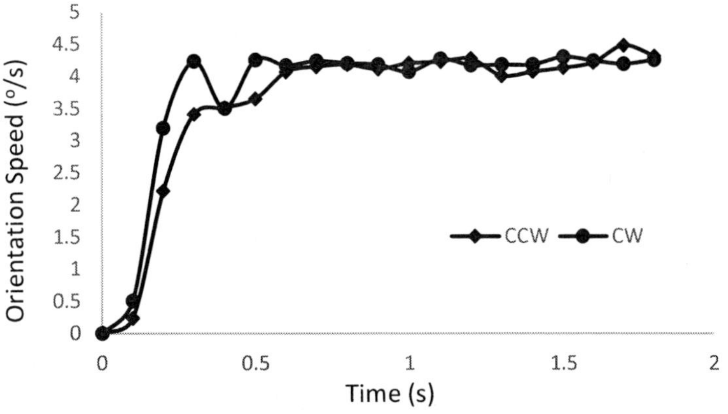

Figure 5. Orientation change.

Based on time response data, the PID controller equation was developed using Ziegler-Nichols method. In spite of all the advances in controller method over the 50th years ago, the PID controller still exists.

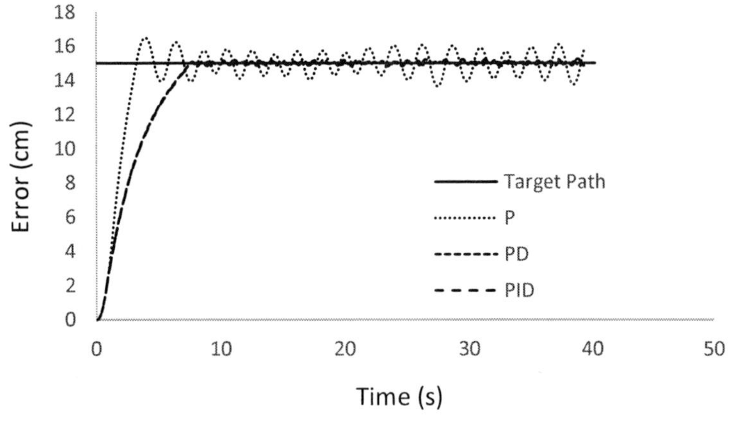

Figure 6. PID equation simulation test.

The Ziegler-Nichols rules was a very influential rule for tuning PID controller. This method developed tuning rules by simulating a large number of different processes and correlating the controller parameters with features of the step response. The first determined the proportional constant then integral and differential constant [6]. The PID constants was obtained for position error $k_p = 0.5$, $k_i = 0.2$, $k_d = 0.5$ and for orientation error $k_p = 0.1$, $k_i = 0.15$, $k_d = 0.01$. The simulation test for P equation (k_d and $k_i = 0$) , PD equation ($k_i = 0$), and PID equation can be shown in figure 6. In simulation test, the model was programmed to reach the straight line target path with distance 15 cm beside of target path.

When position error of combine model was too large, the model approached sharply then slowly horizontal. From the test was known that the P controller had oscillation around the target path but the PD controller and PID controller seen can approach smoothly. Nevertheless, the root mean square error (RMSE) of the PD larger than PID on stable condition. The RMSE error of the PD is 0.65 cm and the PID is 0.07 cm. So that the PID controller is ideal to control this model moving wherever the moving approach in short time and less overshoot so that the constants are used.

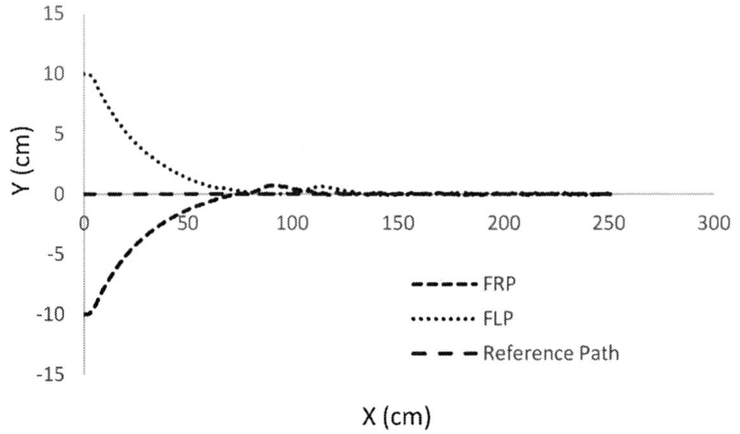

Figure 7. Approaching test.

Approaching test was done to know how the model can reach the target reference path using the controller. In figure 7 can be seen the model approaching from the right path (FRP) and for the left path (FLP). On FRP there is the little overshoot after the model reaches the reference path but on FLP can be seen after the model reaches the reference path little back away. The approaching test can be compared with figure 6 the PID simulation test that shows the approach too but on simulation. The real test on approaching test shows a similar pattern.

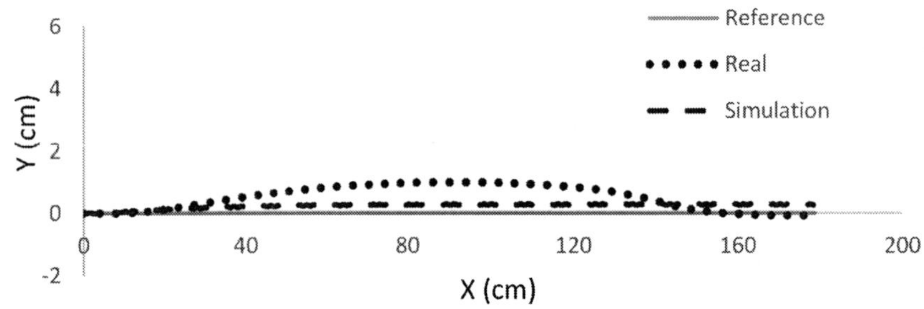

Figure 8. Trajectory straight line test.

In figure 8 shows the result of trajectory straight line test on real condition and simulation. The test was done using average speed 17.8 cm/s on the floor and sampling time 0.1 s. Average error on real condition is 0.67 cm and maximum error 0.98 cm while on simulation the average error is 0.26 cm and the maximum 0.35 cm. In this study was used the odometry method to determined the model position. This method can estimate the position from the speed and direction data. Odometry assume that wheel revolution can be translated into linear displacement relative to the floor. There are several error sources that fit into two categories, systematic errors and nun-systematic errors. The systematic errors consist of unequal wheel diameters, misalignment of wheels, uncertainty about the effective wheelbase, limited encoder resolution, and limited encoder sampling rate. The non-systematic errors consist of travel over uneven floors, travel over unexpected objects on the floor, as well as wheel-slippage [7]. The trajectory straight line test can be seen in figure 9.

In turning test the model was turned by one of track moved maximum speed and the other not moved. In order to turn left, the right track moved forward and the left track did not move as well as vice versa. The turn radius can be determined by ordered the model to make a circle turn. Result of this test, the turn radius of the model is 28 cm.

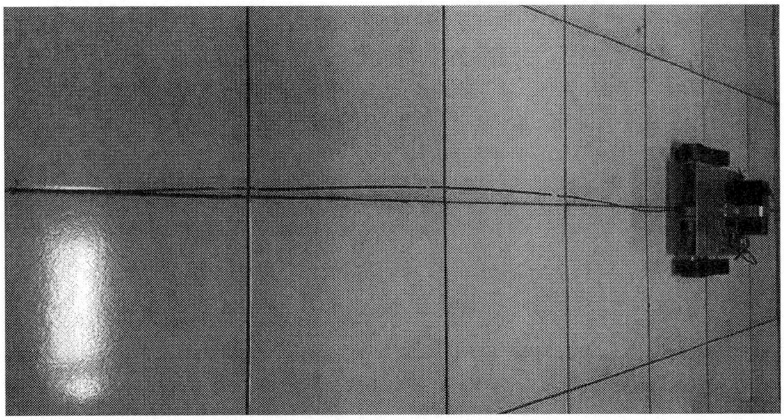

Figure 9. Straight line test in real condition.

4. Conclusions

From this study, it can be concluded that system identification and steering control characteristics should be done to control the combine model, and then can be summarized as follows: (1) The combine model can turn based on the speed difference on each track; (2) The greater speed difference of each track make the turn radius will be smaller; (3) The response time indicates the time required by the track to reach ordered speed. It also affects the turning response; (4) The steering control of model can use PID odometry based on traveling speed of each track and orientation change; (5) Based on the simulation and real test shows this method can make the model trajectory reference path.

References

[1] Miu P 2016 *Combine Harvesters Theory, Modeling, and Design* (New York: CRC Pres)
[2] Cho W, Iida M, Suguri M, Masuda R and Kurita H 2014 Using multiple sensors to detect uncut crop edges for autonomous guidance systems of head-feeding combine harvesters *Engineering in Agriculture, Environment and Food* 7 115-121
[3] Benson E R, Reid J F and Zhang Q 2003 Machine vision-based guidance system for agricultural grain harvesters using cut-edge detection *Biosystems Engineering* **86**(4): 389-398
[4] Li M, Imou K, Wakabayashi K and Yokoyama S 2009 Review of research on agricultural vehicle autonomous guidance *Int J Agric & Biol Eng* **2**(3): 1-16
[5] Modjtahedzadeh A and Hess R A 1993 A model of driver steering control behavior for use in assessing vehicle handling qualities *Journal of Dynamic Systems, Measurement, and Control* **115** 456-464
[6] Astrom K J and Hagglund T 2004 Revisiting the Ziegler-Nichols step response method for PID control *Journal of Process Control* **14** 635-650
[7] Boreinsten J and Feng L 1996 Measurement and correction of systematic odometry errors in mobile robots *IEEE Transactions on Robotics and Automation* **12**(6): 869-880

Economic Ergonomic Approach to Design an Optimal Manpower and Mechanization in Rice Production

Muanah[1], M F Syuaib[2] and Liyantono[2]

[1] Graduate School of Bogor Agriculture University (IPB), Bogor, Indonesia
[2] Department of Mechanical and Biosystem Engineering, Faculty of Agricultural Engineering and Technology, Bogor Agriculture University, Indonesia

E-mail: muanahtmp@yahoo.com

Abstract. Productivity of manpower could be improved by considering the economic and ergonomic aspect. The ergonomic aspect (human factor) is required to design an optimal manpower, while in the economic aspect, the well being of manpower could be evaluated from the amount of received income based on their work capability. This research was conducted on February 2016 to January 2017 in Gapoktan Silih Asih rice field Cigombong, Bogor. This study aims to analyze the income of manpower based on current conditions, increasement of working hours and mechanization addition. The results showed that manpower income based on the current condition and ergonomic consideration was Rp 1,174,030/person.month, by increasing the working hours with ergonomic consideration the obtained income was Rp 1,766,204/person.month. the revenues were based on the existing work system even though the increasement of working hours have not provided optimal income due to low productivity. Therefore the results of analysis with the addition of selective mechanization, showed self-ownership machinery was more profitable the obtained income have more than the Bogor minimum wage standard of Rp 2,969,325/person.month.

1. Introduction

Rice production in the agricultural sector is one of the activities that absorb a lot of manpower. So that manpower conditions affect the productivity greatly. According to [1] which said that several factors that affect the productivity are the variable of wage suitability, work experience and expertise in work. But in this study, the focus was on the suitability of wages to obtain an optimal output. Wage is an acceptance or a compensation from employers to employees (manpower) on work or services that have been done [2].

Generally, in the wage agricultural sector between male and female manpower is different. Male manpower receives a higher wage than the female one. This is different from the wage system that used in the other sectors that is using a minimum wage standards. The minimum wage policy in Indonesia according to Gianie [3] policy is used to look at two aspects namely, (1) as a protection for workers to keep the value of received wage did not decrease in meeting the needs of daily living (2) as a protection for companies to maintain the manpower productivity. However, the policy in the agricultural sector has not been well realized, which caused the manpower's income is still far from the minimum wage standard, while the risk of work is high. The amount of received income determines the welfare of a community. The higher per capita income of a society the higher of the prosperity level [4]

Therefore, based on the description above, in this study an economic ergonomic analysis was conducted. The Economic ergonomic is an ergonomic evaluation to get an optimal wage economically. So, the manpower in addition to the ergonomic conditions will also met an economical conditions (optimal wage) to support the welfare. This research aimed to: (1) to analyze the manpower income based on the existing work system with ergonomic approaches in the study location (2) to analyze the manpower income using scenario by increasing the working hours without changing the existing work systems and performing comparisons with the minimum wage standards, and (3) to analyze the manpower income with scenario of mechanization addition and comparing with the Bogor minimum wage standard.

2. Materials and methods

2.1. Data collection

This study used an ergonomic analysis result data including the number and the distribution of the ideal manpower. So the equipment and subject of research refered to the previous research conducted by [5] While the data of wage information of male and female manpower was obtained through a direct interview with the manpower or farmer.

2.2. Data analysis

2.2.1. Manpower's Income Analysis

1. Manpower's income analysis based on current condition

From the number and distribution of existing manpower and data related to manpower's wage, the manpower income could be determined by using Equation 1, then compared to the Bogor wage standard. The manpower's income depends on the size of productivity called Human Power (HP). Human Power is the amount of daily worker needed for one crop season and one hectare of rice production. The output is measured by the manpower (person) and the area (ha).

$$HI = \frac{[TLW]_m + [TLW]_f}{HP} \tag{1}$$

Description :

HI : Income per person (Rp/person.month)
T : Duration of work (days/month)
L : Manpower needs (person/ha)
W : Manpower's wage (Rp/day.person)
m, f : Male and female manpower index
HP : Human Power (person/ha)

2. Income analysis by increasing working hour (First Scenario)

At the location of this study conducted manpowers work for 4.8 hours/day. Therefore, it can still be increased to 7 hours/day working hour. The increasement of the working hours was based on the standard working hours in Indonesia. The manpower's income was calculated by Equation 1, where the wages is based on working hours and compared to the Bogor wage standard.

3. Income analysis with mechanization addition (Second Scenario)

Based on the existing working system and the increasement of working hours that can not provide optimal income, so that in several work elements the mechanization was added. The addition of mechanization was conducted by selective method which means that the addition of mechanization was conducted on certain working elements based on the needs of existing workforce and with the considerable of available time. Therefore, from the addition of this mechanization, the manpower savings was obtained, then the difference in cost reduction from manpower savings to the machinery costs was an additional advantage of manpower.

3. Result and discussion

3.1. Manpower's Number and Distribution

The economic ergonomic approach is the second stage of research after an ergonomic analysis. In ergonomic analysis, there is an ideal number and distribution of manpower to work on a series of rice production activities. Therefore, from the distribution with the resulting work output the amount of manpower's income could be determined. The number and distribution of manpower based on the ergonomic approach reported by [5] can be seen in Table 1.

Based on work capacity and by using 120 days (4 month) per crop season, the rice production activities required a total manpower force of 2.32 persons consist of 1.85 male manpower and 0.47 female manpower on 100 work days (the minimum wage standar: 25 days/month). The value of 3.32 person/ha means average daily worker required for rice production on a hectare area of crop. From the needs of manpowers, the manpower productivity or called Human Power (HP) could be determined. The term of productivity in this study was obtained by dividing the work area output (ha) with manpower needed (person).

Table 1. Manpower's requirement number and distribution besed on current condition (4.8 hours/day by) [5]

Work Element	Symbol	Required manpower (person.day/ha)
Seedling**	Se	13*
Tillage***	Ti	56
Harrowing	Ha	9
Transplanting	Tr	22*
Fertilizing	Fe	2
Cutting	Cu	37
Paddy transporting	Ev	15
Thressing	Th	38
Unhulled rice tranporting	Tp	9
Drying	Dr	12*
Husking	Hs	9
Polishing	Pl	4
Grading	Gr	2
Packaging	Pc	4
Total		232

Information: *female manpower, **equivalent to provide the requerment of seeds for 1 m^2 of planting area; ***include hoeing the corners of rice field that was not processed by tractor, clean up and tidy up the rice field.

3.2. Manpower's Income Analysis Using Current Condition

Manpower's income in this study is a calculation result based on the amount of remuneration that has been applied in the research location. Where the current wage system is based on gender. Male manpower was Rp 50,000/day while the female manpowerer was Rp 35,000/day with average work time was 4.8 hours/day. Therefore, the calculation results using Equation 1 obtained Rp 1,174,030/person.month income (Table 2). The income was still lower than the Bogor regional minimum wage standard (Rp 2,960,325/person.month) of Bogor Regency in 2016.

Based on the existing work system, manpower income has not met the regional minimum wage standard. Therefore, the optimization was made based on two scenarios in which the existing work-hours are not changed. The first scenario was applied on the same existing work-system but with extended working hour (from 4.5 hours became 8 hours as a daily standard); whilst the second scenario was apply with 'selective mechanization' to increase the daily work's productivity due to increasing the daily wage. The optimization calculation revealed that the scenarios are rational to apply on a 100 ha area paddy field with 9 rotation of crops, so the number of labor expresses a daily routine of workers needed.

Table 2. Manpower's income based on current condition with the UMR standard

Description	Amount
Required Manpower (person/ha)	2.32
Manpower's total cost (Rp/ha.month)	2,723,750
Average wage per person (Rp/person.month)	1,174,030

3.3. Manpower's Income Analysis by Optimizing Working Hours (Scenario 1)

Optimizing the working hours aimed to increase the working hours that have been applied which was 4.8 hours/day, then added to 7 hours/day in accordance to the provisions of working hours in Indonesia. One of the scenarios used in this study aimed to obtained an optimal income, then one way that was conducted was by optimizing the working hours.

Based on the analysis results by 7 working hours per day in a crop season required a total workforce of 1.44 people consisting of 1.16 men and 0.28 women manpowers. Increasement of working hours led to increased the work productivity to 0.69 ha/person because the manpower worked longer than the previous that was only 4.8 hours/day. The increase of working hours aimed to make the manpower more productive in order to get a standard wage. Which made the obtained revenue of Rp 1,766,204/person.month was bigger than before. Increasement of working hours without changing the existing work system proved not provide an income which in accordance to the Bogor minimum wage standards. The manpower income between male and female in this study was assumed to be the same because based on the results of the analysis proved that some elements of work are more productive if conducted by a female manpower, but also in labor regulations in Indonesia that the standard wage for male and female are not different (Table 3).

Therefore, to obtain the income according to the Bogor regional minimum wage, the second scenario must be conducted. The second scenario aimed to make a design by adding selective mechanization to the distribution of existing workforce, so the required manpower could be reduced but the work productivity could be improved to obtain the optimal results.

Table 3. Manpower's income based on the increasement of working hours (7 hours/day)

Description	Amount
Required manpowers (person/ha)	1.44
Manpower's total cost (Rp/ha)	2,543,333
Average wage per person (Rp/person.month)	1,766,204

3.4. Analysis of Manpower's Income with Mechanization Design (Scenario 2)

Based on the workflow that has already existed in gapoktan Silih Asih Cigombong Bogor, which was the place where this research was conducted, one crop season was counted into 134 working days. The results of the analysis showed that the seedling work element (Se) could be conducted simultaneously with the elements of tillage (Ti) and harrowing (Ha) on day-1 and day-14. In the first day, the seed distribution to the seedling area was conducted and the seed removal was conducted in the day of 14th. The work elements Se was conducted parallely with land preparation (Ti and Ha) because they could be done such it was to optimize the work's sequence. On the other hand, the other work elements had to be done in serial sequence, such as: planting was conducted after the tillage and harrowing and so on.

The workflow if conducted only in one round then in one growing season a lot of working time which is still left that can still be optimized was found. This optimization could be conducted by making a cluster with the longest working time limit on the nursery process takes 14 days, more details can be seen in Figure 1 Based on the analysis result that in one work cycle (one series of rice production activities) was divided into 9 clusters with the assumption of work area of 100 ha per crop season. In that cluster, if the works was still conducted manually with the available time, then the target capacity would not fulfilled, so the mechanization input on some work elements was required. In the second scenario with the addition of mechanization inputs selectively could increase the work productivity so that the income earned increased.

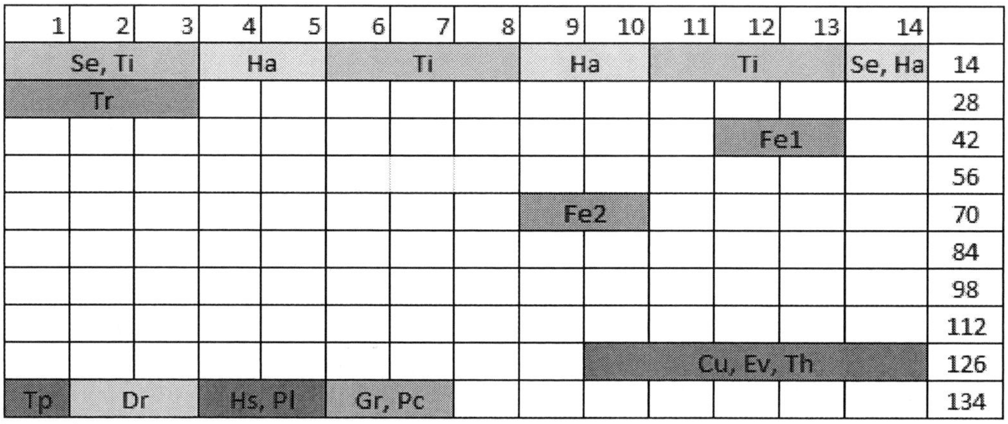

Figure 1. workflow of rice production activity per crop season (100 ha) with total work 134 days

Information:

Se	:	Seedling	Cu	:	Cutting	Hs	:	Husking
Ti	:	Tillage	Ev	:	Evacuating	Pl	:	Polishing
Ha	:	Harrowing	Th	:	Threshing	Gr	:	Grading
Tr	:	Transplanting	Tp	:	Transportation	Pc	:	Packaging
Fe	:	Fertilization	Dr		Drying			

Figure 2. workflow round cluster of rice productivity per crop season (134 work days) in 100ha

Generally, the total work capacity is the resultant of the engine capacity and the total cropping period. The capacity of farm machinery used in this calculation referred to [6] and [7]. In accordance to the observed cropping system, the time-sequence of cropping operations were 14 days of tillage, 5 days of planting, and 5 days of harvesting. So, according to the available time sequence, a walking-type tractor can till for 4.9 ha land area, whereas a riding-type tractor was equivalent for 9.8 ha. As for the planting work-element, a walking-type transplanter was capable for 1.75 ha, and a combine harvester was equivalent for 3.5 ha harvested area. The machine is optimally used if it able to work in one growing season on 9 clusters of 100 ha field area (Table 4).

Table 4. Comparison mechanization work capacity per crop season of rice production (total working time of 134 days)

Mechanization	Mechinery capacity (ha/hour)	Limit of work hour (a)	Machinery work capacity		
			ha/days	ha/(a)	ha/season
Walking-type tractor	0.03*	14	0.2	2.8	25
Riding-type tractor	0.1*	14	0.7	9.8	84
Walking-type Transplanter	0.05**	5	0.35	1.75	42
Combine Harvester	0.1*	5	0.7	4.9	84

Source: *[6], ** [7], one days = 7 hour

Based on the number and distribution of manpower and the available time, the selective mechanization scenario was on 7 elements of work, such as: tillage, transplanting, cutting, transporting and threshing. The addition machinery in the scenario is economically advantageous to the labor wages. As Figure 2 shows that the total labor cost was relatively higher on the first scenario (without additional mechanization) compared to the second ones (with selective mechanization). The use of mechanization was eventually required to completing the tasks in such larger capacity in a given constrain of work time and sequence.

The use of mechanization inputs in this scenario was assumed to use the self-ownership and rental machinery. Each cost that must be spent for both can be seen in Table 5. In this study, costs on the self-ownership machinery used assumptions which was using variable costs that does not include fixed costs because the machine used is a donation machine from the government.

The required manpower due to mechanization addition reduced by 107 people in 100 ha of the crop season. The difference from the manpower requirement could be solved using mechanization. Comparison of total manpower and mechanization cost in the workday increasement and mechanization addition scenarios can be seen in Table 5. The difference of total cost was an advantage as additional income, the amount of profit can be seen in Table 6. The results show that monthly profit using a rental machinery was Rp 680,021/person.month. While the assumption of self-ownership machinery obtained a profit of Rp 1,196,595/person.month. Based on this, it can be concluded that using the self-ownership machinery is more profitable than using the rental machinery. This is because the cost incurred for the self-ownership machinery was less than the rental machinery. The results of this study were then compared to the standard of minimum wage (working for 7 hours/day or 25 days/month). While the previous working hours was 4.8 hours/day so it can still be optimized. Furthermore, the results of the analysis were compared to the standard of minimum wage with the same working hours.

Labour need on first scenario (increasment of working hour) was 144 person not distributed yet smoothly every day (25 days/month). This was caused as practically labour need was different on every element (Table 1). Therefore, it needed to be done an optimation by additional mechanization. With this mechanization, work element that needs many labours can be finished in short time. Thus permanent labour need was 37 person in 100 hectare per crop season can be distributed smoothly to work every day.

Table 5. Total cost of manpower requirement (human and machinery) for rental and own machinery scenario per crop season 100 ha

	Description	Current Condition	Increasment of working hour	Additional mechanization
Rental machinery	Required manpower (person)	193	144	37
	Manpower cost (thousand Rp)	907,917	1,017,333	258,125
	Machinery cost (thousand Rp)	65,968	65,968	553,168
	Total (thousand Rp)	**973,885**	**1,083,301**	**811,293**
own machinery	Required manpower (person)	193	144	37
	Manpower cost (thousand Rp)	907,917	1,017,333	258,125
	Machinery cost (thousand Rp)	27,813	27,813	308,383
	Total (thousand Rp)	**935,729**	**1,045,146**	**566,508**

Table 6 Profit based on the difference of manpower cost with the increasement of working hour and additional mechanization

Assumption of machinery usage	100 ha/crop season	ha/month
Rental machinery (Rp)	272,008,333	680,021
Self-ownership machinery (Rp)	478,637,933	1,196,595

3.5. Changes in Manpower Income with the Addition of Mechanization (Second Scenario)

Total income of manpower after the mechanization design can be seen in Table 7. A work round consisting of 9 clusters with the assumption of 100 ha per crop season proved to increase the income of manpower. The increase of manpower income was obtained from the difference between the manpower cost manually and the cost of using machinery. On the assumption showed that both of the rent machine and the self-ownership machinery could increase the revenue. Revenue of a rental machinery was lower if compared to the self-ownership machinery because the costs incurred of rental machinery was larger than the self-ownership machinery. In addition, the self-ownership machinery used a variable costs because it used a donation machinery from the government. Each income with the assumption that the income of rental machinery usage was Rp 2,446,225/person.month while the self-ownership machinery was Rp 2,962,799/person.month.

Based on these two assumptions, it can be concluded that the assumption by using the self-ownership machinery was more profitable compared to the rental machinery because the income earned already met the Bogor regional minimum wage standard. While the manpower income on assumption with rental machinery still below the Bogor regional minimum wage standard. This happened due to the cost of the existing rental machinery showed that the cost of the rental machinery with manpower costs was almost same, so that the addition of mechanization did not provide the optimum benefit for earning the Bogor regional minimum wage standard income.

Table 7 Total income of manpower using additional mechanization

	Description	Amount
Rental machinery	Manpower income (Rp/person.month)	1,766,204
	Additional income due to addition of machinery (Rp/person.month)	680,021
	Total income (Rp/person.month)	**2,446,225**
Own machinery	Manpower income (Rp/person.month)	1,766,204
	Additional income due to addition of machinery (Rp/person.month)	1,196,595
	Total income (Rp/person.month)	**2,962,799**

*fix costs did not included due to the used machine was a donation from government

4. Conclusion

The result of the ergonomic approach analysis proved that the output of manpower based on the current condition with income of Rp 1,174,030/person.month has not met the Bogor regional minimum wage standard. Further analysis, the scenario by increasing the working hours without changing the existing work system was obtained income of Rp 1,766,204/person.month, where the income was not meet the Bogor regional minimum wage standard. Based on the second scenario analysis which was designing a mechanization with selective method proved to increase manpower income, but the addition of mechanization with the assumption of using a self-ownership machinery (Rp 2,962,799/person.month) was more profitable compared to the assumption of rental machinery (Rp 2,446,225/person.month) due to the obtainrd income with the self-machinery assumption has met the Bogor regional minimum wage standard.

Reference

[1] Fithri P and Regina Y S 2015 Analysis of productivity measurement of agricultural equipment and machinery in Cv. Cerry Sarana Agro (in Indonesian) *Jurnal Optimasi Sistem Industri* **14**: 138-155.

[2] Sumarsono S 2003 *Economics of Human Resource Management and Employment* (in Indonesian). (Jogjakarta (ID): Graha Ilmu).

[3] Gianie 2009 *Effect of Minimum Wages on the acceptance of Low-Education Personnel in the Industry and Trade Sector* (in Indonesian) (Jakarta: Graduate School of University Indonesia)

[4] Findi M and Marshafeni O 2013 Labor acceptance in the agricultural agriculture and the service sector, post wage minimum policy in Banten province for the year period (in Indonesian). *Jurnal Manajement dan Agribisnis*. **10** 2001–11

[5] Fil'aini R 2017 *Ergonomic Study for Modelling Working System on Sustainable Rice Production (Comparative Study of Organic and Conventional Rice Field)* (in Indonesian) (Bogor: Graduate School of Bogor Agricultural University)

[6] Syuaib MF 2016 Sustainable agriculture in Indonesia: Facts and challenges to keep growing in harmony with environmen. CIGR Jurnal. **18** 170-184

[7] Balai Besar Pengembangan Mekanisasi Pertanian 2015 Mesin tanam padi jajar legowo [Internet] [Di unduh 05 September 2017]

IOP Conf. Series: Earth and Environmental Science **147** (2018) 012022 doi:10.1088/1755-1315/147/1/012022

Design and Performance of LPG Fuel Mixer for Dual Fuel Diesel Engine

Desrial[1], W Saputro[1] and P P Garcia[2]

[1] Department of Mechanical and Biosystem Engineering, Bogor Agricultural University, Indonesia
[2] VSU GIS Services Unit, Visayas State University, Philippines

E-mail: desrial_ipb@yahoo.com

Abstract. Small horizontal diesel engines are commonly used for agricultural machinery, however, availability of diesel fuel become one of big problems especially in remote area. Conversely, in line with government policy for conversion of kerosene into LPG for cooking, then LPG become more popular and available even in remote area. Therefore, LPG is potential fuel to replace the shortage of diesel fuel for operating diesel engine in remote area. The purpose of this study was to design mixing device for using dual fuel i.e. LPG and diesel fuel and evaluate its performance accordingly. Simulation by using CFD was done in order to analyze mixture characteristics of LPG in air intake manifold. The performance test was done by varying the amount of LPG injected in intake air at 20%, 25%, 30%, 35%, until 40%, respectively. Result of CFD contour simulation showed the best combination when mixing 30% LPG into the intake air. Performance test of this research revealed that mixing LPG in air intake can reduce the diesel fuel consumption about 0.7 l/hour (without load) and 1.14 l/hour (with load). Diesel engine revolution increases almost 300 rpm faster than when using diesel fuel only. Based on economic analysis, using the fuel combination (diesel fuel – LPG) is not recommended in the area near SPBU where the price of diesel fuel is standard. However, using the fuel combination LPG-diesel fuel is highly recommended in the remote areas in Indonesia where price of diesel fuel is comparatively expensive which will provide cheaper total fuel cost for diesel engine operation.

1. Introduction

In recent decade, Indonesian national fossil fuel consumption has increased significantly and reached 1.5 million barrel per day [1]. Percentage of fuel consumption by type during the year 2000-2014 is gasoline fuel 45.5%, diesel fuel 45.2%, aviation fuel 6.3%, and kerosene 1.5% [2]. Therefore, now the use of alternative fuels, especially on internal combustion engines has become an interesting object to continue to be developed. The use of alternative fuels cannot be separated from the two global problems of crude oil availability continues to decline and the issue of exhaust emissions are increasingly apprehensive. The current effort is to develop alternative fuels that are possible to reduce exhaust emissions and relatively low fuel costs. One effort to answer the problem is by using dual fuel.

The use of dual fuel especially for farmers and fishermen in Indonesia has not been widely used yet. Combination of diesel fuel with LPG (Liquefied Petroleum Gas) for diesel engine is technically practicable. However, the optimum composition of the mixture between LPG and air entering the combustion chamber is not clearly defined yet, especially for single cylinder diesel engine. So far, the comparison of such imprecise composition is carried out by opening the gas regulator valve by trial and error. So that affects the engine performance and efficiency of the engine that will ultimately affect the cost of fuel to be spent.

Based on these problems, the purpose of this research is to design a fuel mixture device for the application of dual fuel (Diesel fuel - LPG) for small diesel engine. The CFD software was used to analyze LPG-air mixing behavior in the mixing device. Cost analysis was also done to determine the ratio of costs incurred when using diesel fuel completely and if using a combination of dual fuel.

2. Material and method

2.1. Material

Research material used in this study were:
- 8 hp single cylinder water cooled diesel engine,
- AC electric generator 3 kW-220V and 2 kW halogen lamp
- LPG in 3 kg tank and regulator valve
- diesel fuel
- LPG-air mixture device,
- instrument for measuring LPG and diesel fuel consumption, and
- workshop tool and measurement equipment

2.2. Method

The method of this research follows typical machine design procedure including problem identification, formulation of design criteria, design analysis, design simulation, design engineering drawing, prototyping, and performance test of the prototype [4]. Design analysis and fabrication of prototype were done in workshop of the Department of Mechanical and Biosystem Engineering Bogor Agricultural University.

2.3. Construction of instrument for measurement of LPG consumption

In this research an instrument for measuring LPG consumption rate was constructed. The instrument was made using pressure loadcell to measure pressure drop in the venturi pipe passed by LPG. This instrument was equipped with microcontroller and LCD display to show real-time consumption rate of LPG as shown in figure 1.

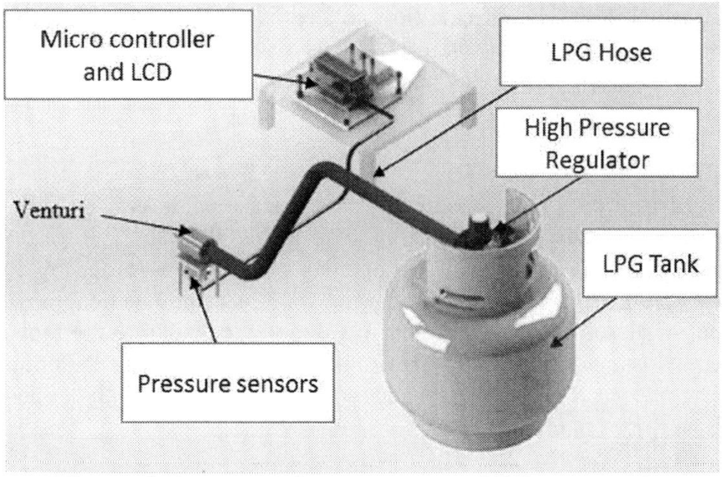

Figure. 1. Instrument for measurement LPG consumption

2.4. Design of LPG-air mixing device

In dual fuel system for diesel engine, LPG is premixed with intake air before entering the combustion chamber. Technical analysis was carried out to determine the speed of the airflow with the design made on the neck of the mixer at the maximum motor speed, so that mixing between air and LPG can take place well. Calculations used include volumetric intake rate, airflow velocity on the mixer's neck, cross-sectional area of the mixer's neck, and discharge to intake port. This LPG mixer has a total length of 50 mm, and its inlet diameter equal with the diameter of air filter outlet, which was 28 mm. The diameter of the outlet is the same as the diameter of the intake manifold was counted 28 mm. The diameter of the input port for LPG was 14 mm which was equal with the diameter of the fuel butterfly cock as an on-off valve of LPG. The engineering drawing process was done by using by Solidworks 2016 software which is shown in figure 3.

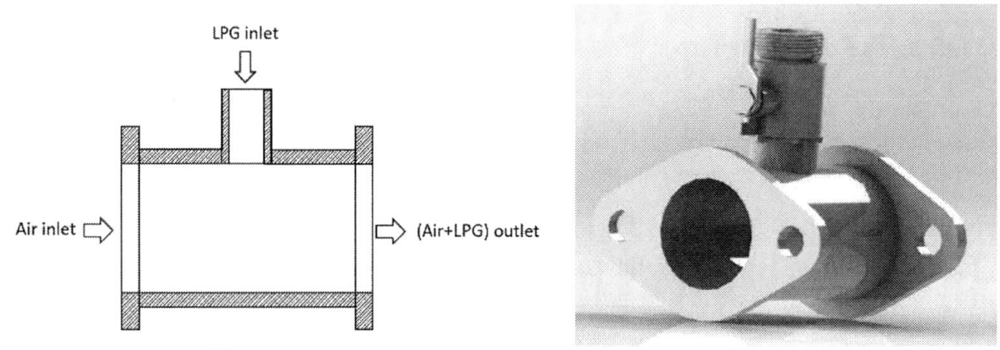

(a) (b)

Figure. 2. Schematic design (a) and engineering drawing (b) of LPG-air mixing device

2.5. Mixing analysis by CFD

Computational Fluid Dynamics (CFD) simulation was performed to analyze the mixed contours of LPG and air from the theoretical result approach at several LPG regulator valve opening positions. This CFD simulation used Solidworks software with its flow simulation menu. In performing the CFD simulation, some input parameters were used as boundary conditions in this simulation. Type of analysis used was internal flow. The internal flow concept is the flow in a container or stream bounded by the surface. The type of flow used was turbulent flow which has Reynold number (Re)> 4000. In the CFD analysis, the target of engineering parameters were pressure, velocity, volume fraction (LPG-air), and density.

2.6. Engine Performance Test

In the engine performance test, fuel consumption was measured in two conditions: 1) test without using load and 2) test with load by using halogen lamp with 2000 watt. The rotational speed of motor was set at 2000 rpm, and LPG regulator valve opening position was varied by 5% step from 5% to 100%. At each step, diesel fuel and LPG consumption were measured and engine rotational speed was noted. Measurement of diesel fuel consumption was done by calculating the time required each drop of 10 ml of diesel fuel using a stopwatch. Measurements of LPG gas consumption were carried out using gas flow rate device with varying range of LPG regulator valve opening. Setup of the performance test is shown in figure 3.

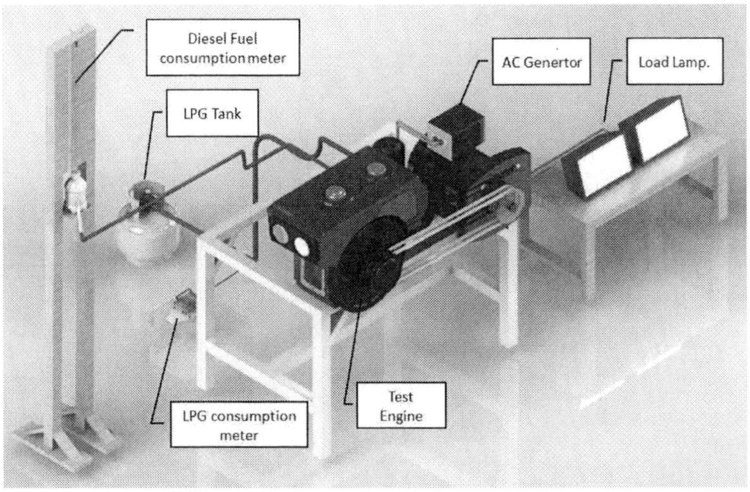

Figure. 3. Setup for performance test

3. Results and discussion

3.1. Computational Fluid Dynamic Simulation Analysis (CFD)

The CFD simulation [3] was use to analyze characteristics of LPG-air mixing which was necessary in design process of mixer device as shown in figure 5. The optimum inlet and outlet diameters of the mixing device was determine through this CFD analysis. In determining the consideration of the boundary conditions, several parameter such as atmospheric pressure, LPG pressure, and the discharge of LPG gas in each variation of valve opening were input to the program. The simulation approach was carried out by inputting the atmospheric pressure of 101325 Pa at the air inlet port and the LPG pressure at the LPG inlet port by 201260 Pa (20%), 218080 Pa (25%), 237239 Pa (30%), 260534 Pa (35%), and 332549 Pa (40%). Further data of LPG discharge inputs were 0.000167 m^3/s (20%), 0.0004 m^3/s (25%), 0.00045 m^3/s (30%), 0.000608 m^3/s (35%), and 0.000872 m^3/s (40%). The CFD simulation approach produces a mixing volume contour of LPG-air as shown in figure 4.

The results of the CFD simulation volume contour show the best mixing of LPG and evenly occur at 30% openings with mixed average yield of 0.064. The CFD simulation approach yields the same mixing result with the theoretical result of 0.064. Meanwhile, the average mixing results at valve openings were 20%, 25%, 35%, and 40% by 0.023, 0.056, 0.081, and 0.108. The mixing of LPG was strongly influenced by the speed and pressure of the LPG and air flow. The CFD simulation produced a pressure contour in the mixer as shown in figure 5.

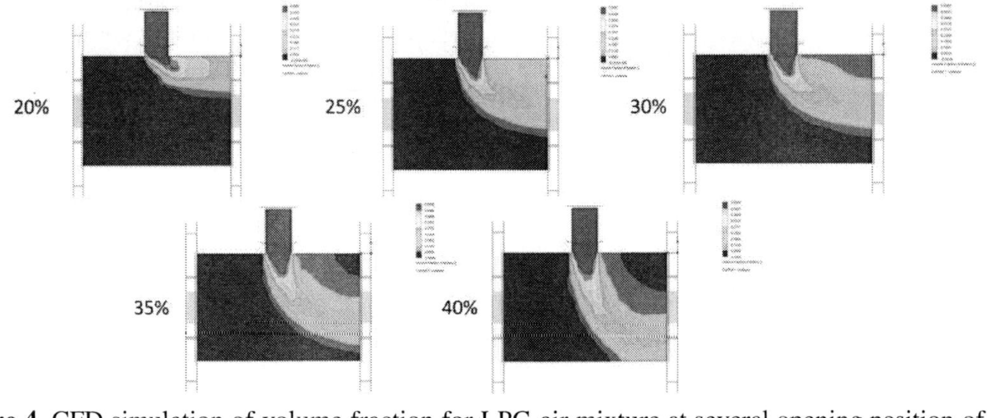

Figure 4. CFD simulation of volume fraction for LPG-air mixture at several opening position of LPG regulator valve

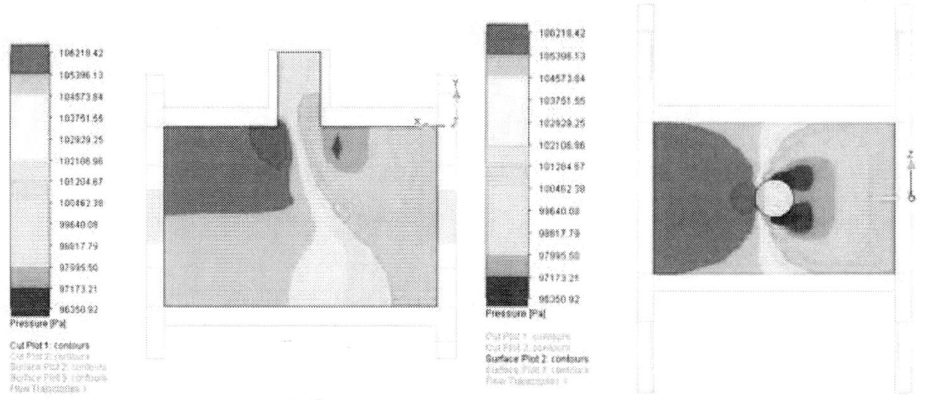

a. Side view b. Top view

Figure. 5. CFD simulation of pressure contour in mixer at 30% valve opening

3.2. Regulator Valve Setting

The regulator valve setting was conducted to determine the amount LPG gas inlet can be used. The test was done with engine speed of 2000 rpm. At the opening of the LPG regulator valve 5%-15%, the gas flow meter could not detect the rate of LPG discharge. Furthermore, at valve opening 20% the LPG flow rate was found to be 10 l/min, at 25% opening was 24 l/min, at 30% was 27 l/min, at 35% opening was 36.57 l/min and at 40% valve opening, the gas flow rate was 52.43 l/min. When the opening of the valve reached 45% there was an explosion in muffler, and this occurs due to the excess of LPG gas composition into the combustion chamber, resulting in the incomplete combustion and then was burned in the muffler. Thus, in this research the regulator valve can be set at the range of 20%-40% only.

3.3. Engine Performance Test

In this study, engine performance was evaluated in term of its dual fuel consumption (LPG and diesel fuel) with and without load applied to the engine. The result of fuel consumption test without using the load is shown in figure 6. It can be seen that the diesel fuel consumption rate decreased with the addition of the amount of LPG gas injected to the diesel engine. It can be explained that the flame speed of LPG in combustion chamber was much higher than the diesel itself, where the flame speed of LPG is 82 cm/s and for diesel only 30 cm/s [4]. With LPG flame speed capability is very fast this causes the combustion in the engine does not require diesel or fuel too much to make engine revolution speed remains stable. This can be shown with the stoichiometric ratio of air and diesel fuel 22: 1[5], in the presence of LPG can enlarge fuel-air ratio (F/A) is lower or in other words, it will reduce diesel fuel consumption. The highest combination of LPG and diesel fuel occurred at regulator valve opening at 40% where diesel fuel and LPG consumption were 0.054 l/h and 3.14 m^3/h, respectively.

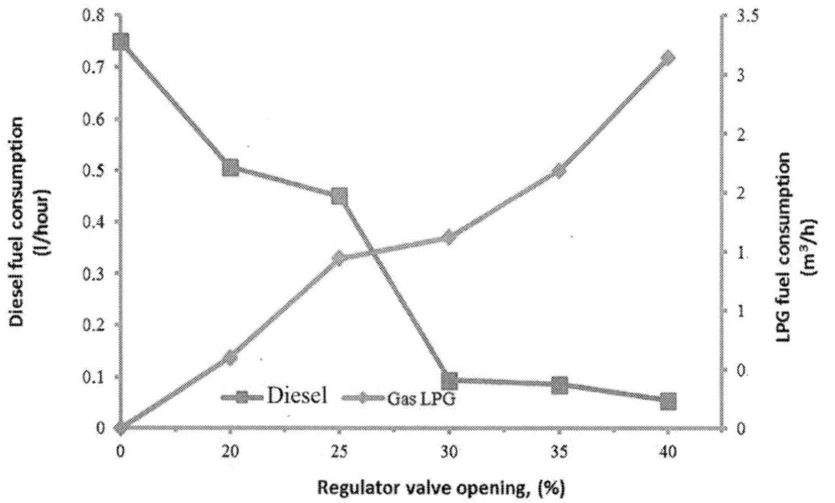

Figure. 6. Fuel consumption rate with no-load at engine speed 2000 rpm

In the engine performance test by applying load, an AC generator with 2000 watts halogen lamps were used. The dual fuel consumption rate results is shown in figure 7. This graph shows that it has similar trend as shown when the engine working without load but with higher amount of fuel consumption. The highest combination between LPG and diesel fuel occurred at regulator valve opening at 40% where diesel fuel and LPG consumption were 1.24 l/h and 3.14 m^3/ h, respectively.

Fuel consumption will increase with the level of engine rotation. In the combustion system of a gasoline engine, when the load is increased then the work of the piston that pushes and pulls will be faster or in other words need an explosion of combustion that can push the piston faster which means need more fuel [6]. This is what causes more savings when using a combination of LPG and diesel fuel at big loads.

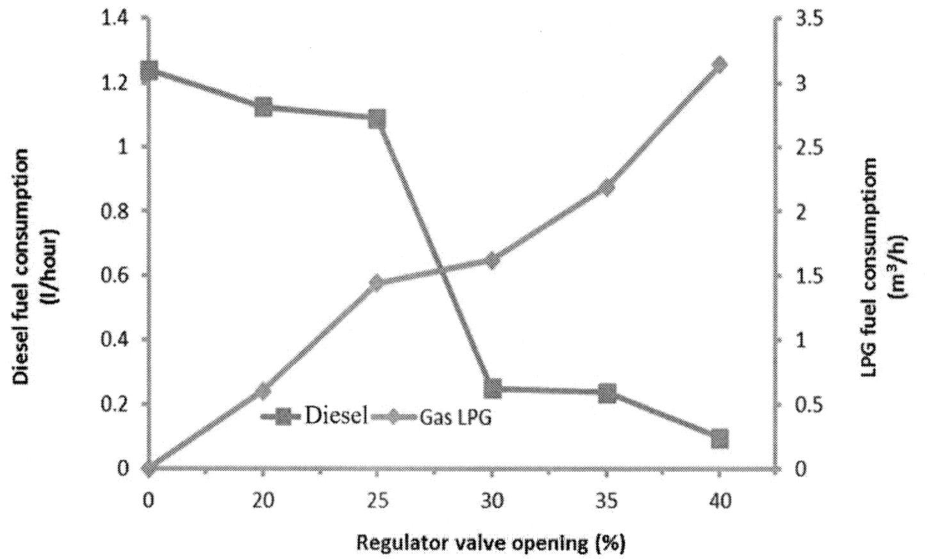

Figure. 7. Fuel consumption rate using load at engine speed 2000 rpm

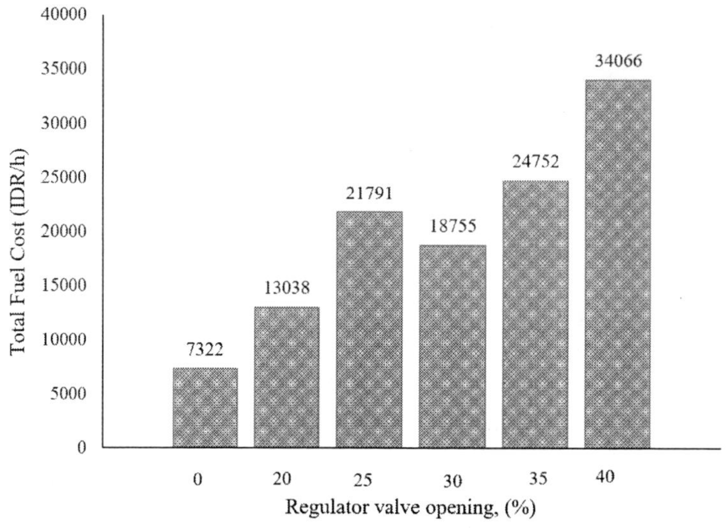

Figure. 8. Relation of total fuel cost with LPG valve opening at standard price of diesel fuel

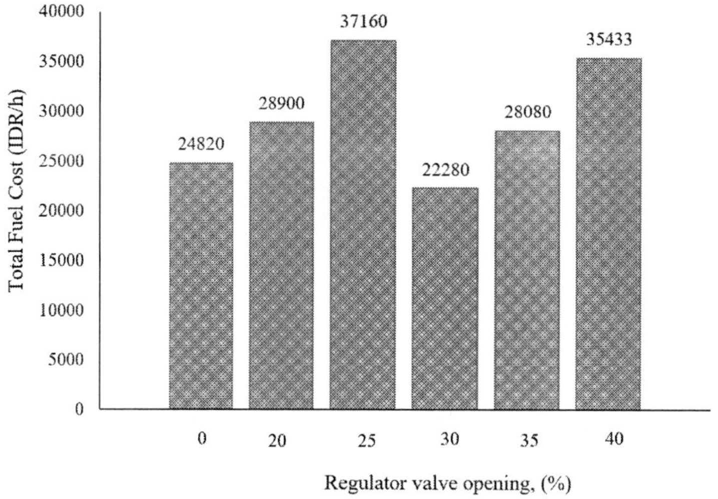

Figure. 9. Relation of total fuel cost with LPG valve opening at retail price of diesel fuel in remote area

3.4. Fuel Cost Analysis

The cost feasibility of the application of LPG in combination with diesel fuel for diesel engine was also considered. Cost analysis was divided into two, that was based on areas using standard government diesel fuel prices that are commonly sold in gas stations and based on retail price at remote areas. Pertamina standard diesel fuel price used in this analysis was Rp 5,900.-/l while diesel fuel price at remote area was Rp 20,000,-/l. This retail diesel price was the retail price used in the Sandaran village, East Kutai regency, East Kalimantan province. The price of LPG used in this analysis was the standard price of Pertamina for 3 kg LPG i.e. Rp 16,000,-/tank. The retail price of LPG is regulated by government, therefore its price is relatively constant everywhere in the country

The results of total fuel cost analysis of engine working with load on the standard price of diesel fuel can be seen in figure 9. The data showed that the lowest fuel cost was achieved when the engine operated by using diesel fuel only, i.e. Rp 7,322,-/h. The total fuel cost is increasing along with the increasing of LPG consumption and reached its maximum at 40% opening valve i.e. Rp 34,066,-/h. It means that in case of using standard price of diesel fuel, the addition of LPG or dual fuel system is not recommended due to its higher total fuel cost.

In case of fuel cost analysis by using retail price of diesel fuel in the remote area, it was found that the combination of application LPG at 30% valve opening with diesel fuel gave minimum total fuel cost of Rp 22,280,-/h at as shown in figure 10. Meanwhile, the fuel cost when using only diesel fuel was found to be of Rp 24,820,-/h with, meaning that the use of a combination of diesel fuel - LPG is more efficient at Rp 2,540,-/h than using diesel fuel only. Therefore the addition of LPG or dual fuel system in remote area is recommended due to its lower total fuel cost.

4. Conclusions

The application of LPG as dual fuel for diesel engine was technically proven. LPG was premixed with intake air in a mixing device that specially designed for this purpose. Based on fuel cost analysis, the application of LPG and diesel fuel only recommended when the price of retail diesel fuel more then three times expensive than its standard price which is commonly found in remote area. The highest fuels cost savings occur at 30% LPG valve opening which was Rp 2,540,-/h. This result also validated the CFD simulation analysis which showed that the best LPG-air mixture occurred at 30% valve openings.

References

[1] SKK Migas. Yearly Report 2015 (in Indonesian). [cited on 2015 November 12]. Available at: http:// www.skkmigas.go.id/publikasi/buletin
[2] BPPT Agency for the Assesment and Application of Technology. 2014. Indonesia Energy Outlook 2014.(in Indonesian) [cited on 2015 November 12]. Available at: http://www.bppt.go.id.
[3] Budiyanto M A 2012 *Simulation of Combustion of Diesel Engine with Dual Fuel* (in Indonesian) Magister thesis (Depok: University of Indonesia)
[4] Sitorus T B 2002 *Study on Development of Gas as Alernative Fuel* (in Indonesian) Magister thesis (Medan: University of North Sumatra)
[5] Dayang 2011 *Effect of Compression Ration on Performance of Diesel engine by Using Gas Fuel* (in Indonesian) Magister thesis (Surabaya: Institut Teknologi Sepuluh November)
[6] Suyanto and Wardan 1989 *Theory of Engine* (in Indonesian) (Jakarta: Depdikbud)

Factors Affecting Planting Depth and Standing of Rice Seedling in Parachute Rice Transplanting

I W Astika[1], I D M Subrata[1] and G Pramuhadi[1]

[1] Department of Mechanical and Biosystem Engineering, Bogor Agricultural University, Indonesia

E-mail: wayanastikaipb@yahoo.co.id

Abstract. Parachute rice transplanting is a simple and practical rice transplanting method. It can be done manually or mechanically, with various possible designs of machines or tools. This research aimed at quantitatively formulating related factors to the planting depth and standing of rice seedling. Parachute seedlings of rice were grown at several sizes of parachute soil bulb sizes. The trays were specially designed with a 3D printer having bulb sizes 7, 8, 9, 10 mm in square sides and 15 mm depth. At seedling ages of 8-12 days after sowing the seedling bulbs were drops into puddled soil. Soil hardness was set at 3 levels of hardness, measured in hardness index using golf ball test. Angle of dropping was set at 3 levels: 0°, 30° and 45° from the vertical axis. The height of droppings was set at 100 cm, 75 cm, and 50 cm. The relationship between bulb size, height of dropping, soil hardness, dropping angle and planting depth was formulated with ANN. Most of input variables did not significantly affect the planting depth, except that hard soil significantly differs from mild soil and soft soil. The dropping also resulted in various positions of the planted seedlings: vertical standing, sloped, and falling. However, at any position of the planted seedlings, the seedlings would recover themselves into normally vertical position. With this result, the design of planting machinery, as well as the manual planting operation, can be made easier.

1. Introduction

Mechanical rice transplanting is still facing some problems in Indonesia. The common nursery mat type transplanter is still facing infrastructure problems, where most of Indonesian rice fields do not have sufficient access road which enables the machine to get into the fields. Another main problem is that the depth of field hard so that the machine undergo many slips during the operation.

The emergence of parachute planting method has been enabling simple ways of planting, both manually and mechanically. In term of mechanical planting, it requires relatively simpler design of machinery, easier to be made by farmers, both man-driven tools or engine-powered machinery, both self-driven and attached to a hand tractor. The productivity of rice planted with parachute transplanting is considerably high. Three transplanting methods were compared: parachute, traditional, and line transplanting, it was found that parachute transplanting gave the highest yield with lowest cost[1]. Another research[2] compared traditional and parachute transplanting methods. Seedlings of parachute transplanting did'not show any stress as the roots are still covered by soil. Furthermore, parachute transplanting gave longer roots and bigger shoots. Other research[3] found that line transplanting gave better yield than parachute transplanting and then followed by random manual transplanting. Parachute method gave the highest population per hectare.

Another possibility of applying parachute rice planting is to transplant young seedling as required by SRI method. SRI method requires young seedling (around 10 days after sowing), and less number of seedling per hill (if possible 1 seedling per hill). It is possible to arrange less number of seedling per hill in parachute seedling by putting less seed per tray holes, and then transplanting the seedling earlier. These will give several advantages at the same time: enabling more number of seedling per tray making less area required for preparing the nursery and requiring less time for nursery the making more use of trays per season.

Parachute transplanting is also potentially applied with minimum tillage in which the untilled soil is just stripped with a furrowed or rotary tiller. A research has been done on strip tillage with rotary tiller and followed by direct seeding. Only the planted area is tilled[4]. As Parachute transplanting only needs soft areas at the places the seedlings are dropped, therefore the area can be untilled inundated soil, strip-tilled, or fully tilled.

This research studied dropping of seedling as the main mechanism in parachute planting, especially for the small size of parachute seedlings. The objective of the research is to observe and to formulate the effect of several factors: height of dropping, soil hardness, the size of seedling parachute, and the angle of dropping to the depth of planting and the standing of the planting.

2. Materials and method

2.1. *Survey on Soil Hardness*

A survey on soil hardness was done at surrounding areas of Bogor City in order to get soil hardness data of farmers rice fields, both untilled soil and pulverized soil. The data was then used as the references in preparing the soil hardness for the main experiment. Soil hardness was measured in term of soil hardness index with golf ball test, in which a golf ball dropped from 1 m height vertical from the soil surface.

2.2. *Preparation of Parachute Seedlings*

Parachute seedlings were prepared at several soil bulb sizes and several seedling ages. The seedling was grown in 4 sizes of trays which were specially designed and manufactured by using 3D printer. The size of tray holes and the mass of the seedlings are shown in table 1, figure 1 and figure 2.

Table 1. Sizes of tray's holes and parachute seedlings at 8-10 days after sowing

No	Sides and depth (mm)	Mass	
		Range (g)	Average (g)
1	7 x 7 x 15	0.1 – 0.3	0.204
2	8 x 8 x 15	0.2 – 0.5	0.322
3	9 x 9 x 15	0.3 – 0.9	0.670
4	10 x 10 x 15	0.3 – 1.3	0.977

Figure 1. Seedlings in the model of tray

Figure 2. Sample of seedlings

2.3. *Preparation of Soil*

The range of soil hardness prepared for experiments was hardness index 1 - 3 as measured in soil hardness index using golf ball test. This range was made based on the result of measurements on soil hardness around Bogor City as shown in table 3. With those data, three levels of soil hardness were prepared: Level 1 (hard) where golf ball test gives the surface of golf ball 1 cm above the soil surface, Level 2 (mild) where the golf ball surface at the soil surface, and Level 3 (soft) where the golf ball surface 1 cm beneath the soil surface as shown in figure 3.

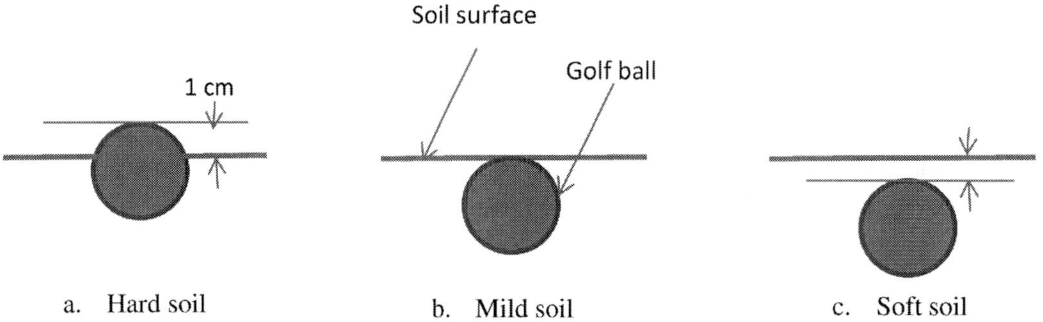

Figure 3. Soil hardnesses indicated with golf ball test, the sinking of a golf ball due to dropping from 1 m height

2.4. *Planting Simulation*

Planting of seedling parachutes were simulated by dropping the seedlings from a certain height to a pulverized soil surface as depicted in figure 4. Four treatments were applied in experiment in order to know the factors affecting the planting depth. The treatments and levels within the treatments are shown in table 2.

Table 2. Treatments and levels of each treatment

No	Treatments	Levels
1	Parachute seedling sizes	4 levels as shown in table 1
2	Dropping height	3 levels : 100 cm, 75 cm, 50 cm
3	Dropping angle	3 levels: 0°, 15°, 30° from vertical line
4	Soil hardness	3 levels : hard, mild, soft as shown in figure 4

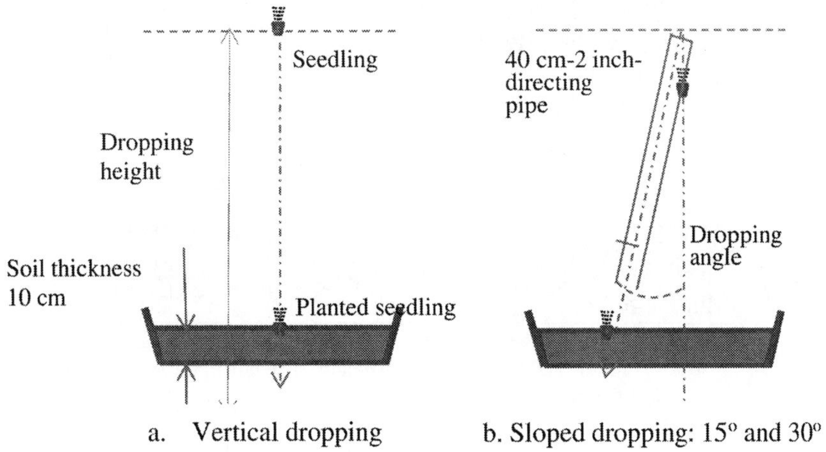

Figure 4. Soil hardness indicated with golf ball test, the sinking of a golf ball due to dropping from 1 m height

2.5. Planting Depth Mathematical Formulation

The relationship between input variables and the planting depth were formulated with and artificial neural network (ANN) structure as shown in figure 5. All input variables are coded in discrete numbers, while the output planting depth are coded as real numbers.

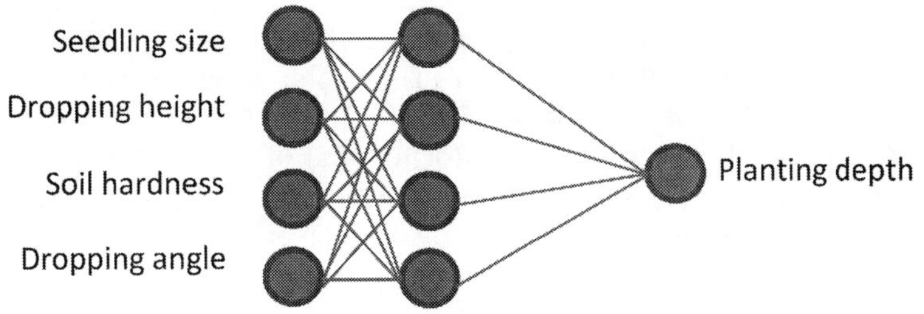

Figure 5. Artificial neural network structure

A number of 108 data were used in the training of the ANN, and no validation step was done as the error of the training were very big indicating that there is no consistency relationship between the input and the output variables.

3. Results and discussion

3.1. Soil Hardness Survey
Survey on soil hardness was done in the beginning of there research. Four sites at Bogor Regency, Bekasi, and Karawang Regency were selected as representatives of pulverized rice field soil ready for planting. The result of soil survey is shown in table 3.

Table 3. Soil hardness of pulverized and inundated soil

Sites	Golf Ball Test (cm*)
Balumbang Jaya Village, Bogor Regency	1.18
Situgede Village, Bogor Regency	0.53
Sukatani Village, Karawang Regency	1.29
Wanajaya Village, Karawang Regency	0.55
Lengah Jaya Village, Bekasi Regency	-0.58
Setia Jaya Village, Bekasi Regency	0.74
Average	0.89
Maximum	-3.00
Minimum	2.10

*) Depth of golf ball sinking due to dropped from 1 m height, average of 10 repetitions. Negative sign means the ball upper surface is on the soil surface

Table 3 shows various values of soil hardness of pulverized and inundated soil. Inundated fields are usually softer than those the pulverized drained ones. Even some of the untilled-inundated fields are also soft as those of pulverized ones. With planting simulations done later, it seems that some untilled-inundated fields are also suitable for parachute planting. This soil condition is suitable for minimum tillage or zero tillage cultivation in which the productivity is similar to tilled soil[5][6][7]. On the other hand, some pulverized fields can also show high hardness if it is drained and exposed to sunlight for a long time.

3.2. Effect of Bulb Size on Planting Depth
The size of seedling soil bulb as shown in table 1 gave a various and inconsistent planting depths. The statistics are shown in table 4. The average of planting depth tends to increase due to the increase in bulb size, but the standard deviation is very high making the insignificant differences among the planting depths among the four level of bulb sizes. There are several possible reasons for this insignificance, among others: 1) the various sizes (or masses) of soil bulb among those in the same level of tray holes, 2) the relatively small mass of soil bulbs as compared to their seedling mass making the parachute effect becomes small. From this result, for parachute planting purpose, any of the size would be suitable for tray design, but agronomic factors such as water and nutrient holding capacity would limit the small sizes. A special research on this aspect would be required.

Table 4. Statistics of planting depth as related to soil bulb size.

Soil Bulb Size	1	2	3	4
Average Planting Depth (cm)	0.422	0.407	0.456	0.563
Standard Deviation (cm^2)	0.568	0.456	0.456	0.818
Maximum (cm)	1.9	1.5	1.4	3.2
Minimum (cm)	0	0	0	0
Mean Difference Significance*)	a	A	a	a

*) The same letters indicate no significance

3.3. Effect of Dropping Height on Planting Depth

Dropping height at 50-100 cm range also did not give a significant effect on the planting depth. The statistics are shown in table 5. The planting depth tends to increase as the dropping height increases, but again the standard deviation is very high. The reasons would probably the same as those for bulb size effect. From this results, design of parachute planters can be flexible in term of dropping height. For manual dropping, the height of 70-90 cm would be comfortable for workers, while for planting machine the height up to 150 cm would be fine.

Table 5. Statistics of planting depth as related to dropping height

Height of Dropping	50 cm	75 cm	100 cm
Average Planting Depth (cm)	0.406	0.464	0.517
Standard Deviation (cm^2)	0.439	0.475	0.804
Maximum (cm)	1.3	1.5	3.2
Minimum (cm)	0	0	0
Mean Difference Significance	a	a	a

*) The same letters indicate no significance

3.4. Effect of Soil Hardness on Planting Depth

Soil hardness apparently gave a significant effect on planting depth, especially the hard soil to mild and soft soil, while the mild and the sot soil did not show a significant difference in planting depth. The statistics are shown in table 6. Hard soil gave a very shallow planting depth that would not give strong standing of the grown plant. The suitable soil hardness for parachute planting is the mild and the soft soil. Fortunately, as shown in table 1, the hardness of pulverized soil are usually mild and soft. In short, soil hardness in the observed area is mainly suitable for parachute rice planting.

Table 6. Statistics of planting depth as related to soil hardness

Soil Hardness	Hard	Mild	Soft
Average Planting Depth (cm)	0.094	0.619	0.672
Standard Deviation (cm^2)	0.188	0.627	0.677
Maximum (cm)	0.6	2.8	3.2
Minimum (cm)	0	0	0
Mean Difference Significance	a	b	b

*) The same letters indicate no significance

3.5. Effect of Dropping Angle on Planting Depth

Slopped dropping would give less parachute effect of the seedling as the vertical force would decrease, and the expected result is that the higher dropping angle would give less planting depth. As shown in table 7, they are not consistently so, and the differences in planting depth are not significantly differ for those 3 levels of dropping angle. The reasons are probably the same as those valid for soil bulb size and dropping angle factors. Certainly, the vertical dropping would give the highest average of planting depth, but slopped dropping would be also applicable for the design of planter in which planting spaces need to be adjusted by setting the angle of directing sliding tube.

Table 7. Statistics of planting depth as related to dropping angle

Dropping angle	0°	15°	30°
Average Planting Depth (cm)	0.542	0.414	0.431
Standard Deviation (cm^2)	0.817	0.430	0.458
Maximum (cm)	3.2	1.4	1.3
Minimum (cm)	0	0	0
Mean Difference Significance	a	a	a

*) The same letters indicate no significance

3.6. Standing of Planted Seedlings

As the parachute seedlings hit the soil, the seedling are immersed into the soil. The standing of the planted seedlings can be vertical or angled. Three categories were defined for the standing direction: vertical for the angle range 0°-30°, sloped for range of angle 30°-60°, and falling for range of angle 60°-90°. From the planting simulation, the standing of planted seedlings is shown in table 8.

Several results of planting simulation were grown in trays in order to have a controlled environment as shown in figure 6. As living things, rice seedlings tend to recover themselves into the proper standings. The angled planted seedling would recover themselves into the normal vertical positions in several days. Therefore, for practical application in planting work, both manually and mechanically, the standing of seedling would not be a problem, making the design of machine/tool would become easier.

Table 8. Standing of planted seedling due to dropping angle

Dropping	Standing of Seedlings (%)		
Angle	Upward	Slopped	Falling
0°	75	25	
15°	63.9	33.3	2.8
30°	36.1	63.9	

a. Sloped and falling seedling
b. Recovered standing of falling and sloped seedling 4 days after planting

Figure 6. Standing of planted seedlings

3.7. Neural Network Formulation

The artificial neural network has been trained. The progress of the training is shown in figure 7. The errors were still very high at the end of the training process, and seemly could not achieve less error significantly if the training is continued. This result is consistent with the fact that almost all of the factors do not give insignificant effect to the planting depth. As the error of artificial neural network is high, the validation process was not done.

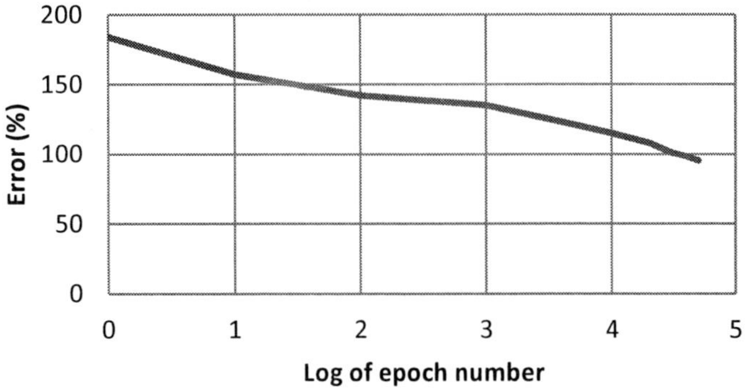

Figure 7. Progress of ANN training

4. Conclusions

Several points can be concluded from this research:

1. Among the four factors: 1) size of parachute seedlings, 2) height of dropping, 3) soil hardness, and 4) angle of dropping, only the soil hardness give significant effect to planting depth. Other factors showed reasonable trends but did not significantly affect the planting depth.
2. Artificial neural network failed to formulate seedling size, height of dropping, soil hardness, dropping angles and planting as the input factors do not consistently affect the planting depth. The training error of the ANN was very big around 90%.
3. The result of planting, aside from planting depth, also characterized by the standing of the seedlings. The standing can be vertical, sloped, or falling. Sloped dropping tends to give sloped planting. However, whatever the position of is, it will be recovered in several days into the normal vertical position.

5. Suggestion

For better result of future researchs, in order to eleminate bias on the factors, it is important to make a relatively uniform sizes in bulb size in the same levels.

References

[1] Akhterii M and Sabar M 2002 Evaluation of rice line-transplanting and parachute planting methods *Proceedings of the National Workshop on Rice-Wheat Systems in Pakistan 11-12 December 2002* (Islamabad) p 38

[2] Nabi G, Akhtar S, Hussan M H and Gill S M 2002 Root growth in parachute and conventional rice transplanting methods *Proceedings of the National Workshop on Rice-Wheat Systems in Pakistan 11-12 December 2002* (Islamabad) p 62

[3] Awan T H, Ali I, Safdar M E, Ahmad A and Akhtar M S 2008 Comparison of parachute, line and traditional rice transplanting methods at farmer's field in Rice growing area *Pak. J. Agri. Sci.* **45** 432

[4] Lee K S, Park S H, Park W Y and Lee C S 2003 Strip tillage characteristics of rotary tiller blades for use in a dryland direct rice seeder *Soil and Tillage Research* **71** 25

[5] Haque M E, Bell R W, Islam M A and Rahman 2016 Minimum tillage unpuddled transplanting: An alternative crop establishment strategy for rice in conservation agriculture cropping systems *Field Crops Research* **185** 31

[6] Erenstein O and Laxmib V 2008 Zero tillage impacts in India's rice–wheat systems: A review. *Soil and Tillage Research* **100** 1

[7] Huang M, Zou Y, Feng Y, Cheng Z, Mo Y, Ibrahim M and Xia B P J 2011 No-tillage and direct seeding for super hybrid rice production in rice–oilseed rape cropping system *European Journal of Agronomy* **34** 278

Application of Natural Air Drying on Shelled Corn in Timor

J Nino[1], L O Nelwan[1] and Y A Purwanto[1]

[1] Departement of Mechanical and Biosystem Engineering, Bogor Agricultural University, Indonesia

E-mail: jefrianusnino@gmail.com

Abstract. A study of the application of natural air drying on shelled corn in Timor using a bed-type dryer has been performed. The study aspects were limited to obtain the suitable air flow rate requirement and duration of the drying operation per day. For each aspect, the treatments were carried out simultaneously. The results showed that at the average ambient air temperature of 30.6°C and relative humidity (RH) of 73.0% the air flow rate of 0.83 L/s-kg provided the highest drying rate. Subsequently, by using the same air flow rate, three scheme of drying operations duration were used, i.e., 8 hours per day (08.00-16.00), 6 hours per day (09.00-15.00) and 4 hours per day (10.00-14.00). The average temperature and RH of ambient air condition at the second experiment were 30.3°C and 73.3% respectively. After 4 days of drying, the 8 hours per day (first scheme) treatment was able to dry the shelled corn from the initial moisture content of 27.24% w.b. to the final moisture content of 14.05% w.b. The specific energy consumption (SEC) of the first scheme was 1.75 MJ/kg. The final moisture content of the second and third schemes were 15.08 % w.b. and 18.45 % w.b. respectively with SEC of 1.41 MJ/kg and 1.21 MJ/kg respectively.

1. Introduction

North Central Timor Regency (TTU) is one of corn production centers in the Province of East Nusa Tenggara (NTT). In 2014, corn production in TTU was 647,072 tons (British Psychological Society Regency TTU 2014). Post-harvest management of corn in Timor is still limited, especially for the drying process. The drying process is usually conducted using smoking method. Such drying method has a disadvantage that is reduced sugar levels of corn. In addition to the smoking method, the drying process is also conducted by directly drying under the sun and spread over on matting and tarpaulin. Such method has weaknesses namely requires large space, more labor, longer time and contamination between the corns seeds and foreign objects so that damage occurs when stored. Artificial drying process is required to overcome these problems. Drying is one of thermal energy saving alternatives because the thermal energy requirement for drying is very high around 90%-95% of the total energy demand [1]. Geographically, Timor region has dry tropical climate that is more dry season than wet season; average rainfall is 265.9 mm, sunlight is 50%-98%, average air temperature is 22°C-34°C and air humidity is 67%-89% [2]. Such natural air condition is very potential to be used as the drying medium. [3] has been conducted a study on natural air drying on shelled corn, equipped with control system. This study used as much as 22.5 kg of corn hybrid varieties with initial moisture content 20.89% d.b and final moisture content of 16.20% d.b. He found that at ambient temperature ranging between 27.9°C-40.6°C and RH environment 39.47%-93.43%, the drying time was 10.517 hours and airflow rate was 0.01 m^3/s is equal 0.4 L/s-kg. For Timor region with dry tropical climate, where during the day the ambient temperature reaches 34°C and RH below 70% are potential as to be used as for natural air drying without control system. Therefore, a drying method is required by determining the appropriate air flow rate and operating time. Airflow rate serves as a heat carrier to evaporate and remove the moisture content. Besides the air flow rate, operating time also greatly affects the drying process. The purpose of this study is to determine appropriate air flow rate and operating time for natural air drying for shelled corn.

2. Materials and methods

The research was conducted in Regency of TTU, District of Kefamenanu City, Laboratory of Agriculture Faculty, University of Timor, and continued at Renewable Energy Laboratory, Department of Mechanical Engineering and Biosystem, Faculty of Agricultural Technology. The material being tested is local white shelled corn from Manusasi Village, District of West Miomaffo. Three bed dryers with each capacity of 30 kg were designed and constructed (figure 1). The shelled corn was put into the drying chamber and then the ambient air was then flowed by a blower into each drying unit. The test was performed in two stages, namely determination of air flow rate (first stage) and operating time (second stage). Both stages were carried out simultaneously using 3 dryer units as presented in figure 2. Air flow rates were measured using orifice plates, that had been calibrated with an anemometer. Measurement of moisture content is done using gravimetric method (oven). Measurement of electrical power for blower was performed using multimeter and clamp meter. In airflow rate test, air flow rate of the three drying units varied and operated continuously at the same time from 08.00-16.00. Testing of operating time is carried out with the air flow rate from previous experimental results. The first drying scheme runs from 08:00 to 16:00 every day (8 hours per day), the scheme II from 09:00 to 15:00 every day (6 hours per day), and scheme III from 10:00 to 14:00 every day (4 hours per day).

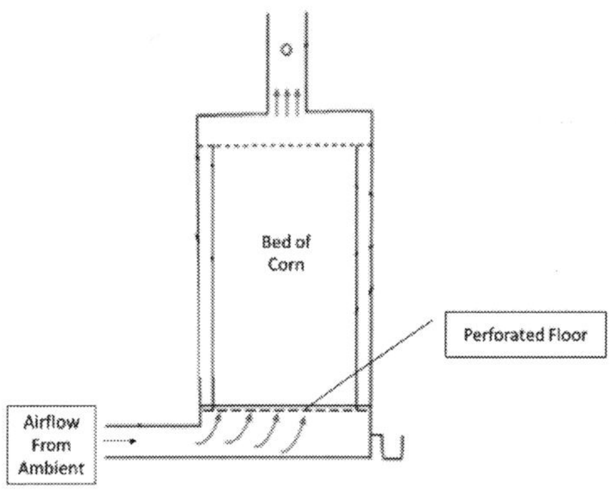

Figure 1. Schematic design of one unit shelled corn bed dryer

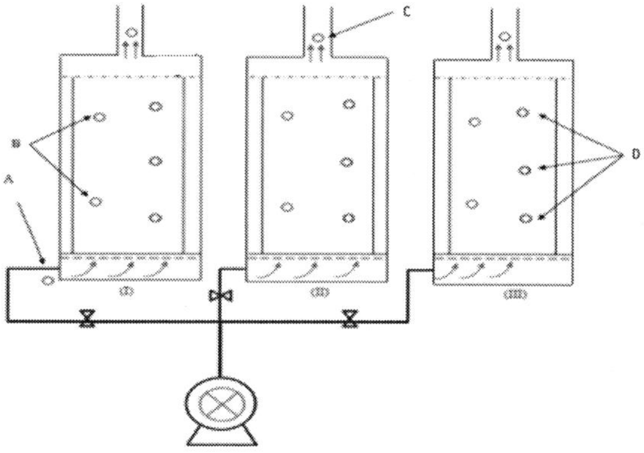

Figure 2. Schematic set up of bed dryer testing (A: air temperature, RH and velocity measurement, B: moisture content measurement, C: air temperature and RH measurement, D: grain temperature measurement)

2.1. *Determination of Air Flow Rate*

Tests on air flow rate were carried out under the same ambient air conditions as presented in figure 1 and figure 2. The air flow rate can be measured using orifice plate equipped with a pipe U manometer and calibrated using an anemometer. The calibration data was used to obtain linear equations of flow rate to the difference of the water column height to obtain equation (1). The air flow rate was obtained using equation 2 where the air velocity air velocity is calculated by equation (1).

$$V = a\,(\Delta h) + b \tag{1}$$

$$Q = V \times A \tag{2}$$

where: air flow rate (m/s), Δh= difference of water height among manometer columns (mm), a,b = calibrated constants, and Q = flow rate (m^3/s), A = cross-sectional area of airway pipe

2.2. *Air Temperature and Relative Humidity (RH)*
Measurement of air temperature was conducted by measuring ambient temperature, inlet, bed and outlet temperature. The measurement point of bed temperature is performed at 2 points i.e., the top and bottom layer. Relative humidity measurement of dryer air is carried out by measuring the temperature of the wet bulb and dry bulb. The measurement point is carried out on ambient air, inlet and outlet.

2.3. *Moisture Content*
Measurements were carried out before and during the drying process with a sample weight of 10 grams at 1-hour interval. Measurements were made by gravimetric method (oven) for 24 hours at 105°C. The sampling point is made at 3 points i.e., the bottom, middle and top

2.4. *Total Fan Efficiency Determination*
Measurement of the total fan efficiency was done by connecting the fan to the pipe equipped with a valve. Valve opening arrangements were performed at 10 levels. The purpose of adjusting the valve opening was to obtain the total fan efficiency at the treatment of various air flow rates. The electric power consumed by the fan is calculated using the equation:

$$W = v \; x \; I \; x \; Cos \; \emptyset \quad (3)$$

where: W = Electrical power (watts), v = Voltage (volt) I = Current (ampere) Cos Ø = Power factor. The total pressure (ΔP_{Total}) (gauge) in Pa, can be obtained by:

$$\Delta P_{Total} = \Delta P_s + \Delta P_d \quad (4)$$

$$\Delta P_d = \rho \; x \; \frac{(\frac{Q}{A})^2}{2} \quad (5)$$

Where ΔP_d = dynamic pressure (gauge) (Pa), ΔP_s = static pressure (gauge) (Pa).

The total fan efficiency can be calculated using the following equation:

$$\eta_{Total} = \frac{Q \; x \; \Delta P_{Total}}{W} \; x \; 100\% \quad (6)$$

where: η_{Total} = Fan total efficiency and W = Fan electrical power (kW).

2.5. *Determination of Electricity Energy Consumption*
Electrical energy consumption was calculated using flow rate (Q), total pressure (ΔP_{total}) and the total fan efficiency specified in equation (11) are expressed as follows:

$$W_{cal} = \frac{Q \; x \; \Delta P_{Total}}{\eta_{Total}} \quad (7)$$

The electrical energy during the drying process is calculated by the equation:

$$E = W_{cal} \; x \; t \quad (8)$$

where: E = Electrical energy drive fan (MJ), W_{cal} = Electric power fan calculation results (Watt), t = Drying time (s)

The specific energy consumption is calculated by the equation:

$$KES = \frac{E}{M_{vap}} \quad (9)$$

where KES = Specific energy consumption (MJ/kg), M_{vap} = Mass of moisture vaporized from corn (kg)

2.6. Three (3) Tons Drying Simulation

Generally, application of shelled corn drying at farmer level is in large scale. In the simulation, application of drying is designed with the assumption of 3000 kg capacity, that is by extends the space, but the bed height remains. The use of such fan shall comply with the pressure and air flow rate for the assumption of 3000 kg capacity. In the fan used, the pressure is the same as the experiment while airflow rate is increased so that the volume per seed mass is met. The fan used is a commercial fan [4]. The performance curve can be seen at http://www.fantech.com.Au/images/PDF /Catalogue/performanceeg. The website may disappear after a certain period.

3. Results and discussion

3.1. Effect of Air Flow Rate on Drying Process

Air flow requirements for grain drying can be expressed as air flow rate required for each kg of dried grain in L/s-kg. In this study, the air flow rate of 0.025 m^3/s is equal to 0.83 L/s-kg, 0.020 m^3/s is equivalent to 0.67 L/s-kg whereas 0.015 m^3/s is equivalent to 0.5 L/s-kg. Figure 3 shows that the greater the rate of air flow, decrease in moisture content will be faster. In the test, the decrease of moisture content at flow rate of 0.83 L/s-kg is reach 16.67% w.b, 0.67 L/s-kg was 18.71% w.b and 0.5 L/s-kg was 20.85% w.b. Air flow rate is very influential on the drying process [5]. Larger air flow rate will result in differences in drying air temperature and moisture content between the bed becomes smaller and shorter drying time. This finding is supported by [6] which state that the higher air flow rate, the higher ability to evaporate water.

3.2. Changes in Temperature and RH on Air Flow Rate Test

The experiment was carried out for 3 days with 24 hours of drying time. Figure 3 shows the graph of ambient temperature fluctuations with the bed temperature over time during drying. During the drying process, the ambient temperature and inlet temperature tend to fluctuate due to fluctuating weather. The average ambient temperature is 30.62°C. In figure 3, it is also seen that during the drying process the lower pile temperature stays higher than at the upper pile temperature. It is because the incoming ambient air will be first coming on the bottom of the pile and before flowing into the upper pile, causing the upper pile temperature tends to be lower. At the same ambient temperature, the air flow rate of 0.83 L/s-kg results in greater drying rate. Higher air flow rate results in more moisture and easier to volatilize. It can be seen from the temperature difference between the lower and upper piles at the highest air flow rate of 0.83 L/s-kg, especially at the end of drying (third day).

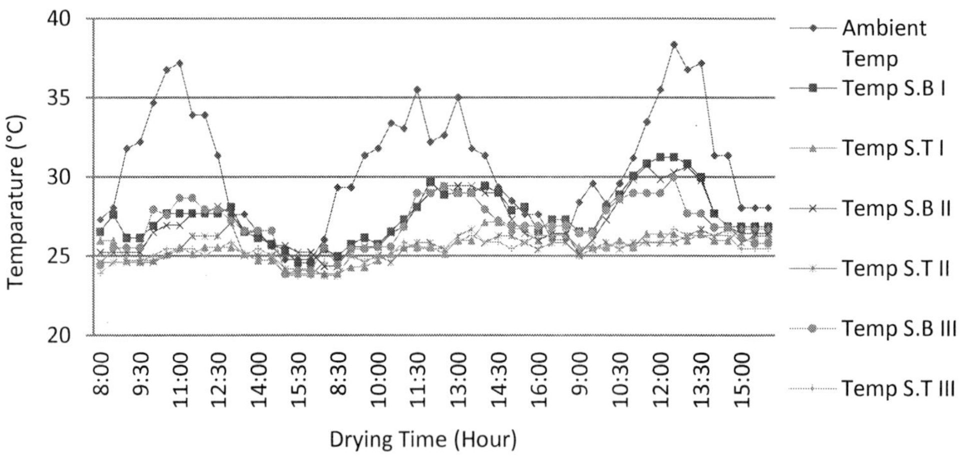

Figure 3. Plots of temperature versus drying time (T refer to upper, B refer to lower layer, while I, II and III refer to 0.83, 0.67 and 0.5 L/s-kg of air flow rate, respectively)

Figure 4 shows a change in RH during the drying process. The graph shows that the difference between RH ambient and RH outlet is very significant, where RH ambient is higher than RH outlet, but RH ambient starting to decrease at 14:30 to 16:00. RH outlets on each drying unit are affected by the air flow rate. It can be seen from the highest RH outlet value at the air flow rate of 0.83 L/s-kg, although the drying rate is the highest since it also contains more air, it results in low RH. The average relative humidities of the outlet portion at the highest to lowest air flow rate are 86.78%, 89.03%, and 90.83% respectively. The greater the temperature and air flow, the lower the relative humidity produced, causing more vapors absorbed by the drying air. The highest air flow rate 0.83 L/s-kg has low relative humidity so that the capability of drying air to absorb vapors can be able to get higher.

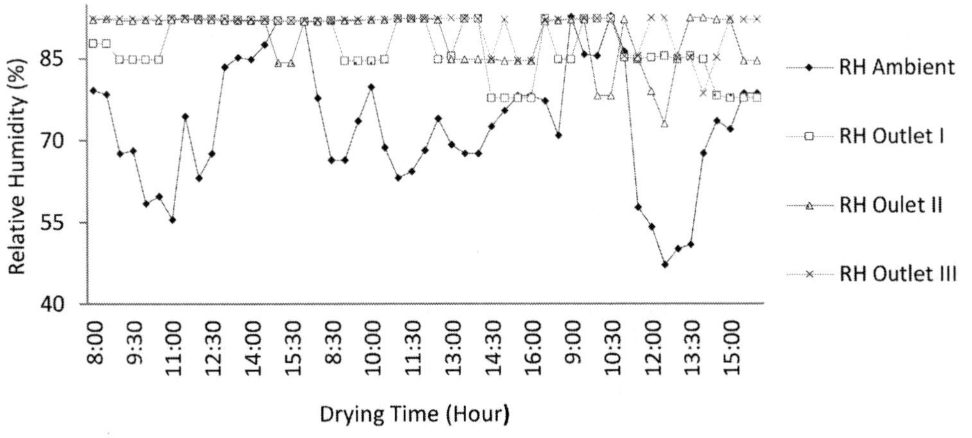

Figure 4. Plots of RH versus drying time (I, II and III refer to 0.83, 0.67 and 0.5 L/s-kg of air flow rate, respectively).

3.3. Changes in the Moisture Content on Air Flow Rate Test

Figure 5 shows the effect of air flow rate on changes in corn moisture content. The decrease of the moisture content of the three air flow rate test is not significant, but the fastest decrease found in higher air flow rates. At the air flow rate of 0.83 L/s-kg, the average of air ambient was 30.62°C, RH was 72.96% with the initial average moisture content 33.11% w.b can decrease the moisture content until the final average moisture content was 16.67% w.b. Figures 5 the decreasing moisture content in the lower layers is always lower for the overall airflow rate treatment. It is because in the lower layers the thickness of the pile closer to the inlet temperature so that faster vapor evaporation occurs than the material resulting in a greater decrease in water content. The thinner or smaller the maize layer, the faster and bigger the decrease of the moisture content of the corn kernels per time unit [7]. However, in the lower, middle and upper layers of the decline, the difference between moisture content is not too large. It is because during the drying process carried out stirring 2-3 times a day so that in the maize layer mixing between corn kernels with different moisture content so that the uniformity of moisture content between layers of corn.

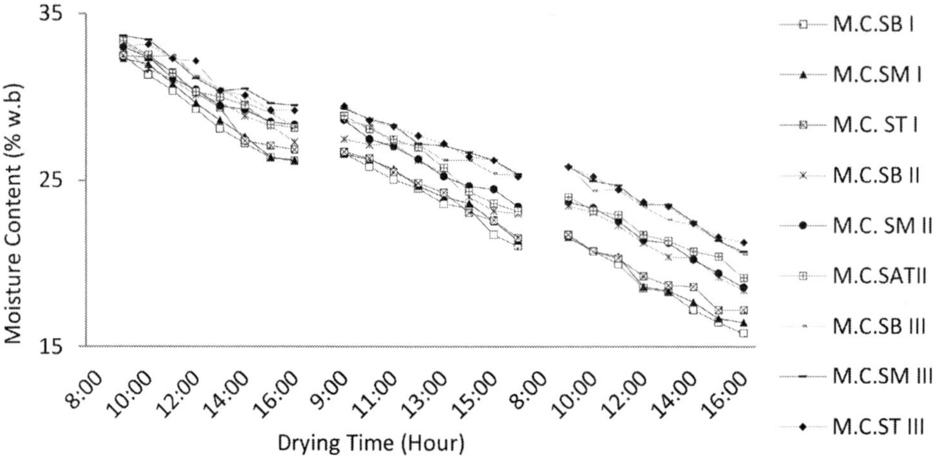

Figure 5. Effect of airflow rates on decrease in moisture content versus drying time (T refer to upper, M to middle, B to lower layer, while I, II and III refer to 0.83, 0.67 and 0.5 L/s-kg of air flow rate, respectively

3.4. Changes in Temperature and RH on Operation Time Test

Figures 6, 7, and 8 show changes in ambient temperature, inlet temperature, pile temperature during the drying process. The average ambient temperature of the three 32-hour operating time schemes (08: 00-16: 00), 24 hours (09: 00-15: 00) and 16 hours (10: 00-14: 00) are 30.03°C, 30.49°C, and 31.03°C, respectively. Based on the graph, it is shown that the potential of ambient temperature as a drying medium is at 10:00 to 14:00, but the rate of decrease in moisture content was slow. It is due to a shorter operating time of 4 hours per day whereas a faster drop in moisture content is an operating time scheme of 8 hours per day. The longer the drying time, the more water vapor is evaporated so that the decrease in moisture content will get higher. The graph also shows that, during the three schemes of the drying process, the lower pile temperature is higher than the upper pile temperature. Based on the temperature data, it can be concluded that the determination of the effective drying time is important so that the drying process can be conducted on the time when the ambient temperature is potential only. It can be seen in the 3rd scheme treatment where only with 4 hours operating time per day (16 hours) the moisture content decrease can reach 18.70% bb, whereas if drying is done with scheme 1 then the drying time can take up to 24 hours.

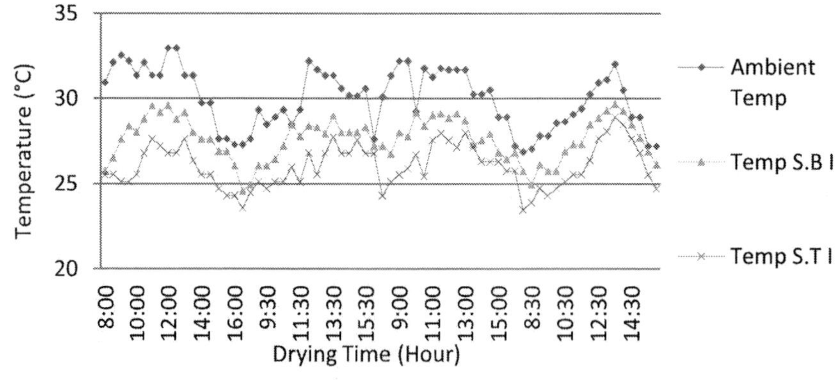

Figure 6. Plots of air ambient temperature and bed temperature versus drying time of the first scheme (32 hours). (T refer to upper, B to lower layer).

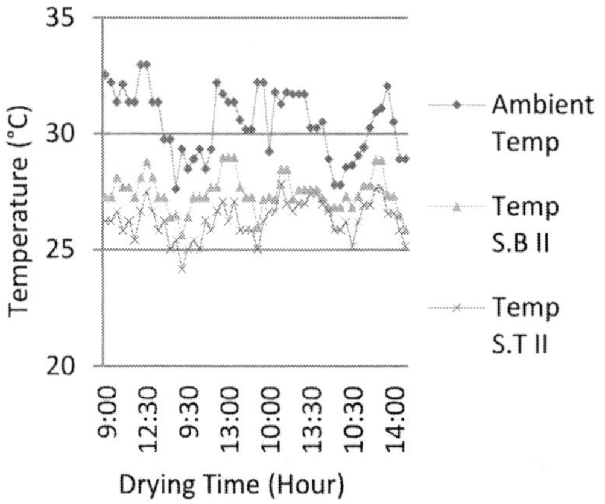

Figure 7. Plots of air ambient temperature and bed temperature difference versus drying time of the second scheme (24 hours) (T refer to upper, B to lower layer).

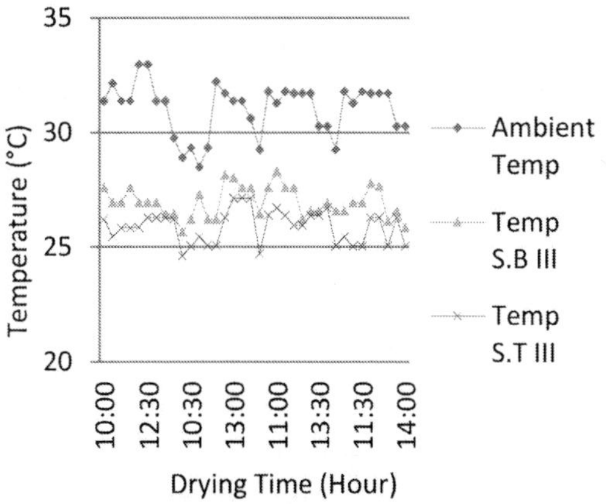

Figure 8. Plots of air ambient temperature and bed temperature difference by time of the scheme versus drying time of the third scheme (16 hours). (T refer to upper, B to lower layer).

Figure 9 shows the change of air ambient RH and outlet RH by time. Due to on the graph, during drying takes place, air ambient RH is lower than the outlet. Air ambient RH was 73.16% on average while outlet RH is 89.95%. Temperature increase due to the length of drying time may affect RH. The higher air temperature of the dryer, air RH is lower. Such low relative humidity will cause higher heat and mass transfer from material to air [8].

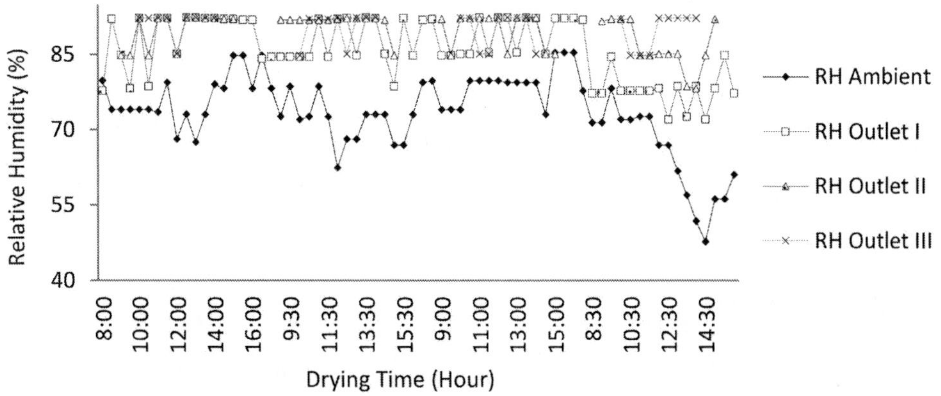

Figure 9. Plots of RH versus drying time (I, II and III refer to the first, second and third operation scheme).

3.5. Changes in Moisture Content on Operation Time Test

Figure 10, 11 and 12 shows the change in moisture content of shelled corn bed. In the testing process, air ambient temperature is 30.32°C on average, and air ambient RH was 75.84% on average. The test was performed on initial moisture content 27.24% w.b until the final moisture content 14.05% w.b on average. During the drying process, moisture content decrease is not significant, but a decrease in moisture content is faster by longer drying time. Operation time at this stage was 32, 24 and 16 hours. The average of the final moisture content of the three drying variables was 14.05% w.b, 16.23% w.b and 18.70% w.b. The decrease in moisture content between the lower, middle and upper beds is not significant. It is because stirring was conducted 2-3 times a day during the drying process. In the stirring process, there will be mixing between dry and wet grains resulting in difference between the lower, middle and top bed. In the testing process, moisture content decrease in the lower bed is faster; the middle and top bed is very slow. It is because the thicker the corn bed, the possibility of heat distribution of drying air is further hampered by the density of the corn grains resulting in a slow decrease in moisture content.

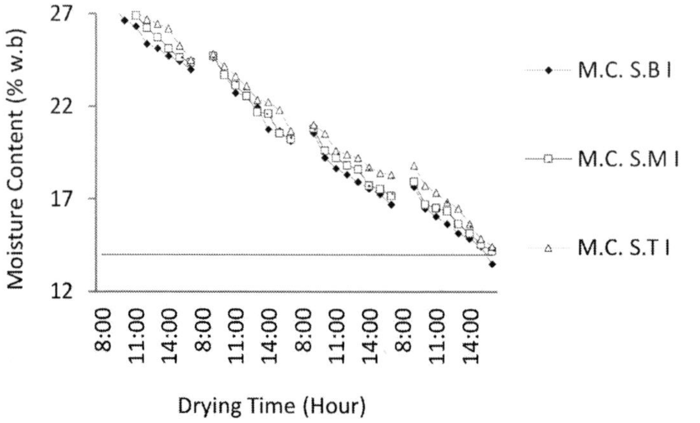

Figure 10. Plots of moisture content versus drying time of the first scheme (32 hours) (T refer to upper, m to middle and B to lower layer).

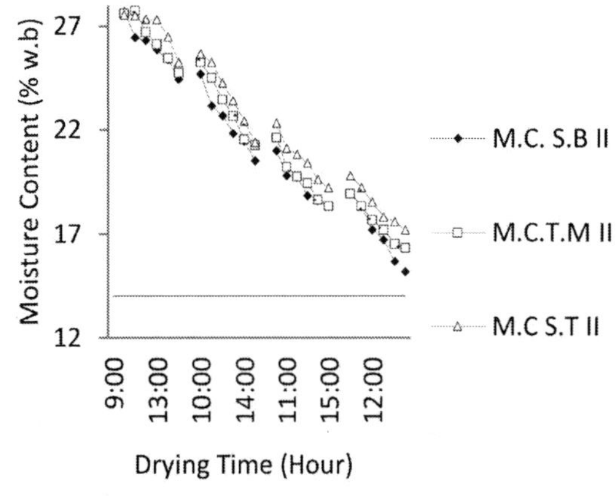

Figure 11. Plots of moisture moisture content versus drying time of the second scheme (24 hours) (T refer to upper, m to middle and B to lower layer).

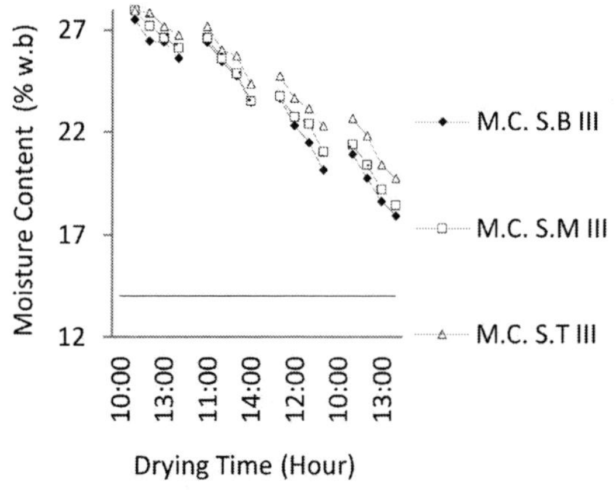

Figure 12. Plots of moisture moisture content versus drying time of the third scheme (16 hours) (T refer to upper, m to middle and B to lower layer).

3.6. Fan Efficiency

Table 1. Shows the average value of total fan efficiency. Due to the table, it can be seen that, in general, the fan test has a maximum total efficiency of 19.29% and an average total efficiency of 14.70%. The value of fan cost Ø was 0.82. From the calculations, the fan has low efficiency.

Table 1. The Average value of total fan efficiency testing

Voltage (V)	Ampere (A)	Power (W)	Cos Ø	$\eta_{Total\ Max}$ (%)	$\eta_{Total\ rata-rata}$ (%)
218.22	1.2	216.4	0.82	19.29	14.70

3.7. Electric Energy Consumption During Drying Process

In the two stages test, the drying process is carried out effectively (fan condition turns on). At the final stage, air flow rate of drying process lasts for 3 days. Electricity consumption during drying process was 4.88 MJ, while specific energy consumption (SEC) was 0.80 MJ/kg. Table 2 shows energy consumption for 8 hours per day operating time which lasted for 4 days. Due to the table, the highest electrical energy consumption is on the first day while the lowest is on the fourth day. It is because the initial corn moisture content is still very high causing the drying process is mostly for water vapor of material surface thereby increasing the use of electrical energy. Electricity consumption during the drying process was 7.99 MJ, while SEC was 1.75 MJ/kg.

Table 2. Electricity consumption of Scheme I with operating time 8 hours per day

Shelled Corn Dryer	Average Initial Water Content (% w.b)	Average Final Water Content (% w.b)	Effective Drying Time (Hours)	Electrical Energy (MJ)	KES (MJ/kg)
	27.24 % bb				
First Day		24.25	8	2.46	2.07
Second Day		20.26	8	2.26	1.50
Third day		17.78	8	1.84	1.69
Fourth day		14.05	8	1.43	1.37
Total				7.99	1.75

Table 3 shows electrical energy consumption for operating time 32, 24, and 16 hours. The table shows that the highest energy consumption is on 32 hours drying treatment. Consumption of electrical energy during the drying process is 8.01, 6.01 and 4.0 MJ, while SEC is 2.42, 1.82, and 1.21 MJ/kg of respectively.

Table 3. Electrical energy consumption

Scheme Shelled Corn Dryer	Average Initial Moisture Content (% w.b)	Average Final Moisture Content (% w.b)	Effective Drying Time (Hours)	Electrical Energy (MJ)	KES (MJ/kg)
	27.24 % bb				
First		14.05	32	8.01	2.42
Second		15.08	24	6.01	1.82
Third		18.45	16	4.00	1.21

3.8. Three (3) Tons Drying Simulation

Table 4 shows electrical energy consumption and commercial fan cost. Due to the analysis, the commercial fan used for 3 ton drying is a fan with an airflow rate of 2.5 m^3/s or equivalent to 0.83 L/s-kg, static pressure 150 Pa and power 650 watt. The total efficiency of the fan was 72% [4]. Electrical energy consumption for 32 hours of operation time was 20.8 kWh. Given the value of electricity tariff was Rp 1.500/kWh then the total cost was Rp 31,200, and the cost was Rp 10/kg. According to [9], the cost of corn dryer per kg was Rp 90.

Table 4. Electric energy consumption and commercial fan cost

Pressure (Pa)	Air Flow Rate (m^3/s)	η Total (%)	Power (W)	Electrical Energy (kWh)	KES (MJ/kg)	Cost (Rp)	Cost Per Kg
150	2.5	57	650	20.8	0.16	31.200	10

4. Conclusions

In natural air drying for shelled corn with 40 cm thick thicknesses conducted at 08:00-16:00, the airflow rate was 0.83 L/s-kg yields the highest drying rate of 1.25% bk/hr compared to the air flow rate 0.67 L/s-kg and 0.5 L/s-kg resulting in an airflow rate was 1.13% bk/h and 1.01% bk/hr respectively. In the operating time scheme test which gives the fastest drying rate of 4 hours per day (10:00-14:00) with a drying rate of 0.93% bk/h while the operating scheme 8 and 6 hours of drying rate is low (0.74% bk/h and 0.64% bk/h). The specific energy consumption (SEC) in the test scheme of operating time of 4 hours per day (10: 00-14: 00) was the lowest of 1.21 MJ/kg compared to the operating scheme of 8 and 6 hours respectively 2.42 MJ/kg and 1.82 MJ/kg. In a specific daily energy consumption (SEC) test of the 8-hour operating time scheme tends to decrease over the drying time was 2.07, 1.50, 1.69 and1.75 MJ/kg. The simulation of 3-ton drying capacity with 32 hours of drying time under ambient air conditions provides a specific energy consumption was 0.16 MJ/kg equivalent to energy cost per kg was Rp 10.

References

[1] Nelwan L O 1997 *Drying Cocoa Using A Dryer Rack With Collector Type Greenhouse Effect* (in Indonesian) (Bogor: Graduate School of Bogor Agricultural University)

[2] [BPS] British Psychological Society Regency TTU Province of East Nusa Tenggara 2014 *Profile of Corn Crops* North Central Timor Regency.

[3] Simbolon L D 2011 *Control Strategy On Drying Pipilan Corn Utilizing Ambient Air* (in Indonesian) (Bogor: Graduate School of Bogor Agricultural University)

[4] Fantech 2016 Axial flow fans component weights. Environmental product declaration of the European plastic. [Internet]. [downloaded 2017 February 02]. Available on: http:www//fantech.com.au/image/pdf/catalogue/section.

[5] Djaeni M 2008 *Energy Efficiency Multistage Zeolite Drying for Heat Sensitive Products*. (Wageningen: Graduate School of Wageningen University) ISBN:9978-90-8585-209-4

[6] Kiranoudis C T, Maroulis Z B and Marinos K D 1996 Drying of solids: Selection of some continuous operation dryer types *Computer and Chem. Eng*. **20** (1): 177-182

[7] Bintoro N, Nugroho J and Maria D 2012 *Development of Mechanical Drying Methods and Equipment for Seeds in Sacks* Proceedings The National Seminar of Perteta (Malang) (Indonesia)

[8] Agusniar A and D Setiyani 2011 Drying Corn With Mixed-Adsorption Drying Method Using Zeolite On Fluidized Bed *Journal of Chemical and Industrial Technology* **2** (2): 103 – 110.

[9] Firmansyah I 2009 *Drying and poultry technology for corn seed quality improvement. Case Study in Tanah Laut District, South Kalimantan*. Proceedings of The 2009 Cereals National Seminar ISBN 978-979-8940-27-9

Simulation of Oil Palm Shell Pyrolysis to Produce Bio-Oil with Self-Pyrolysis Reactor

Fika R[1], L O Nelwan[1] and M Yulianto[1]

[1] Department of Mechanical and Biosystem Engineering, Bogor Agricultural University, Indonesia

Email: fikabayasut@gmail.com

Abstract.A new self-pyrolysis reactor was designed to reduce the utilization of electric heater due to the energy saving for the production of bio-oil from oil palm shell. The yield of the bio-oil was then evaluated with the developed mathematical model by Sharma [1] with the characteristic of oil palm shell [2]. During the simulation, the temperature on the combustion chamber on the release of the bio-oil was utilized to determine the volatile composition from the combustion of the oil palm shell as fuel. The mass flow was assumed constant for three experiments. The model resulted in a significant difference between the simulated bio-oil and experiments. The bio-oil yields from the simulation were 22.01, 16.36, and 21.89 % (d.b.) meanwhile the experimental yields were 10.23, 9.82, and 8.41% (d.b.). The char yield varied from 30.7 % (d.b.) from the simulation to 40.9 % (d.b.) from the experiment. This phenomenon was due to the development of process temperature over time which was not considered as one of the influential factors in producing volatile matters on the simulation model. Meanwhile the real experiments highly relied on the process conditions (reactor type, temperature over time, gas flow). There was also possibilities of the occurrence of the gasification inside the reactor which caused the liquid yield was not as high as simulated. Further simulation model research on producing the bio-oil yield will be needed to predict the optimum condition and temperature development on the newly self-pyrolysis reactor.

1. Introduction

The increase of crude palm oil production in Indonesia results in the increase of waste production, one of which is oil palm shell (OPS), considerably potential to be one of the renewable energy resources due to its woody characteristics similar to coconut shell. Each fresh fruit bunch (FFB) processed at least consists of 6% OPS. OPS can be converted into renewable energy resources through fast pyrolysis process, producing either bio-oil or syngas depending on the temperature and the reaction time. Fast pyrolysis is a biomass decomposition process in the absence of oxygen, with high heating rate and usually takes place ofground feedstock in a constant temperature (±500°C), with short vapor residence time (<2s) [3].

Fast pyrolysis decomposes biomass into parts of gases (condensable and noncondensable), liquid, and char. The production of pyrolysis liquid takes place cooling and condensation of the primary pyrolysis vapor with short residence time, and results in dark-brown liquid able to obtain half of the conventional fuel heating value [4]. This pyrolysis liquid is commonly called bio-oil or bio-crude which is very sensitive to the change of temperature, and able to have calorific value up to 17 MJ/kg (compared to conventional fuel that is able to have calorific value 42-44 MJ/kg) [5] and produced in

medium temperature (around 500°C) [6]. This liquid also called bio-oil, is also able to be utilized as the fuel for gas turbine and diesel machines [7].
Bio-oil production usually utilizes fluidized bed reactor for its process, with the heat sourced from either electric or induction heater [8,9]. The use of fluidized bed reactor with electric heater usually needsa high amount of electricity (~5 kW) [9]. This recent study focuses on the design of fixed bed reactor for oil palm shell pyrolysis forming in the self-pyrolysis heated reactor. Applying the principle of pyrolysis which is heated deformation, the source of the heat in this reactor will come from the stove-like zone at the bottom of the reactor, with fuel coming from the oil palm shell itself, supplying heat to the pyrolysis chamber. Controlled air flow will be used in the combustion zone using a blower to achieve an optimum temperature for the pyrolysis process. The pyrolysis vapor will then be condensed using a cross-flow condenser, resulting in the final product of bio-oil.
The production system will be in batches, aiming at to be able to be used a home-industry effort by the local people of the CPO factories. Expected that this method can reduce the abundant waste of oil palm shell, this reactor also is designed to achieve a total 10% yield of bio-oil (%biomass d.b.).
Numerical simulation will take place to predict the yield from the process, using the developed model from Sharma [1] This model focuses on the volatile composition of a pyrolysis process, which will be the basis of the bio-oil yield prediction.

2. Objectives
This study aims at designing, manufacturing, and evaluating the performance of a reactor by predicting the yield of the bio-oil product from oil palm shell pyrolysis using the developed simulation model by Sharma [1].

3. Research methods

3.1. Materials and sample preparation
Iron pipe with adiameter of 15 cm was utilized to manufacture the reactor, with aniron plate with athickness of 2 mm for its upper end. The cross-flow condenser was made from iron plate with athickness of 2 mm, and copper pipe with adiameter of 0.5 inches.
The performance tests were conducted using ablower, CA-type thermocouples, anemometer, and also AUTONICS recorder. The feedstock was raw oil palm shell (water content of 15% w.b.) without grinding.

3.2. Research procedures
The literature on oil palm shell, pyrolysis, and the manufacturing process are done first before the development of the numerical simulation takes place. The reactor is designed using SolidWorks Premium 2012, manufactured on the local workshop and then tested on the Renewable Energy Engineering Laboratory, Bogor Agricultural University (IPB). The oil palm shell is gathered from state oil palm factory in West Java.

3.3. Experimental set-up
The experimental set up is illustrated in figure 1. The total length of the copper pipe of the condenser is 7 m, with 20 fins. The thermocouples were set apart 15 cm from each, with the main focus on the combustion zone.
As the reactor was aimed to be one of the home-industry efforts, its capacity was set to be only 5 kg as a pilot project. Assuming that the bulk density (ρ_{bulk}) of the oil palm shell was 400 kg/m^3 when designing, so the calculation of the reactor volume (V, in m^3) was obtained through equation (1).

$$V = \frac{m}{\rho_{bulk}} \tag{1}$$

where m is mass of oil palm shell (in kg). Considering that the fast pyrolysis must apply the principle of pressure, was then obtained a ratio of 1:5 for the diameter and the height of the reactor (15 cm and 75 cm).

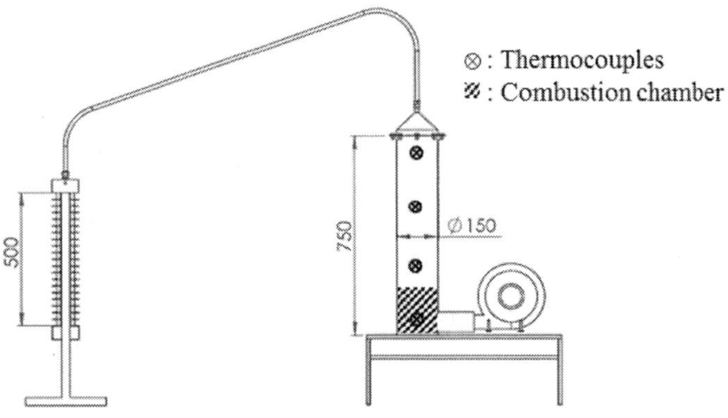

Figure 1. Reactor and condenser assembly during the experiment

3.4. Experimental procedures

The experiment started with putting a netin the middle of the reactor, to provide enough air due to low porosity of the oil palm shell. Three kilograms of oil palm shell was then filled into the reactor, and a bit of them already pyrolyzed (and became char) was put inside the net and burnt as a fuel. After the flame was considered to incinerate enough, the reactor was then closed and sealed using ceramic wool to prevent any vapor leaks. Four CA-type thermocouples were put in places, and thetemperaturewas measured in every five minutes.

The calculation of the bio-oil yields, char, and noncondensable gas was conducted every time the experiment ended. The bio-oil yield (mass fraction), denoted by Y, was contained in a measuring beaker to measure and then weighed. The calculation of each yield was obtained through equations (2-4) [3].

$$Y_{bio-oil} = \frac{m_{bio-oil}}{m_{dryOPS}} \times 100\% \quad (2)$$

$$Y_{ch} = \frac{m_{ch}}{m_{dryOPS}} \times 100\% \quad (3)$$

$$Y_{ncg} = (1 - Y_{bio-oil} - Y_{ch}) \times 100\% \quad (4)$$

where subscript *bio-oil* is for tar, *dryOPS* for dry matter, *ch* for char, and *ncg* for noncondensable gas.

3.5. Simulation model

Sharma [1] developed a model to predict the pyrolysis yield from woody biomass in forms of char, volatile matters, and also pyrolysis heat. This model was based on input data to predict the volatile matters of the oil palm pyrolysis process using theself-pyrolysis reactor, in some temperature levels. The calculation of the developed model was obtained from equation (5) to equation (14).

Due to the dry biomass used, the mass fraction on the dry basis of the biomass was considered to be:

$$Y_{cl} + Y_{hc} + Y_{lg} = 1 \quad (5)$$

The subscript *cl* stands for cellulose, *hc* for hemicellulose, and *lg* for lignin. While the char yield from the pyrolysis was considered to be:

$$Y_{char} = Y_{cl}f_{char,cl} + Y_{hc}f_{char,hc} + Y_{lg}f_{char,lg} \quad (6)$$

where $f_{char,cl}$, $f_{char,hc}$, and $f_{char,lg}$ stand for composition fraction of cellulose, hemicellulose and lignin, respectively. The volatile matters fraction (Y_{vol}) was then obtained from equation (7).

$$Y_{vol} = 1 - Y_{char} \quad (7)$$

Volatile matters from combustion consist of CO, CO_2, H_2, H_2O, light hydrocarbons (ME), and heavy hydrocarbons (tar) [11]. Light hydrocarbons in this studywere then grouped into methane and ethylene, with amass ratio of 3:1 resulting in its empirical formula of $C_{1.16}H_4$. Heavy hydrocarbons (tar) was considered to be levoglucosan [12] which was vapored and condensed atincreasing temperature, with its empirical formula of $C_6H_{6.2}O_{0.2}$ [11,13]. Aiming at achieving the reaction balance where char presence was needed, the char was considered to be pure carbon (C) in this study.

The pyrolytic decomposition was then determined using equation (8).

$$C_6H_{HB}O_{OB} = n_{char,ashfree}C + n_{CO}CO + n_{CO_2}CO_2 + n_{H_2}H_2 + n_{H_2O}H_2O + n_{ME}C_{1.16}H_4 + n_{TAR}C_6H_{6.2}O_{0.2} \quad (8)$$

where n is mole number of the corresponding substance. Characteristics of the oil palm shell [2] were then used to put into the developed model. The characteristic was shown in table 1.

Table 1. Fraction of oil palm shell lignocellulose [2]

Cellulose	Hemicellulose	Lignin	Ash+Other Extractives
0.205	0.223	0.499	0.073

To obtain the volatile matters fraction, char formation characteristic was used as shown in table 2, where each of the C, H, and O balance are obtained through equations (9)-(11), respectively.

$$6n_{DB,ashfree}C - n_{char,ashfree} = n_{CO} + n_{CO_2} + 1.16n_{ME} + 6n_{TAR} \quad (9)$$

$$n_{DB,ashfree}HB = 2n_{H_2} + 2n_{H_2O} + 4n_{ME} + 6.2n_{TAR} \quad (10)$$

$$n_{DB,ashfree}OB = n_{CO} + 2n_{CO_2} + n_{H_2O} + 0.2n_{TAR} \quad (11)$$

where subcript *DB,ashfree* stand for dry biomass. Considering the ratio of water and carbondioxide as one ($H_2O = CO_2$), the ratio of each volatile matters was obtained through equations (12)-(14).

$$\frac{Y_{CO}}{Y_{CO_2}} = e^{-1.845+\frac{7730.3}{T}-\frac{5019898}{T^2}} \quad (12)$$

$$\frac{Y_{H_2O}}{Y_{CO_2}} = 1 \quad (13)$$

$$\frac{Y_{ME}}{Y_{CO_2}} = 5 \times 10^{-16}T^{5.06} \quad (14)$$

where T is temperature of pyrolysis. The value determination from equation (12) to equation (14) was conducted in every five-minutes temperature measured where each of them was put to simulate the volatile matters producing the bio-oil from the experiments. The parameters used in the simulation can be seen in table 3.

Table 2. Char yield fraction.

Composition	Cellulose	Hemicellulose	Lignin	Reference
Char yield fraction	0.05	0.10	0.55	[14]
Formula	$C_6H_{10}O_5$	$C_6H_{10}O_5$	$C_9H_{7.95}O_{2.4}(OCH_3)_{0.92}$	[15]

Table 3. Parameters used in the simulation.

Parameter	Value	Unit
Water Content	15	% (w.b.)
Pyrolysis Rate ($\dot{m}$)	0.023, 0.022, and 0.024	kg/minute
Tar HHV	41,000	kJ/kg

4. Results and discussion

4.1. Experiment results

The designing of the reactor resulted in avolume of 0.0125 m^3, with the total amount of starter fuel was 0.97 kg. The temperature development for every 5 minutes is shown in figure 2.

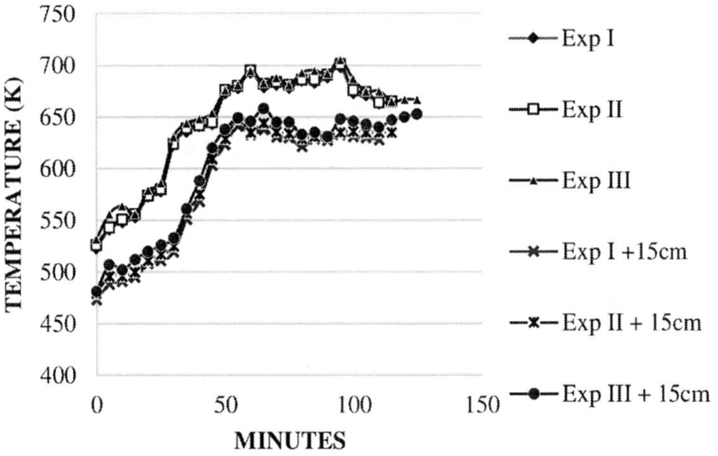

Figure 2. Experiment temperature development

Three experiments showed similar trends of temperature development, as well as their temperature in the second thermocouple spot 15 cm above. These resulted in the bio-oil yield of 275 ml (ρ = 1054.54 kg/m^3), 260 ml (ρ = 1038.46 kg/m^3), and 255 ml (ρ = 1019.60 kg/m^3) respectively. It showed each percentage was 10.23, 9.82, and 8.41 % (d.b.). The char yields were 39.22, 43.14, and 40.34 % (d.b.), with the process time 110, 115, and 125 minutes respectively. The pyrolysis process length using this self-pyrolysis reactor was not one of the controlled variables due to many external influential factors affecting different process conditions, such as ambientairflow velocity, and also environment temperature.

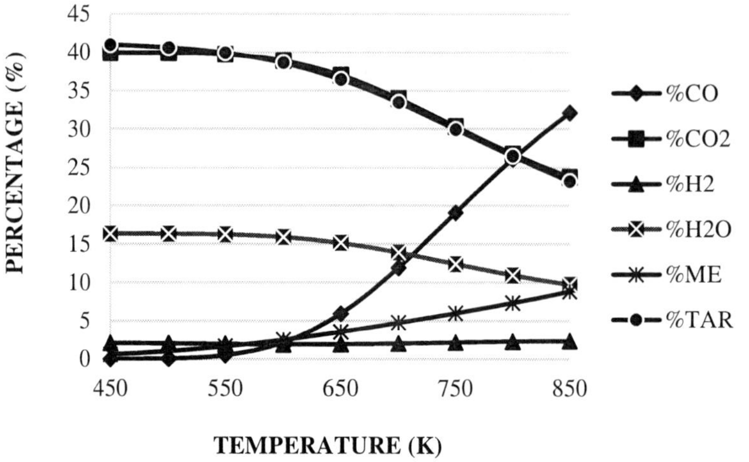

Figure 3. Volatile compositions of oil palm shell

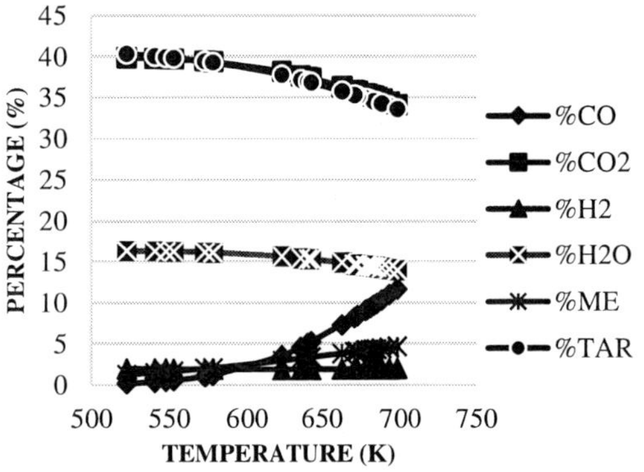

[a] Experiment I

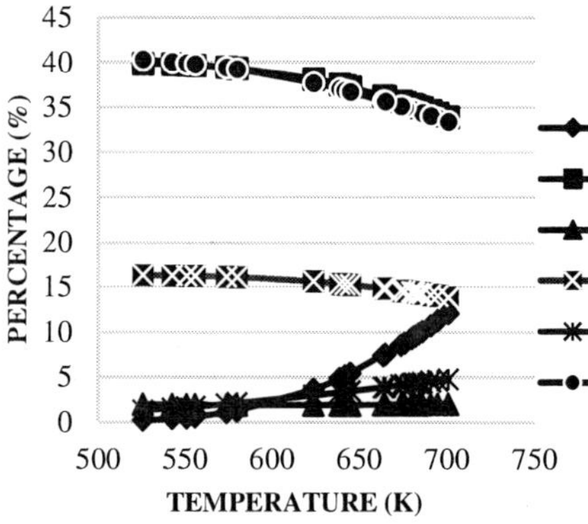

[b] Experiment II

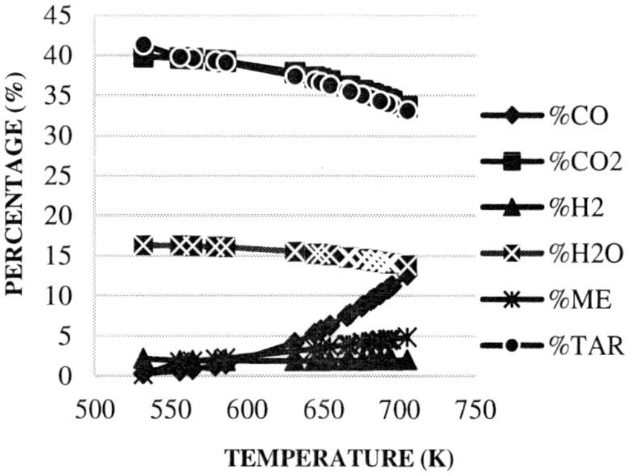

[c] Experiment III

Figure 4. Volatile composition of oil palm pyrolysis experiments

4.2. Combustion zone simulation result

Referring to the developed model by Sharma *et al.* (2016) with the oil palm shell characteristics from Saka *et al.* (2008), the volatile compositions in every temperature level is shown in figure 3.

Plotting the model to the experiment temperatures with assuming constant mass flow will result in volatile composition at every temperature level (every 5 minutes). It is assumed that the bio-oil produced was a combination of levoglucosan and H_2O [16,17]. Figure 3 shows that in the temperature range of 450-850 K, the H_2O content decreases in the increased temperature, while the gases component increases within temperature level. Also, the simulation showed that the hydrogen content of the oil palm shell is rather high, due to its high aromatic compounds [18], even if still in low level.
The volatile compounds for each experiment are shown in figure 4, where the input temperature was the temperature when the bio-oil starts getting produced to when the bio-oil stop coming out of the condenser outlet. Meanwhile, the simulation temperature was the whole process temperature, starts when the reactor was closed,and the temperature was measured every five minutes. It was compared to determine the difference between the real cracking reaction occurring, and the simulated reaction.

4.3. Simulation and experiment validation
There is a significant difference between the result of the simulation and experiment yields. While the experiments resulted in 10.23, 9.82 and 8.41 % (d.b.), the simulation showed in a higher result which was 22.01, 16.36, and 21.89 % (d.b.). The comparison between simulation and experiment is shown in figure 5.

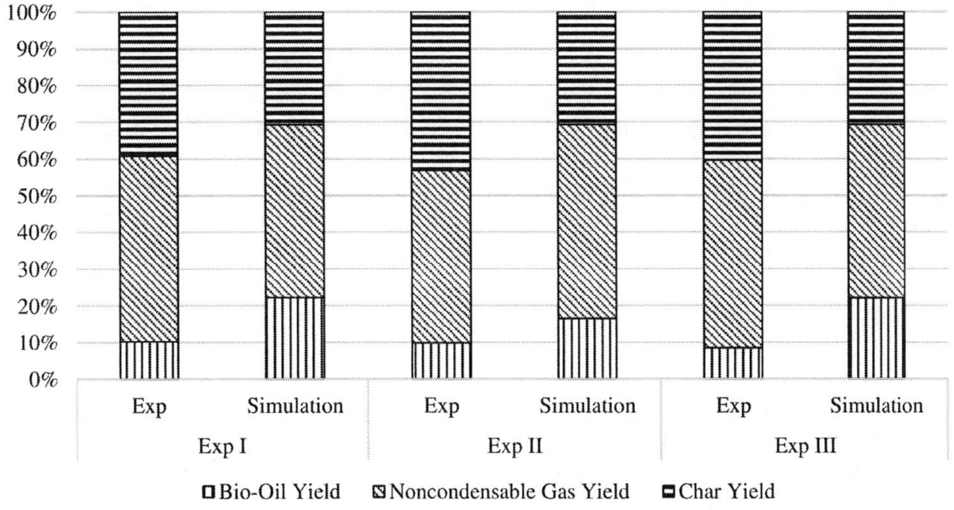

Figure 5. Comparison between simulation and experimental results

It was stated that the char mass fraction could be determined only from the hemicellulose, cellulose, and lignin content of a biomass [1], so the char yield from the simulation resulted in 30.7 % (d.b.), while the experiments show different values which were 39.22, 43.14, and 40.34 % (d.b.) respectively. The reaction or process condition does not count as influential factors [1] while in actual condition, the pyrolysis yield from biomass conversion is dependent to the reactor (shape, size, also optimum capacity), time, and temperature of theprocess.
A significant difference in bio-oil yield also was indicated by the high level of CO_2 content due to slow heating rate, considering that the highest temperatures achieved through the process were 425, 428, and 432 °C respectively. The optimum temperature to produce thehighest yield of oil palm shell bio-oil should range from 500 to 550 °C referring to [8] while usually the lignin decomposition – which contributes the most to the production of bio-oil – occurs at 650 °C [19]. The slow heating rate occasionally causes the organics reaction to occur during the process; this might also be caused by the inconstant air flow to the condenser inducing secondary reactions on the vapor formed, the vaporization of liquid, and also the gasification between the char and water [19].

The optimum size of oil palm shell particle to be pyrolyzed ranged between 1.4 – 2 mm [3], in which way any grinding was not conducted in this research. The minimum pretreatment to the oil palm shell such as grinding and drying was not conducted in this study regarding the aim of feasibility to apply this method towards local people. It also affected the result [19], biomass with very low water content can be very potential to be converted through thermochemical processes including pyrolysis. The heating rate of all three experiments also showed slow pyrolysis mechanism instead of fast pyrolysis, that resulted in higher char yield than the bio-oil yield.
The significant difference between the simulation and actual experiment results show that the process condition affects the final yield of the product including bio-oil, char, and gases formed. The simulation model whichwas only based on the chemical balance couldn't still be able to reflect the real condition of a process depending on the temperature dynamic, heating rate, also the total time of reaction.

5. Conclusions

A developed model from Sharma [1] was used to predict the total yield of oil palm shell pyrolysis using its combustion chamber temperature with a self-pyrolysis method. There was asignificantdifference between the simulation and actual experiments due to different condition applied to both calculations. The simulation did not include external factors such as temperature fluctuation, total time of reaction, and reactor type.

References

[1] Sharma AK, Ravi M R and Kohli S 2017 Modelling product composition in slow pyrolysis of wood. http://web.iitd.ac.in [16 January 2017]
[2] Saka S, Munusamy MV, Shibata M, Tono Y and Miyafuji H2008Chemical constituents of the different anatomical parts of the oil palm (*Elaeis guineensis*) for their sustainable utilization. *JSPS-VCC Group Seminar 2008, Natural Resources and Energy Environment 24-25 November 2008*(Kyoto)(Japan)
[3] Abnisa F, Wan Daud WMA, Husin WNW and Sahu JN2011Utilization possibilities of palm shell as a source of biomass energy in Malaysia by producing bio-oil in pyrolysis process. *Biomass and Bioenergy***35**(5): 1863-72
[4] Bridgwater AV, Meier D and Radlein D1999An overview of fast pyrolysis of biomass*Organic Geochemistry***30**(12): 1479-93
[5] Bridgwater AV and Peacocke GVC2000Fast pyrolysis processes for biomass*Renewable and Sustainable Energy Reviews***4**(1): 1-73
[6] Luo Z, Wang S, Liao Y and Cen K2004Mechanism study of cellulose rapid pyrolysis. *Ind. Eng. Chem***43**(18): 5605-10
[7] Chiaramonti D, Oasmaa A, Solantausta Y. *et al.* 2007Power generation using fast pyrolysis liquids from biomass*Renewable and Sustainable Energy Reviews***11**(6): 1056-86
[8] Asadullah M, Rasid NSA, Kadir SASA and Azdarpour A2013Production and detailed characterization of bio-oil from fast pyrolysis of palm kernel shell*Biomass and Bioenergy***59** 316-324
[9] Muley PD, Henkel C, Abdollahi KK and Boldor D2015Pyrolysis and catalytic upgrading of pinewood sawdust using induction heating reactor*Energy and Fuels***29**(11): 7375-85
[10] [BATAN] Badan Tenaga Nuklir Nasional2012Utilization of Palm Oil Industry Waste for fuel (in Indonesian). http://batan.go.id [1 Oktober 2014]
[11] Thunman H, Niklasson F, Johnsson F and Leckner B2001Composition of volatile gases and thermochemical properties of wood for modeling of fixed or fluidized beds*Energy & Fuels***15**(6):1488-97
[12] Lewellen PC, Peters WA and Howard JB1977Cellulose pyrolysis kinetics and char formation mechanism*Symposium (International) on Combustion***16**(1):1471-80

[13] Ragland KW and Aerts DJ1991Properties of wood for combustion analysis*Bioresource Technology***37**(2):161-68

[14] Reed TB and Das A1981*Handbook of Biomass Downdraft Gasifier Engine Systems*Solar Energy Research Institute. SERIISP-271-3022

[15] Tillman DA, Rossi AI and Bitto WD1981*Wood Combustion*(Amsterdam: Elsevier) ISBN: 978-0-12-691240-1

[16] Okoroigwe EC, Li Z, Onyegegbu SO and Saffron CM2012Maximizing the energy potential of palm kernel shell by pyrolytic conversion to biofuel*The Fourth International Renewable Energy Congress*.http://researchgate.net [6 Maret 2017]

[17] Aramideh S, Xiong Q, Kong SC and Brown RC2015Numerical simulation of biomass fast pyrolysis in an auger reactor*Fuel***15**234-42

[18] Yang H, Yan R, Chen H, Lee DH, Liang DT and Zheng C2006Mechanism of palm oil waste pyrolysis in a packed bed*Energy & Fuels***20**(3):1321-28

[19] Asadieraghi M and Daud WMAW2015In-depth investigation on thermochemical characteristicsof palm oil biomasses as potential biofuel resources*Journal of Analytical and Applied Pyrolysis***115** 379-391

Development of Portable Venturi Kiln For Agricultural Waste Utilization By Carbonization Process

S E Agustina[1], N Chasanah[1], and Eris A P[1]

[1] Dept. of Mechanical & Biosystem Engineering, Faculty of Agricultural Engineering & Technology, Bogor Agricultural University (IPB), Indonesia.

E-mail: endah@perdana-consulting.co.id ; endah859@gmail.com

Abstract: Many types of kiln or carbonization equipment have been developed, but most of them were designed for big capacity and some also having low performance. This research aims to develop kiln, especially portable metal kiln, which has higher performance, more environmental-friendly, and can be used for several kinds of biomass or agricultural waste (not exclusive for one kind of biomass) as feeding material. To improve the kiln performance, a venturi drum type of portable kiln has been designed with an optimum capacity of 12.45 kg coconut shells. Basic idea of those design is heat flow improvement causing by venturi effect. The performance test for coconut shell carbonization shows that the carbonization process takes about 60-90 minutes to produce average yields of 23.8%., and the highest temperature of the process was 441 °C. The optimum performance has been achieved in the 4th test, which was producing 24% yield of highest charcoal quality (represented by LHV) in 65 minutes process at average temperature level 485 °C. For pecan shell and palm shell, design modification has been done by adding 6 air inlet holes and 3 ignition column to get better performance. While operation procedure should be modified on loading and air supply, depending on each biomass characteristic. The result of performance test showed that carbonization process of pecan shell produce 17 % yield, and palm shell produce 15% yield. Based on Indonesian Standard (SNI), all charcoal produced in those carbonization has good quality level.

1. Introduction

Some of biomass waste has been utilized for many purposes such as handicraft and accessories raw materials, active carbon, and also as charcoal for cooking and heating process. But huge of them still have not been utilized well and potential to causing an environmental problem. Carbonization, a thermal decomposition process based on the pyrolysis, is one of the best utilization, especially for woody biomass waste, since it produces economic products. Many types of carbonization equipment have been developed, but most of them were designed for big capacity or has low performance.

Lacross (1985) in [1] explain that high yield of carbonization will be achieved if the pyrolysis process occurred on the primary pyrolysis steps at a temperature range of 150-350 °C, but high temperature will produce higher carbon quality. Implementing venturi effect in the kiln design should be producing higher velocity and temperature of the gas flow inside the kiln or carbonization chamber. Thus carbonization condition will be better and more efficient.

Many designs of kiln have been developed since several years ago. Regular drum type has been chosen since it's easy to build by using regular 'used oil drum' as its main structure, which can be found easily in the rural or urban area [2]. This study aims to develop portable venturi kiln to improve the portable drum kiln performance by implementing venturi effect in the pyrolysis or carbonization chamber.

To increase performance, the kiln should be designed or operated based on each feeding material characteristic. Since biomass or agricultural waste varies in the very wide range, not only in its physical properties but also its composition, some modifications should be done both in kiln design process and in the kiln operation to get optimum performance of the kiln.

2. Method

In general, the kiln development stages are starting with design approach, continuing with design process and construction, then following with performance test and data analysis, and finishing with a recommendation for performance improvement as shown in the flowchart below.

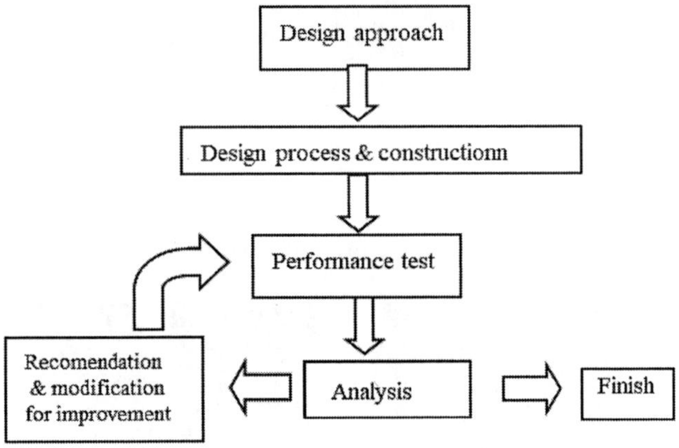

Figure 1. Stages of research activities.

As result of the design approach, kiln was designed as venturi drum type which consists of 6 (six) parts or sections as shown in figure 2. The parts were called as combustion section (d), carbonization section or pyrolysis chamber (f), heating column (c), chimney (a), kiln cover (b), and kiln support (e) as shown in figure 2. The total kiln height is 1763 cm and the wide is 572 cm. Volume (capacity) of carbonization chamber or pyrolysis chamber (f) is 0.3 m^3 or about 13 kg of coconut shell with 2/3 volume occupation, so there is enough space for heat gases which will be produced while pyrolysis process occurred. Combustion section in the bottom part of kiln (d) was designed for initial heat energy supply for the carbonization process, as same as the function of heating column in the upper part of the kiln (c).

Performance of the kiln was evaluated based on several parameters; they were: optimum capacity (kg coconut shell as feedstock), temperature of the process, time or duration of the process, yield and quality of the product. 2 (two) pre-testing was conducted to get the optimum condition of the kiln operation which produced best yield and quality. Main testing was done for 5 carbonization processes. Coconut fiber and coconut shell have been used for initial heat energy supply which was burned (combusted) in the heating column and the combustion section. After performance test by using coconut shell as feedstock, performance test was carried out by using palm shells and pecan shells as feed-stock material. For pecan shell and palm shell, the procedure were same, but some modification has been applied to get the best kiln performance.

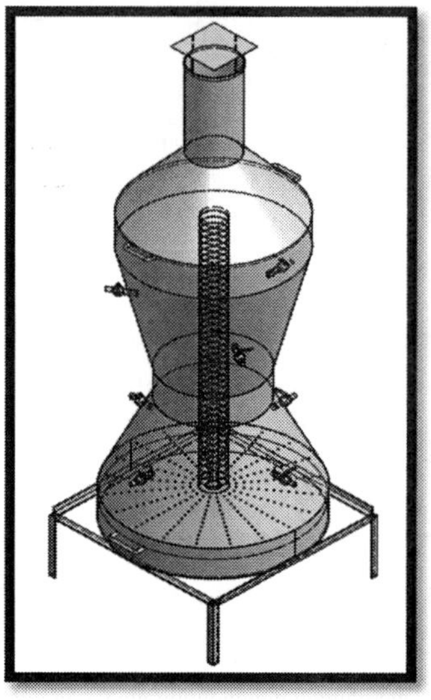

Figure 2. Structure of portable venturi drum kiln

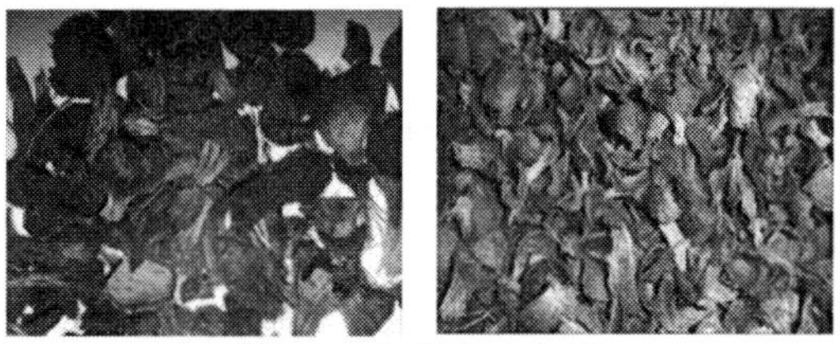

Figure 3. Palm shell

Figure 4. Pecan shell

3. Result and Discussion

Portable venturi drum kiln which has been designed is presented in figure 5 below, and the performance test condition presented in figure 6.

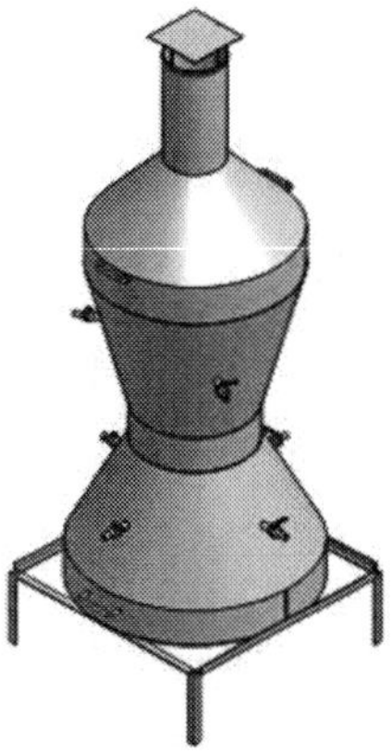

Figure 5. The kiln

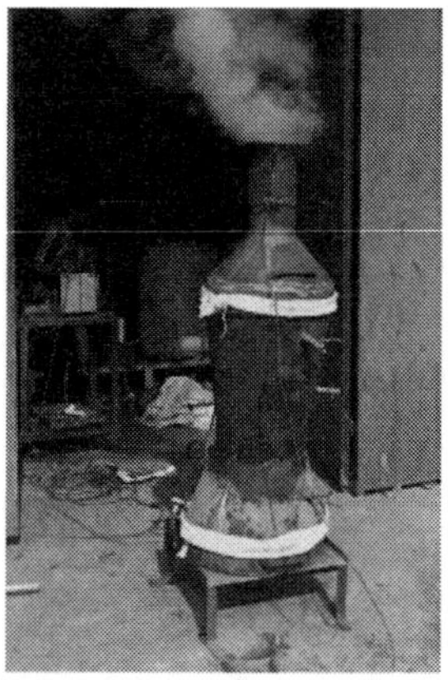

Figure 6. Performance test

3.1. *Kiln capacity*

Pre-testing shows that optimum capacity of the kiln, means the total weight of coconut shell which occupied 75% carbonization chamber, is about 12 - 13 kg, depending on the coconut shell size. The test also shows that the shape and dimension of coconut shell as feedstock, influence the time or

IOP Conf. Series: Earth and Environmental Science **147** (2018) 012026 doi:10.1088/1755-1315/147/1/012026

duration of the process and also the temperature of carbonization process, those influences the yield. Coconut shell with big size and half sphere shape, producing higher yield (with the same quality) compare to the smaller size, due to its higher temperature and longer process (table 1). But, in the real condition, coconut shell found in the municipal waste was not uniform in size and shape, but mostly in small size. So, most of the performance test was done by using small size coconut shell as feeding material.

Performance test by using pecan shell and palm shell, which has better uniform shape and size but smaller than coconut shell, shows that capacity of the kiln was about 34 kg for pecan shell and about 31 kg for palm shell. Higher density of pecan shell and palm shell (compare to coconut shell) causing lower kiln performance (Tabel 2) due to the smaller porous among the shell. Thus, modification should be taken on the pyrolysis chamber to get better performance. Two air column addition has been successful in improving carbonization process but reducing kiln capacity as result of the chamber volume reduction. The kiln capacity became 30-31 kg for pecan shell and about 28 kg for palm shell.

3.2. *Kiln performance for coconut shell carbonization*

Kiln performance for coconut shell carbonization presented as data shown in Tabel 1 and figure 7, figure 8 and figure 9. Based on data in figure 7 and table 1, higher temperature in the 3rd, 4th and 5th test, compared with the 1st and 2nd test, were producing less yield (less amount of charcoal). Those result was matching with pyrolysis theory, which was defined that higher temperature in carbonization process will produce less yield. But this condition (higher temperature) producing better charcoal quality, representing by LHV (Lower Heating Value), as shown in figure 9. Those result also matched with the theory which defined that higher temperature of the process will produce charcoal with higher carbon content so that the charcoal will have higher energy (LHV) content.

Table 1 also shows that big size coconut shell produced higher yield compared with small size. It produced shorter temperature range as shown in figure 7, due to air column among the material layer. While kiln operation by using small size coconut shell shows that best yield achieved in the 1st test. Those process running at average temperature of 270 °C and highest temperature of 552 °C, producing 26% yield and all of coconut shell was 100% successfully pyrolyzed.

Table 1. Kiln performance by using coconut shell as feed stock

Test no	Coconut shell (kg)	MC (%)	Coc.shell size	Process time (minute)	Average temp. (°C)	Highest temp. °C)	Yield (%)	Quality
1	12.45	12.64	Small	65	270	552	26	G
2	12.45	13.03	Big	90	386	541	27	G
3	12.45	12.84	Small	65	554	989	22	G
4	12.45	13.41	Small	70	485	908	24	G
5	12.45	13.41	small	60	558	977	20	G

Note: G= good quality

3.3. *Venturi effect on the carbonization process*

Temperature level during carbonization proses which was presented in figure 7, and temperature distribution inside the kiln (figure 8), were influenced by venturi effect of the throat inside carbonization chamber. Those data showed that venturi throat had successfully reduced the differences temperature between middle and upper level of the kiln as shown in the 1st, 4th and 5th tests. Thus the process temperature could keep in almost same level due to the speed of gas velocity through the venturi tube. In the 2nd and 3rd tests, the venturi effect failed to keep the temperature. It could happen when the throat was interrupted (by material), but it could not recognize or investigate in this study due to the kiln wall was not transparent [2].

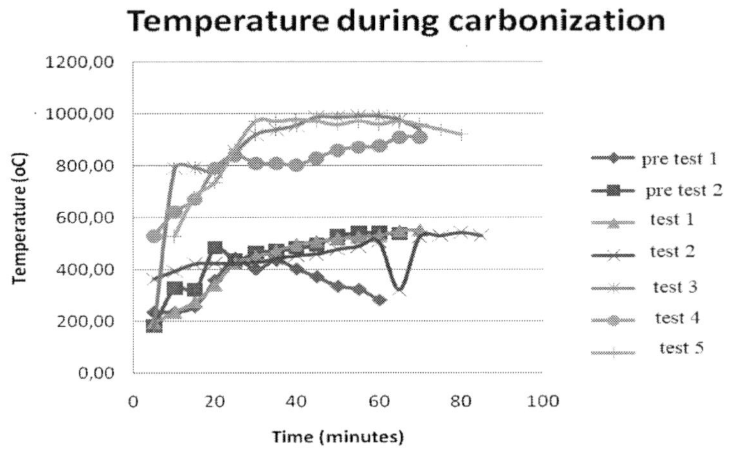

Figure 7. Temperature level during coconut shell carbonization

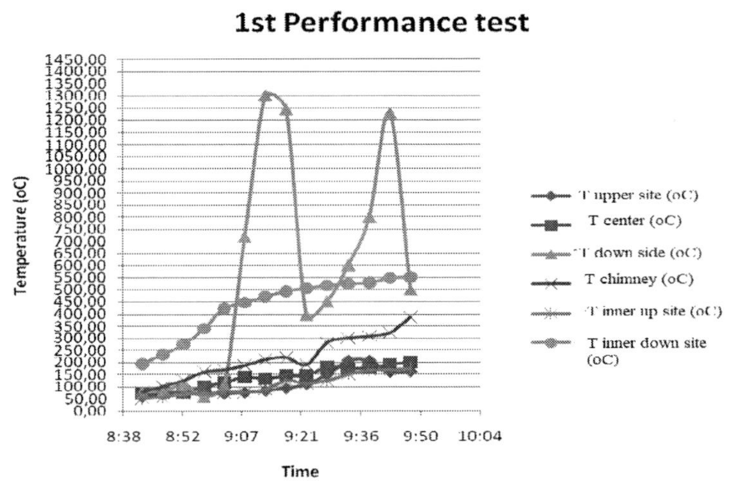

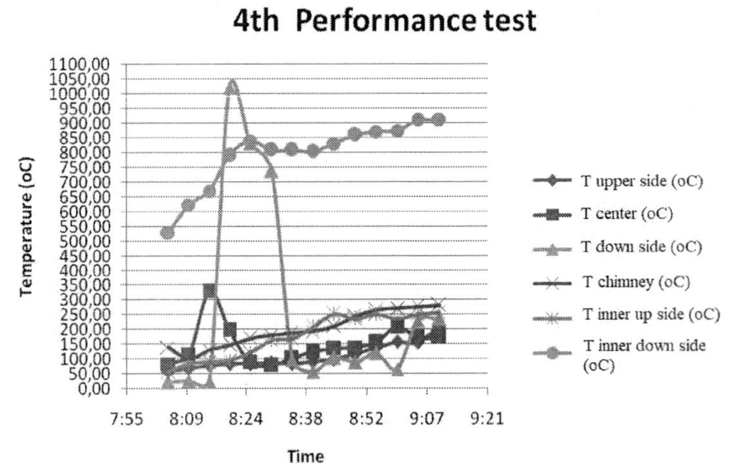

Figure 8. Temperature distribution inside kiln chamber

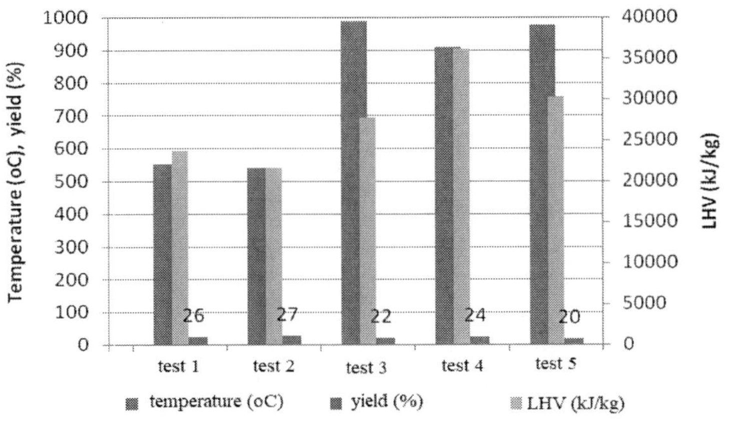

Figure 9. Relationship of temperature, yield, and charcoal quality

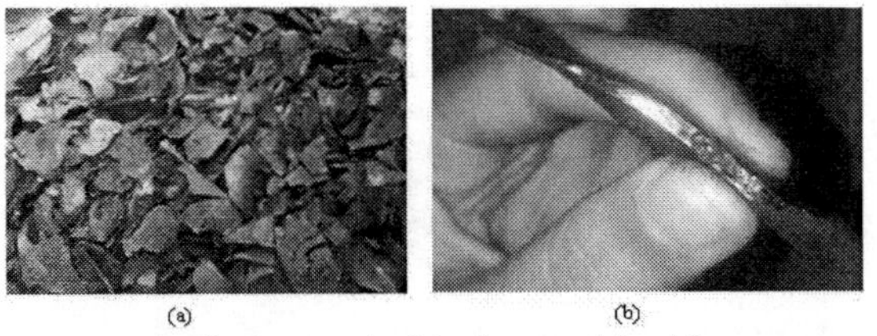

Figure 10. (a) Coconut shell charcoal, (b) charcoal inside, representing the quality

3.4. *Kiln performance for pecan shell and palm shell.*

Result of performance test of the kiln by using pecan shell and palm shell as feeding material presented in table 2, figure 11 and figure 12 below.

Tabel 2. Kiln performance for pecan shell and palm shell

No	Parameter	Palm shell	Pecan shell
1	Optimum capacity (kg)	31.35	34.17
2	Bulk density (kg/m3)	88.57	128.00
3	Carbonization time (minutes)	313.5	346.5
4	Process time (minutes)	362.5	396.25
5	Average carbonization temp (oC)	239.22	226.92
6	Highest temperature (oC)	433	378
7	Yield (%)	15.32	17.07
8	Un-pyrolized raw material (kg)	17.97	16.40

IOP Conf. Series: Earth and Environmental Science **147** (2018) 012026 doi:10.1088/1755-1315/147/1/012026

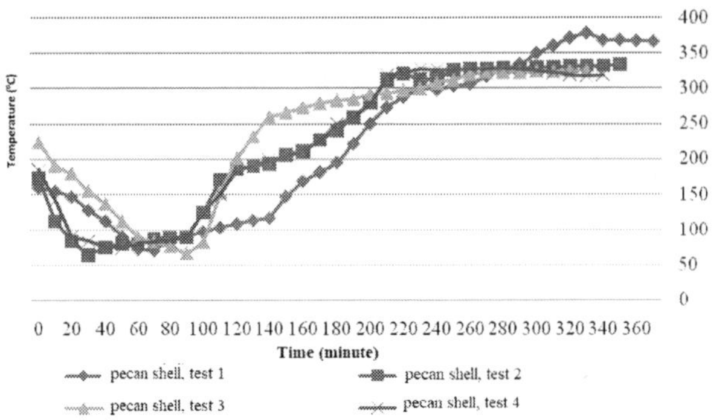

Figure 11. Temperature pattern in pecan shell carbonization

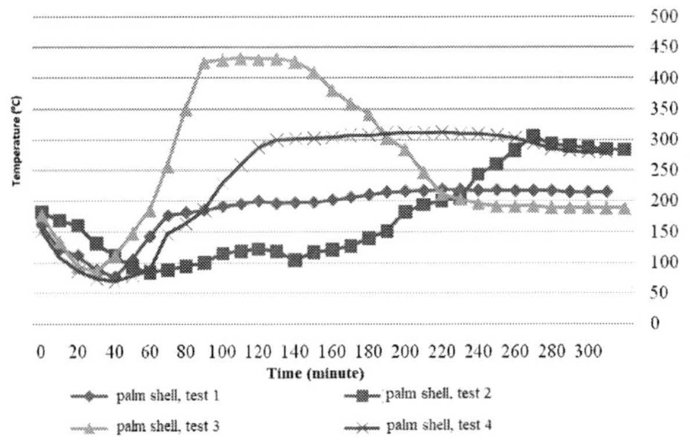

Figure 12. Temperature pattern in palm shell carbonization

The data shows that the kiln performance for pecan shell and palm shell carbonization was poor than its performance for coconut shell. Some modification has been made to improve the kiln performance as shown in table 3. Modification of operation procedure also applied to material feeding/loading. For coconut shell, all material can be loaded to the carbonization chamber as much as kiln capacity, but for pecan shell and palm shell loading should be done step by step to give enough time for material to be pyrolyzed.

Table 3. Kiln modification

No	Component	Original design	Modification
1	Air inlet holes:		
	a.number of holes	7	13
	b.size (inch)	0.75	0.75
	c. position	Center & bottom	Center & bottom
2	Air distribution column	1	3

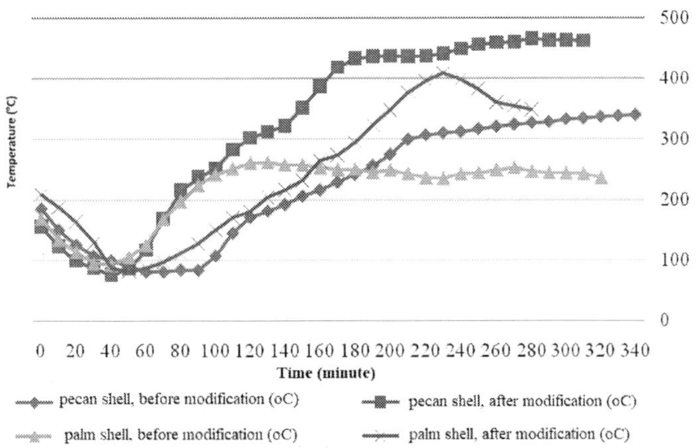

Figure 13. Temperature of pecan shell and palm shell carbonization, before and after modification

Result of the modification presented in table 4 and figure 13. The data shows that modification successfully on improving kiln performance, which is representing by decreasing of carbonization time and un-pyrolyzed material, and increasing carbonization temperature and charcoal yield. But, the kiln capacity was lower due to the space reduction as result of additional "air distribution column".

Table 4. Result of kiln performance test for pecan shell and palm shell, before and after modification

Parameter	Palm shell (original design)	Palm shell (modified)	Pecan shell (original design)	Pecan shell (modification)
Optimum capacity (kg)	31.35	28.03	34.17	30.93
Total carbonization time (minutes)	362.50	333	396.25	360
Temp. of carbonization (°C)	239.2	240.7	226.9	321.
Yield (%)	16.43	26.65	17.07	23.21
Un-pyrolized material (kg)	17.19	0.0	16.40	8.52

4. Conclusion

The venturi drum kiln which has been designed based on the coconut shell characteristic, having total size of 1763 cm height and 572 cm width, and (volume) capacity about 0.3 m3 or about 13 kg coconut shell. The performance test shows that optimum capacity of the kiln is 12.45 kg of coconut shell, and the best yield achieved by the 1st test, which was running in average temperature 270 °C and highest temperature 552 °C , producing 26% yield, and the material was 100% pyrolyzed. Best charcoal quality achieved by the 4th test, which was running in average temperature 485 °C, highest temperature 908°C, and produce 24% yield.

Venturi throat has successfully reduced the differences of temperature between middle and upper level of the kiln as shown in the 1st, 4th and 5th tests. Thus the process temperature could keep in almost same level due to the speed of gas velocity through the venturi tube.

For pecan shell and palm shell carbonization, the kiln has poor performance and some modification should be taken to improve the performance. Modification on air inlet system and material loading has been successful in improving kiln performance, but causing reduction of kiln capacity from about 34 kg to 30 kg of pecan shell, and from about 31 kg to 28 kg of palm shell. The yield was increasing from 17% to 23% for pecan shell, and from 16.43% to 26.6% for palm shell. Even though the modification achieving better carbonization result for palm shell than pecan shell, but design improvement of this venturi kiln still needed to get higher performance, and also availability for other biomass.

References

[1] Abdullah K, Irwanto A K, Siregar N, Agustina S E, Tambunan A H, Yamin M, Hartulistyoso E, Purwanto Y A, Wulandani D and Nelwan L O 1998 Agriculture Energy and Electrification (In Indonesia) (Bogor: Graduate School of Bogor Agricultural University)

[2] Agustina S E and Hasanah N 2013 *Design and performance test of metal kiln venturi drum type for coconut shell carbonization* Proceeding of International Seminar on Agriculture and Biosystem Engineering (ISABE) (Yogyakarta: Gadjah Mada University)

A Preliminary Study on Rock Bed Heat Storage from Biomass Combustion for Rice Drying

L O Nelwan[1,2], D Wulandani[1,2] and I D M Subrata[1,2]

[1] Department of Mechanical and Biosystem Engineering, Bogor Agricultural University, Indonesia

[2] Center for Research on Engineering Application in Tropical Agriculture (CREATA), Bogor Agricultural University, Indonesia

E-mail: lonelwan@ipb.ac.id

Abstract. One of the main constraints of biomass fuel utilization in a small scale rice drying system is the operating difficulties related to the adjustment of combustion/feeding rate. Use of thermal storage may reduce the problem since combustion operation can be accomplished in a much shorter time and then the use of heat can be regulated by simply adjusting the air flow. An integrated biomass furnace-rock bed thermal storage with a storage volume of 540 L was designed and tested. There were four experiments conducted in this study. Charging was performed within 1-2 hours with a combustion rate of 11.5-15.5 kg/h. In discharging process, the mixing of air passing through the rock bed and ambient air were regulated by valves. Without adjusting the valve during the discharging process, air temperature increased up to 80°C, which is not suitable for rice batch drying process. Charging with sufficiently high combustion rate (14 kg/h) within 1 hour continued by adjusting the valve during discharging process below 60°C increased the discharge-charge time ratio (DCTR) up to 5.33 at average air temperature of 49°C and ambient temperature of 33°C.The efficiency of heat discharging was ranged from 34.5 to 45.8%. From the simulation, as much as 156.8-268.8 kg of rice was able to be dried by the discharging conditions.

1. Introduction

The use of biomass fuel in a small-scale drying system has major constraints regarding its operation, especially in regulating the rate of combustion and feeding. Manually combustion rate settings may cause inconvenience for operators by having to set up the feed rate as well as to monitor the combustion process during the drying process. Automatic control can be applied by controlling the motor feeding system, but the use of such a system is quite expensive and often encounters problems of operation.

The use of heat storage in the biomass furnace system can reduce these problems. By applying heat storage system, the operation of combustion can be performed in a much shorter time. Furthermore, during the drying takes place, the heat used is obtained from the heat storage material which in operation is much easier.

Rock is one of the heat storage material that is frequently used for drying. Several studies of rock heat storage using solar collectors for drying have been carried out, for example, the use of rock bed at the bottom of collectors in Jain (2006) [1] and Aboul-Enein et al. (2000) [2] or that separated from collectors in Tambunan et al. (2010) [3]. One of the advantages of rock bed heat storage use is its capability to be used in high temperature compared to other materials such as water. At 1 atm pressure, water can only store heat with temperatures below 100°C, while rock can be used much higher than that temperature. The main problem of use of rock as a heat storage material is its specific heat capacity much lower than water. If the heat storage use is not at sufficiently high temperature, then the total heat that can be stored per unit volume becomes lower than that of water. Therefore, to obtain the real benefit of rock use, heat storage needs to be done at relatively high temperatures. Thermal storage research on high temperature of rock bed is commonly used to drive turbines in concentrated solar power systems [4-7] (Zenangeh et al., 2012; Allen et al., 2014; Allen et al., 2016; Cascetta et al., 2014). The fluids used in these studies include oil, molten salt, and air.Therefore the use of rocks as the heat storage material is considered to be potential from biomass combustion which is also generated at high temperature.

In batch mode grain drying, the desirable air temperature should be moderate, preferably below 55°C. In addition, the air flow rate for drying should also be set at a certain level to avoid inefficiency of the drying process. For this purpose,the heat originated from storage needs to be mixed with the ambient air in a certain ratio to obtain suitable drying air conditions.

Study on rock bed heat storage using biomass energy sources especially applied for drying was still rare, so we conducted a preliminary study to experimentally investigate the performance of the system. Therefore, an integrated heat rock storage – biomass furnace was designed, built and tested.

2. Materials and methods

2.1. Experimental setup

The design of integrated biomass furnace-rock heat storage with a volume of 540 L is schematically described in figure 1. The walls of furnace and storage chamber were made from steel plate and then insulated by ceramic wool and glass wool. The wall of ducting from the heat storage was also insulated by glass wool. The rock used was crushed basalt rock. The rocks were poured into the storage chamber until the bed height was around 1500 mm or equal to 735 kg. The biomass used was teakwood waste piece. The experiments were conducted in two stages, i.e., charging and discharging processes.

The charging process is started by burning the biomass in the furnace. The blower was then switched on, but at the initial period, the valve connecting the ambient air and the chamber above the bed was completely closed so that the negative pressure above the bed will cause the flue gas to pass through the bed of rocks.

To start the discharging stage, the biomass fuel was removed from the furnace and the valves were adjusted in such a way that mixing of the ambient air and the heated air occurred. At both stages of the process, the temperature measurement was carried out by placing a K type thermocouple inside the rock and the air closed to the measured rock representing the bottom, middle and top regions. In addition, temperature measurements were also carried out at air temperature inside the ducting as well as at the ambient. The temperature changes were recorded during the processes with interval of 1 minute. The charging process was stopped when the air (gas) out had reached 60°C, while the discharging process was stopped when the air temperature had reached less than 10°C above the ambient temperature. Airflow rate was obtained from the measurement of air velocity by using an anemometer in the ducting.

The void fraction was determined by putting the rock bed into a cylindrical chamber, so the volume of the rock bed was 20 L and subsequently was filled by water in such a way that the surface was equal to that of the bed. The water volume filled is denoted by V_{water}, while V_{tot} is the 20 L volume. The void fraction (ε) was calculated by:

$$\varepsilon = \frac{V_{water}}{V_{tot}} \quad (1)$$

The true density and the dimension of the rocks involving maximum length, thickness and volume were obtained by measuring 20 samples of rocks. The equivalent diameter was obtained from the measured volume.

2.2. Experimental procedure

Four experiments were carried out in this study with procedures as described in table 1. The biomass fuel feeding was performed manually into the furnace based on the flame condition and the temperature of the furnace. The valves V_1, V_2 dan V_3 (figure 1) were used to adjust the air flow from ambient air as well as from the storage chamber in order to manually control the combustion process (in the charging stage) as well as the temperature of the air flowing in the duct. The outlet air temperature which is used as the parameter to control the charging as well as the discharging stage was measured inside the duct at the middle position.

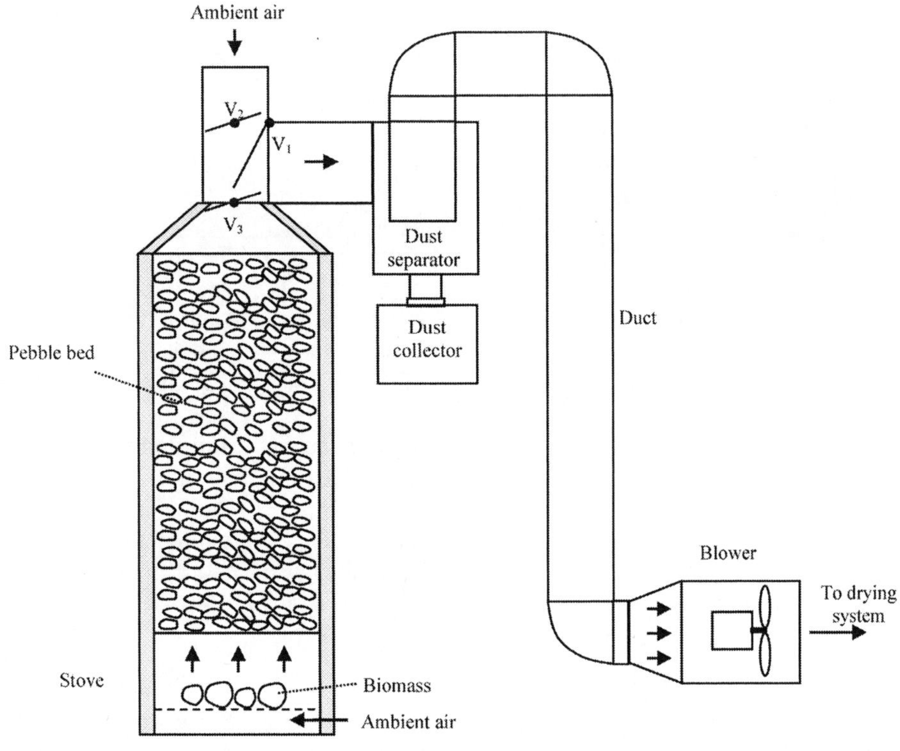

Figure 1. Schematic design of integrated biomass furnace – rock heat storage

2.3. Experimental parameters

The heat used (Q_{disc}, in kJ) to raise the temperature from ambient air (t_i, in °C) to outlet air temperature (t_o, in °C) during the discharging process was calculated numerically by:

$$Q_{disc} = \int_0^{\theta_{disc}} \dot{m} c_p (t_i - t_o) d\theta \quad (2)$$

where $\dot{m}$ and c_p are the mass flow rate (kg/s) and the specific heat capacity (kJ/kg-°C) of flowing air determined at outlet air temperature, respectively.

The overall efficiency (η_{disc}) is the ratio of Q_{disc} to the potential energy of the fuel (i.e heat energy from biomass combustion (Q_b)), which is expressed as:

$$\eta_{disc} = \frac{Q_{disc}}{Q_b} \tag{3}$$

Table 1. Experimental procedures

Procedure		Experiment I	Experiment II	Experiment III	Experiment IV
Valves adjustment		• Only used V_1. • V_2 and V_3 were not installed	• Used V_2 and V_3 • V_1 was not installed	• Used V_2 and V_3 • V_1 was not installed	• Used V_2 and V_3 • V_1 was not installed
Charging	Combustion rate	• Initial period: (± 0.18 kg/min) • End period: (± 0.18 kg/min)	• Initial period: (± 0.18 kg/min) • End period: (± 0.10 kg/min)	• Almost constant (±0.14kg/ menit)	• Almost constant (±0.23kg/ menit)
	Final charging condition	• Outlet air temperature of 70°C.	• Temperature of top region of bed at 120°C. V_2 was adjusted so the outlet air temperature less than 70°C.	• Temperature of top region of bed at 120°C. V_2 was adjusted so the outlet air temperature less than 65°C	• Charging time was set for 60 minute
Discharging	Outlet air temperature target	• Initially V_1 was adjusted to reach initial outlet air temperature of 60°C, and subsequently, V_1 was not altered.	• Outlet air temperature was kept below 60°C by continuously adjusting V_2 dan V_3	• Outlet air temperature was kept below 60°C by continuously adjusting V_2 dan V_3	• Outlet air temperature was kept below 60°C by continuously adjusting V_2 dan V_3

where Q_b is defined as

$$Q_b = \dot{m}_b h_b \theta_{ch} \tag{4}$$

with $\dot{m}_b$ denotes the average of fuel combustion rate (kg/s), h_b denotes the *lower heating value* of the biomass used (kJ/kg) and θ_{ch} is the charging time (s).

The ratio of discharging-charging time (DCTR) was used as the convenience parameters in thermal storage operation, which is expressed by:

$$DCTR = \frac{\theta_{disc}}{\theta_{ch}} \tag{5}$$

where θ_{disc} is the discharging time (s).

3. Results and discussion

3.1. Physical characteristics of the crushed basalt rock

The physical characteristics of the crushed basalt rock were described in table 2. The length was defined as the maximum length of the piece, while the thickness was measured at the position of half-length. The rock density was 2.63 kg/m^{3}, and their sizes (length and thickness) varied widely. In general, the pieces have oblong shape with average equivalent diameter of 30 mm. By such the distribution of shape and size, the obtained porosity of the rock bed was 0.47.

Table 2. Physical characteristics of the crushed basalt rock

Parameter	Value
Density (kg/m^3)	2.63
Length (mm)	51±11
Maximum thickness (mm)	28±6
Minimum thickness (mm)	19±4
Porosity (-)	0.47

3.2. Heat storage performance

In experiment I and II, the fuel feeding rate at the first 30 minutes were set to be higher than the rest, while in experiment III and IV the feeding rates were almost constant during the charging state. Table 3 summarizes the experimental parameters in both charging and discharging processes.

The change of bed temperature in bottom region during the charging process (not shown in this paper) from each experiment exhibited different patterns. In experiment I, the temperature of the bottom region increased rapidly up to 350°C and fluctuated at 300-350°C. However in experiment II after the bottom temperature reaches 450°C, the temperature slowly decreased as the feeding rate reduced. The experiment III exhibited continuous increase up to 500°C at the end of charging while in the experiment IV the temperature stable at around 500°C after 35 minutes discharging started due to the higher feeding (and then combustion) rate of fuel. However, change of bed temperature in middle and top region almost similar for all experiments. In addition, the temperature less fluctuated due to the flue gas that had passed through the high heat capacity of rock bed.

Discharging process was started when the heat supply from furnace had been stopped. The temperature at the bottom region was instantaneously decreased for all experiments. For the middle region, the change was closely related to the condition of end of the charging process. At the earlier period, experiment I and III showed that the middle region temperature still increased while that of experiment II had decreased with respect to time. In experiment II, due to the lower combustion rate in the end period of charging, the middle region temperature had reached its peak before enter to the discharging process. In the experiments I, II and IV the heat transfer from the bottom region to the middle region were still occurred. In the top region the temperature change pattern was similar to that in the middle region.

The air coming out of the bed was mixed with ambient air entering from the chimney inlet so that the temperature could be set by adjusting the valve V_1 (in experiment I) or V_2 and V_3 (in experiments II, III dan IV). In experiment I, the valve V_1 was adjusted only at the initial period of discharging to set the outlet temperature at 60°C. Since the bed temperature at the top region still increased, the outlet temperature increased as well since the valve was not adjusted anymore. The outlet temperature reached 83°C and subsequently decreased when the bed temperature in the top region had reduced.

Such the high temperature is not suitable for rice drying in batch mode. In addition, the discharging time was short, i.e. only 170 min.

Table 3. Performance of furnace-heat storage system

Stage	Parameter	Experiment			
		I	II	III	IV
Charging	Charging time (min)	72	92	115	60
	Total fuel (kg)	11.5	12.5	15.5	14
Discharging	Discharging time (min)	170	282	360	320
	Discharging final temperature (°C)	42.4	38	40.6	41.1
	Average outlet air temperature (°C)	69.4	51.4	52.8	49.6
	Average ambient temperature (°C) (°C)	36.0	33.7	34.3	33.7
	Volume flow rate or air (m^3/s)	0.263	0.245	0.278	0.293

In experiment II, III and IV, the valves $V_2$and V_3were adjusted so the outlet temperature below 60°C. The air volume flow rate used were 0.245, 0.278 and 0.293 m^3/s, respectively. By adjusting the valves the average temperature were 51.4, 52.8°C and 49.6oC with discharging time reached 282, 360and 320minutes.It is important to note that in this study the discharging time was defined as the time started from the fuel was removed to a condition where the difference between the outlet and the ambient temperature was still higher than 10°C. If the final discharging is defined by using a lower difference of those temperatures, then the discharging time can be sufficiently higher.

3.3. *Energy obtained during discharging process*

Table 3 shows the energy obtained during discharging process, discharging efficiency and DCTR. The discharging energy of experiment I, II, III and IV were 89.6, 73.4, 111.1 and 89.5 MJ with discharging efficiencies were 45.8, 34.5, 42.2 dan 37.6%, respectively. Such the low discharging efficiencies were contributed by combustion process and heat loss from the system. The heat loss from the system involved heat loss through walls during the charging and the discharging process as well as convectively through exhaust gas. The heat loss through exhaust gas was not sufficiently high, however, due to its temperature was kept below 60 or 70°C during the charging process. The highest heat loss was probably the heat penetrating the furnace door since it was made of an uninsulated iron plate.

The discharging efficiency could be influenced by discharging rate and charging method. In experiment I, discharging process was performed in relatively short time, therefore, the heat stored in the rock can be used more efficiently than the others. However, the experiment III, due to its final temperature in the charging period was higher, had higher efficiency than the experiment II. The slightly lower discharging efficiency in Experiment IV might be caused by higher energy loss from the furnace due to its highest combustion rate.

The high rate of fuel combustion in Experiment IV provided sufficient heat charging process in a relatively short charging time, i.e., 60 minutes. By discharging time similar to that of Experiment II and III, the DCTR of this experiment reached 5.33 while its efficiency was still higher than that of Experiment II. This shows that at a higher combustion rate, charging process was able to be performed effectively. Charging process was able to be carried out sufficiently fast since the bed pores providing high heat transfer rate. The higher combustion rate is still able to be performed to provide higher DCTR, despite a possible reduction of discharging efficiency to occur.

Table 4. Discharging energy and DCTR

Parameters	Experiment			
	I	II	III	IV
Discharging energy (MJ)	89.6	73.4	111.1	89.5
Energy stored in the fuel (MJ)	187	204	263.5	238
η_{disc}	45.8%	34.5%	42.2%	37.6%
DCTR	2.36	3.07	3.13	5.33

3.4. *Drying simulation*

Experiments II-IV yielded a suitable drying air temperature for rice drying. The determination of the amount of rice able to be dried by those conditions was performed by deep bed rice drying simulation which is developed by Nelwan, et al. 2013 [8]. The simulation results showed that the air having conditions resulted from experiments II, III and IV could be used to dry 156.8, 268.8 and 201.6 kg of rice, respectively, from initial moisture content of 23% w.b. to final moisture content of 14% w.b. By the definition discharging time mentioned before, the final discharging temperature was still in range of 38-41.1°C. Such the condition of air is still to be very potential to be used for drying process, for example the drying air could be used until it decreases up to 35°C, so that the drying load can still be increased.

4. Conclusions

A rock bed heat storage system with heat source from biomass combustion for heating of drying air has been developed in this study. The basalt rock piece was used and they had oblong shape with equivalent diameter of 30 mm and porosity of 0.47. By using 735 kg of rock bed, the charging process could be accomplished within 1-2 hours. Without adjusting the air flow rate in discharging process, from an initial temperature of 60°C, the air drying increased continuously up to 83°C which is not suitable applied for rice drying. By using sufficiently high combustion rate of 14 kg/h in 1-hour charging, the obtained DCTR was 5.33 at outlet and ambient air temperatures of 49°C and 33°C, respectively. However, the high combustion rate could reduce the discharging efficiency. The efficiency of heat discharging for all experiments were ranged from 34.5 to 45.8%. From the simulation, as much as 156.8-268.8 kg of rice was able to be dried by the discharging conditions. The performance improvement could still be performed by lowering the limit of temperature to be used for drying.

Acknowledgement

This research is funded by PTUPT, year 2017. The authors also thank Rustam, Jefrianus and Iqbal who assisted in data collection.

References

[1] Jain D 2007 Modeling the performance of the reversed absorber with packed bed thermal storage natural convection solar crop dryer *Journal of Food Engineering* **78** (2007) 637–647

[2] S Aboul-Enein A A El-Sebaii, M R I Ramadan and H G El-Gohary 2000 Parametric study of a solar air heater with and without thermal storage for solar drying applications *Renewable Energy* **21** (2000): 505-522

IOP Conf. Series: Earth and Environmental Science **147** (2018) 012027 doi:10.1088/1755-1315/147/1/012027

[3] Tambunan A H, L P Manalu and Kamaruddin A 2010 Exergy Analysis on simultaneous Charging and Discharging of Solar Thermal Storage for Drying Application *Drying Technology: An International Journal* **28**(9): 1107-12
[4] Zanganeh G, A Pedretti S. Zavattoni M. Barbato A and Steinfeld 2012 *Solar Energy* **86**: 3084–98
[5] Allen K G, T W von Backström and D G Kröger 2014 Packed rock bed thermal storage in power plants: design considerations. SolarPACES 2013 *Energy Procedia* **49** 666 – 675
[6] Allen K G, T W von Backström, E Joubert and P Gauché 2016 *Rock bed thermal storage: Concepts and costs* AIP Conference Proceedings 1734
[7] Cascetta M, Giorgio Cau, Pierpaolo Puddu and Fabio Serra 2014 Numerical investigation of a packed bed thermal energy storage system with different heat transfer fluids *Energy Procedia* **45**:598-607
[8] Nelwan L O, I D M Subrata, D Wulandani, L T Mulyantara and M J Djafar 2013 *The Development of Energy Stand-alone Rice Drying System by Using Absorption Heat Pump and Controlled Operation Fueled by Rice Husk Gasification* KKP3N Report Ministry of Agriculture (in Indonesian).

IOP Conf. Series: Earth and Environmental Science **147** (2018) 012028 doi:10.1088/1755-1315/147/1/012028

Characteristics of Microactive Carbon from Bamboo Var. Petung as Adsorbent

I P S Wirawan[1], Sutrisno[2], K B Seminar[2] and L O Nelwan[2]

[1] Agricultural Engineering Division, Faculty of Agricultural Technology, Udayana University, Bali, Indonesia
[2] Departement of Mechanical and Biosystem Engineering, Bogor Agricultural University, Bogor, Indonesia

E-mail: putusurya2014@gmail.com

Abstract. Bamboo has unique characteristics, such as in the carbonization process at a temperature of 500^{0}C, the carbon characteristics is homogeneous. The characteristics of bamboo have great potential as a future bio-energy resource. Apart from being a bio-energy source of bamboo can also be used as an adsorbent material in the form of activated carbon. Activated carbon is the most inexpensive and easy to produce adsorbent material. One of the activated carbons of bamboo materials used is the micro-active carbon from bamboo. Microactivated carbon bamboo has a pore structure which is good for adsorption because of its surface area being much better than the other adsorbent, mainly on mesopore and micropore pore size. The purpose of this research is to make micro-activated carbon adsorbent bamboo var. petungand to analyze their characteristics. The characteristic of microactivecarbon was analyzed by SEM EDS and Iod number. The result showed a variation in pore size from 1μm to 11.157μm. The surface area of micro-active carbon of 200 mesh and 80 mesh is 1954.95 m^2g^{-1} and 1516.34 m^2g^{-1}.

1. Introduction

Charcoal and activated carbon made from bamboo is a very promising product. Bamboo charcoal and bamboo activated carbon is a good product for moisture control because it has a specific surface area and large pore volume. Bamboo is composed of cellulose and lignin. Cellulose in petung bamboo has a homogeneous fiber and its lignin is a pore matrix. The characteristics of this bamboo have great potential as a future bio-energy resource in China [1]. The carbon characteristics of petung bamboo is homogeneous and the highest calorific value produced on bamboo petung by using argon gas at carbonation process with a temperature of 500°C are 10924 $Calg^{-1}$[2].

Activated charcoal is an amorphous carbon material having a very large surface area of 200 to 2000 m^2 g^{-1}. Activated charcoal has better absorption and surface area than other adsorbents. The pore structure of the active charcoal causes the ability to absorb water [3]. The activated charcoal consists of two types: the activated charcoal of the liquid phase and the gas phase. The liquid phase activated charcoal is usually a very fine powder, whereas gas phase one is usually granular or hard pelleted. The main characteristic that distinguishes these two types of adsorbents is the distribution and size of the pores. The liquid phase carbon has a surface area almost equal to the gas phase but the liquid phase carbon has larger number of pores in the transition area (mesopore), with the pore volume between 0.2-10 mL g^{-1}while in the gas phase is 0.15-0.5 mL g^{-1} (micropore) [4]. The materials typically used commercially for liquid phase carbon are charcoal, sawdust, lignite and fly ash. The active bamboo charcoal pore structure looks like figure 1[5].

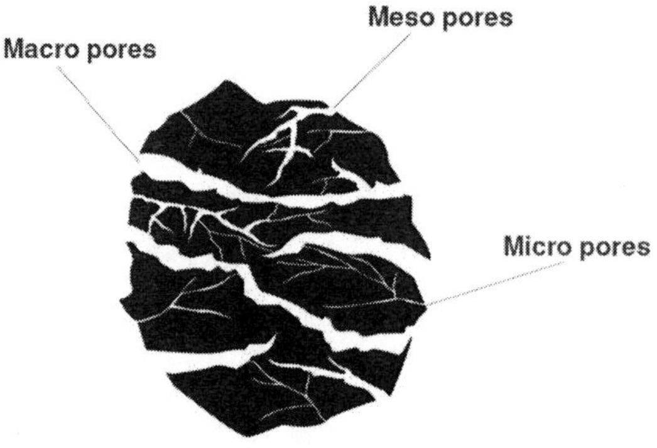

Figure 1. Activecharcoal pore structure

Most of the pores in activated charcoal in the form of pellet or granular pore structure are macro pore while they in the form of powder structure are usually mesopore and micropore. The activated charcoal pore structure of macropore is not well used in alcohol purification because of its small surface area and relatively short contact time. The process of carbonation and activation greatly affects the quality of activated carbon produced [4]. The purpose of this research is to study the making and characteristic of activated charcoal bamboo charcoal with carbonation and activation temperatures of 300°C and 600°C, respectively.

2. Materials and Methods

2.1. Materials
The material used in this study was bamboo petung.

2.2. Preparation of charcoal
The bamboo was cut into 1-cm cubes, placed on a porcelain plate and then put into the furnace with temperature of 300°C for 3 hours. Subsequently, the charcoal produced was washed by aquades and then dried in drying oven at temperature of 110°C for 1 hour.

2.3. Activation of charcoal
Activated charcoal making consists of three stages of process, namely pretreatment, carbonation and activation [6]. The purpose of activation is to enlarge the size of the active charcoal pore and the wider surface area so that the absorption improves. The activation process was performed at an atmospheric fixed bed reactor. The reactor is equipped with N_2 gas inlet and heater. The reactor schematic is shown in figure 2.

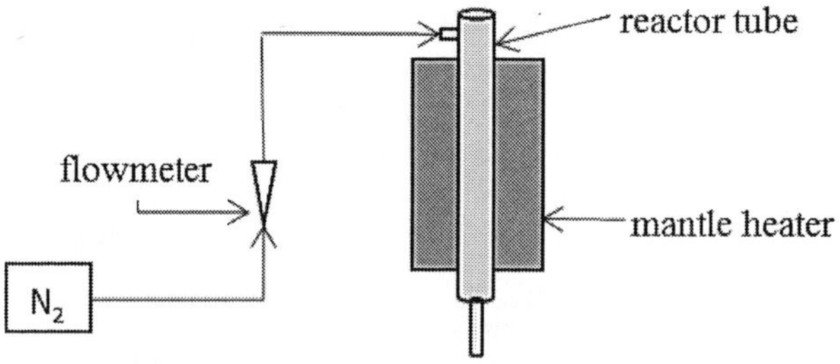

Figure 2. Atmospheric fixed bed reactor

2.4. Scanning electron microscopy (SEM)

Structure and area of pore diameter of adsorbent using scanning electron microscope (JEOL, Ltd, Japan, Model JSM 6510LA). The active charcoal sample will be scanned in the form of active charcoal powder. Sample is inserted into vacuum tube with voltage 20kV with distance of 10 mm. For SEM that has a vacuum condition, the resulting image contrast depends on the topography; by varying the orientation of the specimen to the signal detector a dark light effect can be produced to form a three-dimensional image.

2.5. Iod-Number Test

The surface area is measured by iodometric titration method. The principle of this titration is the number of milligrams of iodine adsorbed by one gram of activated charcoal. The absorption of activated charcoal against iod indicates the ability of activated charcoal to adsorb low molecular weight components [7]. The resulting iod number is a representation of the surface area of the activated charcoal. The correlation of the linear regression equation refers to ASTMD-4607-94[8] with the following equation:

$$\text{activated charcoal surface area} = \frac{\text{Iod number} - 174{,}34}{0{,}6366} \tag{1}$$

The process of the carbonization and activation process to the analysis phase is shown in figure 3.

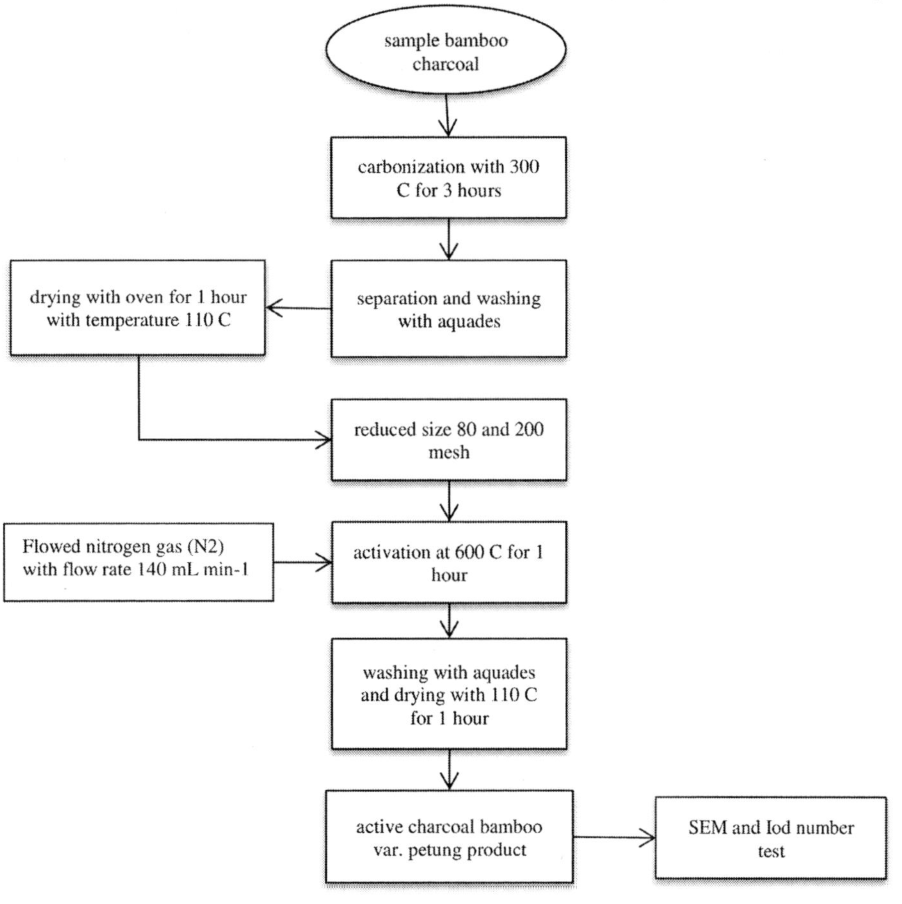

Figure 3. Carbonization and activation charcoal bamboo process

3. Results and Discussion

3.1. Carbonization and activation

The process of carbonizing bamboo samples was performed at 300°C for 3 hours. Bamboo charcoal product averages 30% of the sample's initial weight. This mass loss occurs due to several factors such as moisture content and ash content of the samples used in the process of carbonization in the furnace. The mass loss increases with increasing carbonization temperature [4]. The carbonized bamboo is then washed with purified water and then dried in 110°C oven for 1 hour. Subsequently, they are crushed and sieved manually by using 80 and 200 mesh sievers. A sample of 10 g of each charcoal powder was put into the reactor. Then, for activation process the procedure described in table 1 was performed.

Table 1. Action and observation during activation

Temperature (°C)	Time (minute)	Action	Observation
30	5	N_2 started to flow into the reactor with a rate of 140mL min^{-1}	-
30-200	10	Heater was turned on with a rate of 10°C min^{-1}	water vapor started to come out marked by the exit of smoke
200-400	20	Heating rate of 10°C min^{-1}	smoke came out increased
400-600	20	Heating rate of 10°C min^{-1}	more smoke and stopped at temperature of 600°C
600	60	Keeping the temperature at 600°C for 1 hour	No smoke
600-200	30	Turning off the heater and constantly draining nitrogen gas	Temperature decreased No smoke
200-30	30	Turning off the reactor and discharging sample	-

Table 2 shows the mass reduction due to activation process. The mass reduction was related to the volatile substances which was still exist despite experienced the carbonization process. The washing process after activation process was suspected to influence the mass reduction, since the dirt occupying the pores was removed in this process.

Table 2 Result of activation process

Size(mesh)	Mass before activation(g)	Mass after activation(g)	Mass reduction (%)
80	10.0	7.5	25
200	10.0	4.0	60

3.2. Surface area

The surface area of activated charcoal as an adsorbent is very important. Activated charcoal is said to be a good adsorbent when it has a large surface area. The surface area is measured using the iodine number method. The calculation result obtained by the amount of iod from the active charcoal size 80 and 200 mesh was 1139.64 mgg^{-1} and 1418.86 mg g^{-1}. The surface area of activated charcoal is calculated by referring to the linear regression equation in ASTM D-4607-94[8]. The surface area of each sample size is 1516.34 m^2 g^{-1} and 1954.95 m^2 g^{-1}[9,10]. Activated carbon that has high absorbency to iodine means having a large surface area and a larger micropore and mesopore structure [11]. The carbon content of the sample used is quite high, between 86.1 to 95.4% with diameters between 1μm to 11 μm. The pore structure image and carbon element content by SEM test are presented in figure 4.

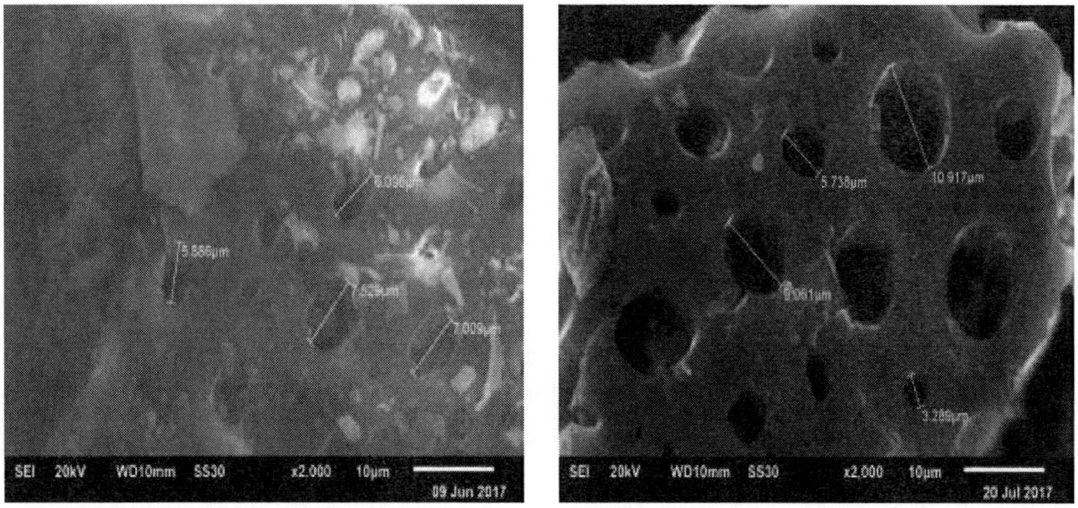

(a) (b)

Figure 4. Activated charcoal pore structure of: (a) 200 mesh, (b) 80 mesh

The sample obtained was mostly rich in carbon followed by oxygen, which indicates that pyrolysis enriches the carbon content by eliminating non-carbon species [10].

4. Conclusions

This study demonstrated that bamboo var. petung may be used to prepare micro-active carbon via pyrolysis and activation. Optimum conditions for preparing micro-active carbon from char include activation temperature of 600° Cactivation time of 60 min. This conditioning can produce a high carbon micro-active surface area of 1954.95 m^2g^{-1}. Subsequent activation at 600°C for 60 min is recommended.

References

[1] Liu Z and Jiang Z 2013 Effects of carbonization conditions on properties of bamboo pellets *Journal Renewable Energy* **51** 1-6

[2] Salihati F G and Ardhyananta H 2013 Study of Making Black Carbon from Bamboo var. Ori (Bambusa Arundinacea) and Bamboo var.Petung (Dendrocalamus asper) *Journal of Pomit Technique* 1(2)

[3] Sudibandriyo M 2003 *Generalized Ono-Kondo Lattice Model for High Pressure on Carbon Adsorben* (Oklahoma: Oklahoma State University)

[4] Rijali A, Usman M and Zuikarnain 2015 Making and Characteristics of Activated Carbon from Bamboo Var. Petung with Activation using H_2O Activating Agent **2**(1)

[5] Strand G 2001 Actived Carbon For Purification Of Alcohol. **202** (12):1-28

[6] Marsh H, Rodriguez and Reinoso F 2006 *Activated Carbon* (Amsterdam: Elsevier)

[7] Novitasari D and Kusumaningrum D 2012 Purification of Bioethanol Using Adsorption Process and Adsorption Distillation Using Zeolite Adsorbent *Journal of Chemistry and Industry* **1**(1): 11-19.

[8] ASTM D,4607-94. 2011 *Standard Test Method for Determination of Iodine Number of Activated Carbon* (West Conshohocken: ASTM International)

[9] Ip A W M, Barford J P and Mc kay G 2008 Production and Comparison of High Surface Area Bamboo Derived Active Carbons *Journal Bioresource Technology* 8909-16

[10] Zhang Y J, Xing Z J, Duan Z K, Li M and Wang Y 2014 Effects of Steam Activation on The Pore Structure and Surface Chemistry of Activated Carbon Derived from Bamboo Waste *Journal Applied Surface Science* :279-286

[11] Jankowska H, Swatkowski A and Choma J 1991 *Active Carbon* (New York: Ellis Horword)

CFD-based Analysis of Non-Premixed Combustion Model in Biomass Grate Furnaces

M Hafiz[1], L O Nelwan[1] and M Yulianto[1]

[1] Department of Mechanical and Biosystem Engineering, Bogor Agricultural University, Indonesia

E-mail: mohamadhafiz.eng@gmail.com

Abstract. Biomass grate furnace is widely used as heat source for various uses including grain drying. In this study, a CFD simulation using Fluent 18.0 academic was performed on a biomass ladder grate furnace, which can be used later to improve the design as well as the operation technique of the furnace. A downscaled overfeed type furnace with size of 15 x 30 x 50 cm was built to validate the model. The turbulence model used in this study was k-epsilon while the combustion model of non-premixed combustion was used. The simulation was performed with the biomass feed rate of 4 kg/h and air flow velocity of 7.5 m/s at 3.81 cm inlet diameter. The simulation result at outlet temperature was 673 ^{0}C and inside temperature were 775 and 717 ^{0}C, while the composition of gases was 0.18 for CO, 0.2 for CO2, 0.001 for CH4, 0.09 for H2O, 0.51 for N2 and 0.029 for other gases. Test results from a biomass ladder grate furnace were used to validate the model and the results are 646 ^{0}C for the outlet temperature, 712 and 582 ^{0}C for inside temperature. Comparison between simulation and measurement results shows good value with average percentage of deviation 12.12%.

1. Introduction

Biomass grate furnace is one of simple technologies for heating that is frequently used by small scale industries. One way to develop a detailed design of furnace with is by using CFD simulation. Furthermore the simulation can increase the understanding of the processes occurring in the biomass furnace systems. The virtual view inside the furnace delivers a comprehensive overview of the system behavior and can help to solve combustion and operation problems. [1]. Wang and Yan [2] summarizes in an article about researches in CFD for biomass thermochemical conversion and conclude the CFD can predict accurately biomass conversion from thermochemical aspect. CFD can analyze distribution of product, flow, temperature, ash, until emission such as NO_x. even though it still need model approaching and assumption and resulting a little error.

Some authors have proposed efficient furnace with various biomass fuels. Rene et al [3] developed furnace for boiler with capacity 176 kW with fuel from corn and wheat straw in round bales shape with feeder type batch. Gomez et al [4] developed mathematic model for wood boiler stove with underfeeding system and Nesiadis et al. [5] optimized small scale boiler with two combustion chamber using logwood as fuel. Combustion simulation by fossil fuels has widely used with non-premixed or mixture fraction approaches but it rarely used in biomass combustion appliances so far. Prieler et al. [6] and Mayr et al. [7] showed the high accuracy of the non-premixed in case of methane-air and oxy-fuel combustion. They predicted temperatures, heat flux and fractions of species of combustion in a very good way. Albrecht et al. [8] and Venturini et al. [9] presented biomass combustion in grate furnaces with mixture fraction models. Both used the GRI 3.0 methane combustion mechanism for chemistry calculation. Without tars represented in the studies. The mixture fraction approach unites the benefits of the classical species transport models. The advantage of this modelling approach is that complex chemical reaction mechanisms can be implemented at low computational effort.

The main objectives of the present work were to evaluate furnace performance with CFD by non-premixed combustion model with turbulence model and make scenarioes for new design to optimize furnace performance.

2. Model description

The present paper using non-premixed combustion as species model with non-adiabatic energy treatment. Fuel and oxidizer enter combustion chamber as separated streams. Mixing and burning take place simultaneously. The reaction chemistry in the Probability Density Function (PDF) method for solving turbulent chemistry interaction is determined using mixture fraction model. The equilibrium model is used and it assumes that the chemistry is rapid enough for chemical equilibrium to always exist at the molecular level [10]. Basing on the simplifying assumptions, the instantaneous thermochemical state of the fluid is related to the mixture fraction f. Species mole fraction is computed based using minimization of Gibbs free energy concept. The mixture fraction f is defined in terms of the atomic mass fraction as Equation 1.

$$f = \frac{m_i - m_{i,ox}}{m_{i,fuel} - m_{i,ox}} \tag{1}$$

Where m_i is the mass fraction for element i. The subscript ox denotes the value at the oxidizer stream inlet and fuel the value at fuel stream inlet. Knowing f we can also calculate mass fraction of fuel and oxidizer at any place x,y,z.

Radiation model P1 used in this research because the model is quite steady to predict temperature and gas composition [11]. For turbulence model, three k-epsilon models were used. The three model were k-ε standard, k-ε renormalization group (RNG) and k-ε realizable. The major differences in the models are as follows, the method of calculating turbulent viscosity, the turbulent Prandtl numbers governing the turbulent diffusion of k and ε and the generation and destruction terms in the ε equation. [10]. The k-epsilon model for k-ε standard, k-ε renormalization group (RNG) and k-ε realizable are expressed in equation 2, 3, and 4.

$$\frac{\partial}{\partial x_i}\rho u_i \varepsilon = \frac{\partial}{\partial x_i}\left[\left(\mu + \frac{\mu_t}{\sigma_\varepsilon}\right)\frac{\partial \varepsilon}{\partial x_i}\right] + C_{1\varepsilon} G_k \frac{\varepsilon}{k} - C_{2\varepsilon} G_k \rho \frac{\varepsilon^2}{k} \tag{2}$$

$$\frac{\partial}{\partial x_i}\rho u_i \varepsilon = \frac{\partial}{\partial x_i}\left[\alpha_\varepsilon \mu_{eff} \frac{\partial \varepsilon}{\partial x_i}\right] + C_{1\varepsilon} G_k \frac{\varepsilon}{k} - C_{2\varepsilon} G_k \rho \frac{\varepsilon^2}{k} \tag{3}$$

$$\frac{\partial}{\partial x_i}\rho u_i \varepsilon = \frac{\partial}{\partial x_i}\left[\left(\mu + \frac{\mu_t}{\sigma_\varepsilon}\right)\frac{\partial \varepsilon}{\partial x_i}\right] + \rho C_1 S\varepsilon - \rho C_2 \frac{\varepsilon^2}{k+\sqrt{v\varepsilon}} + C_{1\varepsilon}\frac{\varepsilon}{k} - C_{3\varepsilon} \tag{4}$$

In the derivation of the k-ε model, k-ε standard assumption is that the flow is fully turbulent, and the effects of molecular viscosity are negligible. The standard k-ε model is therefore valid only for fully turbulent flows. While the standard k-ε model is a high-Reynolds number model, the RNG theory provides an analytically-derived differential formula for effective viscosity that accounts for low-Reynolds number effects. K-ε realizable is a modified transport equation for the dissipation rate, ε, has been derived from an exact equation for the transport of the mean-square vorticity fluctuation [12], [13].

Solution control method was used to choose interpolation scheme in mesh nodes. The pressure-velocity coupling scheme used is the SIMPLE scheme. In this scheme selected Least Squares Cell Based for gradient, Pressure standard and First-Order Upwind Scheme for Momentum, Turbulent Kinetic Energy, Turbulent Dissipation Rate, Energy, Mean Mixture Fraction and Mixture Fraction Variance. Where the First-Order Upwind Scheme is the lightest and fastest converging interpolation scheme.

IOP Conf. Series: Earth and Environmental Science **147** (2018) 012029 doi:10.1088/1755-1315/147/1/012029

3. Methodology

Material used in the research was chopped corncob as fuel, furnace build from concrete (brick, cement and sand) and steel for feeder, ladder and outlet. Equipment for record temperature data was Thermocouple Type-K and recorder, and air flow data used anemometer. The main research was simulation for the furnace with non-premixed model with steady state condition until convergent, valid with test performance and then make scenario for better furnace.

3.1. Furnace configuration

The size of the furnace was 30 cm (l) x 15 cm (w) x 50 cm (h) with diameter of screw feeder 7.62 cm, inlet air diameter 3.81 cm and outlet diameter 9 cm. Wall of the combustion and air chamber made by concrete and the combustion ladder, screw feeder, outlet and inlet cylinder made by steel. The schematic diagram of the performance test is presented in figure 1. Chopped corncob used in the test have average diameter and length 0.04 m and 3 cm. The combustion begins by burning corncob on the combustion ladder until the flame steady, and then the corncob insert continuously by the feeder with capacity 4 kg/h. The blower used to flow the fresh air with velocity ± 7.5 m/s or equal to mass flow ± 36.73 kg/h. Thermocouple position for outlet was in the middle of outlet cylinder, about 10 cm (x-axis) and 4.5 cm (y-axis), for combustion chamber the position is 10 cm and 20 cm (x-axis) from outlet wall with 10 cm (y-axis).

3.2. CFD Modelling Configuration

The CFD software used for simulation was ANSYS Fluent R18.0 Student version. Mesh 3D model was done with mesh program from ANSYS itself with curvature advance size function, medium relevance center, medium smoothing, fast transition and medium span angle center. Nodes for the simulation were 33697 nodes with 153988 elements. With boundary condition velocity inlet for inlet air, pressure outlet for gas outlet, mass flow inlet for fuel inlet and wall for furnace wall. Furnace model used in this simulation as same as the furnace for test performance. The position of the fuel inlet is different from performance test like at figure 2, this is because the combustion happened in ladder of the furnace, not in the feeder.

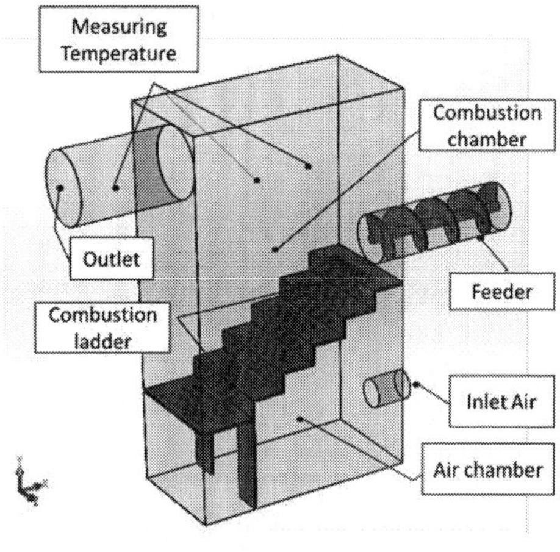

Figure 1. Schematic view of the performance test used in this study

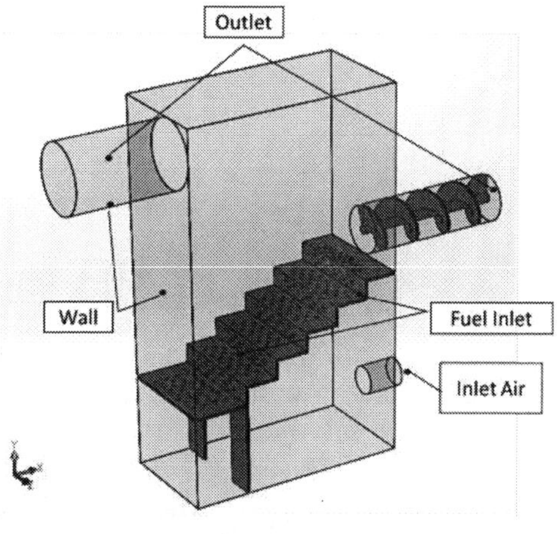

Figure 2. Schematic view of the computational study used in this study

Fuel used during the experiments was corncob. Table 1 presents properties data (proximate and ultimate) of the fuel used; the value was input in coal calculator facility. Wall used in the simulation are concrete and steel. The density of the concrete and steel used are 1300 kg/m^3 and 8030 kg/m^3 with specific heat of 750 J/kg.K and 502.48 J/kg.K and thermal conductivity of 1.7 W/m.K and 16.27 W/m.K.

IOP Conf. Series: Earth and Environmental Science **147** (2018) 012029 doi:10.1088/1755-1315/147/1/012029

Table 1. Corncob composition and properties [14]

Parameter	Value	Unit
Proximate analysis		
Moisture	7.96	w.b.wt%
Ash	1.49	d.b.wt%
Volatiles	85.08	d.b.wt%
Fixed Carbon	13.43	d.b.wt%
Ultimate analysis		
Carbon (C)	48.1	d.b.wt%
Hydrogen (H)	6.02	d.b.wt%
Nitrogen (N)	0.36	d.b.wt%
Oxygen (O)	43.96	d.b.wt%
Sulphur (S)	0.05	d.b.wt%
Calorific Value	18.4	MJ/kg

Thermal conditions of the wall boundary are convection. Heat transfer coefficient for concrete was 15.7 W/m^2.K, free stream temperature as same as environment temperature 31 ^{o} C, internal emissivity 1 and wall thickness 0.12 m, Heat transfer coefficient for steel 10.46 W/m^2.K, and wall thickness 0.003 m.

3.3. Model Validation

Validation was conducted by comparing temperature of the experimental and simulation data. Accuracy of experimental and simulation result expressed as average percentage of deviation. Value of average percentage of deviation defined as Equation 5.

$$Average\ percentage\ of\ deviation = \frac{|T_{simulation} - T_{experimental}|}{T_{experimental}} \times 100\% \quad (5)$$

4. Results and discussion

The combustion chamber temperature of the test can be seen at figure 3. The test was conducted for 1 hour and 30 minutes, the temperature of the furnace hard to steady because of the problem of the furnace design, where the flame come out from the feeder so the biomass in feeder is combusted and stuck in the middle of the screw feeder. Steady flame was observed in the last 20 minutes, with average outlet temperature of 646 ^{o} C, the middle furnace temperature was 581 ^{o} C and 711 ^{o} C. average inlet temperature of air flow was 39 ^{o} C with environment temperature of 31 ^{o} C. The inlet air temperature was higher than ambient temperature because the position near feeder with lack of flame so the heat is sucked by the blower.

Figure 4 until figure 7 shows contours and vectors of the velocity. Figure 4 show primary air injection vectors that pass through the chamber, figure 5 and figure 6 show the vector of velocity in outlet furnace and inside the screw feeder. In the screw feeder the air flow entered to the feeding biomass area.

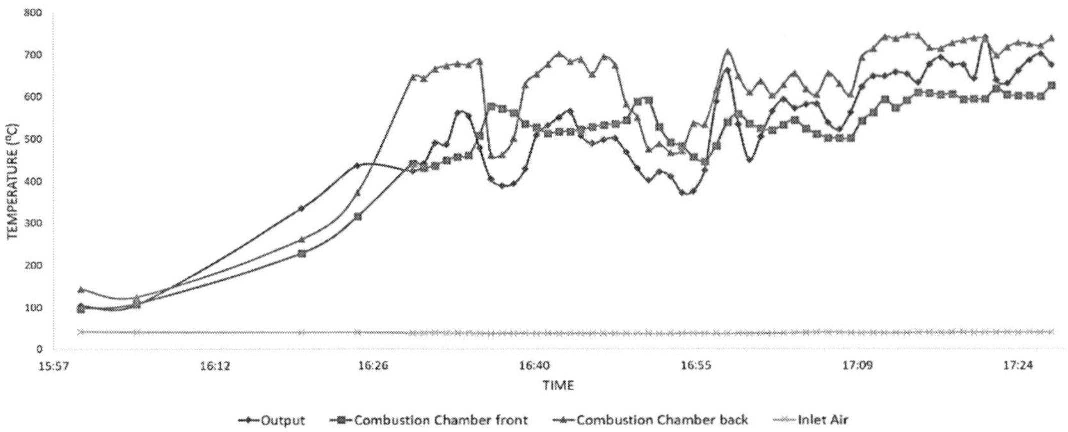

Figure 3. Temperature of furnace test performance

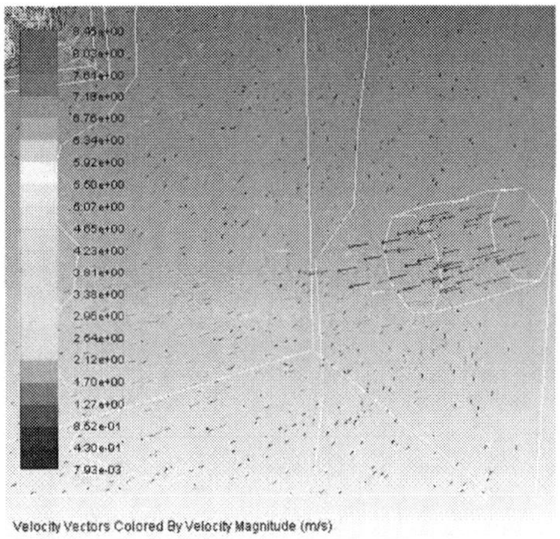

Figure 4. Velocity vectors of primary air

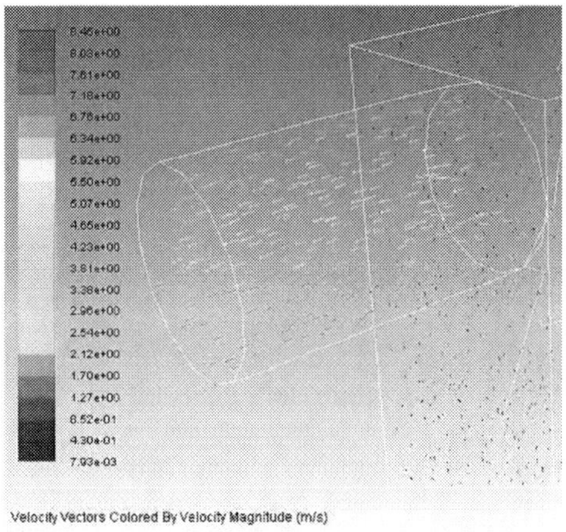

Figure 5. Velocity vectors of primary outlet

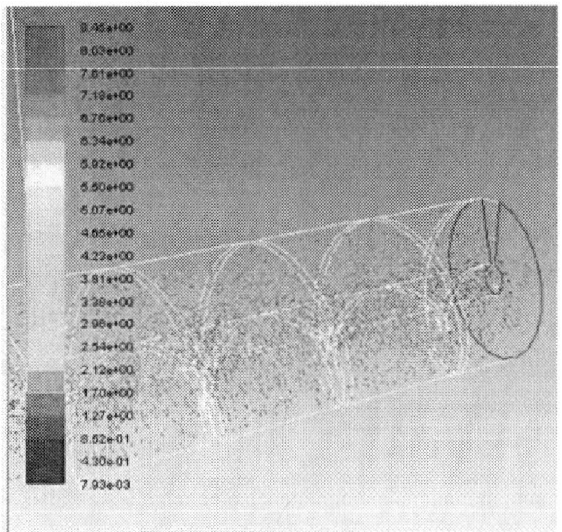

Figure 6. Velocity vectors of feeder

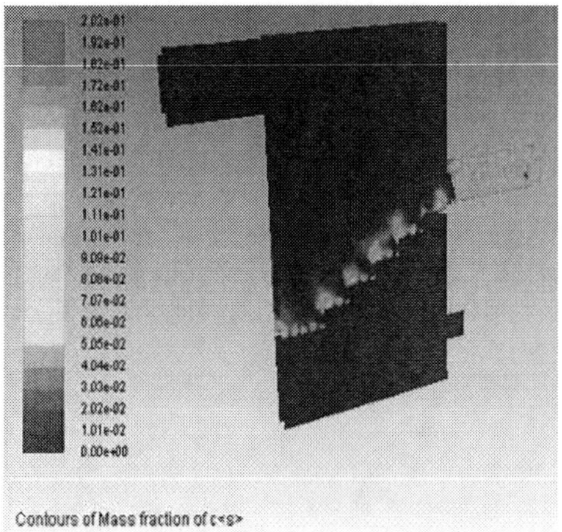

Figure 7. Contours of char in the furnace

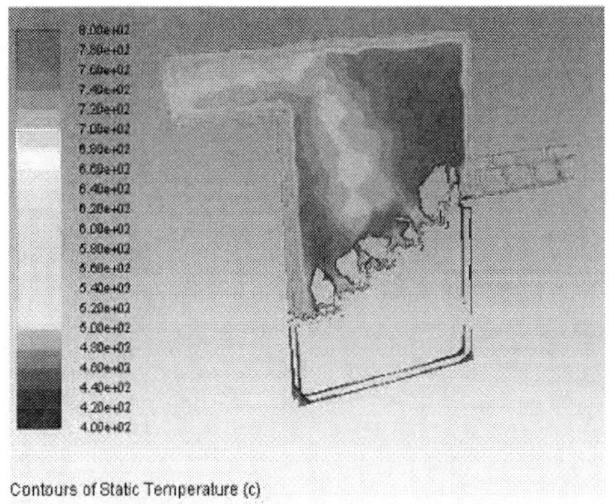

Figure 8. Detail contours of temperature in combustion chamber

In the simulation, the biomass was assumed to come out through the ladder. The biomass was injected from the floor by discrete phase model. Figure 7 shows carbon solid (c<s>) or char of the biomass in the bed region. The char was fly in the furnace and not burned out all. Some of char was come through the feeder. Figure 8 shows contour of the temperature in which the flame rising from the ladder

Table 2. Deviation and accuracy between simulation and experimental

Turbulent model	Position	Simulation (°C)	Experimental (°C)	Deviation (%)	Accuracy (%)	Average Accuracy (%)
	Outlet	642	646	0.61	99.39	
k-ε Standar	Back	747	712	5.02	94.98	92.25
	Front	684	582	17.62	82.38	
	Outlet	746	646	15.38	84.62	
k-ε RNG	Back	982	712	37.99	62.01	72.27
	Front	755	582	29.82	70.18	
	Outlet	674	646	4.35	95.65	
k-ε Realizable	Back	783	712	10.07	89.93	87.26
	Front	720	582	23.79	76.21	

Table 2 shows the comparison of experimental result vs model. These data represent the time averaged values once steady state had been achieved. The experimental values were collected for approximately 20 minutes from 17:07 – 17:27 at experimental time. From the three models of turbulence used it can be seen the accuracy of the outlet and rear of the furnace is very good for k-ε standard and k-ε realizable, above 90%, but for k-ε RNG only 84% at the output and 62% in part back. The front for the standard k-ε is 82%, 70% for k-ε RNG and 76% for k-ε realizable. To illustrate the visualization of the temperature ratio obtained between the test and the three models of turbulence can be seen in figure 12. From the data it can be seen that the k-ε RNG turbulence model is not suitable for this furnace case, since this model is more suitable in the case of turbulence with the number Reynold is low, and the best turbulence model for combustion is a standard k-ε with an average accuracy of 92.25%, of that value then the simulation is acceptable, so the boundary conditions used in this model will be used in the next furnace design.

IOP Conf. Series: Earth and Environmental Science **147** (2018) 012029 doi:10.1088/1755-1315/147/1/012029

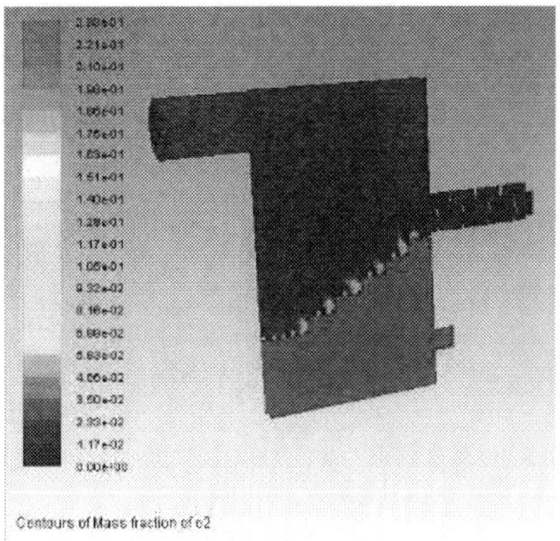

Figure 9. Contours of gas O_2

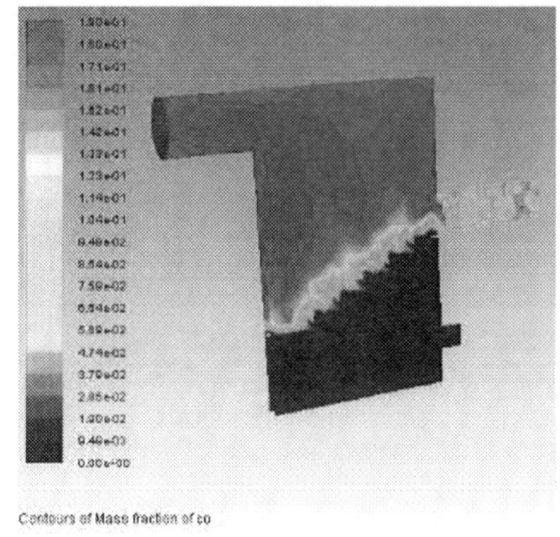

Figure 10. Contours of gas CO

Gas contour in the simulation was varied, with domination of N_2 for 51%, the O_2 had zero value after pass through the ladder, shown in figure 9 and composition of other was 18% for CO, 20% for CO_2, 0.1% for CH_4, 8% for H_2O, and 2.9% for other gases. Gas CO_2 and CO emission in the simulation quite high, especially for gas CO shows in figure 10, it still not confirmed for the validation in the test performance but it shown the gas still not combusted completely.

There was one issue in the furnace that can disturb the combustion stability of the fuel in the combustion chamber there was the flame come out from the feeder. When the flame came out from feeder, the fuel pyrolysis in feeder and made the corncob sticky and stuck in pitch of feeder. A new design of the furnace is a must and in this study, one alternative design of the furnace is evaluated. The alternative design made change the position of the feeder to upper side, add one secondary air inlet and insert partition between the combustion chamber and feeding chamber. The combustion ladder is smaller in the 2nd furnace with just four ladders. Figure 12 shows flow of air in the furnace, the primary inlet air same was 7.4 m/s and secondary air 1.5 m/s, average outlet flow of the furnace was 6.2 m/s and flow air in the feeder 1.2 m/s.

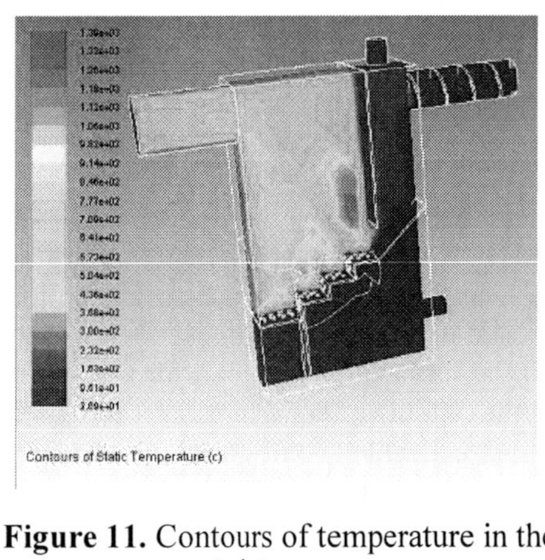

Figure 11. Contours of temperature in the 2nd furnace

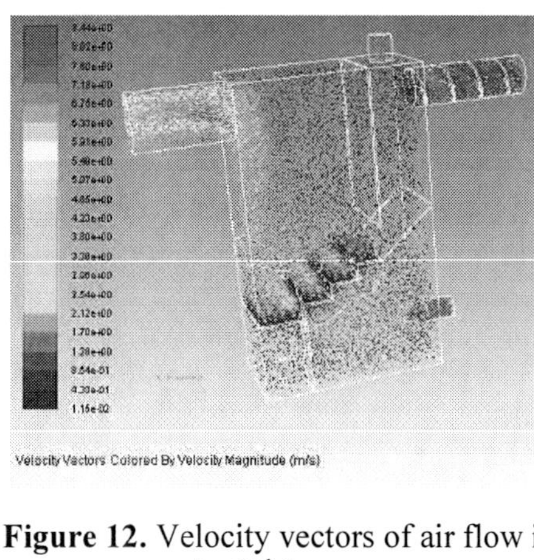

Figure 12. Velocity vectors of air flow in the 2nd furnace

IOP Conf. Series: Earth and Environmental Science **147** (2018) 012029 doi:10.1088/1755-1315/147/1/012029

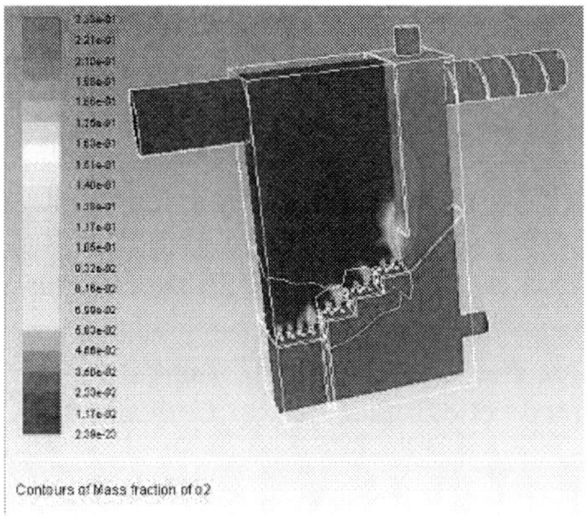

Figure 13. Contours of O_2 in the 2nd furnace

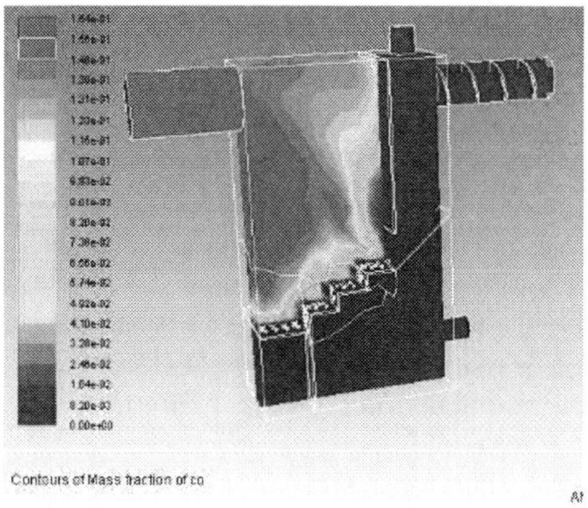

Figure 14. Contours of CO in the 2nd furnace

New design of the furnace solved issue of the furnace and help the biomass did not pyrolysis or burned in the feeder, figure 11 show the temperature in the feeding chamber just around 95 ^{o} C and temperature in the feeder still around 75 ^{o} C. the temperature enough to prevent the corncob as fuel to be pyrolysis or burned. Average outlet temperature of the new design was 700 ^{o} C, with zero concentration of O_2 in the combustion chamber (figure 13) and concentration of gas CO (figure 14) quite high but smaller than original design with composition 0.15.

Based on comparison between original design (1st design) and improved design (2nd design) we can see different value of temperature produced both, the outlet temperature produced by 2nd design higher ± 25 ^{o} C with higher gas outlet rate due to additional air, the sensible enthalpy value for both fuels is almost similar and the sensible gas outlet enthalpy is higher in the 2nd design because the outlet temperature is higher so that the resulting heat is higher 2192 J/s or about 21%. Detailed comparison between the two furnaces can be seen in table 3.

Table 3. Comparison of sensible enthalpy and heat of furnace simulation

Furnace design	1st design	2nd design
Biomass mass flow (kg s-1)	0.00113	0.00113
Biomass sensible enthalpy (J kg-1)	102494	102500
Gas flow (m s-1)	5.2	6.2
gas outlet mass flow (kg s-1)	0.0099	0.0118
gas sensible enthalpy (J kg-1)	824491	876848
Heat of furnace (J s-1)	8063	10254

5. Conclusions

This research shown a good CFD simulation of combustion with non-premixed combustion model for corncob fuel, the average accuracy of the model is 92.25% using k-ε standard. Simulation process of furnace can analyze gas composition of the combustion (CO, CO_2, O_2, etc.) and velocity of air in the combustion. A new design of the furnace has been designing in this research with higher heat of stove and solve issue of the flame come out from the feeder area.

Acknowledgement
This research is funded by Indonesia Endowment Fund for Education (LPDP) year 2017.

References

[1] M Buchmayr, J Gruber, M Hargassner and C Hochenauer 2016 A computationally inexpensive CFD approach for small-scale biomass burners equipped with enhanced air staging *Energy Convers. Manag.* **115**: 32–42.

[2] Y Wang and L Yan 2008 CFD Studies on Biomass Thermochemical Conversion *Int. J. Mol. Sci.* **9**(6): 1108–30.

[3] R Morissette, P Savoie and J Villeneuve 2013 Corn Stover and Wheat Straw Combustion in a 176-kW Boiler Adapted for Round Bales *Energies* **6**(11): 5760–74.

[4] M A Gómez, J Porteiro, D Patiño and J L Míguez 2015 Eulerian CFD modelling for biomass combustion. Transient simulation of an underfeed pellet boiler *Energy Convers. Manag.* **101**: 666–680.

[5] N Athanasios, N Nikolaos, M Nikolaos, G Panagiotis and E Kakaras 2015 Optimization of a log wood boiler through CFD simulation methods *Fuel Process. Technol.* **137**: 75–92.

[6] R Prieler, M Demuth, D Spoljaric and C Hochenauer 2015 Numerical investigation of the steady flamelet approach under different combustion environments *Fuel* **140**: 731–743.

[7] B Mayr, R. Prieler, M Demuth, M Potesser and C Hochenauer 2015 CFD and experimental analysis of a 115 kW natural gas fired lab-scale furnace under oxy-fuel and air-fuel conditions *Fuel* **159**: 864–875.

[8] B A Albrecht, S Zahirovic, R J M Bastiaans, J A van Oijen and L P H de Goey 2008 A premixed flamelet-PDF model for biomass combustion in a grate furnace *Energy and Fuels* **22**(3): 1570–80.

[9] P. Venturini, D Borello, C Iossa, D Lentini and F. Rispoli 2010 Modeling of multiphase combustion and deposit formation in a biomass-fed furnace *Energy* **35** (7): 3008–21.

[10] Ansys *ANSYS Fluent Theory Guide* 2010

[11] X Lu and T Wang 2012 Effect of Radiation Models On Coal Gasification Simulation *ASME 2012 Int. Mech. Eng. Congr. Expo* **6**: 1175–92

[12] T H Shih, W. W. Liou, A Shabbir, Z Yang and J Zhu 1995 A new k-e eddy viscosity model for high reynolds number turbulent flows. *Comput. Fluids* **24** (3):227–238

[13] R Prieler, M Demuth, D Spoljaric and C Hochenauer 2014 Evaluation of a steady flamelet approach for use in oxy-fuel combustion *Fuel* **118**: 55–68

[14] S Melin 2013 *Considerations for Grading Agricultural Residue*

Solar Powered Automated Pipe Water Management System, Water Footprint and Carbon Footprint in Soybean Production

Satyanto K S[1], Abang Z E[1], C Arif[1] and M Yanuar J P[1]

[1] Department of Civil and Environmental Engineering, Bogor, Indonesia

E-mail: saptomo.sk@gmail.com

Abstract. An automatic water management system for agriculture land was developed based on mini PC as controller to manage irrigation and drainage. The system was integrated with perforated pipe network installed below the soil surface to enable water flow in and out through the network, and so water table of the land can be set at a certain level. The system was operated by using solar power electricity supply to power up water level and soil moisture sensors, Raspberry Pi controller and motorized valve actuator. This study aims to implement the system in controlling water level at a soybean production land, and further to observe water footprint and carbon footprint contribution of the soybean production process with application of the automated system. The water level of the field can be controlled around 19 cm from the base. Crop water requirement was calculated using Penman-Monteith approach, with the productivity of soybean 3.57t/ha, total water footprint in soybean production is 872.01 m^3/t. Carbon footprint was calculated due to the use of solar power electric supply system and during the soybean production emission was estimated equal to 1.85 kg of CO^2.

1. Introduction

Soybean water saturation cultivation is one of the cultivation techniques that has one objective of increasing sub-optimal wetland productivity while preserving it from degradation due to over drainage. It is important to provide appropriate field water level as the root may not be submerged under the water for a long time but in the other hand it may not be too low that will stop the capillary rise to keep water available in the soil. One method that we are proposing is by using perforated pipes that is laid below the surface. The perforated pipe will be water line for drainage and/or irrigation depends on the case.

Irrigation automation refers to the operation of system that minimized manual intervention in irrigation, which can be conducted in small area or in a large area segmented into smaller parts. The segments are called irrigation block and receive water regarding the available discharge as mention in Anitha [1]. The performance of automatic irrigation had been increased by the introduction of microcontroller which reduces the complicity of circuit design and the cost while upgrading the system is become easier [2]. There are many additional sensors and components that were designed for direct interfacing with microcontroller that ease the design and modification [3].

The use of electricity in electromechanical control system like automatic irrigation however causes carbon footprint due to equivalent CO2 emission for every unit usage of electricity that added in soybean production at the implemented land. Life cycle assessment (LCA) produces complete figure from input and output of components: air pollution emission, water utilization, wastewater generation, energy consumption and greenhouse gasses emission. To determine carbon footprint, LCA estimate the emitted or produced greenhouse gasses gas in every step identified for product life cycle, technically known as greenhouse gases calculation [4].

Water footprint is a measure of humanity's appropriation of fresh water in volumes of water consumed and/or polluted in the production process. Water footprint of a province consist of internal and external. Internal means the annual volume of water sourced from the province that is consumed to produce crop for the local population. While external means if the water was originated from outside of the province as mentioned by Hoekstra and Chapagain [5]. Consumers water footprint is also divided into direct and by consumption of agricultural and industrial product. In Indonesia only 3% of water footprint from industrial product, and greater portion is from agriculture. Water footprint and carbon footprint can describe how well we use and conserve resources for our production activity, and the method we choose will add particular water footprint and carbon footprint to our product.

The objective of this study are to develop a solar-powered automation control applied to pipe irrigation and drainage system which was set below the soil surface using perforated pipe and use it in trial and to estimate the contribution of water footprint and carbon footprint originated from soybean production related to automatic water management.

2. Material and methods

2.1. *Water management setting*

The field for soybean was made into beds having width of 4 m and height of 30 cm from the base. Each bed was separated by small 30 cm width and depth channel where perforated PVC pipe was laid at the bottom. The pipes connected to unperforated collector pipe which will transmit water to outlet. Figure 1 shows the water management setting of the field and figure 2 shows pipes installation with coconuts fiber to tighten the pipes to the bottom of the channel.

At the inlet an electric motorized valve was installed to control flow of water into the field. The operation of the valve was done automatically with controller. With perforated pipes position below the bed, they also functioned as drainage. Both inlet and outlet were attached with flowmeter to observe the amount of inflow and outflow.

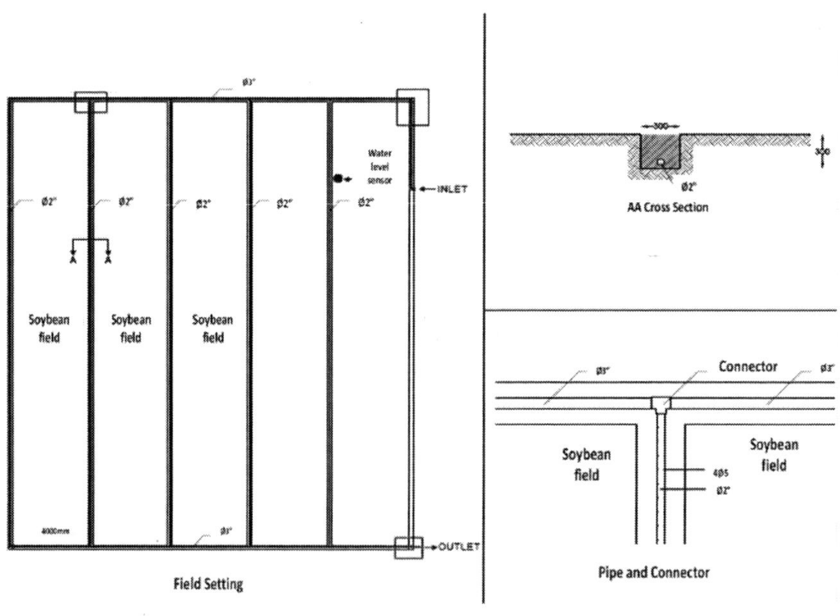

Figure 1. Field Setting

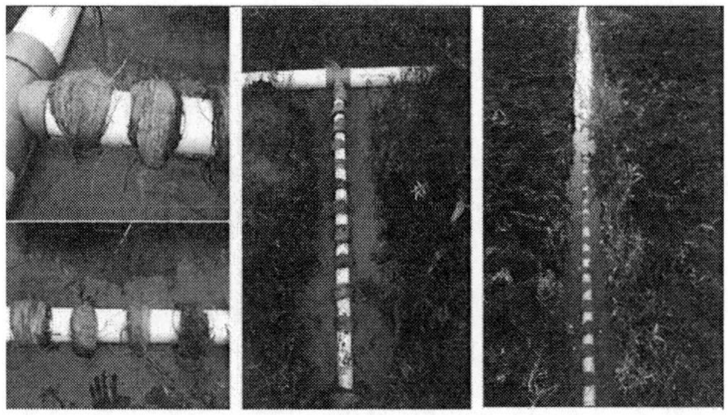

Figure 2. Pipes installation

2.2. *Solar power automated irrigation system*

Automatic irrigation system that was implemented in the study is the solar powered automatic irrigation system that was developed for rice field (figure 3), with modification in water delivery system to the field by using perforated pipe. The system is fully operated by using battery continuously 24 hours. During daytime the battery is charged by solar panel to ensure the availability of energy.

Solar power generated electricity can save energy consumption of more than 35% [6]. Another modification that was made is the use of Raspberry Pi minicomputer, instead of microcontroller. The system is responsible to read water level with water level sensor and to open or close irrigation by operating motorized valve automatically.

There are two setpoints that were used, which are upper setpoint and lower setpoint. Water from reservoir will flow through the pipes as water level fall below lower setpoint and irrigation will be stopped if water level reach the higher setpoint. This method was chosen considering its simplicity and effectiveness in providing appropriate level for the land.

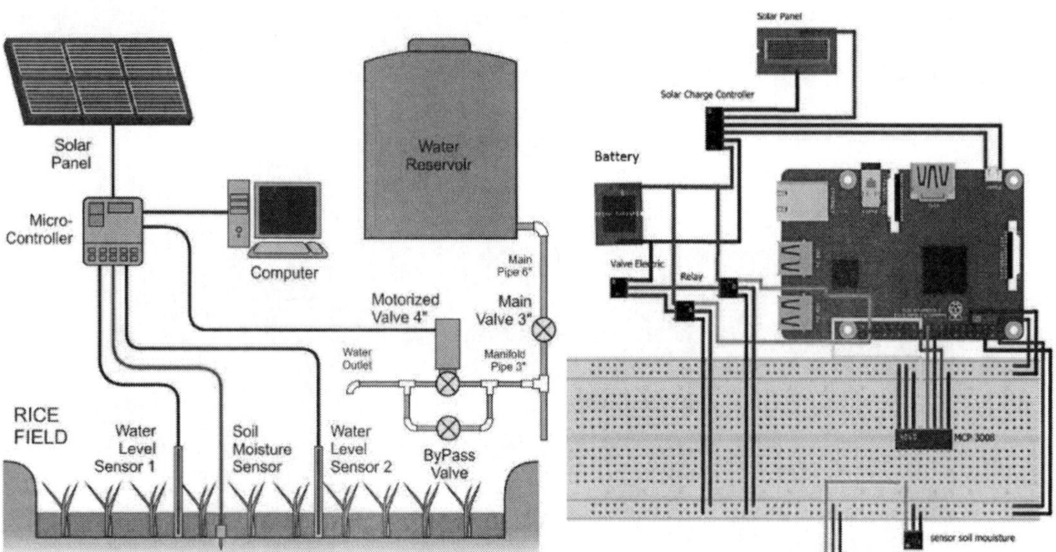

Figure 3. Solar powered automatic irrigation system setting and Raspberry Pi controller prototyping schematic

IOP Conf. Series: Earth and Environmental Science **147** (2018) 012030 doi:10.1088/1755-1315/147/1/012030

The components of the automatice controller are:

1. Raspberry Pi
2. Project Board
3. Water Level sensor
4. Soil Moisture Sensor
5. Solar Panel
6. Solar Charge Controller
7. Magnetic Relays
8. Electric Valve

Raspberry Pi is a credit card sized miniPC device having system of Broadcom BCM2835 chip (SoC), which includes ARM1176JZF-S 700 MHz to 1 GHz, VideoCore IV GPU [7]. All sensors and actuators are connected and controlled by this device. Figure 4 shows the installation of sensors and valve and figure 5 shows the experimental land during soybean cultivation experiment.

Figure 4. Sensors and Valve

Figure 5. Experimental land

2.3. *Water Footprint Calculation*

As it was described by Hoekstra et al. [8], water footprint is total water use to produce a product. The calculation of water footprint is divided into green water footprint, blue water footprint, and grey water footprint. They describe the amount of rainwater and irrigation water that were used and amount of water to clean the pollution resulted during the production activity, in volume of water per unit of product mass (m^3/t). This can be calculated using equation 1.

$$WF = WF_{green} + WF_{blue} + WF_{grey} \quad (1)$$

Where :

- WF : *Water footprint* (m^3/ton)
- WF_{green} : Green *Water footprint* (m^3/ton)
- WF_{blue} : Blue *Water footprint* (m^3/ton)
- WF_{grey} : Grey *Water footprint* (m^3/ton)

Each green, blue and grey water footprint can be calculated with equations 2, 3 and 4. In this study, crop water use (CWU) was estimated by at first determining reference evapotranspiration (ETo). This was done by using FAO standardize method to calculate ETo [9]. ET_{green} and ET_{blue} are the consumptive water use of the plants regarding to it source of water: green (rainwater) and blue (surface/groundwater).

$$WF_{green} = \frac{CWU_{green}}{Y} \quad (2)$$

$$WF_{blue} = \frac{CWU_{blue}}{Y} \quad (3)$$

$$WF_{grey} = \frac{(\alpha \times AR)(C_{max} - C_{nat})}{Y} \quad (4)$$

where

- WF_{green} : green water foot print (m^3/ton)
- WF_{blue} : blue water foot print (m^3/ton)
- WF_{grey} : green water foot print (m^3/ton)
- CWU_{green} : crop water use/ET_{green} (m^3/ha)
- CWU_{blue} : crop water use/ET_{blue} (m^3/ha)
- α : times the leaching-run-off fraction (-)
- AR : chemical application rate to the field (kg/ha)
- C_{max} : maximum acceptable concentration (kg/m^3)
- C_{nat} : natural concentration for the pollutant considered (kg/m^3)
- Y : crop yield (ton/ha)

2.4. *Carbon Footprint Calculation*

Carbon footprint is the total greenhouse gas that is emitted or produced from production process of a product. We can distinguish the calculation into primary carbon and secondary carbon footprint. In this study, only the later mentioned was implemented because there was no fossil fuel being used in water management process. Instead, electricity was used which CO2 emission is calculated as equation 5 [10].

$$CO_2 = EF \text{ x electricity consumption} \quad (5)$$

where:

CO_2 = emission of CO_2 (kg CO_2)
EF = emission factor (kg CO_2/kWh)
Electricity consumption = Quantity of used electrical energy (kWh)

3. Result and discussion

3.1. *Control performance*

Result of automatic irrigation operation is shown in figure 6 which depicts upper and lower setpoints, fluctuation of actual water level and the state of irrigation (valve). Valve's state was given value 0 or 1, which mean the valve is closed to stop irrigation or is opened to apply irrigation. It is clearly seen that valve's state became 1 when water level below lower setpoint and was kept open until water level reaches the upper setpoint. After upper setpoint was reached or exceeded valve's state turn to 0 which meant irrigation was stopped and water level will gradually decrease.

There were events when water level suddenly increases above upper setpoint, which happened as rain fallen. This sudden event would trigger the control system to close irrigation valve. In this case water level was not gradually decrease but rather drop due to drainage.

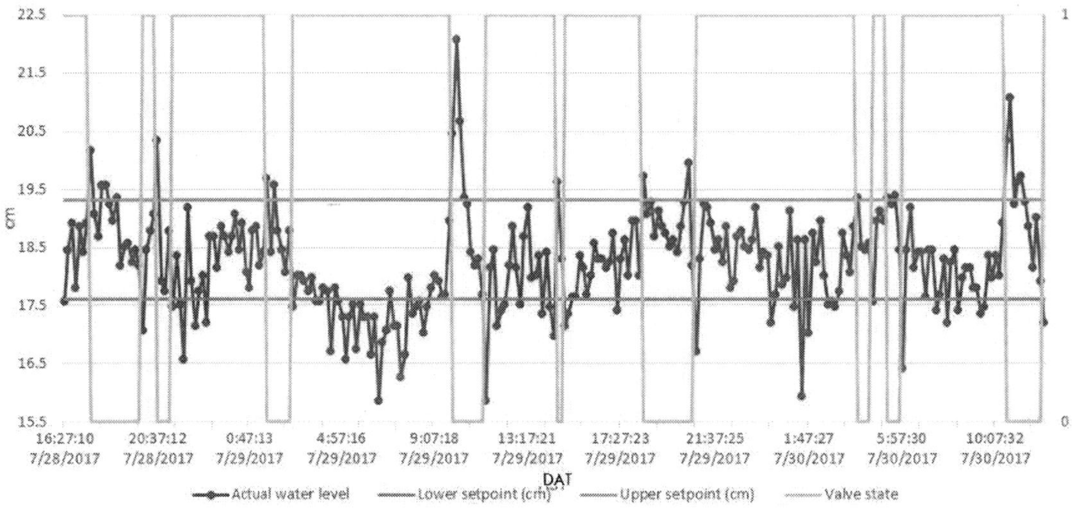

Figure 6. Automatic irrigation control performance

3.2. *Water Footprint*

Water footprint was obtained from the amount of water applied to provide crop water requirement which originated from seepage, rainfall and irrigation. Effective rainfall was used as part of rainfall that can be held by soil pores and available for crops, which are always less than total rainfall. Seepage is determined as water that comes out from the soil, although it was originated from the adjacent irrigated field due its higher elevation. The amount of seepage is in accordance to rainfall, the greater rainfall drove greater seepage inflow. This stage of study was conducted during rainy season, and water input to the land through seepage was dominant. Figure 7 shows monthly evapotranspiration, effective rainfall and seepage during the observation.

Table 1. Calculation of crop water use (CWU)

Month	Min Temp °C	Max Temp °C	RH %	Wind Speed km/day	Sunshine hours	Radiation MJ/m^{2}month	ETo Mm/mo nth
March	18.80	34.00	85.35	26.13	4.86	529.00	108.89
April	18.00	34.30	87.67	20.73	3.63	420.90	87.65
May	22.20	42.90	87.42	1.42	5.73	484.70	101.09
June	22.30	34.00	85.40	1.5	6.40	157.70	31.73

IOP Conf. Series: Earth and Environmental Science **147** (2018) 012030 doi:10.1088/1755-1315/147/1/012030

Due to unavailability harvest data during preparation of this report, soybean productivity was assumed following Ghulamahdi et al. [11], for water saturated soybean production: 3.57 t/ha. Therefore, water footprint added to soybean production from this stage are as shown in table 2 and 3 which shown grey water footprint. Water footprint of seepage was difficult to be categorized as it came from the adjacent field through subsurface flow. The origin of water itself was either from irrigation or rainfall, thus for the moment we let it not be defined as green or blue. Here green water footprint was found 433.74 m^3/t, and water footprint for seepage is 303.68 m^3/t.

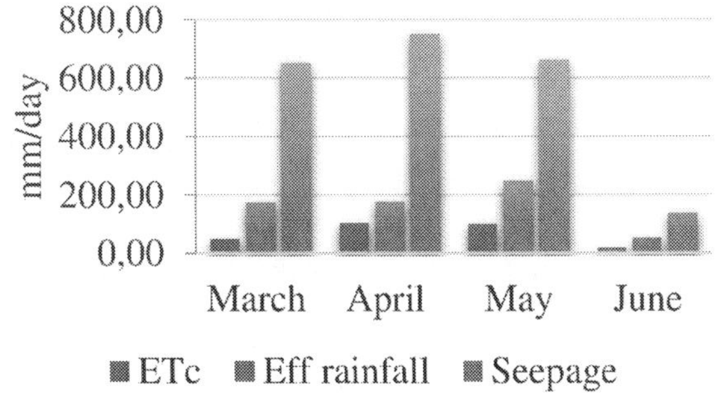

Figure 7. ET, Rainfall and Seepage

Table 2. Calculation of green and blue water footprint

Month	ETc mm/day	CH eff mm/day	Seepage mm/day	CWU green m^3/ha	CWU seepage m^3/ha	Yield t/ha	WF green m^3/t	WF seepage m^3/t
March	48.33	174.14	651.76	342.12	141.14	3.57	95.83	39.54
April	100.80	175.64	749.41	599.61	408.36	3.57	167.96	114.30
May	98.27	245.60	662.65	511.71	470.98	3.57	143.33	131.93
June	15.87	50.24	135.94	95.00	63.65	3.57	26.61	17.83

Table 3. Calculation of grey water foot print

Mineral	AR (kg/ha)	α	Cmax (μg/L)	Cmax (kg/L)	Cnat (mg/L)	Cnat (kg/m^3)	GWfp (m^3)
N	0.3	$1.42x10^{-1}$	13000	$1.30\ x10^{-2}$	0.26	$2.60\ x10^{-4}$	3.35
P	0.2	$2.34\ x10^{-1}$	20	$2.00x\ x10^{-5}$	0.01	$1.00\ x10^{-5}$	467.73

Grey water footprint of soybean was estimated from Nitrogen and Phosphorus content in manure and fertilizer. Greywater footprint for N and P are 0.96 m^3/t dan 133.64 m^3/t and total grey water footprint is 134.59 m^3/t. In summary, water footprint for soybean from during cultivation period was 872.01 m^3/t.

Arena et al. [12] presented soybean water footprint in Argentina with and without irrigation are 2,572.2 m^3/t and 2,440.7 m^3/t with productivity 3.2 t/ha and 2.8 t/ha. In America, Mubako dan Christoper [13] had shown that soybean water footprint is between 651 m^3/t to 2,281 m3/t. The difference of water footprint in different country is influenced by geographical condition, climate, the length of cultivation period and the productivity. In this case, our study result is close to that in America. However since this was done only in one trial, and this study only considers the period after planting to harvest, and the exact harvest had not been quantified yet, we will need more trial and extensive data before we can conclude the right figure of water footprint of soybean in water-saturated soil cultivation.

3.3. *Carbon footprint*

Carbon footprint calculation was based on the amount of electricity of the whole automation system with solar electricity generation. Postnote [15] suggested that solar panel emits 58 g CO_2/kWh it produced, this is based Lifecycle CO_2 emissions for UK photovoltaic power systems with current technology. During non-irrigation period, when automatic system was not operating, only monitoring function controlled by the controller, total daily consumption for the whole system is 0.31 kwh per day and thus emission for 102 days soybean cultivation 1.85 kg CO_2.

Table 4. Electricity consumption of automatic system components

Device	Voltage (V)	Current (A)	Power (watt)	Operation	Time (hour)	Energy Consumption (kWh)
Raspberry pi 3 model B	5	2.5	12.5	1 day	24	$3.00x10^{-01}$
Sensor E tape			0.5	1 day	24	$1.20x10^{-02}$
Sensor VH400 Vegetronix	3.5	0.012	0.042	1 day	24	$1.01x10^{-03}$
Relay (1)			0.15	1x valve opening	$8.33x10^{-03}$	$1.25x10^{-06}$
Relay (2)			0.15	1x valve closing	$8.33x10^{-03}$	$1.25x10^{-06}$
Valve Valworx 561086	12	2.1	25.2	1x valve opening or closing	$8.33x10^{-03}$	$2.10x10^{-04}$

By analyzing control performance data shown in figure 6 electricity that was used by actuators (relays and valve) can be estimated. The time that was required by water level to decrease to lower setpoint and increase again to upper setpoint is 2 hours 30 minutes, where valve opened and closed once. During 102 days, valve opened and closed up to 1958 times. One rotation of electric emitted carbon emission of 24.1 g. Assuming automatic irrigation would operate each day the same as it' performance in the trial, during the production period, the system will produce 1.88 kg of CO_2 equivalent emission. Table 4 shows the detail energy consumption and carbon emission of the automatic irrigation.

Irrigated land produced emission more than the non irrigated one due to its larger electricity consumption for operation of relays and valve. Esteves et al [15] revealed that soybean irrigation in Brazil produced 6.04 kg CO_2. The larger figure compared to our case is due to the use of diesel fuel, not solar electricity. Therefore carbon footprint added to soybean production can be minimized by using solar-powered automatic irrigation system.

4. Conclusions

Automatic water management system had been developed and water level of the field was controlled within the setpoints range (around 20 cm from base). With productivity of soybean 3.57t/ha, total water footprint in soybean production is 872.01 m^3/t. Carbon footprint due to the use of solar-powered system was estimated equal to 1.85 kg of CO_2 when irrigation is not operating and 1.88 kg of CO_2 if the field automatically irrigated.

References

[1] Anitha K 2016 Automatic Irrigation System. Journal Innovative Trend in Science *Engineering and Management* **2**(8): 301-309

[2] Al Smadi T A 2011 Low-cost smart sensor design *Journal Engineering and Applied Sciences* **4**(1): 162-168

[3] Fisher D K and Hirut K 2010 A low-cost microcontroller based system to monitor crop temperature and water status *Journal Computers and Electronics in Agriculture* (74): 168-173
[4] Pandey D, Madhoolika A and Jai S P 2010 Carbon Footprint: Current Methods of Estimation (National Environmental Engineering Research Institute (NEERI)) (Nagpur: Springer)
[5] Hoekstra A Y and Chapagain A K 2007 The Water Footprint Of Coffee and Tea Consumption in the Netherlands *Journal Ecological Economics* **64**(1): 109118
[6] Alam N and Naseem A 2014 Solar powered auto irrigation system *Journal Sci. Int* **26**(4) : 1515 – 17. ISSN: 1013 – 5316.
[7] Horan B 2013 Practical Raspberry Pi, Tecnology In Action [Internet]. https://thepiratebay.sx/torrent/8606141/Practical_Raspberry_Pi_V413HAV.
[8] Hoekstra A Y, Ashok K C, Maite M A and Mesfin M M 2011 The Water Footprint Assessment Manual, Setting the Global Standard (London: Earthscan)
[9] Allen R G, Pereira L S, Raes D and Smith M 1998 Crop evapotranspiration: guidelines for computing crop water requirements (Rome: FAO Drainage and Irrigation) Paper 56
[10] Prihatmaji Y P, Akhmad F, Syaiful R and Feris F 2016 Carbon Footprint Analysis of Central Library Building, Rectorate, MIPA Laboratory of UII based on Existing Vegetation as Greenhouse Gasses Emision Reduction *Journal of Innovation and Entrepreneurship* **1**(2): 148-155 (in Indonesian)
[11] Ghulamahdi M, Sandra A A and Abdul K M 2012 Water-saturated Cultivated Technology Implementation for Paddy and Soybean to Increase Plantation Index in Tidal Land. *Proceeding of Symposium and Joint Seminar PERAGI-PERHORTI-PERIPI-HIGI Suporting Sustainable Food and Energy Sovereignty* (Bogor) pp 113-118
[12] Arena A P, Roxana P and Barbara C 2011 *Water Footprint of Soybean Production in Argentina* (Mendoza: Grupo CLIOPE, CONICET)
[13] Mubako S T and Christopher L L 2013 Agricultural Virtual Water Trade and Water Footprint of U.S. States *Annals of the Association of American Geographers* **103**(2): 385-396
[14] [Postnote] Parliamentary Office of Science and Technology 2006 *Carbon Footprint of Electricity Generation* (London: Postnote)
[15] Esteves V P P, Elisa M M E, Davi J B, Daniel G, Daniel D, Luiz EV, Ofella and Claudia 2016 Land Use Change (LUC) Analysis and Life Cycle Assessment (LCA) of Brazilian Soybean Biodiesel *Journal of Clean Techn Environ Policy*

Sustainability Challenge of Micro Hydro Power Development in Indonesia

Didik H[1], Bambang P N[2], Asep S[3] and Y A Purwanto[2]

[1] Study Program of Natural Resources and Environmental Management, Bogor Agricultural University, Indonesia
[2] Department of Mechanical and Biosystem Engineering, Bogor Agricultural University, Indonesia
[3] Department of Civil and Environmental, Bogor Agricultural University, Indonesia

E-mail: masdidikhadiyanto@gmail.com

Abstract. Rural electrification using renewable energy is the best choice for many locations that far away from national grid. Many renewable energy project have been built for rural electrification such as micro hydro power plant (MHPP) and solar photovoltaic (SPV). Sustainability still the main challenge of off-grid renewable energy development for off-grid rural electrification in Indonesia. The objective of this paper is to review sustainability of micro hydro power development in Indonesia. The research method was done by field observation, interview with MHPP management, and reviewing some research about MHPP in Indonesia. Sustainability issues include various aspects that can be classified into 5 dimensions: technical, economic, socio-cultural, institutional, and environmental. In technical factors that lead to sustainability problem are: improper MHPP design and construction, improper operation and maintenance, availability of spare parts and expertise. In the economic dimension are generally related to: low electricity tariff and utilization of MHPP for productive use. In the social dimension are: the growth of consumer's load exceeding the capacity, reduced number of consumers, lack of external institutional support. In the institutional side, it is generally related to the ability of human resources in managing, operating and maintaining of MHPP. Environmental factors that lead the sustainability problems of MHPP are: scarcity of water discharge, conflict of water resources, land conversion over the watershed, and natural disasters.

1. Introduction

Electrical energy is one of the infrastructure to realize sustainable development, encourage economic growth, and increase prosperity, health and educations. Electricity is one of critical inputs for overall development of country and is one of main infrastructure requirements for agricultural, industrial and socio-economic development [1].

Electricity development in Indonesia aims to ensure the sufficient electricity supply, good quality electricity supply with reasonable price in order to improve the welfare and prosperity [2]. Today, there are many Indonesian people who have not been able to access electrical energy. The electrification ratio of Indonesia in 2014 is about 84.4%, there are 15.6% of Indonesian people who have not yet electrified [3].

Electrical energy supply for rural communities or rural electrification is an authority of both central and local governments. The demand for electricity in a remote area can be satisfied by either extending the coverage of the national power grid (the on-grid approach) or constructing and operating stand-alone power system (the off-grid approach) such as micro hydropower plant, solar photovoltaic, wind turbine, biogas and biomass [4]. The selection of rural electrification methods (on-grid or off-grid) depends on several factors such as distance from existing grid, voltage and capacity of existing grid, load capacity and density, rough terrain, investment cost, operation and maintenance costs, willingness and ability to pay of electricity, and renewable energy potential [5].

The rural electricity program in remote areas by national power grid expansion is not economic. In technical side will create another problem related to power quality i.e. voltage drop, in economic side it is requires a large investment cost. Diesel power plants utilization for rural electrification, is an unfavorable choice today, due to the high fuel costs and environmental problems. Diminishing sources of fossil fuel and environmental concerns are encouraging factor to switch Indonesian power generation sources to renewable energy [7].

Utilization of local renewable energy potentials for rural electrification in the areas that far away from national power grid is the most realistic solution. Indonesia has abundant renewable energy potential, but the utilization is still low. Renewable energy potential includes hydro, wind, solar, biomass, biogas, and geothermal. Indonesia's renewable energy potential data can be seen in table 1 [3].

Table 1. Renewable Energy Potential and Installed Capacity [3]

No	Renewable Energy Sources	Potential (MW)	Installed Capacity (MW)	Utilization (%)
1	Hydro	75.091	4.826,7	6,4
2	Geothermal	29.544	1.438,5	4,9
3	Minihydro/Microhydro	19.385	197,4	1,0
4	Bioenergi	32.654	1.671,0	5,1
5	Solar	207.898	78,5	0,04
6	Wind	17.989	0,3	0,002
7	Air Laut	443.208	8.215,5	1,9

Renewable energy is widely built in Indonesia today i.e. micro hydro power plants (MHPP) and solar photovoltaic (SPV). Indonesia's renewable energy technology data that has been built by central government (Ministry of Energy and Mineral Resources/MEMR only) can be seen in table 1 [8]. Some ministries also have built a lot of MHPP in Indonesia, such as Ministry of Home Affairs, Ministry of Village, Ministry of Forestry, Ministry of Public Works etc. The government, both central and local government, has spent a lot of money every year to develop renewable energy to power up unelectrified villages. By 2017, MEMR has budgeted more than 100 billion rupiahs for renewable energy development for rural electrification [9].

Table 2. Renewable Energy Technology Development by MEMR [8]

Year		RE Technology	
		MHPP	SPV
2011	Unit	8	0
	Capacity (kW)	216.4	0
2012	Unit	0	123
	Capacity (kW)	0	4828
2013	Unit	11	121
	Capacity (kW)	1301	5275
2014	Unit	14	87
	Capacity (kW)	651	2650
2015	Unit	14	163
	Capacity (kW)	1193	8795

MHPP is the best choice of renewable energy technology if a location has hydro power potential. There are several reasons why it is very interesting to be developed in Indonesia today. From a technical point of view, MHPP technology is relatively simple, almost every component can be produced in Indonesia. In terms of economy and finance: investment, operating, and maintenance costs are relatively cheap compared with other renewable energy technologies. Water energy is a clean, cheap and environmental friendly of power generation source [10].

MHPP is a hydropower plant with a capacity below 1 MW that come from irrigation channels, rivers or natural waterfalls by utilizing the height of head and water discharge [11]. MHPP is a small power plant that uses hydropower with a capacity below 200 kW [12]. According to UNIDO, MHPP is a power generation system using energy sources from hydropower with a capacity of 500 W - 100 kW [13]. According to JICA, MHPP is a hydropower plant with capacity up to 200 kW [14].

In general, MHPP is a run-off river hydroelectric or has no dam (reservoir) that can be used to store water for a certain period of time (daily, weekly or monthly). The availability of water discharge, at that time used to generate electricity. In MHPP, head obtained is not by constructing a dam, but by diverting some stream flow to one side of the river and dropping it into the same river at a place where the required head has been obtained naturally. The water is deployed to rotate the turbine inside the powerhouse via penstock. The mechanical energy from turbine rotation is converted to electrical energy by a AC generator.

Sustainability is the main issues in the development of off-grid MHPP in Indonesia. Many MHPP projects have failed to address sustainability from the start [15]. The sustainability is determined to start from the planning, construction until operation and maintenance. It is very important to identify sustainability problem of MHPP. The objective of this paper is to review sustainability of MHPP development in Indonesia.

2. MHPP development for rural electrification

Many considerations in the development of MHPP so it can provide optimal benefits and sustainable. The identification of many factors starts from the planning stage: a potential survey followed by a feasibility study of technical, economic, social-cultural, institutional and environmental aspect. The next stage is detailed engineering design (DED) based on feasibility studies and budgets. In the other hand, preparation of community development and management must be done [12].

Social problems sometimes arise at the planning stage. The problem must be solved first before step to the next stage. Coordination of each stakeholder should start from the beginning to establish commitments and formulate the rules of the game among parties. Society should be the subject in every stage of development and not just as beneficiaries. The stages of MHPP development are broadly comprised of: potential surveys; feasibility study; technical design; and preparation of development; implementation of development; commissioning; operation, maintenance, and management. MHPP development process can be seen in figure 1 [12].

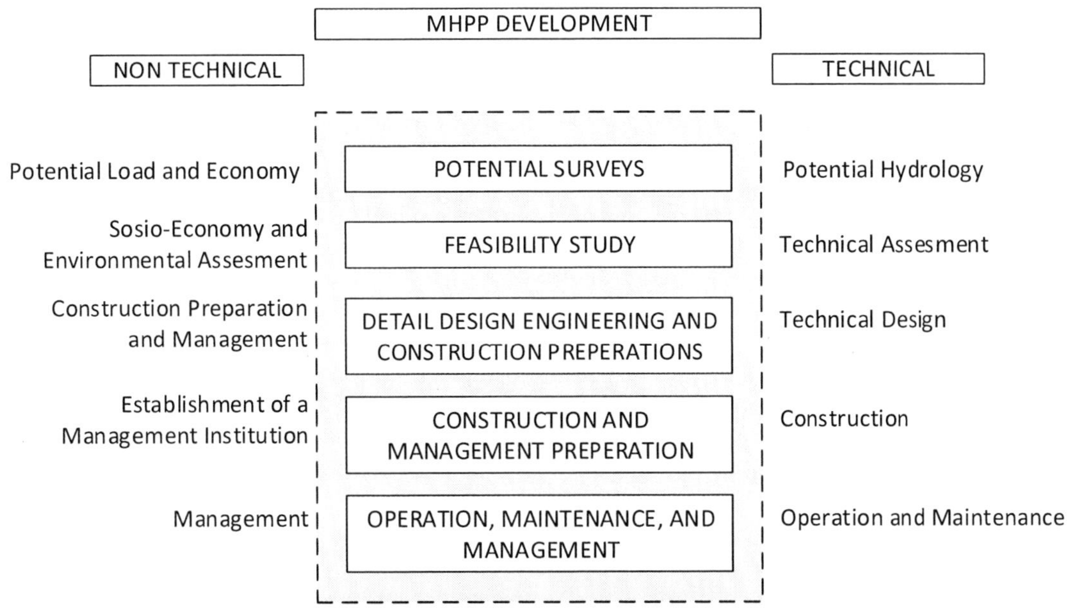

Figure 1 MHPP Development [12]

3. Methodology

There are several research methods undertaken in preparing this paper. The first study was conducted by reviewing various studies that have been done related to MHPP and the problem. Several studies have examined many MHPP locations in Indonesia such as Sumatra, Java, and Sulawesi.

In addition to reviews of various studies that have been conducted, research is also based on field observation at several locations of existing MHPP in Indonesia. There are several locations that have been surveyed such as North Sumatra, West Sumatra, West Java, Yogyakarta, West Nusa Tenggara, Gorontalo, Central Sulawesi, West Sulawesi, and Papua. Surveys are conducted by observing directly the condition of existing MHP and its problems and interviews with management institutions.

The third research method is through in-depth interviews of the management institutions. Interviews with the management institutions of both operators and managers are conducted simultaneously with MHPP training for management institutions in 2017. There are more than 100 MHPP managers and operators from various provinces in Indonesia such as: North Sumatra, Riau, West Sumatera, Lampung, West Java, West Kalimantan, West Nusa Tenggara, East Nusa Tenggara, South Sulawesi, West Sulawesi, Central Sulawesi, Gorontalo, Southeast Sulawesi, West Papua and Papua follow the training.

4. MHPP sustainability review in Indonesia

After the project is completed by the government or donor agencies, the MHPP is usually handed over to the village community. Management, operation and maintenance are carried out independently by the community through MHPP management institution. Operation maintenance and management should be well done so that the supply of electrical energy can be sustainable. The sustainability is not only responsibility of the management institution. Consumers, governments, contractors and manufacturers play an important role in maintaining MHPP sustainability.

The Ministry of Public Works conducts research about sustainability of the MHPP with capacities ranging from 2.5 kW to 120 kW in Lampung, West Java, Central Java and East Java provinces. There are several issues related to the sustainability. Low electricity tariff is ranged between Rp 10.000, - Rp. 20.000, - per month, so it is not enough for operation and maintenance cost. Conflict of water resources for other purposes such as raw water needs for the daily needs of the community. The expansion of national power grid by National Electricity Company (PLN) has caused MHPP customers move to PLN's customers [17].

In 2012, Energising Development Indonesia (EnDev Indonesia) undertook research about MHPP performance in Indonesia through a key performance indicator (KPI) survey of 20 sites in Sulawesi and 27 sites in Sumatra. The capacity factor of MHPP in Indonesia is relatively low about 6.1%, while electricity availability factor is around 63%. The low capacity factor indicates that the potential of electrical energy that can be produced is very large, but its utilization is very low. The low availability factor indicates a limited operating time. Another result of this study, there are many problems about MHPP sustainability: some MHPP components are broken and not repaired, financial administrations report not yet booked, lack of water supply during the dry season due to land conversion in the upstream of the watershed, and the availability of MHPP spare parts [18].

In 2013 EnDev Indonesia conducted a sustainability study of 32 MHPP units in Sulawesi. There are several problems related to MHPP sustainability: social conflicts between villagers and villagers with MHPP management caused MHPP damage and stop operation, conflict of water resource utilization for agriculture and plantation cause MHPP not optimal in operation. The Expansion of national electric grid will reduce MHPP's consumers, natural disasters such as floods and landslides caused massive damage to the civil components. Another problem is poor quality of civil construction and electro-mechanical components will make MHPP low in reliability [19].

Murni et al. in 2013 conducting sustainable MHPP research at two sites in East Kalimantan. The sense of ownership and community participation in management is key factors that determine the sustainability. Sufficient electric tariff will ensure the availability of operation and maintenance fund. The management has a significant role in operating maintaining and managing the MHPP. Feasibility studies and good technical design produce a reliable MHPP which will be 'the initial capital' of sustainability [20].

Ranzanici in 2013 undertook the research of MHPP sustainability on several sites in Sulawesi Island. High investement costs are required to produce a good quality of MHPP that has a significant impact on sustainability. Problems on social and economic dimensions have a fatal effect on micro-hydro sustainability. The sustainability is a very complex interaction among various aspects: technical, economic, social, institutional and environmental [21].

Purwanto and afifah conduct the research sustainability of 2 unit MHPP in Java. The aim of the research is evaluate MHPP sustainability using sustainable development indicators. The results obtained that both projects performed poorly in the economical dimension and positively in other dimensions. The economic sustainability is facing great difficulties, as the project has no financial scheme, low electricity tariff and also low utilization of electrical energy for productive enterprises [22].

5. Discussion

Development of MHPP has been spending a lot of money either sourced from government budget or grants and loans from various donor agencies. Evaluation of technical, economic, socio-cultural and environmental condition is very important to ensure the sustainability of rural electricity supply. There are many factors affecting the sustainability, which can be classified into 5 dimensions: technical, economic, socio-cultural, institutional and environmental [16].

5.1. Technical Dimension

There are many MHPP built without feasibility studies and proper technical design, it will result interruption in MHPP operation is more often. Improper head and water discharge design, causing MHPP not working optimally.

MHPP should be operated and maintained using proper operating and maintenance procedures. Improper operation and maintenance will cause damage of MHPP component more often. It may disrupt the electric supply for rural communities. Standar operation and maintenance procedures shall be provided by contractor or manufacturer. The operator must conform with standar operating and maintenance procedure.

Another technical factor is maintenance guarantees and availability of spare part. The damage of some components can not be fixed by local operators, so maintenance guarantees from the contractor and manufacturer is one important factor for electricity supply continuity. Every micro-hydro component have a life time, therefore spare part availability is absolutely needed to ensure sustainability. In general, location of MHPP is in remote areas, so the availability of spare parts becomes the main problem for MHPP.

5.2. Economical Dimension

In many cases the MHPP electricity tariff is set without correct calculation and only by community agreement. Tariff setting is only base on willingness to pay of community. so that the money earned is not sufficient for the operation and maintenance. Finally, the operation of the MHPP will be disrupted, because there are no funds available for component replacement.

The utilization of MHPP for productive use is one solution of low electricity tariff. Electrical energy is mostly used for consumptive activities at the night. In general, during the day the potential of electrical energy is not utilized. The utilization of MHPP energy for productive activities will increase the income of MHPP and community.

5.3. Social Dimension

Socio-cultural issues contribute to the sustainability. Lack of ownership and community participation is one of problem in MHPP. If some component or civil works having damaged, how the community responds to provide support both funds and workforce to repair it. Community compliance with MHPP rules and tariffs is a social capital that needed to maintain the sustainability.

Uncontrolled growth in consumer load often causes MHPP is overload. Overload operation will lead damage of turbine, generator and other components more frequent. It is necessary to adjust the consumer's load during peak load at 18:00 to 22:00. This rule must be followed by every consumer.

National power grid expansion to the location will provide an alternative energy for the community. Quantity and quality of electrical energy usually become a consideration for consumers. In the other hand low electricity tariff will be an attraction for the consumer. If consumer move to PLN, it is important to create an alternative utilization of MHPP. Productive use of MHPP for the small and medium enterprise (SME) is the best solution to keep it still running. SME is is one of the best ways to improve the welfare of the community.

Sustainability is strongly influenced by external institutions support. Policy, empowerment and funding support from external institutions plays important role to determine sustainability. Funding support can be sourced from village funds (Dana Desa), special allocation funds (DAK), as well as from local government revenue budgets (APBD). Government funds should only for large maintenance costs such as repair of civil works due to natural disasters, turbine replacement, and other major maintenance. Other institutions such as manufacturers or workshops play a significant role in maintaining the sustainability. The support of the workshops generally involves the availability of spare parts and experts. In west Sumatera, the workshop sometimes give a soft loan to MHPP for component replacement.

5.4. Institutional Dimension

Institutional issues that determine the sustainability of the MHPP are the capacity of the management. The institutional capability of management includes human resource competency, both operator, and administrative personnel in performing their duties and functions of managing, operating and maintaining the installation. Education and training for managers and operators is one way to improve the capability of MHPP management institutions.

Completeness of a set of rules and infrastructure for management, operation and maintenance will support the sustainability. The set of rules in the organization can be electricity tariff decree, agency decree, articles of association and bylaws, operation and maintenance manual, standing operation procedures (SOP). The infrastructure facilities organization can be office, safety equipment, work equipment, measuring equipment and financial administration report.

5.5. Environmental Dimension

Environmental is great issues in MHPP sustainability. The availability of adequate water discharge throughout the year is a critical point. Well-maintained forest will ensure sufficient water resource availability throughout the year. Land conversion in the watershed will cause decrease of water discharge during dry season.

Conflict of water resources can disrupt the operation MHPP. In general, utilization of water resources for other purposes besides MHPP, such as agriculture, plantation, fishery, industry, and water supply. Regulating the use of water resources is necessary to ensure continuity electricity supply to the community.

Natural disasters is one of factor that can influence the operation of MHPP. In general, MHPP is located in hilly areas so that natural disasters such as floods, earthquakes, forest fires, and landslides often occur. Flood and landslide cause severe damage to civil components. The damage caused by natural disasters generally require high maintenance costs. The availability of operating and maintenance funds sometimes insufficient to repair the severe damage caused by natural disasters.

6. Conclusions

Sustainability is still an important issue in the development of renewable energy for rural electricity in Indonesia today. The sustainability of an MHPP installation is determined from the first time, from the planning and construction stage until operation and maintenance. Feasibility studies and good technical design at the planning stage provide reliable and good quality of MHPP. Involving the community in every stage of development is a necessity that can increase ownership and participation in MHPP management.

In the operation and maintenance phase, there are several factors determining the MHPP sustainability. The main factors are the availability of operation and maintenance fund, the availability of water discharge during the year, and the ability of the management institution to manage the MHPP and external institutional support such as local government and MHPP workshop.

References

[1] Kumar A, Mohanty P, Palit D and Chaurey A 2009 Approach for Standardization of Off-Grid Electrification Projects *Renewable and Sustainable Energy Reviews* **13** 1946-56

[2] The Law of the Republic of Indonesia No30 2009 *About Electricity* (in Indonesian) (Jakarta: Presiden of Republik Indonesia)

[3] Regulation of the President of the Republic of IndonesiaNo22 2017 *About the General Plan of National Energy (RUEN) (*in Indonesian) (Jakarta: Presiden Republik Indonesia)

[4] Poudel R C 2013 Quantitative decision parameters of rural electrification planning:A review based on a pilot project in rural Nepal. *Renewable and Sustainable Energy Reviews* **25** 291-300

[5] Rahman M M, Paatero J V and Lahdelma R 2013 Evaluation of choices for sustainable rural electrification indeveloping countries: A multicriteria approach. *Energy Policy* **59** 589–599

[6] The World Bank 2008 *Designing Sustainable Off-Grid Rural Electrification Projects: Principles and Practices* (Washington D.C.: The World Bank)

[7] Erinofiardi, Gokhale P, Date A, Akbarzadeh A, Bismantolo P, Suryono AF, Mainil A K and Nuramal A 2017 A Review On Micro Hydropower In Indonesia *Energy Procedia* **110** 316 – 321

[8] Direktorat Jenderal Energi Baru, Terbarukan dan Konservasi Energi (Ditjen EBTKE). 2016. *Statistic book of EBTKE 2016* (In Indonesia). (Jakarta)

[9] https://finance.detik.com/energi/3529716/jonan-minta-anggaran-rp-65-t-tahun-depan-buat-apa-saja (accessed 29.09.17).

[10] Date A and Akbarzadeh A 2009 Design And Cost Analysis Of Low Head Simple Reaction Hydro Turbine For Remote Area Power Supply *Renew Energy* **34** 409-415.

[11] Regulation of the Minister of Energy and Mineral Resources of the Republic of Indonesia No 03

IOP Conf. Series: Earth and Environmental Science **147** (2018) 012031 doi:10.1088/1755-1315/147/1/012031

2016 *Technical Guidelines for the Using Allocation Fund for Small Scale Energy Schemes in Fiscal Year 2016* (in Indonesian) (Jakarta: ESDM)

[12] [DJLPE] Direktorat Jenderal Listrik dan Pemanfaatan Energi 2006 *Manual program of Microhydro Power Plant Development (MHPP) for Rural area Electricity* (in Indonesian) (Jakarta: DJLPE)

[13] [IMIDAP] Integrated Microhydro Development and Application Program 2009 *Modul Pelatihan Operator Mikrohidro* (Jakarta: IMIDAP)

[14] [JICA] Japan International Cooperation Agency 2003 *Study On Rural Energy Supply With Utilization Of Renewable Energy In Rural Areas In The Republic Of Indonesia* (Tokyo: JICA)

[15] Mainali B and Silveira S 2015 Using A Sustainability Index To Assess Energy Technologies For Rural Electrification *Renewable and Sustainable Energy Reviews* **41** 1351–65

[16] [IMIDAP] *Integrated Microhydro Development and Application Program 2010 Pedoman Studi Kelayakan Komprehensif Berkelanjutan* (Jakarta: IMIDAP)

[17] [PU] Kementerian Pekerjaan Umum 2011 *Research and Development of Community-Based Microhydro Technology Management* (in Indonesian) (Jakarta: PU)

[18] [EnDev] Energising Development Indonesia 2012 *Survey on Key Performance Indicators for Indonesian Micro-Hydro Power Sites* (Jakarta: Energising Development (EnDev) Indonesia)

[19] [EnDev] Energising Development Indonesia 2013 *EnDev2 Indonesia: Impact on Sustainability* (Jakarta: Energising Development (EnDev) Indonesia)

[20] Murni S 2014 *The Implementation of Micro Hydro Projects in Remote Villages in Developing Countries: An Interdisciplinary Approach* (Western Australia: Graduate School of Murdoch University)

[21] Ranzanici A 2013 *Sustainability Comparison Between EnDev and non-EnDev Micro-Hysro Power (MHP) in Indonesia* (Madrid: Universidad Politécnica De Madrid)

[22] Purwanto WW and Afifah N 2016 *Assessing the Impact of Techno Socioeconomic Factors on Sustainability Indicators of Microhydro Power Projects in Indonesia: A Comparative Study* Renewable Energy 93: 312-322

Anaerobic Co-digestion of Cow Dung and Rice Straw to Produce Biogas using Semi-Continuous Flow Digester: Effect of Urea Addition

A Haryanto[1], B P Sugara[1], M Telaumbanua[1] and R A B Rosadi[1]

[1] Agricultural Engineering Department, Faculty of Agriculture, The University of Lampung. Soemantri Brojonegoro Road, Bandar Lampung, Indonesia

E-mail: agus.haryanto@fp.unila.ac.id

Abstract. The objective this research was to investigate the effect of urea addition on the biogas yield from co-digestion of rice straw and cow dung using semi-continuous anaerobic digester. The experiment was conducted by using self-made semi-continuous anaerobic digester having a working volume of 30 L. Cow dung was provided from Department of Animal Husbandry, University of Lampung; while rice straw was collected from farmer at Way Galih, Tanjung Bintang, South Lampung. Rice straw was sun-dried to about 12% of moisture content and then ground into fine particles. Cow dung and ground straw were mixed at a dung-to-straw ratio of 3:1 based on total solid (TS) and four different urea additions (0, 0.25, 0.65, and 1.30 g/L) were applied to have a C/N ratio between 20 and 30. The mixture was diluted with water to create TS content of 10%. As much as 30 L of the substrate mixture was introduced into the digester as a starting load. The same substrate was added daily at a loading rate of 0.5 L/d. The experiment was made in triplicate and observation was performed for two months. Total and volatile solids of influent and effluent and daily biogas production were observed. The biogas quality was measured by its methane content using gas chromatography. Results showed that urea addition influenced the biogas yield and its quality. Substrate mixture with urea addition of 0.25 g/L (C/N ratio of 27.3) was the best in terms of biogas yield (434.2 L/kgVSr), methane content (50.12%), and methane yield (217.6 L/kgVSr).

1. Introduction

Biogas is produced through anaerobic digestion process of organic substances. The biogas is mainly composed of CH_4 (45-70%) and CO_2 (30-45%) and traces of H_2, water vapor (H_2O), ammonia (NH_3), and hydrogen sulfide (H_2S) [1, 2]. Biogas can be used to fuel several applications, from cooking stove to generating electricity. Biogas technology is now an ecologically sound option to reduce environmental burden by decomposing organic material and producing not only energy but also good quality organic fertilizer [3, 4]. Application of small biogas digester provides economic and environmental benefits to the society [5]. Biogas is one of renewable fuel that can be an important source of Indonesia's energy in the near future. This is accentuated by a fact that Indonesia is bestowed with enormous biomass both fresh matters and as wastes from agro-industrial processing. During biogas process, which is carried out under anaerobic conditions, volatile solids are decomposed and converted into methane and carbon dioxide.

Rice straw can be a promising substrate for biogas production. Straw is produced during paddy harvesting. In the past time, when paddy was harvested manually simple cutter and then sickle, straw is left on the plant. Nowadays, rice straw is separated from the grains after the plants are threshed either manually or using stationary threshers. Recently, mechanical harvesting is conducted by using combine harvesters that is facilitated with the threshing unit so that straw is produced just after harvesting. According to IRRI, the amount of straw is roughly 0.7–1.4 kg for every kg of milled rice, depending on varieties, cutting-height of the stubbles, and moisture content during harvest [6]. As one of the largest rice producing countries in the world, Indonesia has abundant rice straw. The potential of rice straw in Indonesia in 2012 was about 91.753 million tons which is equivalent to approximately 50.974 million tons of coal with potential energy around 382,305 GWh of electricity, and the electric power potential around 43,642 MW [7]. According to the report of the Agriculture Ministry of Indonesia, about 79.4 million tons of milled rice is produced in 2016 [8]. It can be calculated that Indonesian rice straw is 49.63–99.26 million ton. It is an abundant waste that can be used to generate energy to substitute fossil fuels [7, 9].

The most common utilization of rice straw in Indonesia is for animal feed, either directly or after fermentation, even though it is classified as a poor feed for the animals due to high silica content [10]. Rice straw is also used for fuel in brick, roof tile, and pottery industries. The very little amount is used for cooking fuels. Significant amounts, however, remain unused in the fields. One common managing practice of is incorporating the rice straw into the soil during plowing to decompose and provide fertilizer for the next crop. Buresh and Sayre, however, noted that incorporation of rice straw in the soil may have detrimental effects because of the initial immobilization of soil N, decreasing Zn availability, and increasing methane emission [11]. This practice is supposed to reduce harvesting yields due to foliage diseases [12].

Another common practice is barely burning *in-situ* the straw on the fields. One of the widely accepted reasons of burning rice straw in the field is to accelerate soil preparation and to provide minerals. Research indicates, however, that open burning has negative effects such as nutrient loss, removal of soil organic matter, and reduction of beneficial soil insects and microorganisms [10]. Open burning has also been observed to contribute to emissions of harmful air pollutants. During uncontrolled burning, pollutants such as CO_2, nitrous oxide (N_2O), CH_4, CO, non-methane hydrocarbons, NO_x, SO_2, and particulate matter are emitted [13]. This emission not only does pollute the environment but also causes serious impacts on human health due to polycyclic aromatic hydrocarbons [14] which have significant toxicological properties and are notably potential carcinogens. In addition, open burning of rice straw also threats security problems of fire disasters [15].

Anaerobic co-digestion of animal manure and agricultural byproducts has drawn increased attention [16]. Studies on biogas production by co-digestion of animal wastes with rice straw have attracted special interest. Biogas production has shown to be one of the key technologies for sustainable utilization of rice straw as renewable energy source [15]. Rice straw has such high organic matter with cellulose content of 25.4–35.5%, hemicelluloses of 32.3–37.1%, and lignin of 6.4–10.4% [17].

Rice straw is potential for biogas production because of high organic matter. The problems encountered in biogasification of rice straw are mainly related with high C/N ratio or low hydrolysis performance and digestibility because of high lignin content and its complex, stable and recalcitrant lignocellulosic structure [18], which needs a further balance of nutrients and destructive pretreatments if rice straw is used as substrate for biogasification [17, 19]. The objective of our research is to evaluate the effect of urea addition on biogas yield resulted from anaerobic co-digestion of cow dung and rice straw using semi continuous digester.

2. Materials and method

2.1. Digester preparation

Biogas production was carried out using semi-continuous anaerobic digester. The digester vessels were made of two 5-gallon drinking water containers as depicted in figure 1. The two containers were cut at their bottom and then combined by using fiber resin and let to dry for 24 hours. A hole was made on the digester to deliver biogas into a storage balloon trough a plastic tube. The tube was facilitated with a stop valve to close the piping for biogas volume measurement and biogas sampling.

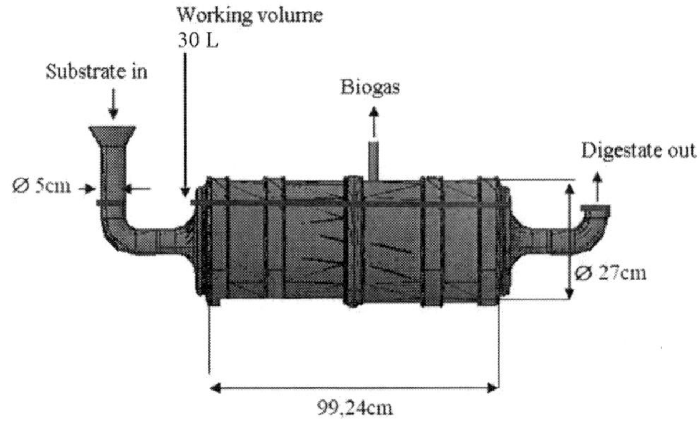

Figure 1. A self-designed semi-continuous digester prepared for the experiment

2.2. Substrate Preparation

Fresh cow dung, taken from the Department of Animal Husbandry, the University of Lampung, was used as a microbial seed source. Rice straw (figure 2) was of Ciherang variety and was collected from farmer field in Way Galih, Tanjung Bintang, Regency of South Lampung. Straw was sundried until its moisture content is about 12% (wet basis). The dried straw was chopped and ground to fine particles. Biomass size reduction is important to enhance biogas production [20]. Samples of straw and cow dung were analyzed for total solids (TS), volatile solids (VS), and carbon (C) and nitrogen (N) contents. Table 1 shows characteristics of each substrate. Urea granule with a nitrogen content of 46% was purchased from a local supplier and was used as external nitrogen source.

Figure 2. Rice straw: sun dried (left) and ground (right)

Table 1. Fresh substrate characteristic

Characteristic	Cow dung	Rice straw
Water content (%, wet basis)	71.0	11.0
Total solid (TS) (%, wet basis)	29.0	89.0
Ash (% TS)	25.04	28.48
Volatile solid (VS) (% TS)	74.96	71.52
C (%)	39.87	38.55
N (%)	1.42	0.58
C/N Ratio	28.08	66.46

2.3. Treatments and Loading

Anaerobic digesters can be classified into mainly three types based on the feeding strategy, namely batch, semi-continuous and continuous modes. In the semi-continuous system, the digester is periodically loaded with substrate according to a specific rate [21]. Most family scale biogas digesters installed in Indonesia use cow dung as substrate with semi-continuous loading mode.

The experiment was designed with TS content of about 10% of substrate mixture and TS ratio of 1:3 (straw:dung). With digester working volume of 30 L, and referring to table 1, then the composition of substrate for each digester is equivalent to 8.03 kg cow dung, 0.83 kg ground straw, and 21.10 L tap water. Initially, fine straw and cow dung were thoroughly mixed. Rice straw has so high C:N ratio that external nitrogen such as urea is required [22]. For this experiment, we prepared four level of urea addition, namely 0 (P1), 0.25 g/L (P2), 0.65 g/L (P3), and 1.3 g/L (P4). Table 2 shows substrate compositions along with their TS, VS, and C/N ratios. The experiment was conducted with three replications and 12 digesters were prepared.

Table 2. Treatment and substrate composition

Characteristic	P1	P2	P3	P4
Water content (%, wet basis)	89.48	89.49	89.51	89.50
Total solid (TS) (%, wet basis)	10.52	10.51	10.49	10.50
Ash (% TS)	25.66	25.69	25.70	25.61
Volatile solid (VS) (% TS)	74.34	74.31	74.30	74.39
C (%)*	39.75	39.75	39.75	39.75
N (%)*	1.34	1.59	1.98	2.64
C/N ratio*	30	27.3	24.3	20.5

*) Calculated based on Equation (4)

2.4. Analysis and Calculations

For determining the total solid (*TS*), a sample with a certain weight (W_1) was placed in ceramic vessels and dried in an oven (Memmert, type UM 500, Germany) at 105°C for 24 hours until constant weight. After cooling in the desiccator, the sample was weighed (W_2) for TS measurement. A part of this sample (W_3) was taken and burnt in a furnace (Barnstead International model FB1310M-33, USA) at 550°C for 3 hours for volatile solids (*VS*) determination. Five samples were taken for both fresh and spent substrate with an interval of one week. The average values were compared.

Total solid and *VS* are calculated by using equation (1) and (2), respectively:

$$TS\ (\%,\ wb) = \frac{W_2}{W_1} \times 100 \qquad (1)$$

$$VS\ (\%\ TS) = \frac{W_3 - \text{Ash}}{W_3} \times 100 \qquad (2)$$

In order to evaluate the digester efficiency, the destroyed or removed VS (VSr) was calculated using equation developed by Koch [23]:

$$VSr\ (\%) = \left[1 - \frac{VS_{out}(1 - VS_{in})}{VS_{in}(1 - VS_{out})}\right] \times 100 \tag{3}$$

where VS_{in} and VS_{out} refer to VS of fresh substrate (input) and spent substrate (output), respectively.

Carbon and nitrogen contents of each substrate were measured using element analyzer (Elementar Vario EL Cube, Germany). Carbon to nitrogen (C:N) ratio of the mixture is calculated using equation (4):

$$C:N = \frac{(C_c \times m_c) + (C_s \times m_s)}{(N_c \times m_c) + (N_s \times m_s) + 0.46 \times m_{Urea}} \tag{4}$$

where m is dry mass and subscripts c and s denote for cow dung and rice straw, respectively.

To evaluate process condition, temperature and pH of the substrate during the experiment were also checked daily. The temperature was monitored using a thermocouple inserted into the digester. The thermocouple tip was positioned around the center of digester as seen in figure 1. pH values were determined using pH meter (PHMETER, PH_009(I), China). The pH of fresh substrate was measured before its loading into the digester, while pH of spent substrate was measured just after it exits from the outlet of the digester.

Biogas production was determined using simple water displacement method. Initially biogas quality was simply observed daily by burning the biogas in a simple burner. If the biogas can be burnt it means that biogas contains enough methane. Around two weeks since the biogas can be burnt for the first time, the biogas is assumed to have a stable composition. At this time, the biogas was sampled using sampling bag to be analyzed its composition. The analysis was performed using gas chromatograph (Shimadzu GC 2014, Japan) with thermal conductivity detector (TCD), 4-m length of shin-carbon column, and Helium gas as carrier gas with flow rate 40 ml/min. Biogas yield (BY) was calculated from biogas production (BP) and VSr :

$$BY = BP/VSr \qquad (L/kg\ VSr) \tag{5}$$

3. Results and discussion

3.1. ANOVA test

Table 3 showed results from ANOVA test for some parameters to evaluate digester performance, including average pH, average temperature, total biogas yield, average daily biogas yield, and day at which biogas can be burnt for the first time. The explanation for each variable is incorporated in the following discussion.

Table 3. Summary of ANOVA test for six parameters to evaluate digester performance*).

Treatment	Average pH	Average T (°C)	Outlet VS (% TS)	Total Biogas (L)	Daily Biogas (L)	Day Biogas Burnt**)
P1	6.81 a	30.56 a	67.5 d	259.1 b	4.3 b	23.0 b
P2	6.84 a	30.42 a	67.1 c	297.0 c	5.0 c	15.3 a
P3	6.86 a	30.62 a	64.9 a	288.1 c	4.8 c	18.3 ab
P4	6.77 a	30.29 a	68.2 b	198.4 a	3.3 a	31.3 c

Note: *) numbers followed by the same letter in the same column is not significantly different at $\alpha = 5\%$
**) day at which the biogas can be burnt for the first time.

3.2. Operation condition

Temperature and pH are among the important factors influencing digester performance. Figure 3 presents daily pH values of substrate from all treatments. Abbasi *et al.* [1] noted that optimum pH range for anaerobic degradation processes satisfying the requirement for both activities and cell growth of anaerobic microorganisms is 5.5–8.5. In our experiments, digesters for co-digestion of cow dung and rice straw have initially basic pH value of 8.1 (P2 and P4) and 8.2 (P1 and P3). This actually was in the good range for anaerobic digestion process. During the first week, the pH decreased to a value of 6.5. In the second week the pH still fluctuated from around 6.5 to 7.0. Starting from the end of week two, the pH was practically stable between 6.7 and 7.0 with average value of 6.7 for P4 and 6.8 for the others. As presented in table 3, there was no significant difference on average pH value of all treatments. According to de Mes *et al.*, methanogenesis proceeds when the pH is close to neutral, and outside pH values of 6.5–7.5, the rate of methane production is low [24].

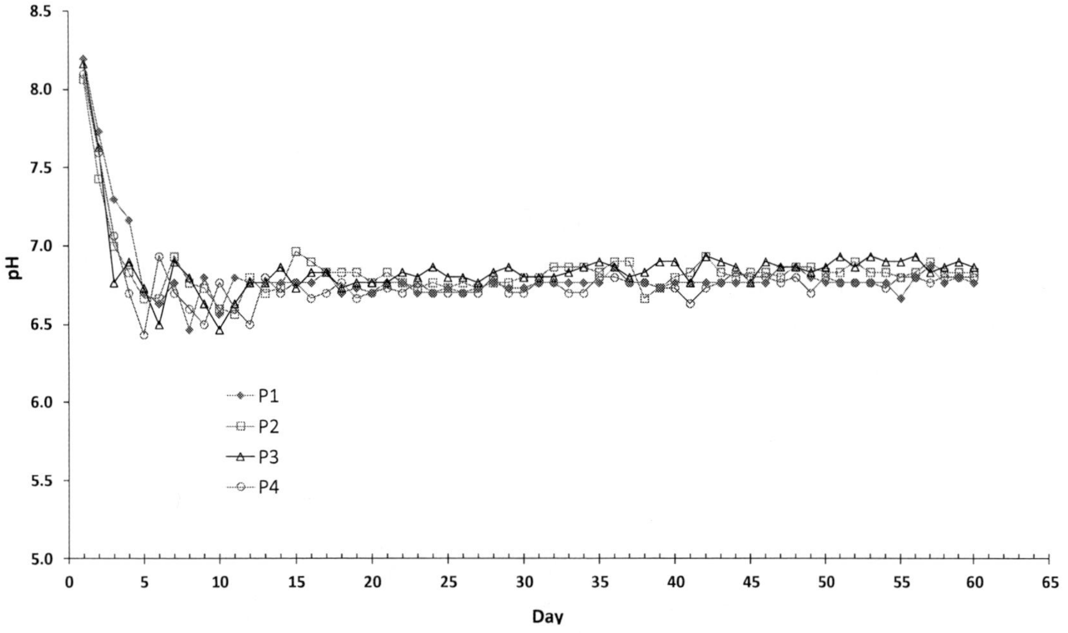

Figure 3. Change of daily pH of the substrate for different treatments

Figure 4 showed daily temperature of the digesters. All digesters operated in the mesophilic temperature region with average value of 30.56°C (P1), 30.42°C (P2), 30.62°C (P3), and 30.29°C (P4). It appears that the operating temperature of all digesters is very close to each other. There was no statistical difference of average daily temperature between all treatments. Daily temperatures for all digesters had a pattern similar to ambient temperature. This indicates that the digester temperature is strongly influenced by environmental conditions. In general, all digesters work at temperatures slightly higher than ambient temperatures (28.61°C). This is understandable because the overall reaction to the biogas process is a slightly exothermic, that is, producing heat [24].

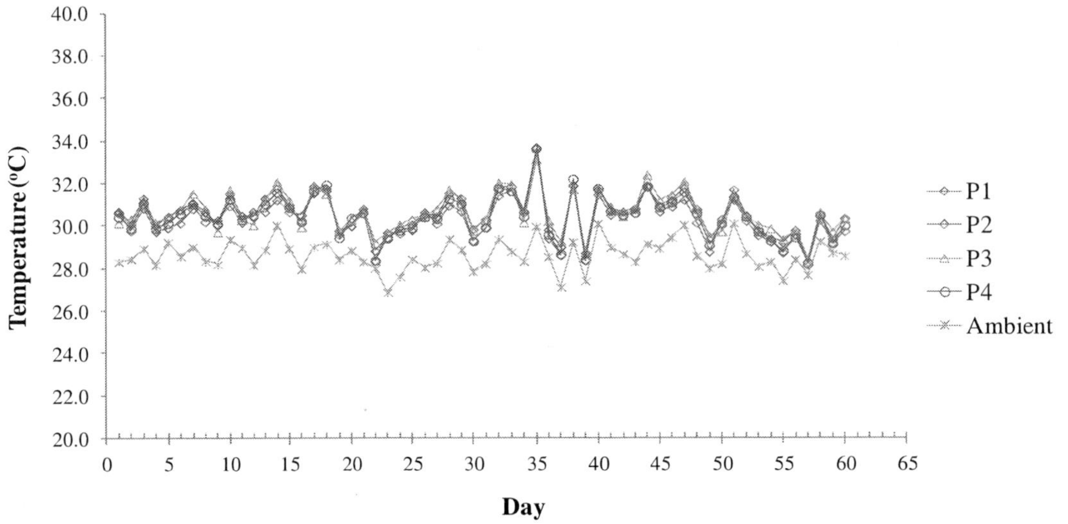

Figure 4. Comparison of daily digester temperature and ambient air.

Figure 5 shows that the digester temperature changed with time. During noon and afternoon, the digester temperature was higher than the temperature in the morning. The same was also observed for ambient temperature. It strengthens that the working temperature of the digester is greatly influenced by the ambient temperature.

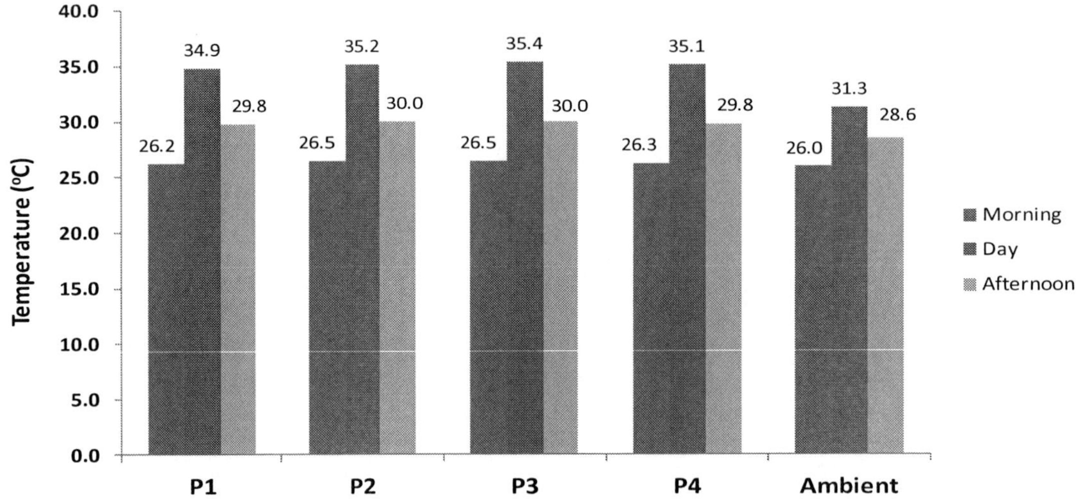

Figure 5. Average working temperature of digesters and ambient air: morning, noon, and afternoon

During anaerobic digestion process, substrate decomposition occurs. Table 4 presents characteristics of spent substrates for each treatment. The table also includes VS removal that is calculated from equation (3) and biogas yield calculated from equation (5). The addition of urea has resulted in significant difference on VS_{out} as well as *VS* removal. Organic material degradation (presented by *VS* removal) increased with increasing urea addition and achieved the highest (36% or 14 g VS/d) at urea addition of 0.65 g/L (treatment P3), and then decreased with more increase in urea addition. Recently, family size digesters using cow dung substrate have been reported to have average organic material removal of 51.32% [5]. This means that *VS* removal from current experiment is significantly lower than that of actual field practices. The addition of rice straw could be responsible for this low degradation of organic matter. Therefore, other pretreatments should be applied to the straw prior to use for biogas substrate.

Table 4. Average volatile solid (*VS*) removal

Treatment	VS_{in} (%, TS)	VS_{out} (%, TS)	VSr (%)	VSr (gVS)	BY (L/kgVSr)
P1	74.34	67.5	28	11.1	388.4
P2	74.31	67.1	29	11.5	434.2
P3	74.30	64.9	36	14.0	341.7
P4	74.39	68.2	26	10.2	322.9

Note: *VS* values are average of five measurements.

3.3. Biogas production

Figure 6 presents daily biogas production resulted from different treatments. During the first week, the digesters showed decreasing trend of biogas production. Initially, gas was produced from respiration of the substrates due to the existence of air filling void space in the digester. Starting day 6th digester with 0.65 g/L urea addition showed increase in biogas production; while other treatments practically start to increase by day 9th. As presented in table 3, addition of urea had significantly influenced average daily biogas production. Treatment with urea addition 0.25 and 0.65 g/L produced the highest daily biogas production (5.0 L/d for P2 and 4.8 L/d for P3) compared to those of other treatments. Table 3 also confirms that amount of urea addition affects biogas quality. Initially biogas is not able to be combusted for all treatments, indicating that biogas quality is still very poor (low methane content). Slowly, the quality increases and eventually the biogas containing high enough methane so that it can be burnt. The day at which biogas can be burnt for the first time is significantly affected by amount of urea addition with P2 treatment shows the fastest (day 15) followed by P3 (day 18), P1 (day 23), and P4 (day 31). Table 4 shows that P2 produced the highest biogas yield in term of biogas production per unit of degraded organic material (434.2 L/gVSr.d). This further strengthens that the addition of urea of 0.25 g/L is the best treatment for biogas production from a mixture of cow dung with rice straw.

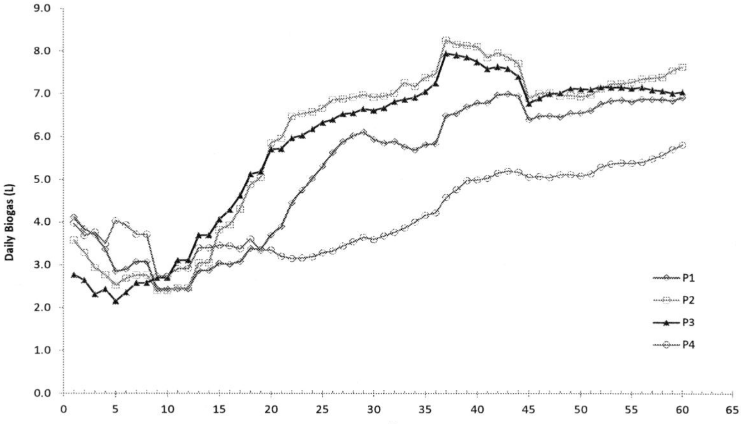

Figure 6. Daily biogas production from different treatments using 8-day moving average

Figure 7 presents cumulative biogas production. During the first three weeks, all digesters showed comparable cumulative biogas production. After that, the effect of urea addition appeared more clearly. Treatments P2 and P3 lead the biogas production with a total production of 297.0 L and 288.1 L. Addition more urea resulted in detrimental effect on biogas production. Treatment P4 with 1.3 g/L urea addition produced the lowest biogas production (198.4 L) with daily average of 3.3 L. This correlates to low C/N ratio (20.5) and the low VS removal (26%).

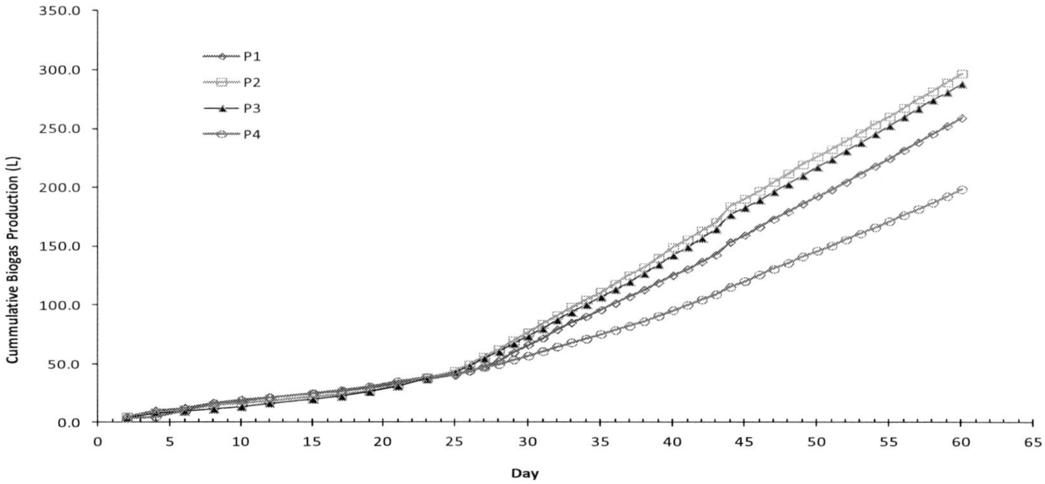

Figure 7. Cumulative biogas production from different treatments.

3.4. Methane yield

Table 5 presented biogas composition from different treatments. The methane content of P2 (urea addition 0.25 g/L) was the highest. Increasing more urea addition resulted in lower methane content. Treatment P4 with urea addition of 1.3 g/L produced biogas with lowest methane content (37.04%). Based on this composition we have calculated methane yield that was also presented in table 4 (last column). It is obvious that treatment P2 with urea addition of 0.25 g/L gave the highest methane yield; whilst P4 with urea addition of 1.3 g/L produced the lowest methane yield. Table 6 compares methane yield of our work with the values already reported by other research. Our work is a half of the maximum value reported by Lei *et al.* (2010) [25]. This meant that there is possibilities to increase methane (biogas yield) by, for example adding other pretreatments.

Table 5. Biogas composition (%) and methane yield (L/kg *VSr*)

Treatment	CH_4	CO_2	N_2	CH_4 Yield
P1	49.82	40.43	9.75	193.5
P2	50.12	39.97	9.91	217.6
P3	46.49	43.15	10.36	158.9
P4	37.04	35.45	27.56	119.6

Note: Biogas composition was measured around two weeks after the biogas can be burnt for the first time.

Table 6. Comparison of methane yield (L/kgVSr)

Condition	CH_4 Yield	Reference
Cow dung 3: Straw 1, TS 10%, Urea 0.25 g/L; 30.6°C	217.6	This work
Batch, 37°C; hydrothermal pretreatment 5% NaOH	132.7	Chandra *et.al.* (2010) [26]
Phosphate addition 115 mg/L; 25°C	440.0	Lei *et al.* (2012) [25]
Batch, kitchen waste : pig manure : straw = (0.4) : (1.6) : (1); 37°C	383.9	Ye *et.al.* (2013) [27]
Straw 10 mm, preheat 110°C, ammonia 2%; 35°C	247.2	Zhang & Zhang (1999) [28]

4. Conclusions

Rice straw is promising for biogas co-substrate. The addition of rice straw as co-substrate potentially improve total biogas yield. Urea addition influenced biogas production and biogas quality from co-digestion of rice straw and cow dung using semi-continuous digester. Substrate mixture with urea addition of 0.25 g/L at which C/N ratio is 27.3 was the best treatment in terms of biogas yield (434.2 L/kg *VSr*/d), day at which the biogas can burnt for the first time (day 15), as well as its methane content (50.12%) and methane yield (217.6 L/kg *VSr*/d).

Acknowledgement

This work was financially supported by DGHE (Directorate General of Higher Education), Ministry of National Education though Strategis Nasional (STRANAS) research scheme. Contract number: 071/SP2H/LT/DRPM/IV/2017585.

References

[1] Abbasi T, Tauseef S M and Abbasi S A 2012 *Biogas Energy* (New York: Springer) pp 1–10
[2] Gomez C D C 2013 Biogas as an energy option: an overview *The Biogas Handbook: Science, Production and Applications*, ed A Wellinger *et al.* (Cambridge: Woodhead Publishing Limited) pp 1–16
[3] Higgins K T 2008 Anaerobic gas and cash *Food Engineering* **80**(4) 119–20
[4] Petis M 2006 Anaerobic digestion plant yields nutrients and power *BioCycle* **47**(12) 61–62
[5] Haryanto A, Cahyani D, Triyono S, Murdapa F and Haryono D 2017 Economic benefit and greenhouse gas emission reduction potential of a family-scale cow dung anaerobic biogas digester *Int. Journal of Renewable Energy Development* **6**(1) 29–36
[6] IRRI (International Rice Research Institute) 2017 Rice Straw. http://ricestraw.irri.org/ (October 29, 2017)
[7] Makhrani 2014 Potential analysis of rice straw as an alternative energy source for substitute coal in Indonesia *Applied Mechanics and Materials* **554** 276-280
[8] Sulaiman A A, Simatupang P, Las I, Jamal E, Hermanto, Kariyasa I K, Syahyuti, Sumaryanto S, Suwandi and Subagyono K 2017 Success Self-Sufficiency: Indonesia Becomes a World Food Baron 2045(in Indonesian) (Jakarta: Ministry of Agriculture)
[9] Sidjabat O 2011 Prospect of rice straw potential for bioenergy in indonesia as substituted gasoline and reducing of environmental problem *Ketenagalistrikan dan Energi Terbarukan* **10**(2) 113-122
[10] Mandal K G, Misra A K, Hati K M, Bandyopadhyay K K, Ghosh P K and Mohanty M 2004 Rice residue- management options and effects on soil properties and crop productivity *Food, Agriculture & Environment* **2**(1) 224-231
[11] Buresh R and Sayre K. 2007. Implications of straw removal on soil fertility and sustainability *Expert Consultation on Biofuels* (Los Banos: IRRI) p 34
[12] Zhang R and Zhang Z 1999 Biogasification of rice straw with an anaerobic-phased solids digester system *Bioresource Technology* **68** 235–245
[13] Gadde B, Menke C and Wassmann R 2009 Rice straw as a renewable energy source in India, Thailand, and the Philippines: Overall potential and limitations for energy contribution and greenhouse gas mitigation *Biomass and Bioenergy* **33** 1532–46
[14] Korenaga T, Liu X and Huang Z 2001 The influence of moisture content on polycyclic aromatic hydrocarbons emission during rice straw burning *Chemosphere - Global Change Science* **3**(1) 117-122
[15] Lei Z, Chen J, Zhang Z and Sugiura N 2010 Methane production from rice straw with acclimated anaerobic sludge: Effect of phosphate supplementation *Bioresource Technology* **101** 4343–48
[16] Song Z, Yang G, Han X, Feng Y and Ren G 2013 Optimization of the alkaline pretreatment of rice straw for enhanced methane yield. *BioMed Research International* 1-9

[17] Jin S and Chen H 2006 Structural properties and enzymatic hydrolysis of rice straw *Process Biochemistry* **41** 1261–64

[18] Chen Y, Cheng J J and Creamer K S 2008 Inhibition of anaerobic digestion process: A review *Bioresource Technology* **99** 4044–64

[19] Ward A J, Hobbs P J, Holliman P J and Jones D L 2008 Optimisation of the anaerobic digestion of agricultural resources *Bioresource Technology* **99** 7928–40

[20] Kratky L and Jirout T 2011 Biomass size reduction machines for enhancing biogas production *Chemical Engineering & Technology* **34**(3) 391–399

[21] Aboudi K, Álvarez-Gallego C J and Romero-García L I 2015 Semi-continuous anaerobic of sugar beet byproduct and pig manure: Effect of the organic loading rate (OLR) on process performance *Bioresource Technology* **194** 283–290

[22] Malik R K, Singh R and Tauro P 1987 Effect of inorganic nitrogen supplementation on biogas production *Biological Wastes* **21** 139–142

[23] Koch K 2015 Calculating the degree of degradation of the volatile solids in continuously operated bioreactors *Biomass & Bioenergy* **74** 79–83

[24] de Mes T Z D, Stams A J M, Reith J H and Zeeman G 2003 Methane production by anaerobic digestion of wastewater and solid wastes *Bio-methane & Bio-hydrogen: Status and Perspectives of Biological Methane and Hydrogen Production* (Dutch: Dutch Biological Hydrogen Foundation)

[25] Lei Z, Chen J, Zhang Z and Sugiura N 2010 Methane Production from Rice straw with acclimated anaerobic sludge: Effect of phosphate supplementation *Bioresource Technology* **101** 4343–48

[26] Chandra R, Takeuchi H and Hasegawa T 2012 Hydrothermal pretreatment of rice straw biomass: A potential and promising method for enhanced methane production *Applied Energy* **94** 129–140

[27] Ye J, Li D, Sun Y, Wang G, Yuan Z, Zhen F and Wang Y 2013 Improve biogas production from rice straw by co-digestion with kitchen waste and pig manure *Waste Mangement* **33** 2653–58

[28] Zhang R and Zhang Z 1999 Biogasification of rice straw with an aerobic-phased solids digester system *Bioresource Technology* **68** 235–245

IOP Conf. Series: Earth and Environmental Science **147** (2018) 012033 doi:10.1088/1755-1315/147/1/012033

The Integration of Gasification Systems with Gas Engine to Produce Electrical Energy from Biomass

K Siregar[1], R Alamsyah[2], Ichwana[1], Sholihati[3], S B Tou[4] and N C Siregar[2]

[1] Department of Agricultural Engineering, Syiah Kuala University, Indonesia
[2] Center for Agro-Based Industry, Ministry of Industry, Indonesia
[3] Department of Agricultural Industry Technology, Serambi Mekkah University, Indonesia
[4] Ministry of Forestry and Environment, Aceh Province, Indonesia

E-mail: ksiregar.tep@unsyiah.ac.id

Abstract. The need for energy especially biomass-based renewable energy continues to increase in Indonesia. The objective of this research was to design downdraft gasifier machine with high content of combustible gas on gas engine. Downdraft gasifier machine was adjusted with the synthetic gas produced from biomass. Besides that, the net energy ratio, net energy balance, renewable index, economic analysis, and impact assessment also been conducted. Gas engine that was designed in this research had been installed with capacity of 25 kW with diameter and height of reactor were 900 mm and 1000 mm respectively. The method used here were the design the Detailed Engineering Design (DED), assembly, and performance test of gas engine. The result showed that gas engine for biomass can be operated for 8 hours with performance engine of 84% and capacity of 25 kW. Net energy balance, net energy ratio, and renewable index was 30 MJ/kWh-electric; 0.89; 0.76 respectively. The value of GHG emission of Biomass Power Generation is 0.03 kg-CO2eq/MJ. Electrical production cost for Biomass Power Generation is about Rp.1.500,/kWh which is cheaper than Solar Power Generation which is about of Rp. 3.300,-/kWh.

1. Introduction

In 2015, Indonesia set a target to realize 23% of primary energy supply from modern renewable energy by 2025 [1]. Indonesia is the largest country in ASEAN, so the need for energy continues to increase. The role of renewable energy is needed to meet the growth of energy demand. There is a large difference in the ratio of electrification throughout Indonesia, where for Papua the electricity ratio is less than 40% while in the Capital of Jakarta is close to 100%. The Indonesian state electricity company PT PLN (Persero) is seeking to increase the electricity ratio by 98% in 2022 [2].

The government has committed to realize 35,000 MW of electricity supply within 5 years (2014-2019). The contribution of new and renewable energy-based power plants is still quite low, i.e. 9.9% Hydropower (PLTA), Geothermal Power Plant (PLTP) of 2.6% and other renewable energy (EBT) below 0.5%.

The process of gasification consists of four stages of the process based on the difference of temperature range , which are drying ($T > 150$ °C), pyrolysis (150 °C $< T <$ 700 °C), oxidation (700 °C $< T <$ 1500 °C), and the reduction (800 °C $< T <$ 1000 °C). The results of this system produce combustible gases (CO, H_2, CH_4, etc). Gasification technology is suitable for the area which has large quantities of biomass, which currently have not been yet traversed by electrical network system such as PT.PLN (Persero) and private companies. In this area the price of diesel fuel is very expensive with price is about Rp. 50,000 per liter. This condition caused the price of electrical energy per hours would be very expensive for all of them e.g. in the Island of Papua.

The gasification is designed to produce combustible gas (CO, H_2, CH_4) [3, 4]. These combustible gas are produced to replace fossil fuel as engine fuel to generate electricity. However, the properties and characteristics of these combustible gas should meet the characteristic of the engine as every engine has its own specific characteristic. Some engines are designed working more properly with methane gas (CH_4) but some work more properly with carbon monoxide gas (CO). Biomass Power Generation (BPG) designed in this research adjusted with combustible gas produced that is CO and H_2. Therefore the objective was to design biomass gasification system that suits the dominant gas produced from biomass. High level of tar contained in produced combustible gas contaminates the engine filter rapidly. Therefore, the machine cannot be operated at a longer time. The objectives of this research is to design gasification system with high content of combustible gases (CO, H_2, CH_4, etc). The gasification system will be connected to gas engine to produce electrical energy from biomass. Besides that, the net energy ratio, net energy balance, renewable index, economic analysis, and impact assessment also been conducted.

2. Methods

2.1. Tools and materials research

The biomass power plant (PLTBm) consists of biomass tank, reactor, cyclone, tar wet scrubber, gas filter and gas engine. Gasification engine was made using downdraft type in order to produce fewer tar values. The specifications of materials, materials and gasification machine capacity produced are shown in table 1.

Table 1. Specification gasification engine/Biomass power plant

No	Description	Technical specifications	Quantity
1	Gasification type	*Downdraft* gasifier	1 set
2	*Carbon steel* tabung reactor	OD 15 – 20 inchi x 1 mm thickness	1 pcs
3	*Engine*	25 kW	1 pcs
4	*Blower*	Daya ¼ - ½ HP	1 pcs
5	*H-Beam*	Steel, 2 – 4 inchi	20 m
6	*Steel plat*	50 mm width x 5 mm *thickness*	20 m
7	*Pipe from carbon steel*	OD 1 – 2 inchi, 1 mm *thickness*	30 m
8	Power cable	Merk ETERNA	50 m
9	Temperature indicator	Merk SIKA	6 pcs
10	*Pressure indicator*	Merk SIKA	4 pcs
11	*Control panel*	ASTM Standard	1 set
12	Portable temperature	Termometer laser dan digital	1 pcs
14	Heast resistant hoses	Material tahan panas, ± 100 °C	5 m

2.2. Research method

2.2.1. Analysis of synthetic gases. The method used in this research was to analyze the gases produced from the downdraft gasifier system which was designed and made. The biomass utilized in this research was wood. Wood was fed on the top of downdraft gasifier and then enters biomass storage in gasifier. The biomass would enter pyrolysis phase about 200 - 700°C, and then combustion and reduction phase about 1000 - 1400°C. The gases produced passed the wet tar scrubber and gas filter to ensure the tar limit meet the requirement and clean gases was used to run the gas engine. Engine performance was measured by test the effectiveness power produced for specific time given.

2.2.2. Analysis of energy, econonmy, and life cycle assessment. The important parameter regarding analysis of energy is Net Energy Ratio (NER), Net Energy Balance (NEB), and Renewable Index (RI). It can be stated as formula as below :

$$NER = \frac{E_o}{E_i} \tag{1}$$

$$NEB = E_o - E_{pr} \tag{2}$$

$$RI = \frac{E_{rn}}{E_{pr}} \leq 1 \tag{3}$$

The determination of the value of Input Energy (E_i), Process Energy (E_{pr}), Renewable Energy (E_{rn}), Output Energy (E_o) are shown on equations 4, 5, 6 and 7.

$$E_i = E_{bm} \tag{4}$$

$$E_{rn} = E_{bm} \tag{5}$$

$$E_{pr} = E_f + E_{bm} + E_{el} \tag{6}$$

$$E_o = E_{ge} \tag{7}$$

Where E_i is input energy of biomass (E_{bm}); E_{rn} is renewable energy of biomass (E_{bm}); E_{pr} is process energy which is the sum of fossil energy (E_f), biomass energy (E_{bm}), and electrical energy (E_{el}), while E_o is output energy of electric power generated by gas engine (E_{ge}).

Impact Assessment was measured by using Life Cycle Assessment (LCA) method which is described below :

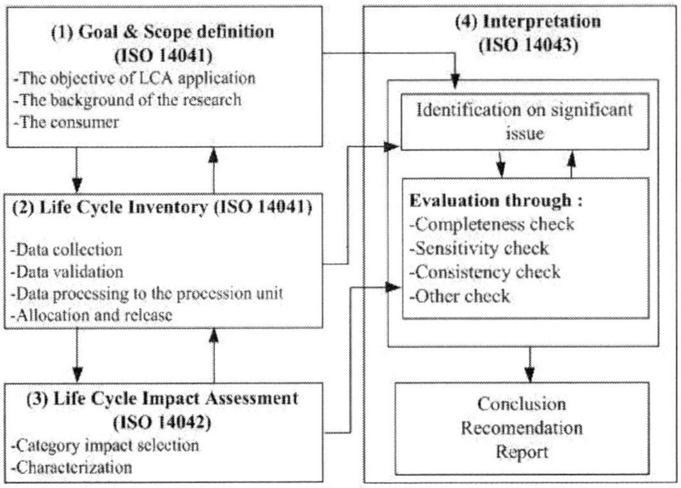

Figure 1. Impact Assessment by Life Cycle Assessment (LCA) Method

To estimate the economic feasibility, the parameter used is to compare electrical production cost frorm Biomass Power Plant and Diesel Power Generation in Rp/kWh unit.

3. Results

3.1. Biomass Power Plant

The biomass power plant produced has been connected to the gas engine which in its series already uses tar wet scrubber and gas filter. Overall the biomass power plant engine consists of tanks of biomass, reactor, cooler, cyclone, tar wet scrubber, gas filter, blower, and valve system control. The series of PLTBm engines generated in this study are shown in figure 2. This gasification technology was aimed to produce combustible gas (CO, H_2, CH_4). This combustible gas is used to replace fossil fuel as an engine fuel used to generate electricity. The biomass fuel used in the research is wood. With relatively uniform shape and size, congestion can be avoided [5], so in this study, woodchopper was also used as a cutting tool that can produce uniformity of biomass sizes.

The type of downdraft gasifier designed in this study is a fixed bed downdraft gasifier with a direction of air flow from the bottom up as shown in figure 3a and figure 3b. The advantages of this type of downdraft are it is not too sensitive to tar and can easily adapt to the amount of biomass feed. Another advantage was that it appears to be cleaner because the tar and oil will burn . This is reinforced by the tar data generated from the various types of gasifiers produced as shown in table 2.

IOP Conf. Series: Earth and Environmental Science **147** (2018) 012033 doi:10.1088/1755-1315/147/1/012033

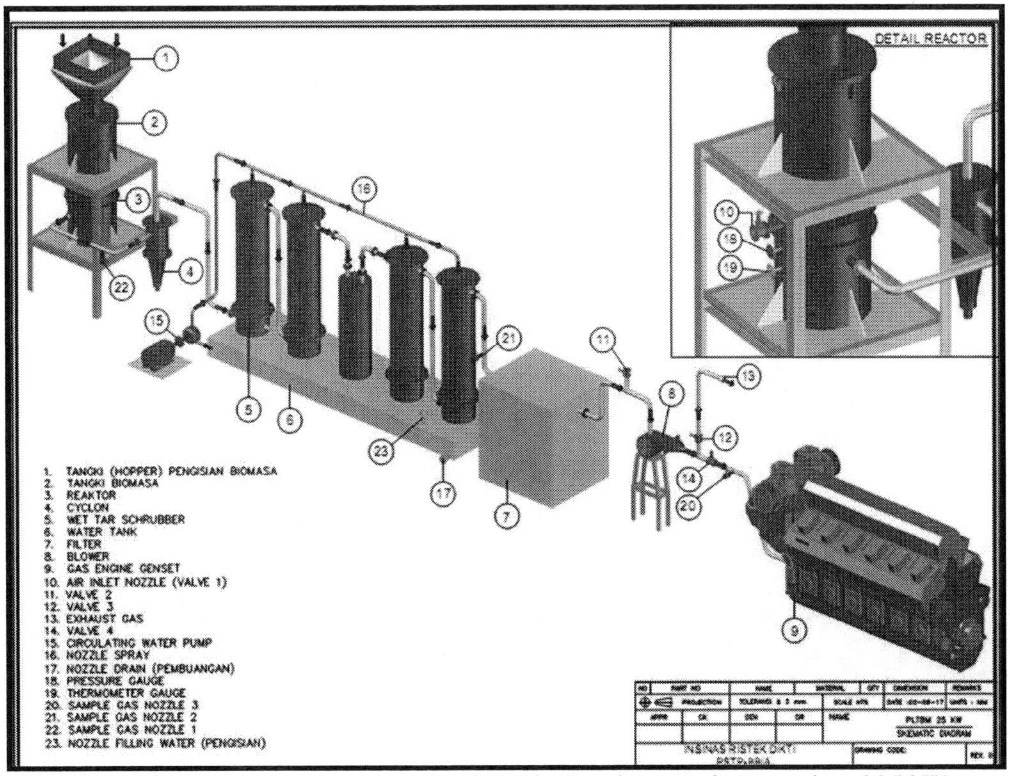

Figure 2. A system gasification type downdraft and gas engine capacity of 25 kW

Figure 3a. Gasifier type downdraft designed and used in this research

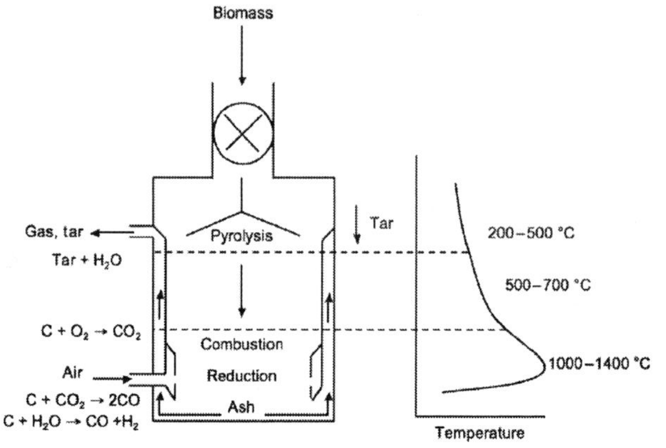

Figure 3b. Commonly phase on Gasifier type downdraft [4]

Table 2. Tar content resulting from various types of gasification produced [6]

Type of gasification	The average fuel gas concentrations in the tar is produced (g/Nm^3)	Tar percentage of biomass used
Downdraft	< 1	< 2
Fluidized bed	10	1 – 5
Updraft	50	10 – 20
Entrained flow	ignored	

3.2. Design and Main Components of Biomass Power Plant with Capacity of 25 kW

The gasifier used in this study has a total dimension of 1340 mm diameter and a height of 2455 mm as shown in figure 3 above. Cyclon has a length of 580 mm, a width of 426 mm with a height of 1766 mm. Gas filter has a length of 700 mm, a height of 700 mm and a width of 700 mm. Tar wet scrubber consists of 5 tubes (300 mm tube diameter) which are strung together with the aim of capturing the tar still contained in the combustible gas produced from the gasification reactor. Gasification reactor itself has total dimensions of 1750 mm width and 1300 mm height.. The design has considered resident time of biomass in gasifier reactor in order to produce synthetic gas optimally. The comparison of height to diameter of 1.8 has been chosen. Overall this series of Biomass Power Plant with Engines is shown in figures 4 and 5.

Figure 4. Biomass Power Generation with Capacity of 25 kW

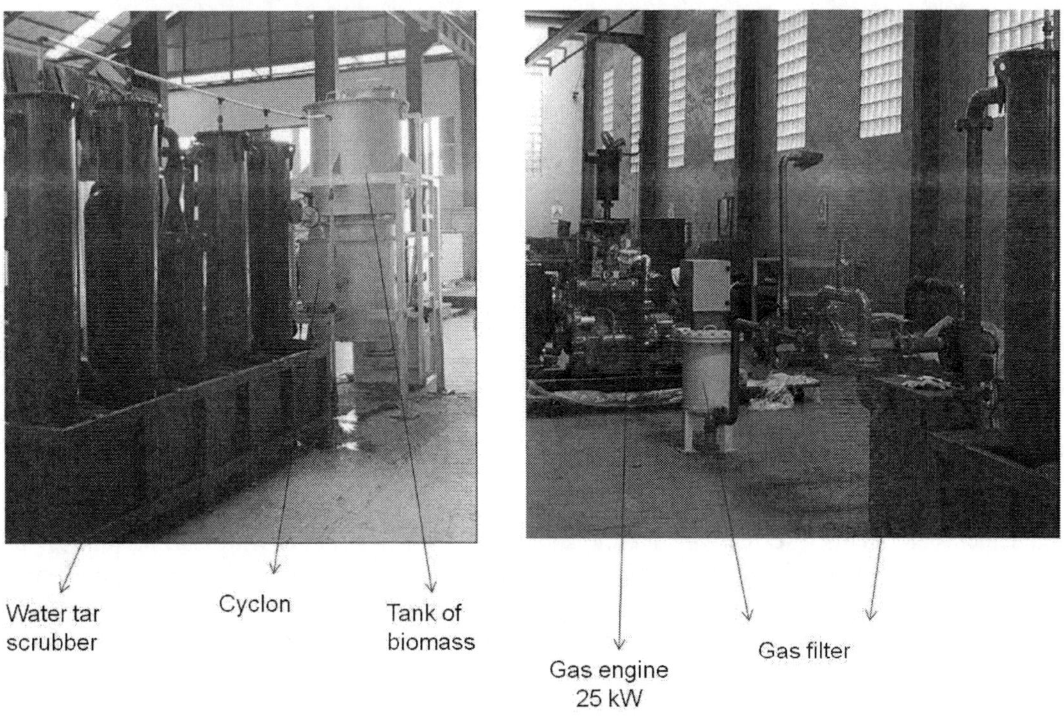

Figure 5. Biomass Power Generation with Capacity of 25 kW

3.3. Engine Performance Test

To ensure the composition of the fuel gas content, the laboratory testing was carried out and the results is shown on table 3. Laboratory test result that the gases produced contained combustible gas those are CO and H_2 in sufficient concentration. In case of biogas in which dominant gas was CH_4, then the gas engine designed is which is suitable for CH4.

Table 3. The average composition of the combustible gases

No	Component name	Concentration (% -vol)	Normalization 100 %
1	Hydrogen (H_2)	11.22	11.43
2	Nitrogen (N_2)	48.23	51.45
3	Carbon monoxide (CO)	23.16	24.86
4	Methane (CH_4)	1.72	1.85
5	Carbon dioxide (CO_2)	10.26	11.02

The gas engine with capacity of 25 kW was operated in specified operating condition and run for 8 hours. The result was that gas engine from biomass-based gas (synthetic gas) can give power with effectiveness of 84%. The net energy ratio obtained in this research was 0.89. This value less than 1 because much of energy losses into the environment because the material of reactor body and gas piping have not been coated with insulators.

The renewable index (RI) on this research was 0.76 which is close to 1. The closer the value to 1 the better the renewable index is. This value assumed that the use of unrenewable fuel on this research was very little. The use of diesel / kerosene is only on the starting point, so are the electricity on blower as the initial trigger to pull the combustible gas from the reactor. Net energy balance on this research was 30 MJ/kWh-electric.

3.4. *Economic feasibility*

Economic feasibility studies has been conducted. The assumption used was the average price of biomass about Rp.700/kg and the rate of biomass use on electricity production about 1.4 - 2 kg/kWh-electrical. Therefore, the cost of electricity produced about Rp.1.650 per kWh. This cost is cheaper than electrical production cost from Diesel Power Generation which is about Rp. 3.300,-/kWh.

3.5. *Analysis of Impact Assessment*

The calculation of greenhouse gas emission (ERK) was approached using Life Cycle Life Assessment Method ISO 14040. The GHG emission value for Biomass Power Plant is 0.04 kg-CO2eq / MJ, and the Diesel Power Plant engine is 0.111 kg-CO2eq / MJ or in the other words the PLTBm GHG emission value was almost 3 times lower than the Diesel Power Plant. Siregar et al. (2015) states that the production of biodiesel in catalyst can reduce global warming emissions 37,83% compared to fuel oil (diesel). In the national arena this activity will support the government target to reach renewable energy mix up to 23% by 2025. This activity is also in line with the achievement of 35,000 MW electricity production until 2019.

4. Conclusions

The conclusion of this research was that the Biomass Power Generation with capacity of 25 kW has already been designed and operated with performance test of 84 % from installed capacity. It resulted *net energy balance*, *net energy ratio*, and *renewable index* was 30 MJ/kWh-electric; 0.89; 0.76 respectively. The value of GHG emission of Biomass Power Generation is 0.03 kg-CO_2eq/MJ. Electrical production cost for Biomass Power Generation is about Rp.1.500,/kWh, whereas electrical production cost for Solar Power Generation is about Rp. 3.300,-/kWh.

Acknowledgement

The authors say thank you to KEMRISTEKDIKTI for its support under joint research INSINAS RISET PRATAMA KEMITRAAN 2017.

IOP Conf. Series: Earth and Environmental Science **147** (2018) 012033 doi:10.1088/1755-1315/147/1/012033

References

[1] Department of Energy and Mineral Resources (DESDM) 2005 Blueprint of National Energy Management (in Indonesia) (*Workshop Sosialisasi Blue Print Pengelolaan Energi Nasional 28-29 June 2005.*)
[2] IRENA 2017 Renewable Energy Prospects: Indonesia, a REmap analysis, International Renewable Energy Agency (IRENA) (Abu Dhabi) accassed: www.irena.org/remap.
[3] Pranolo H 2013 The Potential of aplication technolgy gasification with corn biomass as energy alternative in the village *National Seminar Nasional of Renewable Energy in Indonesia* (Purwokerto: Jendral Sudirman of University)
[4] Reed T B and Das A 1988 *Handbook of Biomass Downdraft Gasifier Engine Systems* (Colorado: Solar Energy Research Institute)
[5] Higman C and Van der Berg M 2003 *Gasification* (Amsterdam: Elsevier Science)
[6] Milne T A and Evans R J 1998 *Biomass Gasifier "Tars": Their Nature, Formation, and Conversion National* (Missouri: Midwest Research Institute)

IOP Conf. Series: Earth and Environmental Science **147** (2018) 012034 doi:10.1088/1755-1315/147/1/012034

Analysis of potency and development of renewable energy based on agricultural biomass waste in Jambi province

W H Devita[1], A M Fauzi[1] and Y A Purwanto[2]

[1] Department of Agroindustrial Technology, Bogor Agricultural University, Indonesia
[2] Department of Mechanical and Biosystem Engineering, Bogor Agricultural University, Indonesia

E-mail: wildaharliadevita@gmail.com

Abstract. Indonesia has the big potency of biomass. The source of biomass energy is scattered all over the country. The big potential in concentrated scale is on the island of Sumatera. Jambi province which is located in Sumatra Island has the potency of biomass energy due to a huge area for estate crop and agriculture. The Indonesian government had issued several policies which put a higher priority on the utilization of renewable energy. This study aimed to identify the conditions and distribution of biomass waste potential in Jambi province. The potential biomass waste in Jambi province was 27,407,183 tons per year which dominated of oil palm residue (46.16%), rice husk and straw (3.52%), replanting rubberwood (50.32%). The total power generated from biomass waste was 129 GWh_{th} per year which is consisted of palm oil residue (56 GWh_{th} per year), rice husk and straw (3.22 GWh_{th} per year), rubberwood (70.56 GWh_{th} per year). Based on the potential of biomass waste, then the province of Jambi could obtain supplies of renewable energy from waste biomass with electricity generated amount to 32.34 GWh_e per year.

1. Introduction

Indonesia today is currently facing energy problems. Indonesian energy condition that occurs due to the growth in energy consumption that has not been matched by sufficient energy supply, the increased of energy consumption by an average of 3.46% per year [1]. Energy supply crises that have been facing by Indonesia has pushed the government to issue a policy on the utilization of renewable energy to meet energy needs. One of the policies that have issued attached to the Presidential Decree No. 2 of 2015 on the National Medium Term Development Plan (RPJMN) 2015-2019. The policy is explicitly state that the issue of 2015-2019 science and technology development strategy was the increase of science and technology capacity of the ability to give a real contribution to the competitiveness of the production sector, sustainability and utilization of natural resources, and the preparation of Indonesian society to meet the advanced and modern global life. Furthermore, the research focused on the mandated areas of Long-Term Development Plan (RPJPN) 2005-2025, which is new and renewable energy as it mentioned in the regulation.

Biomass is a renewable resource, and the energy from it called renewable energy [2]. The most suitable biomass as energy resources is biomass with the low economic value. An ideal type of biomass to be utilized is biomass that has lost the primary product. Currently, biomass potential in Indonesia that can be used for energy generation is available in a waste of agricultural, plantation, forest and livestock [3]. Biomass for energy has the highest value because it's characteristics. Some of the characteristics are renewable and sustainable, can be stored, replaced, and transported, high of availability, and carbon neutral. Biomass from waste (agricultural, plantation, forest and livestock) and garbage waste can be an environmentally friendly source of energy because biomass is derived from non-fossil organic materials that produced non-pollutants CO_2 from it burning results [2].

Jambi province has the biggest biomass potential sources in some area. The potential sources came from agricultural, plantation, forest, livestock and municipal waste. Each region in Jambi province produces biomass from waste by palm oil, coconut, rice, cassava, maize, rubber, cow liquid waste, municipal waste, pulp, industrial forest, and wood industry. The information from data that accessed through the website from Directorate General of Estate Crops (Ditjenbun) and Central Agency of Statistics (BPS) of Jambi province. It is known that solid waste from plantation and processing industry of palm oil, rubber and rice milling waste have the most potential biomass source to be developed into energy generation. In 2014, Jambi province had a palm oil plantation area fixed a total of 692,967 ha, and the agricultural area had a rice harvesting fixed a total of 145,990 ha [4, 5]. Rubber plantation area fixed a total of 664,704 ha in 2015 [12].

Renewable energy development potency in Jambi province is quite large to produce energy from biomass, hence the objectives of this study were to identify the potential condition and distribution of agricultural solid waste biomass that can be utilized in each region. As is known the land area that has the highest value in Jambi province are palm oil with total production by 1.773.735 tons per year, rice by 664,720 tons per year and rubber by 328,581 tons per year [4, 5, 12]. The identification result of the potential condition and agricultural waste distribution are expected to be used as information for utilization in renewable energy development.

1.1. Production and total area of crops

Palm oil production increased the amount to 774,116 tons from 2006 to 2015. The increase in palm oil production by an average of 8.78% per year in eight times increased within ten years. At the same period, an increase occurred average palm oil area total 8.89%, i.e. 46,437 ha in seven times. The highest increase of palm oil production occurred in 2012, with an increase 459,449 tons also corresponded to 32.22% of the production total 1,426,081 tons. The highest increase in palm oil area totals up to 29.23%. Despite production decreased 7.21% in 2013 and area total decreased average 2.55% in 2010 and 2013, palm oil production tends to increase. The production changes more clearly are shown in figure 1.

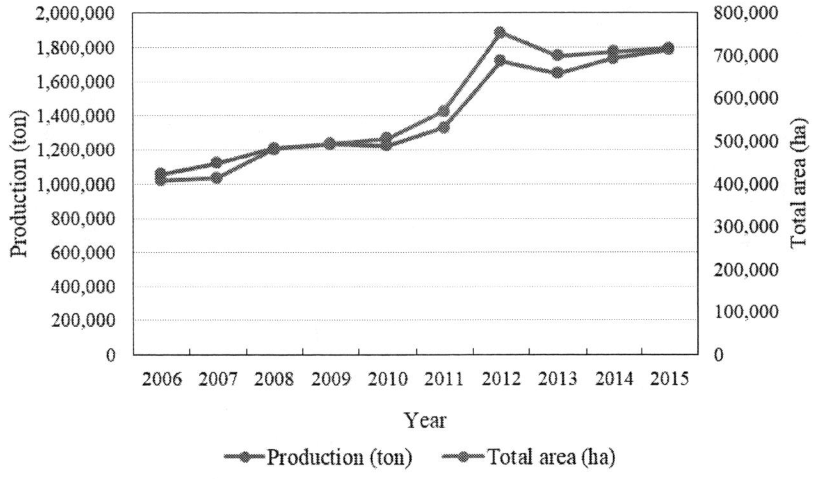

Figure 1. Trend of palm oil production and plantation area in 2006-2015.
Source: [11, 12, 14, 18, 19]

From the results rice production data, known 2005 to 2015 the change annually. Rice production decreased the amount to 38,149 tons (6.58%) from 2005 to 2015. Changes in production were averaged 5.90% per year with an average increase of 5.55% from 579,635 tons to 541,486 tons in five times increased within ten years. The highest increase of production amount occurred in 2009 by 63,243 tons, corresponded to 10.87% of the total production 581,704 tons in 2008. At the same time of production changes, total rice area decreased the amount to 32,727 ha (21.12%) from 2005 to 2015. The lowest production decreased down to 18.54% in 2015. Annually an average change 7,343 ha (5.03%) with an average increase up to 5.10% from 154,941 ha to 122,214 ha in four times increased. The production changes more clearly are shown in figure 2.

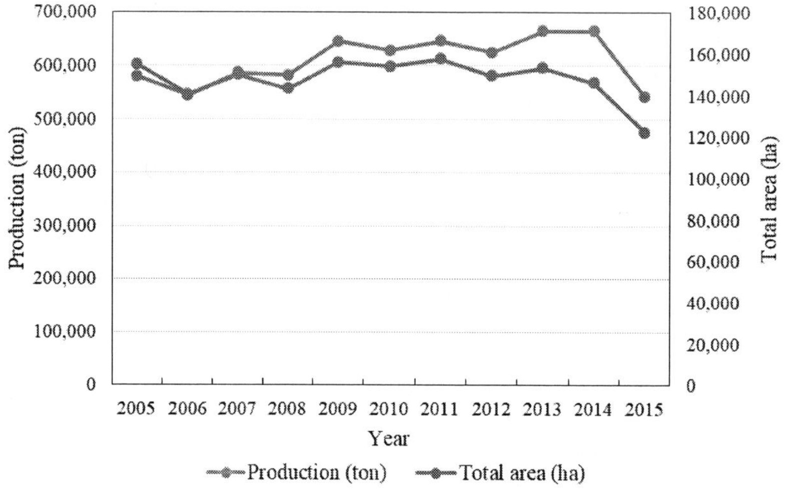

Figure 2. Trend of rice production and rice harvesting area in 2005-2015.
Source: [5, 12, 14, 15, 16, 17]

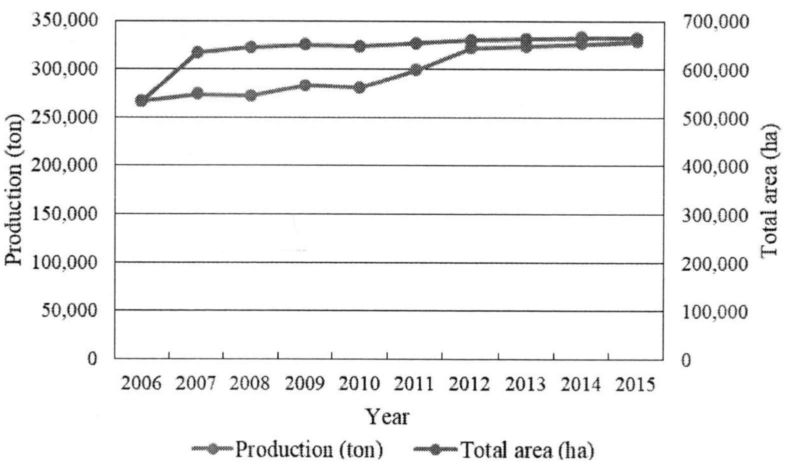

Figure 3. Trend of rubber production and rubber plantation area in 2006-2015.
Source: [12]

The total increase in production and a rubber plantation area from 2006 to 2015 by 22% and 24%. Amount of production increase and plantation area is 62,318 tons and 131,455 ha. Increased production of the total production of 266.263 tons in 2006 and the total production of 328.581 tons in

2015. While the increase in rubber plantation area of the total area of 533.249 ha in 2006 and 664.704 ha in 2015. The average number of changes in rubber production each year is 2.40% with production value average is 6.924 tons and an average number of changes in rubber plantation area per year is 2.63% with plantations area value average is 131.455 ha in seven times increased. An increased average amount of rubber production is 3.28% meanwhile plantations area 3.48%. The production changes more clearly are shown in figure 3.

1.2. Waste biomass potency for a renewable energy source

The biomass potency of palm oil is produced by during activity in the palm oil processing industry and oil palm plantation. The waste composition of biomass production solid waste during palm oil processing industry activity are 5-7% shell, 11-12% fiber, 20-23% empty fruit bunch (EFB) of fresh fruit bunch (FFB) and biomass production from palm frond during activity in oil palm plantation is 8.4 ton/ha per year [8]. During harvest, harvested paddy is cut off, and the biomass from husk 8.4 % and 58% straw are produced during rice mill activity [8]. While biomass from rubberwood is produced during replanting activity in rubber plantation 21.8 ton/ha per year [9].

Table 1. The calorific value of potential waste

Waste	Calorific value (kcal/kg)
Shell	4105-4802
Fiber	2637-4554
EFB	4492
Frond	3757
Rubberwood	4400
Husk	3350
Straw	2800

Source: [7, 8]

The direct combustion process is the most convenient and conventional in the thermal conversion technology of biomass. Biomass is burned directly without any material process. Using additional generators to obtain electricity. This method is mostly using in factories that use waste as fuel [6]. The efficiency of biomass energy is approximately 20-25 percent to generate electricity [13].

2. Methodology

This study consists of a literature study and secondary data collection. Secondary data collection based on Ministry of Agriculture's data that is obtained by Directorate General of Estate Crops (Ditjenbun) database and Central Agency of Statistics (BPS) database of Jambi province. The statistics that were collected is the area of oil palm plantation, rubber plantation area, palm oil production, and rice harvest area. The waste biomass calculated from the statistic data. Descriptive analysis was used to explain and describe the condition and potential agricultural waste biomass of plantation and palm oil processing, rubber and rice mill. Descriptive analysis used tabulation to providing the detailed explanation. Forecasting method was used to estimate the availability of production and total area to produce the biomass from waste to generate power of thermal and electricity to 2025. The technique used in this method was time series analysis, mathematical equation by double exponential smoothing. Data analysis of biomass waste potency in Jambi Province was carried out to identify the potency, location of waste biomass for a renewable energy source.

3. Result and discussion

3.1. Potential of agricultural waste biomass

Agricultural waste biomass potential that has calculated are a shell, fiber, EFB, frond, husk, straw, and rubberwood with waste total volume 27,407,183 tons per year. As based on table 2, the rubberwood waste biomass obtained from replanting activity in rubber plantations was 13,790,997 tons also corresponded to 50.32% of the agricultural waste production total volume. The calculated rubber waste biomass is rubberwood of replanting. Rubber trees in plantations are replanted 3-4% per year and produce an average rubberwood of 35 m^3 per year or 21,5 ton/ha per year [9]. Palm oil waste biomass production was 12,651,965 or 46.16%, then 3.52% waste biomass by rice. Among the seven kinds of waste, rubberwood was the highest value with power reaching 70,559 MWh_{th} (54.55%) then followed by frond 40.83%, straw 2.12%, EFB 1.25%, fiber 0.53%, husk 0.37%, and the smallest by shell 0.69%, from the power total potency up to 129,351 MWh_{th} and 32,339 MWh_e.

Table 2. The annual average of agricultural waste potency in Jambi province 2015

Production	Volume (ton)	Waste (ton)		Thermal (MWh_{th})	Electrical (MWh_e)
Palm oil	1,439,359	Shell	86,362	447	112
		Fiber	165,526	692	173
		EFB	309,462	1,616	404
		Frond	12,090,615	52,820	13,205
Rice	609,899	Husk	121,980	475	119
		Straw	842,241	2,742	686
	Area (ha)				
Rubber	641,442	Rubberwood	13,790,997	70,559	17,640

Source: Processed from secondary data based on [4, 5, 11, 12, 14, 15, 17, 18, 19]

The potency calculation result detailed of waste biomass by Regency/city in Jambi province can be seen in table 2. From table 2 is known, a waste biomass obtained from Merangin regency is the area that has the highest biomass production value. The biomass production value was 3,356,5748 tons or 16.75% of the total volume of agricultural waste biomass potential with power total potency up to 105,438 MWhth and 26,361 MWhe. Second place of the highest biomass production value is Sarolangun regency with biomass production value 3,214,597 tons. Followed continuity Batanghari regency 3,036,193 tons, Tebo regency 2,867,981 tons, Bungo regency 2,770,642 tons, Muaro Jambi Regency 2,495,561 tons, Tanjung Jabung Barat 1,422,126 tons, Tanjung Jabung Timur 568,678 tons, Kerinci regency 240,149 tons and the lowest biomass production value are Sungai Penuh regency 62,811 tons and Jambi city 6,429 tons. Sungai Penuh regency and Jambi city are areas that have the lowest potential value caused the area did not have oil palm and rubber plantations and effected to waste biomass production total.

3.2. Projection Biomass Production as Renewable Energy Source

From forecasting results, the increased trend of palm oil, rice, and rubber production several years to come was estimated. Estimation result shows that there will be an increase in agriculture production 2,814,153 ton in 2025. Increase in production will be related to the increased area, therefore Jambi province will continue to produce biomass as a source of renewable energy materials. Each agricultural production increased graphics more clearly are shown in figure 4, 5, and 6. The potential production waste from palm plantations and industry, rice mills and rubber in the future is huge. Based on table 4, it is known that the estimated amount of agricultural waste biomass production from palm plantations and industry, rice mills and rubber, annually from 2016-2025 in Jambi province.

Table 3. The annual average of agricultural waste in Jambi province 2006-2015 by regency/city

Regency/city	Production	Waste biomass (ton)		Thermal (MWh_{th})	Electrical (MWh_e)
Batanghari	Palm oil	Shell	12,654	66	16
		Fiber	24,254	101	25
		EFB	45,344	237	59
		Frond	668,925	4,091	1,023
	Rice	Husk	7234	28	7
		Straw	49,950	163	41
	Rubber	Rubberwood	2,227,832	11,398	2,850
Total			3,036,193	16,084	4,021
Muaro Jambi	Palm oil	Shell	18,332	95	24
		Fiber	35,136	147	37
		EFB	65,689	343	86
		Frond	1,030,982	6,306	1,576
	Rice	Husk	7941	31	8
		Straw	54,828	179	45
	Rubber	Rubberwood	1,282,653	6,562	1,641
Total			2,495,561	13,663	3,417
Bungo	Palm oil	Shell	9,994	52	13
		Fiber	19,154	80	20
		EFB	35,810	187	47
		Frond	568,075	3,475	869
	Rice	Husk	8020	31	8
		Straw	55,374	180	45
	Rubber	Rubberwood	2,074,215	10,612	2,653
Total			2,770,642	14,617	3,655
Tebo	Palm oil	Shell	6,871	36	9
		Fiber	13,170	55	14
		EFB	24,623	129	32
		Frond	374,222	2,289	572
	Rice	Husk	6299	25	6
		Straw	43,495	142	35
	Rubber	Rubberwood	2,399,301	12,276	3,069
Total			2,867,981	14,952	3,737
Merangin	Palm oil	Shell	10,575	55	14
		Fiber	20,269	85	21
		EFB	37,894	198	49
		Frond	466,187	2,851	713
	Rice	Husk	13502	53	13
		Straw	93,230	304	76
	Rubber	Rubberwood	2,715,091	13,891	3,473
Total			3,356,748	17,437	4,359
Sarolangun	Palm oil	Shell	8,456	44	11
		Fiber	16,207	68	17
		EFB	30,300	158	40
		Frond	476,180	2,913	728
	Rice	Husk	11176	44	11
		Straw	77,168	251	63
	Rubber	Rubberwood	2,595,110	13,277	3,319
Total			3,214,597	16,755	4,189
Tanjung Jabung Barat	Palm oil	Shell	17,282	89	22
		Fiber	33,123	138	35
		EFB	61,925	323	81
		Frond	882,942	5,400	1,350
	Rice	Husk	13584	53	13
		Straw	93,795	305	76
	Rubber	Rubberwood	319,475	1,635	409
Total			1,422,126	7,943	1,986
Tanjung Jabung Timur	Palm oil	Shell	2,198	11	3
		Fiber	4,212	18	4
		EFB	7,875	41	10
		Frond	257,103	1,573	393
	Rice	Husk	18616	73	18
		Straw	128,539	419	105
	Rubber	Rubberwood	150,135	768	192
Total			568,678	2,903	725
Kerinci	Palm oil	Shell	1	0.003	0.001
		Fiber	1	0.005	0.001
		EFB	2	0.012	0.003
		Frond	646	4	1
	Rice	Husk	26848	105	26
		Straw	185,380	604	151
	Rubber	Rubberwood	27,271	140	35
Total			240,149	853	213
Sungai Penuh	Rice	Husk	7946	31	8
		Straw	54,865	179	45
Total			62,811	210	53
Jambi	Rice	Husk	813	3	1
		Straw	5,616	18	5
Total			6,429	21	6
Total			20,041,915	105,438	26,361

Source: Processed from secondary data based on [4, 5, 11, 12, 14, 15, 17, 18, 19]

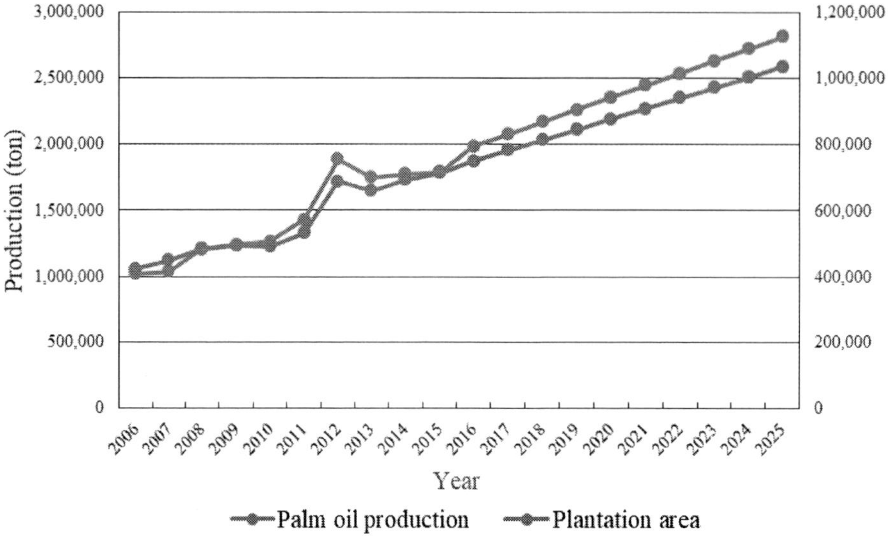

Figure 4. Palm oil production projected graphics in 2006-2025.
Source: Processed secondary data based on [4, 5, 11, 12, 14, 15, 16, 17, 18, 19]

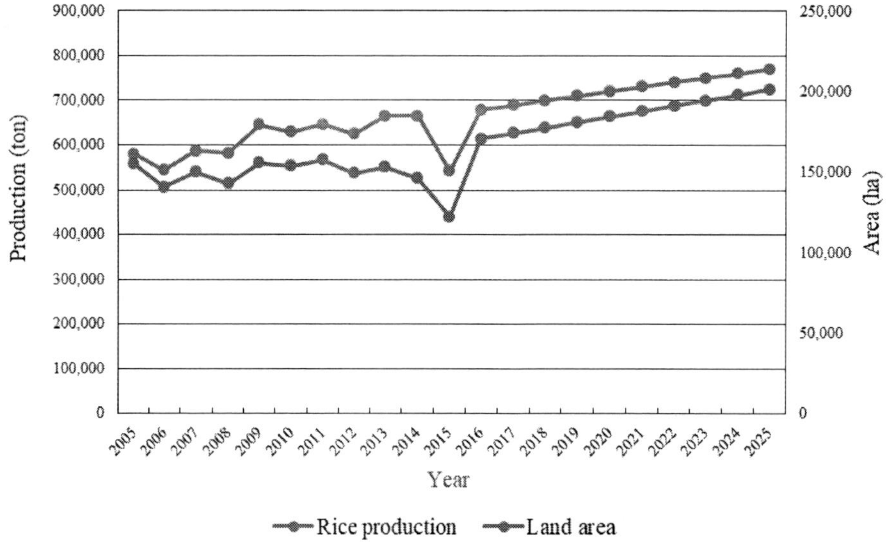

Figure 5. Rice production projected graphics in 2005-2025.
Source: Processed from secondary data based on [4, 5, 11, 12, 14, 15, 16, 17, 18, 19]

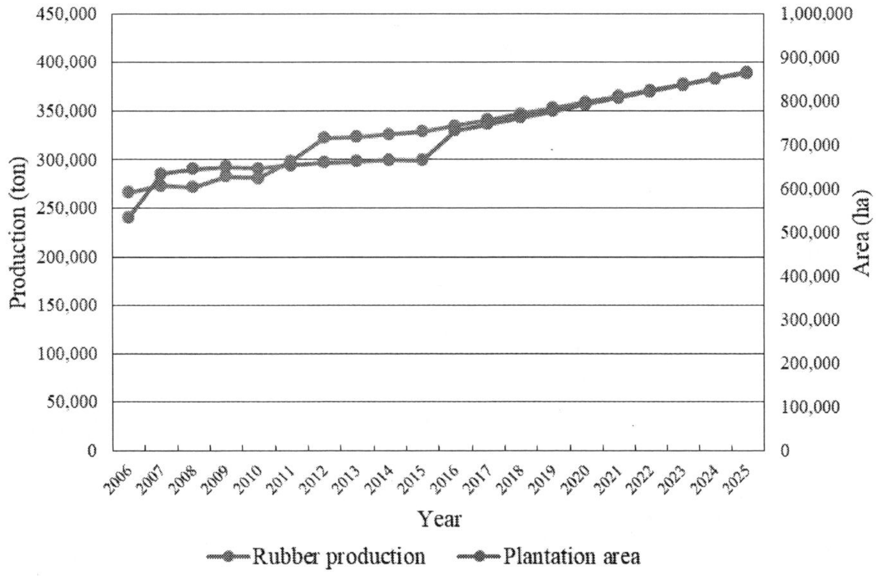

Figure 6. Rubber production projected graphics in 2006-2025.
Source: Processed from secondary data [4, 5, 11, 12, 14, 15, 16, 17, 18, 19]

Table 4. Estimated production of biomass from agricultural waste in 2016-2025.

Year	Palm oil (ton)				Rice (ton)		Rubber (ton)
	Shell	Fiber	EFB	Frond	Husk	Straw	Rubberwood
2016	119,111	228,297	426,816	6,300,609	135,753	937,341	16,126,559
2017	124,638	238,889	446,619	6,566,439	137,802	951,493	16,806,957
2018	130,164	249,481	466,422	6,832,269	139,852	965,643	17,487,356
2019	135,691	260,074	486,225	7,098,100	141,901	979,795	18,167,755
2020	141,217	270,666	506,028	7,363,930	143,951	993,947	18,848,154
2021	146,744	281,258	525,831	7,629,763	146,001	1,008,099	19,528,560
2022	152,270	291,851	545,634	7,895,593	148,050	1,022,251	20,208,958
2023	157,796	302,443	565,437	8,161,423	150,100	1,036,403	20,889,357
2024	163,323	313,035	585,240	8,427,253	152,149	1,050,555	21,569,756
2025	168,849	323,628	605,043	8,693,084	154,199	1,064,707	22,250,155

Source: Processed from secondary data based on [4, 5, 11, 12, 14, 15, 16, 17, 18, 19]

Table 5. Estimated waste biomass potency for a renewable energy source in 2016-2025.

Year	Palm oil (MWh_{th})				Rice (MWh_{th})		Rubber (MWh_{th})
	Shell	Fiber	EFB	Frond	Husk	Straw	Rubberwood
2016	617	954	2,229	27,525	529	3,052	82,509
2017	645	999	2,333	28,686	537	3,098	85,990
2018	674	1043	2,436	29,848	545	3,144	89,471
2019	703	1087	2,540	31,009	553	3,190	92,952
2020	731	1132	2,643	32,170	561	3,236	96,433
2021	760	1176	2,747	33,332	569	3,282	99,914
2022	789	1220	2,850	34,493	577	3,328	103,395
2023	817	1264	2,953	35,654	585	3,374	106,877
2024	846	1309	3,057	36,816	593	3,420	110,358
2025	874	1353	3,160	37,977	601	3,467	113,839

Source: Processed from secondary data based on [4, 5, 11, 12, 14, 15, 16, 17, 18, 19]

Table 6. Estimated waste biomass potency for electricity in 2016-2025

Year	Palm oil (MWh_e)				Rice (MWh_e)		Rubber (MWh_e)
	Shell	Fiber	EFB	Frond	Husk	Straw	Rubberwood
2016	154	239	557	6,881	132	763	20,627
2017	161	250	583	7,172	134	774	21,497
2018	169	261	609	7,462	136	786	22,368
2019	176	272	635	7,752	138	798	23,238
2020	183	283	661	8,043	140	809	24,108
2021	190	294	687	8,333	142	821	24,979
2022	197	305	713	8,623	144	832	25,849
2023	204	316	738	8,914	146	844	26,719
2024	211	327	764	9,204	148	855	27,589
2025	219	338	790	9,494	150	867	28,460

Source: Processed from secondary data based on [4, 5, 11, 12, 14, 15, 16, 17, 18, 19]

Results of analysis on biomass potency show that there is going trend in raw material for renewable energy. Based on table 5 and 6, it is known that the estimated amount of thermal and electrical annually from 2016-2025 will increase every years.

Conclusion

Waste that has the highest potential to be developed into a source of renewable energy raw materials from waste biomass in Jambi province is replanting rubberwood which has 68.81% biomass production percentage of waste biomass potential in Jambi province. There are 11 areas in Jambi province consist of 10 regencies and 1 city that have potential to be developed into a source of renewable energy. The highest percentage of waste biomass potential is Merangin regency. The total raw materials from potential waste in Jambi province consisted of: Merangin (16.75%), Sarolangun (16.04%) Batanghari (15.15%), Tebo (14.31%), Bungo (13.82%), Muaro Jambi (12.45%), Tanjung Jabung Barat (7.10%), Tanjung Jabung Timur (2.84%), Kerinci (1.20%), Sungai Penuh (0.31%), and Jambi city (0.03%).

References

[1] BPPT 2015 *Indonesia Energy Outlook 2015,* ed A Sugiyono et al. (Jakarta: Center for Energy Resources Development Technology) p 11
[2] Yokoyama S 2008 *Asian Biomass Handbook,* ed S Yokoyama et al. (Jakarta: The Japan Institute of Energy) p 16
[3] Hermawati W 2014 *Biomass Conversion for Alternative Energy in Indonesia: A Review of Resources, Technology, Management, and Policy,* ed H Abimanyu and S Hendarana (Jakarta: LIPI Press) p 14
[4] Ditjenbun 2015 *Tree Crop Estate Statistics of Indonesia 2014-2016,* ed M E Subiyantoro et al. (Jakarta: Directorate General of Estate Crops) pp 18-51
[5] BPS 2016 *Jambi in Figures 2016,* ed Division of Data Processing Integration and Statistical Dissemination (Jambi: Statistic of Jambi Province) pp 325-327
[6] Kong GT 2002 *Biomass Conversion for Alternative Energy in Indonesia: A Review of Resources, Technology, Management, and Policy,* ed H Abimanyu and S Hendarana (Jakarta: LIPI Press) pp 88-89
[7] Budiarto R and Agung A 2009 *Potential of Energy from Palm Oil Mill Waste* (Malang: Brawijaya University) p 6
[8] Tajali A 2015 *Guide to Assessing Biomass Potential as an Alternative Energy Source in Indonesia* (Jakarta: Penabulu Alliance) pp 7-11

IOP Conf. Series: Earth and Environmental Science **147** (2018) 012034 doi:10.1088/1755-1315/147/1/012034

[9] Sudrajat R 2004 *The Potential of Biomass Energy Resources in Indonesia for the Possible Development of Clean Technology Process (CTP)* (Jakarta: Center for Energy Resources Development Technology)

[10] Directorate General of New Renewable Energy and Energy Conservation Ministry of Energy and Mineral Resources 2013 *Map of Biomass Resources in Indonesia 2012-2013* (Jakarta: Directorate General of New Renewable Energy and Energy Conservation)

[11] Ditjenbun 2016 *Tree Crop Estate Statistics of Indonesia 2015-2017,* ed W K Zuraina et al. (Jakarta: Secretariate of Directorate General of Estate Crops) p 21

[12] BPS 2017 *Plantation Crop Production by Regency/City* (Jambi: Statistic of Jambi Province)

[13] BERC 2009 *Biomass Energy: Efficiency, Scale, and Sustainability* (Montpelier: Biomass Energy Resource Center) p 1

[14] BPS 2016 *Agricultural Surveys: Rice Production of Jambi Province 2015* (Jambi: Statistic of Jambi Province) pp 14-49

[15] BPS 2010 *Jambi in Figures 2010,* ed Division of Data Processing Integration and Statistical Dissemination (Jambi: Statistic of Jambi Province) p 303

[16] BPS 2012 *Jambi in Figures 2011,* ed *Regional Account and Statistical Analysis Division* (Jambi: Statistic of Jambi Province) p 303

[17] BPS 2012 *Jambi in Figures 2012,* ed *Division of Data Processing Integration and Statistical Dissemination* (Jambi: Statistic of Jambi Province) p 303

[18] Ditjenbun 2014 *Tree Crop Estate Statistics of Indonesia 2013-2015,* ed I Y Soependi et al. (Jakarta: Directorate General of Estate Crops) pp 21-51

[19] Ditjenbun 2013 *Tree Crop Estate Statistics of Indonesia 2012-2014,* ed B S Juga et al. (Jakarta: Directorate General of Estate Crops) pp 21-51

IOP Conf. Series: Earth and Environmental Science **147** (2018) 012035 doi:10.1088/1755-1315/147/1/012035

Evaluation of Working Fluids for Organic Rankine Cycle Based on Exergy Analysis

D Setiawan[1], I D M Subrata[1], Y A Purwanto[1] and A H Tambunan[1]

[1] Department of Mechanical and Biosystem Engineering, Bogor Agricultural University, Indonesia

E-mail: ahtambun@yahoo.com

Abstract One of the crucial aspects to determine the performance of Organic Rankine Cycle (ORC) is the selection of appropriate working fluids. This paper describes the simulative performance of several organic fluid and water as working fluid of an ORC based on exergy analysis with a heat source from waste heat recovery. The simulation was conducted by using Engineering Equation Solver (EES). The effect of several parameters and thermodynamic properties of working fluid was analyzed, and part of them was used as variables for the simulation in order to determine their sensitivity to the exergy efficiency changes. The results of this study showed that water is not appropriate to be used as working fluid at temperature lower than 130 °C, because the expansion process falls in saturated area. It was also found that Benzene had the highest exergy efficiency, i.e. about 10.49%, among the dry type working fluid. The increasing turbine inlet temperature did not lead to the increase of exergy efficiency when using organic working fluids with critical temperature near heat source temperature. Meanwhile, exergy efficiency decreasing linearly with the increasing condenser inlet temperature. In addition, it was found that working fluid with high latent heat of vaporization and specific heat exert in high exergy efficiency.

1. Introduction

In most of industrial processes, energy is often lost as waste heat and released to the atmosphere directly. Le et al. [1] reported that about 20-50% of energy to the industry is lost as waste heat. Furthermore, BCS Inc. [2] estimated that 60% of the industrial waste heat has temperatures below 230 °C. Palm oil plant, which is one of the most important industries in Indonesia, has several parts of it that potentially generate waste heat, such as boiler's flue gas with temperature around 290 °C. For palm oil plant with capacity of 5 tons fresh fruit bunch per hour, the flow rate of the flue gas is around 2.57 kg/s [3]. Conversion of the waste heat into electrical energy is indispensable in the midst of energy crisis and environmental problem. In practice, the most widely used technology for generating electrical energy from heat is the Rankine cycle (steam power cycle). Rankine cycle uses water as working fluid, so the ideal turbine inlet temperature has to be above 350 °C [4]. Using Rankine cycle is considered to be inefficient when using waste heat with lower temperature as heat source. Therefore, it is necessary to find working fluids that can give comparable performance at lower temperature.

One of the working fluid group that may have comparable performance with water is organic fluid. Therefore, the thermodynamic cycle that used organic fluid is called Organic Rankine Cycle (ORC). Studies related to the determination of working fluid on ORC have been done by several researchers. Drescher and Bruggemann [5] reported that alkylbenzene had better thermal efficiency than Toluene and OMTS as working fluid based on biomass combustion (temperature 250 – 350°C) as heat source. Liu et al. [6] found that n-pentane had the highest thermal efficiency for heat source at temperature range of 100 – 140°C. Lakew and Bolland [7] concluded that R227ea produced the highest power for heat source temperature range of 80 –160 °C, while R245fa gave higher work output for temperature greater than 160 °C. Although there have been many studies on the selection of working fluids, no single fluid has been identified as the most optimal for the ORC. It is due to the large number of working fluid variations, the difference of heat sources temperature and working conditions, and the difference in indicator performance [8].

Many of the studies evaluating the appropriate working fluids for ORC were based on thermal efficiency. However, exergy efficiency is a more rational criterion to evaluate the working fluid performance based on waste heat recovery. Therefore, the purposes of this study are: (1) to simulate the performance of several organic fluids, including water, based on exergy analysis with heat source from waste heat recovery (2) looking for the parameters that influence the exergy efficiency level, so it can be useful as guidelines for selecting working fluids.

2. Methods

This research is a simulative study for investigating the properties of working fluid by using Engineering Equation Solver (EES).

2.1. *System description*

The ORC system being studied is a simple cycle, which consists of a pump, evaporator, turbine and condenser (figure 1). The working fluid absorbs heat from the heat source (waste heat recovery), evaporates in the evaporator (process 2-3). The vapor from the outlet of the evaporator will expands in the turbine for doing useful work (process 3-4). The vapor enters the turbine at saturated vapor condition (quality = 1). Then, the working fluid enters condenser to dissipate heat out (process 4-1) before returning to the pump to complete the cycle (process 1-2). The phase of working fluid is saturated liquid at the exit of the condenser. The characteristics of the heat sources (flue gas) and cooling water are preset by fixing the inlet temperature, inlet pressure, mass flow rate. Performance of the heat transfer process is determined by fixing the pinch point temperature. Pinch point temperature is the minimum temperature difference between the saturation temperature of the working fluid and the temperature of the flue gas or cooling water. Temperature of the evaporator and the condenser in this research is the working fluid temperature at saturation condition on the evaporator and the condenser. Assumptions used in this research are: the system operates under steady state, there are no pressure drop and heat loss. The Temperature-Entropy diagram (T-S) of this cycle is shown in figure 2.

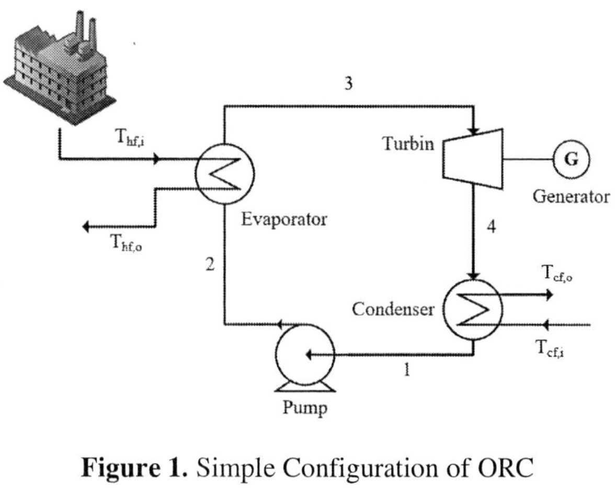

Figure 1. Simple Configuration of ORC

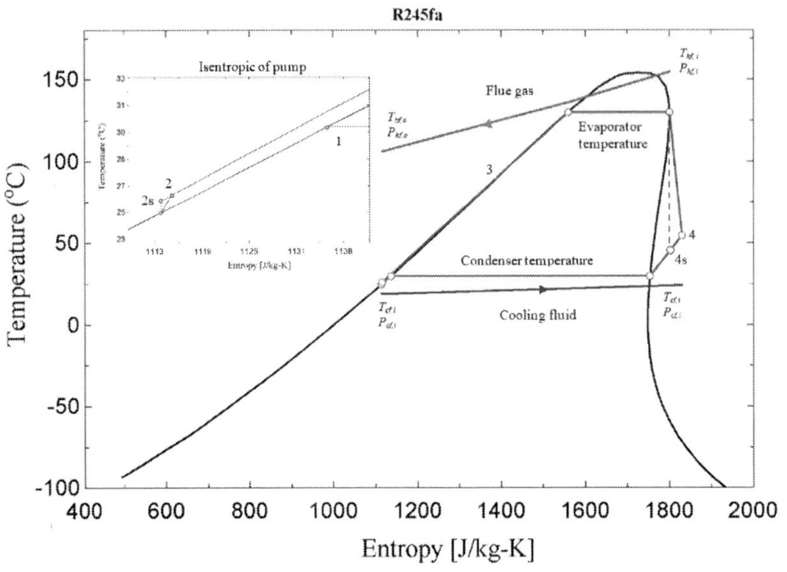

Figure 2. T-S Diagram

2.2. *Thermodynamic analysis*

Using the First and Second Law of Thermodynamics, the performance of working fluid for ORC can be evaluated under different condition. Based on figure 1 and 2, the mathematical expression for each of the components are given as follows:

a. Pump (process 1-2)

The work consumption in the pump is given in equation 1.

$$W_{p,act} = \frac{W_{p,ideal}}{\eta_p} = \frac{\dot{m}_f\,(h_{2s}-h_1)}{\eta_p} \tag{1}$$

where $W_{p,act}$ is the actual power of the pump (W), $W_{p,ideal}$ is the ideal power of the pump (W), $\dot{m}_f$ is the working fluid mass flow rate (kg/s), η_p is the isentropic efficiency of the pump (%), while h_1 and h_{2s} are the enthalpy of the working fluid at the inlet and at isentropic condition of the pump's outlet (J/kg), respectively. The exergy destruction rate (irreversibility) for uniform flow conditions can be expressed as in equation 2.

$$\dot{I} = T_0\frac{ds}{dt} = T_0\dot{m}_f\left[\sum s_{out} - \sum s_{in} + \left(\frac{ds_{sistem}}{dt}\right) + \sum_j\left(\frac{q_j}{T_j}\right)\right] \tag{2}$$

Here, $\frac{ds_{sistem}}{dt} = 0$ for steady state conditions, T_0 is ambient temperature (°C). So, the exergy destruction in the pump can be expressed as in equation 3.

$$\dot{I}_p = T_0\dot{m}_f(s_2 - s_1) \tag{3}$$

Here, s_1 and s_2 are the specific entropy of the working fluid at the inlet and outlet of the pump (J/kg.K), respectively.

b. Evaporator (process 2-3)

Heat transfer in the evaporator can be expressed as in equation 4.

$$Q_e = \dot{m}_f\,(h_3 - h_2) \quad (4)$$

Where h_3 and h_2 are the enthalpy of the working fluid at the outlet and inlet of the evaporator (J/kg), respectively. Using equation 2, the exergy destruction in the evaporator can be expressed as in equation 5.

$$\dot{I}_e = T_0[\dot{m}_f(s_3 - s_2) - \dot{m}_{hf}(s_{hf,o} - s_{hf,i})] \quad (5)$$

Where s_3, $\dot{m}_{hf}$, $s_{hf,o}$ and $s_{hf,i}$ are the specific entropy of the working fluid at the inlet of evaporator (J/kg.K), mass flowrate of heat source (kg/s), specific entropy of the heat source at outlet and inlet of evaporator (J/kg.K), respectively.

c. Turbine (process 3-4)

The turbine power is mathematically described by equation 6.

$$w_{t,act} = w_{t,ideal}\eta_t = \dot{m}_f\,(h_3 - h_{4s})\eta_t \quad (6)$$

Here, $W_{t,act}$ is the actual power produced by turbine (W), $W_{t,ideal}$ is the ideal power of turbine (W), h_{4s} and η_t are the enthalpy of the working fluid at the outlet isentropic condition of turbine (J/kg) and isentropic turbine efficiency (%), respectively. Meanwhile, the exergy destruction of the turbine can be expressed as in equation 7.

$$\dot{I}_t = \dot{m}_f\,T_0(s_4 - s_3) \quad (7)$$

where s_4 is specific entropy at the outlet of the turbine (J/kg.K).

d. Condenser (process 4-1)

Heat transfer in the condenser can be expressed as in equation 8.

$$Q_k = \dot{m}_f\,(h_4 - h_1) \quad (8)$$

The exergy destruction in condenser can be expressed as in equation 9.

$$\dot{I}_k = T_0[\dot{m}_f(s_4 - s_1) - \dot{m}_{cf}(s_{cf,o} - s_{cf,i})] \quad (9)$$

Where h_4, $\dot{m}_{cf}$, $s_{cf,o}$ dan $s_{cf,i}$ are enthalpy of working fluid at inlet of condenser (J/kg), mass flow rate of cooling fluid (kg/s), specific entropy of the cooling fluid at outlet and inlet of condenser (J/kg.K), respectively.
The total of exergy destruction can be expressed as in equation 10.

$$\dot{I}_{total} = \dot{I}_p + \dot{I}_e + \dot{I}_t + \dot{I}_k \quad (10)$$

The net output work of the cycle:

$$W_{net} = \dot{m}_f\ (w_t - w_p) \tag{11}$$

Meanwhile, the cycle efficiency can be expressed as:

$$\eta_{th} = \frac{w_{t,act} - w_{p,act}}{Q_e} \tag{12}$$

The exergy efficiency is defined as:

$$\eta_{ex} = \frac{w_{t,act} - w_{p,act}}{Ex_{hf,i}} \tag{13}$$

Where the inlet exergy of the heat source is:

$$Ex_{hf,i} = \dot{m}_{hf}[(h_{hf,i} - h_o) - T_0(s_{hf,i} - s_o)] \tag{14}$$

As shown in figure 3, working fluids can be classified into wet, isentropic, or dry, with respect to the slope of saturation curve in T-S diagram, i.e negative, infinite, or positive. The type of the working fluid, then, can be determined by using equation 15.

$$\frac{ds}{dT} = \frac{S_1 - S_2}{T_1 - T_2} \tag{15}$$

where T_1 and T_2 are saturation temperature at evaporator and condenser, respectively. Meanwhile, s_1 and s_2 are specific entropy at 1 and 2 conditions, as shown in figure 3.

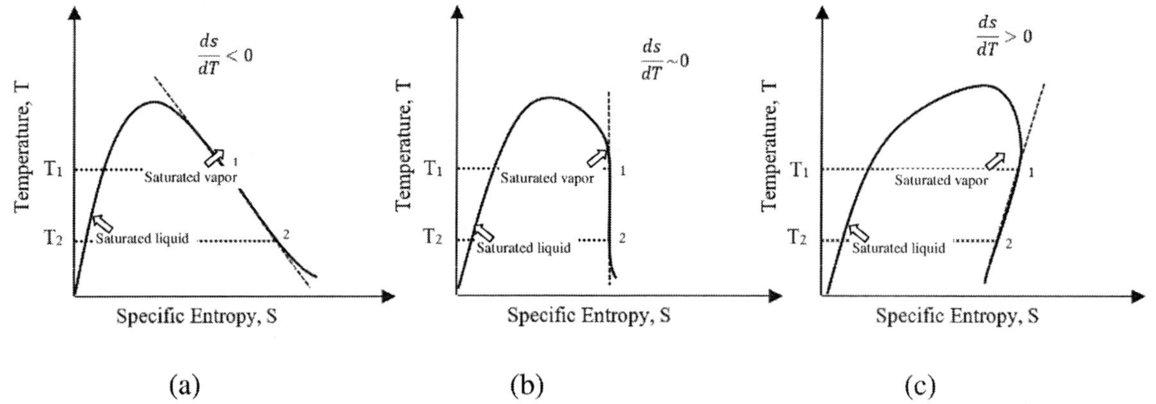

Figure 3. T–s diagram for (a) wet, (b) isentropic and (c) dry working fluids

2.3. *Operation parameters*

The operation parameters used for the simulation process are presented in table 1. In order to know the effect of parameter changes on exergy efficiency and exergy destruction, the turbine and condenser inlet temperature were used as variables.

Table 1. Operation Parameters

Parameters	Value	Note
Environmental pressure [kPa]	101.3	
Environmental temperature [°C]	25	
Working fluid mass flowrate [kg/s]	0.5	
Evaporator temperature [°C]	130	
Condenser temperature [°C]	30	
Cooling fluid pressure [MPa]	0.2	
Cooling fluid inlet temperature [°C]	18	
Water cooling mass flowrate [kg/s]	5	
Pinch point evaporator [K]	6	
Pinch point condenser [K]	6	
Overheating [°C]	0	
Sub cool [°C]	5	
Flue gas inlet pressure [MPa]	0.4	[3]
Flue gas inlet temperature [°C]	170	
Flue gas mass flowrate [kg/s]	2.57	[3]
Isentropic efficiency of pump [%]	80	
Isentropic efficiency of turbine [%]	80	

3. Results and discussion

3.1. *Analysis of working fluid performance*

In order to meet the objectives of this study, 33 organic fluid were evaluated, and their effect on the performance of the ORC are summarized in table 2. Based on net output work, cycle efficiency, and exergy efficiency, the use of water as working fluid by utilizing waste heat recovery has better performance than organic working fluids. The significant difference between water and organic fluids is the highest latent heat, where at the evaporator temperature, latent heat of evaporation of water is 2174 kJ/kg while the organic fluids are ranged in 57.8 – 405.7 kJ/kg. Accordingly, water needs more thermal energy to change its phase from saturated liquid to saturated vapor, yields in lower mass flow rate required to absorb the same amount of thermal power from heat source than organic fluids. A higher mass flow rate leads to higher power consumption by pump and a higher piping system diameter required to overcome pressure losses.

In real conditions, the use of conventional Rankine cycle by utilizing waste heat recovery is not efficient both in terms of thermodynamics and economic view. It can be explained through the T-S diagram of the simulation result presented in figure 4. The expansion process occurs in the saturated area (figure 4(a)), which will produce droplets in the turbine. It does not only make the expansion process fails to occur but the droplets could also be corrosive to the turbine [9]. In case of organic fluid, in as shown in figure 4(b), the expansion process occurs in superheated region.

In addition, condensation process of organic fluids occurs at saturation pressure above atmospheric, while water occurs below atmospheric pressure. Condensation above atmospheric pressure is advantageous to avoid air intrusion to the system which can decrease the cycle's efficiency. Previous works recommend the minimum condensing pressure of ORC between 0.05 – 0.1 MPa [5], [9], [10]. In terms of evaporator pressure, most of organic fluids have higher pressure than water, which will require more safety concern. Some studies suggest that the maximum pressure on the evaporator is between 2 - 2.5 Mpa [5], [11], [12]. The two parameters can be limited factor to determine the best working fluid.

Table 2. Simulation result of working fluid performance

Name	T_c (°C)	P_c (Mpa)	W_t (Watt)	W_p (Watt)	$W_{\_net}$ (Watt)	η_{th} (%)	η_{ex} (%)	$\dot{I}_{total}$ (Watt)	P_{cd} (Mpa)	P_{ev} (Mpa)	P_{ratio}	Type
Acetone	235	4.7	53522	573.3	52949	16.35	11.51	51912	0.038	0.758	19.912	Wet
Benzene	288.9	4.894	45943	259.7	45683	16.51	10.49	44287	0.016	0.379	23.802	Dry
Cis-2-butene	162.6	4.226	38617	2259	36358	14.74	9.724	40730	0.250	2.475	9.889	Dry
Cyclohexane	280.5	4.081	43314	279.7	43035	15.89	10.26	43567	0.016	0.363	22.321	Dry
Diethylether	193.6	3.644	38939	1011	37929	14.68	9.829	42318	0.086	1.231	14.265	Dry
Dimethylcarbonate	284.2	4.909	45158	176	44982	16.45	10.43	44139	0.009	0.310	33.503	Dry
HFE7100	195.3	2.229	14103	251.5	13851	12.82	4.471	17680	0.031	0.642	20.401	Dry
HFE7200	209.8	2.006	13891	181.9	13709	13.05	4.384	17046	0.020	0.434	21.535	Dry
Isobutene	146.1	4.005	34291	2881	31411	14	9.088	37532	0.344	3.057	8.876	Isentropic
Isopentane	187.2	3.37	37325	1223	36102	14.37	9.564	41234	0.109	1.311	12.024	Dry
mm	245.5	1.939	25492	179.2	25313	12.96	7.203	31825	0.008	0.225	29.730	Dry
M-Xylene	343.7	3.535	44523	55.94	44467	16.35	10.31	43469	0.001	0.078	52.716	Dry
N-butane	152	3.796	36363	2562	33801	14	9.516	40215	0.284	2.632	9.270	Dry
N-heptane	267	2.727	40994	209.8	40784	14.8	10.04	44626	0.008	0.236	30.335	Dry
N-hexane	234.7	3.058	40357	438.8	39919	14.66	9.944	44253	0.025	0.486	19.455	Dry
N-nonane	321.4	2.281	41050	48.39	41002	14.76	9.982	44855	0.001	0.056	70.378	Dry
N-octane	296.2	2.497	41306	100.7	41205	14.84	10.06	44863	0.002	0.115	47.098	Dry
N-pentane	196.5	3.364	39619	1026	38593	14.57	9.909	43308	0.083	1.101	13.326	Dry
R11	198	4.408	18298	574.9	17723	15.77	5.384	18293	0.125	1.483	11.831	Isentropic
R113	214.1	3.392	16386	312.2	16074	15.03	5.008	17428	0.054	0.835	15.386	Dry
R114	145.7	3.289	12890	950.4	11940	13.33	4.022	14870	0.250	2.464	9.864	Dry
R123	183.7	3.668	17912	576.8	17335	15.05	5.396	18839	0.110	1.461	13.316	Dry
R1233ZD(E)	165.6	3.573	19168	867.6	18300	14.52	5.7	20780	0.155	1.908	12.347	Dry
R141B	204.2	4.249	23348	585.5	22763	15.52	6.628	23903	0.094	1.251	13.287	Dry
R236EA	139.3	3.429	14794	1154	13639	13.08	4.614	17436	0.245	2.879	11.751	Dry
R245Fa	154	3.651	18783	1010	17773	14.03	5.655	21009	0.177	2.339	13.202	Dry
R365MFC	186.9	3.266	20981	533.8	20447	14.18	6.197	23654	0.069	1.143	16.511	Dry
R600A	134.7	3.64	30179	3363	26816	13	8.553	35059	0.404	3.364	8.317	Dry
Toluene	318.6	4.126	45085	120.6	44964	16.49	10.4	43595	0.005	0.171	35.014	Dry
Trans-2-butene	155.5	4.027	37553	2452	35101	14.44	9.611	40327	0.273	2.622	9.594	Dry
Isohexane	224.6	3.04	38616	552.6	38063	14.5	9.714	42730	0.035	0.608	17.585	Dry
N-decane	344.6	2.103	41148	24.11	41123	14.81	9.997	44801	2.56E-4	0.028	110.65	Dry
N-dodecane	385	1.817	41204	6.141	41198	14.8	10	44902	2.68E-5	0.007	274.2	Dry
Water	374	220.6	238713	166.6	238546	18.24	18.22	221679	0.004	0.27	67.5	Wet

Note: T_c = critical temperature, P_c = critical pressure, P_{cd} = condenser pressure, P_{ev} = evaporator pressure, $P_{ratio} = \frac{P_{ev}}{P_{cd}}$

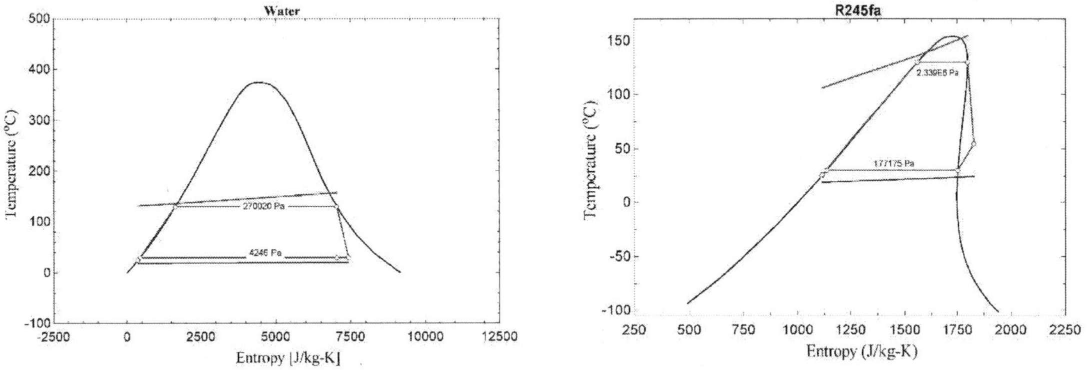

(a) (b)

Figure 4. Process line of ORC using (a) water and (b) R245fa on T-S diagram

3.2. *The effect of operation parameters*

Figure 5 shows the variation of the exergy efficiency of the system with the turbine inlet temperature, using selected organic fluids to represent wet, isentropic, and dry types. It was found that highest exergy efficiency for dry fluid type is given by Benzene, while for wet type organic fluid it is given by Acetone. Wet fluids like Water and Acetone need to be superheated, while dry or isentropic type of organic fluids does not need superheating [13]. The figure shows that the exergy efficiency of organic fluids can be related to the evaporation temperature in polynomial relationship until it approaches the critical temperature. Organic fluids with low critical temperature, such as HFE7100, R11, and R245fa, reveals the polynomial relationship clearly in the figure. Thereby, the selected organic fluids to be used in an ORC should not be operated at evaporation temperature near its critical temperature.

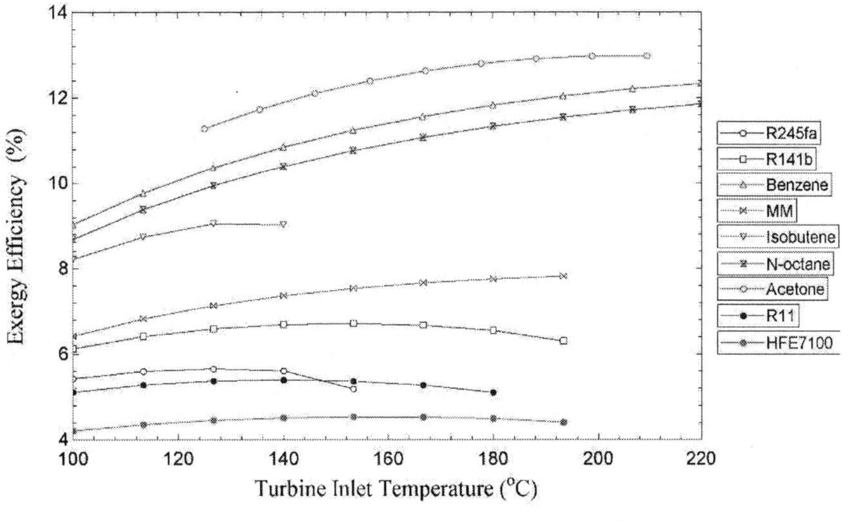

Figure 5. Variation of the exergy efficiency with the turbine inlet temperature

Figure 6 shows the variation of exergy destruction with the turbine inlet temperature. It can be observed that exergy destruction of several organic fluids increases with the increment of the turbine inlet temperature. The largest gradient of exergy destruction occurs in the n-octane, while acetone, R141b, R11, and isobutene show a decreasing trend of exergy destruction to the turbine inlet temperature. It can be concluded that turbine inlet temperature does not significantly contribute to the changing of exergy destruction for certain organic fluids, such as Acetone, R141b, R11, and Isobutene.

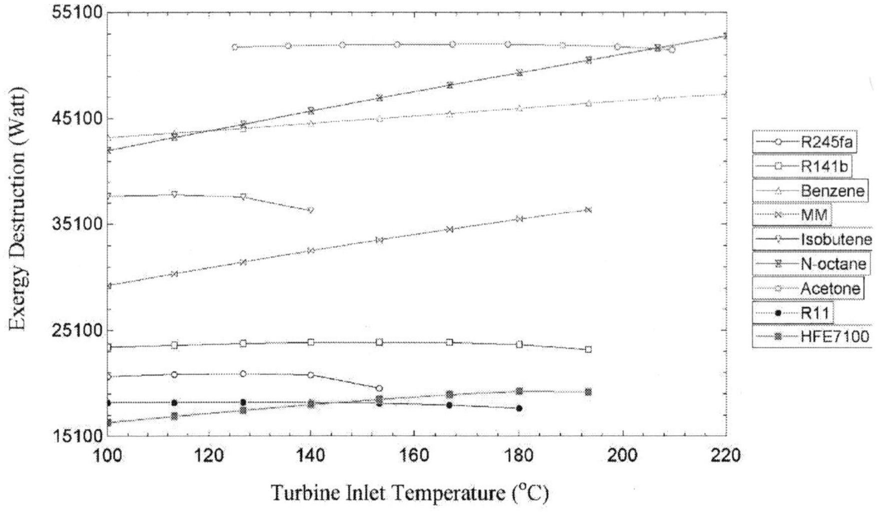

Figure 6. Variation of exergy destruction with the turbine inlet temperature

Condensation temperature is one of the important parameters that affects ORC performance. The limitation of condenser temperature depends on the condenser pressure and freezing point of the working fluid. figure 7 shows the variation of exergy efficiency with the condenser inlet temperature. It can be observed that for all of the organic fluids, exergy efficiency decreases linearly with the increasing condenser inlet temperature. These results indicate that ORC will be more beneficial if operated at low ambient temperatures. The trend observed in this figure is consistent with the results shown in other figures where benzene shows the best exergy efficiency while HFE7100 the worse among the eight evaluated dry type organic fluids.

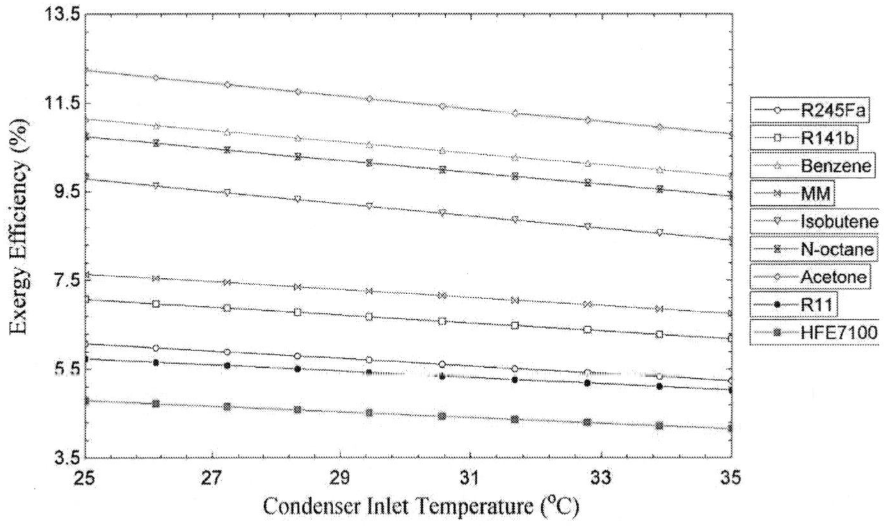

Figure 7. Variation of the exergy efficiency with the condenser inlet temperature

3.3. *The influence of selected thermodynamic properties to exergy efficiency*

Organic fluids that have high latent heat of vaporization, high density, and low specific heat are preferable. It absorbs more energy from heat sources, thereby reducing flow rate, pump power consumption, and the installation size [11]. However, Yamamoto et al. [14] suggested that low latent heat of vaporization provides better performance because the working fluid is in superheating conditions when entering the turbine. Based on exergy efficiency, latent heat of evaporation (turbine inlet temperature 130 °C at saturation pressure) has significant effect on the exergy efficiency, as shown in figure 8. It can be observed that the exergy efficiency of organic fluids increases with the latent heat. In addition, figure 9 illustrates the relationship between specific heat (Temperature =25 °C and Pressure = 1 atm) with exergy efficiency. The higher specific heat of the organic working fluid, there is a tendency to increase its exergy efficiency.

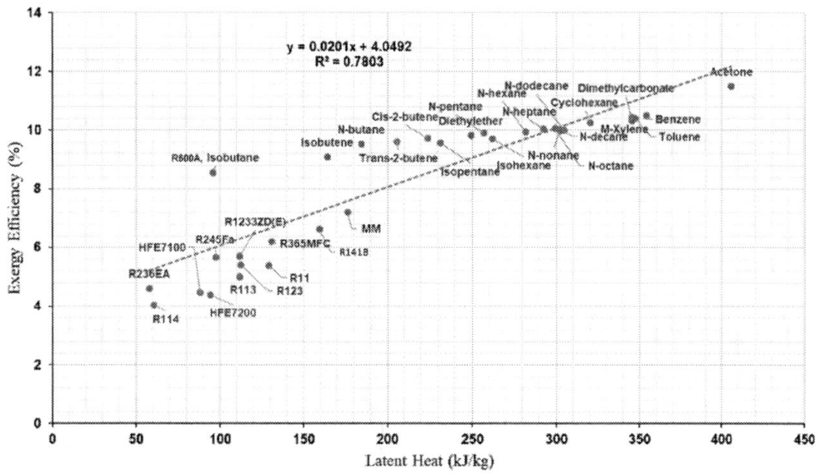

Figure 8. Variation of the exergy efficiency with the latent heat

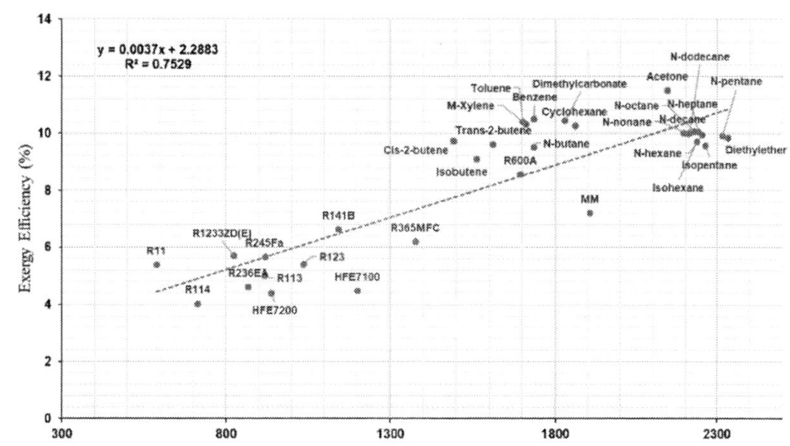

Figure 9. Variation of the exergy efficiency with the specific heat

A good working fluid should have low vapor and low liquid specific volumes. These properties affect the rates of heat transfer in the heat exchangers. The vapor specific volumes relate directly to the size and cost of the cycle components. Moreover, a high vapor specific volume leads to larger volumetric flows. The specific volume of the liquid at the condenser pressure should be as small as possible to minimize the required pump work [15]. While the relationship of specific volume (Temperature: 130 °C at saturation vapor) with the exergy efficiency does not show good relationship, which is the correlation value of $R^2 = 0.0681$. It can be seen clearly in figure 10.

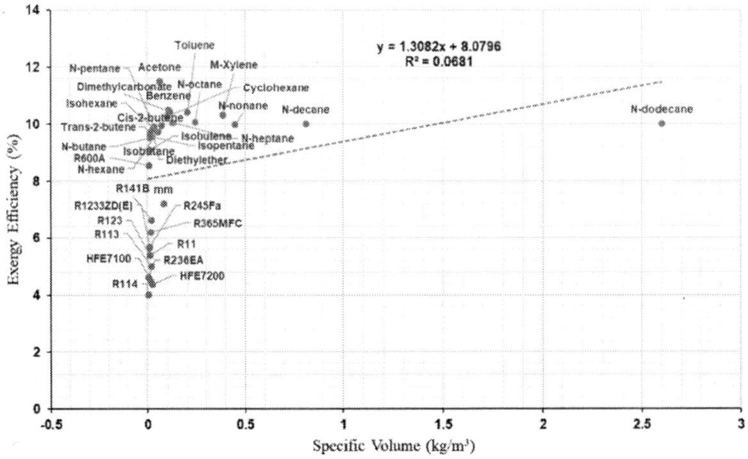

Figure 10. Variation of the exergy efficiency with the specific volume

Any fluid with temperature and pressure above its critical point shows no distinct phases of liquid and gas. When the organic fluid operates above the critical point, it will have a very high work potential due to increased pressure and temperature. Baik et al. [16] found that the output power produced by transcritical ORC using R125 as working fluid is 10% greater than subcritical ORC using R134a, R245fa, and R152a as working fluids. The study by Chen et al. [13] found that transcritical ORC using CO_2 as working fluid has higher cycle efficiency than using subcritical ORC using R123 as working fluid. Meanwhile, the selection of working fluids based on critical temperature shows a positive tendency, even though the correlation is quite low, i.e. R^2=0.2765, as shown in figure 11.

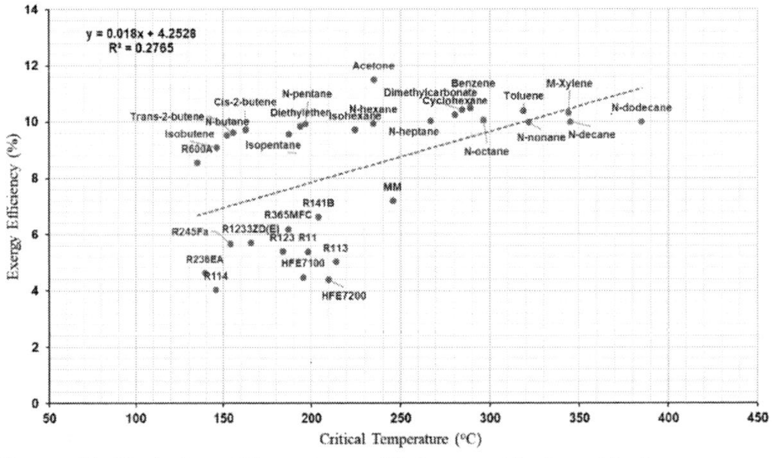

Figure 11. Variation of the exergy efficiency with the critical temperature

4. Conclusions

Based on the analysis that has been done, it can be concluded that the utilization of heat source with temperature lower than 130°C was inappropriate for a Conventional Rankine cycle since the expansion process takes place in the saturated area. Higher exergy efficiency was found in wet type organic working fluids, but it should be noted that the use of wet working fluids is not considerable for ORC. It was also found that Benzene had the highest exergy efficiency of about 10.49% for dry type organic working fluid. Meanwhile the highest exergy destruction of dry type organic working fluid was N-dodecane. The increasing turbine inlet temperature did not lead to the increasing exergy efficiency of organic working fluids with critical temperature near a heat source temperature. Exergy efficiency decreases linearly with the increase of condenser inlet temperature. In addition, it was also found that working fluid with high latent heat of vaporization and specific heat exert in high exergy efficiency. Meanwhile the parameters such as specific volume at saturation vapor and critical temperature do not show good relationship to the exergy efficiency.

References

[1] Le V L, Kheiri A, Feidt M and Pelloux-prayer S 2014 Thermodynamic and economic optimizations of a waste heat to power plant driven by a subcritical ORC (Organic Rankine Cycle) using pure or zeotropic working fluid *Energy*
[2] BCS Inc 2008 Waste Heat Recovery: Technology and Opportunities in U.S *Industry (US)*
[3] Situmorang H 2007 Utilization of exhaust gas from palm oil mill as heating feedwater woth shell and tube heat exchanger (Medan: North Sumatera University)
[4] Bajaj S S, Patil H B, Kudal G B and Shisode S P 2016 Organic rankine cycle and tts working fluid selection-A review **4**(4): 20–26
[5] Drescher U and Bru D 2007 Fluid selection for the Organic Rankine Cycle (ORC) in biomass power and heat plants **27** 223–228
[6] Liu H, Liu Y and Li J 2011 A biomass-fired micro-scale CHP system with Organic Rankine Cycle (ORC) - Thermodynamic modelling studies *Biomass and Bioenergy* **35**(9): 3985–94
[7] Lakew A A and Bolland O 2010 Working fluids for low-temperature heat source *Appl. Therm. Eng.* **30**(10): 1262–68
[8] Bao J and Zhao L 2013 A review of working fluid and expander selections for Organic Rankine Cycle *Renew. Sustain. Energy Rev.* **24** 325–342
[9] Tchanche B F, Lambrinos G, Frangoudakis A and Papadakis G 2011 Low-grade heat conversion into power using Organic Rankine Cycles - A review of various applications *Renew. Sustain. Energy Rev.* **15**(8): 3963–79
[10] Katsanos C O, Hountalas D T and Pariotis E G 2012 Thermodynamic analysis of a Rankine cycle applied on a diesel truck engine using steam and organic medium *Energy Convers. Manag.* **60** 68–76
[11] Maizza V and Maizza A 1996 Working fluids in non-steady flows for waste energy recovery systems *Appl. Therm. Eng.* **16**(7): 579–590
[12] Lai N A, Wendland M and Fischer J 2011 Working fluids for high-temperature organic Rankine cycles *Energy* **36**(1): 199–211
[13] Chen H, Goswami D Y and Stefanakos E K 2010 A review of thermodynamic cycles and working fluids for the conversion of low-grade heat *Renew. Sustain. Energy Rev.* **14**(9): 3059–67
[14] Yamamoto T, Furuhata T, Arai N and Mori K 2001 Design and testing of the organic rankine cycle *Energy* **26**(3): 239–251
[15] Badr O, Probert S D and O'Callaghan P W 1985 Selecting a working fluid for a Rankine-cycle engine *Appl. Energy* **21**(1): 1–42
[16] Baik Y, Kim M, Chang K, Lee Y and Yoon H 2013 A comparative study of power optimization in low-temperature geothermal heat source driven R125 transcritical cycle and HFC organic Rankine cycles *Renew. Energy* **54** 78–84

An Environmental Impact Analysis of Semi-Mechanical Extraction Process of Sago Starch: Life Cycle Assessment (LCA) Perspective

M A Yusuf[1,2], M Romli[1], Suprihatin[1] and E I Wiloso[3]

[1] Department of Agroindustrial Technology, Bogor Agricultural University, Bogor, Indonesia
[2] Department of Agricultural Engineering, Musamus University, Merauke, Indonesia
[3] Research Center of Chemistry, Indonesian Institute of Science (LIPI), Serpong, Indonesia

E-mail: megaayu.yusuf@gmail.com

Abstract. Industrial activities use material, energy and water resources and generate greenhouse gas (GHG). Currently, various regulations require industry to measure and quantify the emissions generated from its process activity. LCA is a method that can be used to analyze and report the environmental impact of an activity that uses resources and generates waste by an industrial activity. In this work, LCA is used to determine the environmental impact of a semi-mechanical extraction process of sago industry. The data was collected through the sago industry in Cimahpar, Bogor. The extraction of sago starch consists of stem cutting, rasping, mixing, filtration, starch sedimentation, washing, and drying. The scope of LCA study covers the harvesting of sago stem, transportation to extraction site, and the starch extraction process. With the assumption that the average transportation distance of sago stem to extraction site is 200 km, the GHG emission is estimated to be 325 kg CO2 eq /ton of sundried sago starch. This figure is lower than that reported for maize starch (1120 kg CO2 eq), potato starch (2232 kg CO2 eq) and cassava starch (4310 kg CO2 eq). This is most likely due to the uncounted impact from the use of electrical energy on the extraction process, which is currently being conducted. A follow-up study is also underway to formulate several process improvement scenarios to derive the design of sago starch processing that generates the minimum emissions.

1. Introduction

Sago (sago palm) plants grow naturally in humid tropic regions, such as in Southeast Asia including Indonesia, Thailand, Malaysia, The Philippines, Vietnam, and Papua New Guinea. Starch contained in a sago stalk is a source of carbohydrates. [1] reported a rapid increase in sago starch production in recent years. Sago has advantages over other carbohydrate source plants, namely its high productivity that reaches 25-30 tons/hectare.year [2]. Compared to other carbohydrate sources such as rice and wheat flour, sago has also higher carbohydrate content, namely 84.7%. Sago starch is an intermediate product that can be processed further into various food and industrial products, such as liquid sugar for beverage industry [3] and bioethanol [4]. The by-products of sago processing in the forms of solid and liquid can be converted into value-added products [5].

Life Cycle Assessment (LCA) is a method to analyze resource issues across the life cycle of a product. It can systematically identify key areas to improve environmental performance, and can be applied to agroindustrial systems especially in sago industry. A standardized LCA methodology for agroindustries will help practitioners undertake LCA studies and greatly increase their value by providing results that are comparable between product and industries.

Increased production of sago starch causes an increase in sago processing wastes, both liquid and solid. Handling of sago processing waste is still very limited, generally, waste is discharged directly into the river without treatment. Phang et al. [6] stated that each plant produces between 10-22 tons of wastewater containing carbon and nitrogen at a ratio of 105: 0.12.

This study analyses the environmental impact of sago processing by using a Life Cycle Assessment (LCA) method. The environmental impact of each stage of sago starch production process is analyzed by calculating the environmental loads based on inventory analysis of resource, energy, water, fuel and other materials.

2. Methodology

This study uses the LCA methodology to evaluate the inputs, outputs, and the potential environmental impacts of a product system throughout the sago starch's life cycle (ISO 14040:2006). The LCA methodology consists of four stages: goal and scope determination, inventory analysis, impact analysis and interpretation.

2.1. Determine the purpose and scope (Goal Definition and Scoping)

In this work, LCA is used to determine the environmental impact of a semi-mechanical extraction process of sago industry. Goal and scope definition is the first step in conducting LCA analysis. At this stage the objectives of the LCA study and the limits or scope of LCA analysis are determined. The parameters associated with the analysis include a functional unit

2.2. Inventory analysis (Life Cycle Inventory/LCI)

The LCA method is used to explore the sources that used to obtain the required data. Process flowchart of semi-mechanical sago processing that displays the sources of waste generation and emissions is depicted in figure 1.

2.3. Impact analysis (Life Cycle Impact Analysis / LCIA)

The impact analysis aims to determine the likely impacts that may occur during the life cycle of a product. Impact estimates are intended as a careful and in-depth study of the quality of the environment, indicated by their magnitude of impact and importance. In this study, a mathematical model is a formal method commonly used to estimate the magnitude of impacts on environmental components due to the calculated data. In this study, the data is calculated by using SIMAPRO software.

2.4. Interpretation (Interpretation / Life Cycle Improvement Analysis)

This step is the last step in doing the LCA, namely by interpreting the results, evaluating, and then analyzing the things that can be done in order to improve the products, processes or services related to environmental impact reduction. The results of analyzes that have been carried out in the inventory and impact assessment stage are manifested in actions that will benefit industry and the environment.

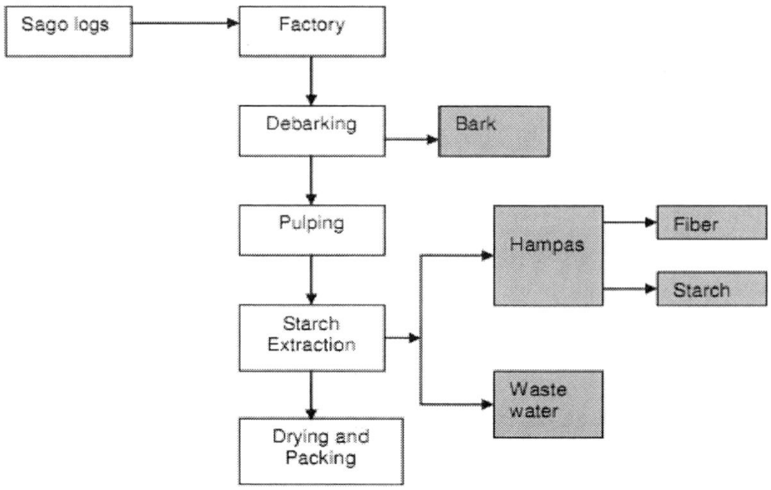

Figure 1. Process diagram and the formation of waste in sago starch agroindustry

3. Result and discussion

3.1. Semi-Mechanical Sago Starch Extraction

Semi-mechanical type of sago starch production is in principle the same process as the traditional one, except that tools or machines are involved in some process steps. For example, the use of the rasping machine in the process of pith destruction; the use of tank equipped with a mechanical stirrer in the process of dissolving sago starch; and the use of pumping system in the filtering process of sago starch separation. This semi-mechanical process is chosen in this study because most of the production of sago in Indonesia (especially in Riau, West Java, and South Sulawesi) uses this type of technology.

The process begins with cutting stems of sago plants after the trees are felled in length of 0.5-1 meter. These pieces are then peeled, split, and shredded. The grated materials are collected in a wooden tub equipped with a mechanical agitator. Agitation is usually done in two stages to make sure that all starch is extracted from the fibers. Afterwards, a mixture consisting of fibers, starch, and water is pumped to a series of filters. The slurry of the starch solution is collected in other wooden tubs for the process of starch deposition. The starch deposition is then washed in a tub or tank equipped with a stirrer and deposited further. The wet sago starch is then sun dried and ground with a grinder. Finally, the milled starch is stuffed into burlap sacks, and ready to be marketed.

3.2. Boundary System of Semi-Mechanical Extraction Process

The purposes of this LCA study are to identify sago starch production process, to determine the emission hotspots, and to calculate the environmental impact of the sago starch processing. Sago industry that being studied is just one sago industry in Cimahpar, Bogor as the case study. The scope of LCA study covers the harvesting of sago stem, transportation to extraction site, and the starch extraction process. The functional unit used is 1 ton of sundried sago starch. Functional unit is the value or services provided from a sago starch production system. The system boundary of sago starch production is defined in figure 2. The type of environmental impact category used in this study is global warming potential by quantifying the emission generated from the entire sago starch life cycle in the form of specific CO_2 eq.

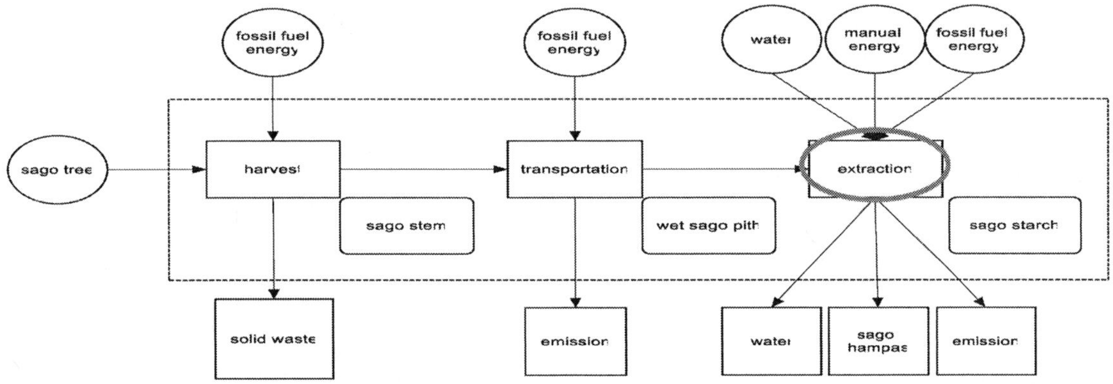

Figure 2. Boundary system scenario of semi-mechanical sago starch agroindustry

3.3. Life Cycle Inventory

For the inventory analysis, the data related to the extraction process is obtained from direct measurement in the industry, while for harvesting and transportation the secondary data is obtained from the Ecoinvent LCA database. The data collected by directly counting on the field and make a calculation through it. In this study, it is assumed that the distance of sago trunk transport to the extraction site is 200 km. Data used for inventory analysis is in accordance with pre-determined LCA goals and scopes. Inventory analysis consists of data collection and data calculation. In the data collection, each process unit will be described by the flow diagram of input and output. Process flowchart of sago starch extraction is given in figure 3. Mass balance calculation of each extraction process is presented in table 1. These figures combined with the secondary data from the Ecoinvent LCA database form the basis for the estimation of the environmental impact.

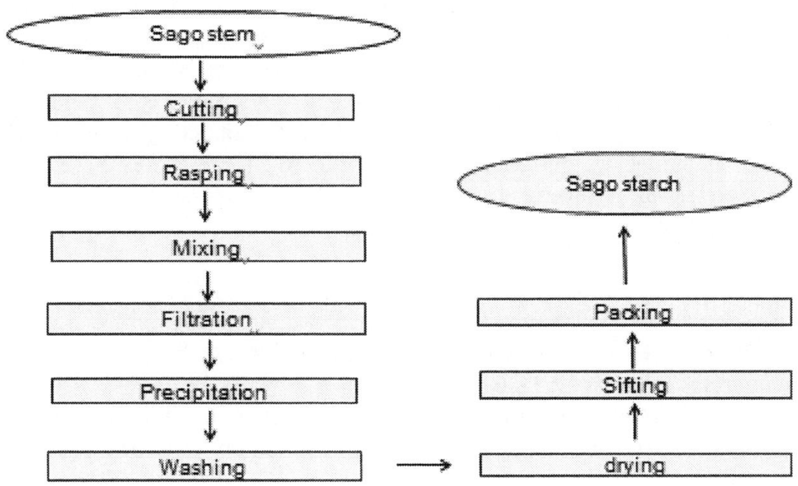

Figure 3. Flowchart of sago starch processing

Table 1. Mass balance calculation of each extraction process

Process	Input		Output	
	Material	Quantity (kg)	Material	Quantity (kg)
Cutting	Sago tree	7000	Sago stem	6965
			Solid waste	35
Sifting	Sago stem	6965	Sago powder	8
Mixing	Water	118430	Mixed with sago powder and water	125395
Filtration	Mixed with sago powder and water	125395	Wet hampas sago and water	17250
			Starch liquid	108145
Sedimentation	Starch liquid	108145	Water	106595
			Wet sago starch	1550
Washing	Wet sago starch	1550	Waste water	142316
	Water 132316	132316	Wet sago starch	1450
Drying	Wet sago starch	1450	Evaporate water	245
				1200

*this study has not accounted for the energy balance in the extraction process

With the assumption that the average transportation distance of sago stem to extraction site is 200 km, the GHG emission is estimated to be 325 kg CO_2 eq /ton of sundried sago starch. This lower figure for sago starch from this study is most likely due to the uncounted impact from the use of electrical energy on the extraction process, which is currently being conducted. A follow-up study is also underway to formulate several process improvement scenarios to derive the design of sago starch processing that generates the smallest emissions.

4. Conclusions

In this research work, LCA is used to determine the environmental impact of a semi-mechanical extraction process of sago industry. Using data from direct measurement of the extraction process and the secondary data obtained from the Ecoinvent LCA database for harvesting and transportation, and assuming the distance of sago trunk transport to the extraction site is 200 km, the GHG emission is estimated to be 325 kg CO2 eq /ton of sundried sago starch. This figure is lower than that reported for maize starch (1120 kg CO2 eq), potato starch (2232 kg CO2 eq) and cassava starch (4310 kg CO2 eq). This is due to the uncounted impact from the use of electrical energy on the extraction process. A follow-up study is underway to formulate several process improvement scenarios to get the design of sago starch processing that generates minimum emissions.

References

[1] Singhal R S, Kennedy J F, Gopalakrishman S M, Kaczmarek A, Knill C J and Akmar P F 2008 Industrial production, processing, and utilization of sago palm-derived products *Carbohydrate Polymers* **72** 1-20
[2] Bintoro H M H, Asmona D and Erwin 2009 Some efforts to rehabilitate sago palm plantation at Meranti District Riau Province Indonesia *1st ASEAN Sago Symposium* 2009 pp 1-4
[3] Purwaningsih I W 2016 *Life cycle assesment of PG. Subang Sugar Cane, West Java* Magister thesis (Bogor: Bogor Agriculural University)
[4] Jong F S and Widjono 2007 The sustainability of sago palm (*Metroxylon spp.*) cultivation on deep peat in Sarawak *Sago Palm Studies* **3** 13-20
[5] Adeni D S, Abd-Aziz S, Bujang K and Hassan M A 2010 Bioconversion of sago residue into value added products *African Journal of Biotechnology* **9**(14) 2016-21
[6] Phang S M, Miah M S and Yeoh B G 2000 Spirulina cultivation in digestes sago starch factory wastewater *Journal of Applied Phycology* **12** 395-400

IOP Conf. Series: Earth and Environmental Science **147** (2018) 012037 doi:10.1088/1755-1315/147/1/012037

The Influence of the Density of Coconut Fiber as Stack in Thermo-Acoustics Refrigeration System

E Hartulistiyoso[1], M Yulianto[1] and L Sucahyo[1]

[1] Department of Mechanical and Bio-system Engineering, Bogor Agricultural University (IPB), Indonesia

E-mail: edyhartulistiyoso@ipb.ac.id

Abstract. An experimental study of using coconut fiber as stack with varying density in thermo-acoustics refrigeration system has been done. Stack is a device which is described as the "heart" in thermo-acoustics refrigeration system. The length of stack is a fix parameter in this experiment. The performance of the coconut fiber was evaluated from the density of stack (varied from 30%, 50% and 70%), position of stack (varied from 0 to 34 cm from the sound generator), and frequency of sound generator (varied from 150 Hz, 200Hz, 250Hz and 300Hz). The inside, outside, and environment temperatures were collected every second using Data Acquisition (DAQ). The result showed that the increase of stack density will increase the performance of thermo-acoustics refrigeration system. The higher density produced temperature differences in cold side and hot side of 5.4°C. In addition, the position of stack and frequency of sound generator have an important role in the performance of thermo-acoustics refrigeration system for all variations of the density.

1. Introduction

Refrigeration system is generally used for air conditioning and cold storage, which is important to maintain the quality of agricultural products. Many refrigeration systems have been developed in many levels of temperature based on the application, such as vapor compression, absorption, thermoelectric, and pulse tube refrigeration. Recently, some researchers develop thermo-acoustics refrigeration (TAR) as alternative refrigeration system. TAR is purposed as an alternative refrigeration system, which is considered as environment-friendly, low energy consumption, and there is no moving part in the system [1,2,3]

TAR has been studied since several years ago. First observation of heat-induced acoustic oscillation was informed by Byron Higgins in 1777 and then basic theoretical for thermo-acoustic phenomena was established in1969 until 1980 by Roth [4]. Actually, TAR is a system that uses thermoacoustic effect to achieve energy conversion between thermal and acoustic power. The thermoacoustic effect is either the energy transformation of acoustic work absorbed to transport heat (thermoacoustic refrigeration) or the energy conversion of the heat supplied to produce acoustic work (thermoacoustic engine) [5]. TAR system is consisting of resonator, loudspeaker and stack. Resonator is a part of TAR to resonance the process, in which the material should be compact, lightweight, and strong. The resonator tube should be chosen by considering the lowest energy loss [6,7]. The length of resonator in $\lambda/4$ has better energy dissipation than $\lambda/2$. Loudspeaker produces an acoustic standing wave at the fundamental frequency which interacts with a porous medium known as stack. The resonance frequency in thermo-acoustic refrigeration has a linear function; therefore the frequency is the key in performance of TAR. Stack is part of TAR with a porous and this part known as the "heart" of the temperature gradient created in this part. The stack acts as a boundary for the compression and the expansion of the fluid (gas) [5].

The experiment in TAR system is divided into two main categories that are geometrical and operating parameters for each component of TAR and working fluid. The experiment to evaluate the influence of the length, diameter and the shape of resonator was done [8,9,10], concluding that the length and diameter of resonator is depend on the frequency. The shape of resonator consists of cylindrical, conical, tubular, Helmholtz, two sides tube and irreducible tube. The experiment for stack in TAR system is about the shape, material, length, porosity, position and acoustic frequency. The shape of stack has been studied by several researchers consist of parallel plates, honeycomb, corning celcor, spiral and pin array [5,6]. The material of stack has also been tested for stainless steel and cooper [11], random steel wool, copper scourers, carbon foam, mylar sheet [12], ceramic substrates [5], as well as organic ingredient such as drying squash [13,14], rice straw, pineapple fiber [14]. The experiment for working fluid, which is declared by researcher are using air, argon, helium, nitrogen [15].

Based on the explanation above, Indonesia as one of agricultural countries has a potency to develop a stack using agricultural waste. This paper will discuss about the use of coconut fiber as stack, which is potential to be used in Indonesia. The production of coconut in Indonesia reaches the number of 15.814 ton in 2015. Coconut fiber has also low thermal conductivity of 0.0488 W/m.K [16], smaller than that of mylar or PVC, which is about 0.2 W/m.K [17].

The objectives of this research are to identify the specific resonance of parallel stack from coconut fiber with random type which have acoustic effect, to measure temperature differences at steady state for all different frequency and position, as well as to describe the node and antinode region for stack from coconut fiber.

2. Materials and methods

2.1. *Material and Apparatus*

The discussion on this paper will focus on the performance of coconut fiber as stack with constant dimension but different position and frequency, therefore no heat exchanger was used. The geometry of stack is described as in table 1 below:

Table 1 the parameter of stack made of coconut fiber

Parameter		Value
Geometry	*Thermal penetration depth* (δ_k)	0.18 mm
	Length (L_s)	6 cm
Thermal Properties	Thermal conductivity(K)	0.0488 W/mK

The density of stack varied from 30%, 50% and 70% with the form as shown in figure 1 and table 2 below:

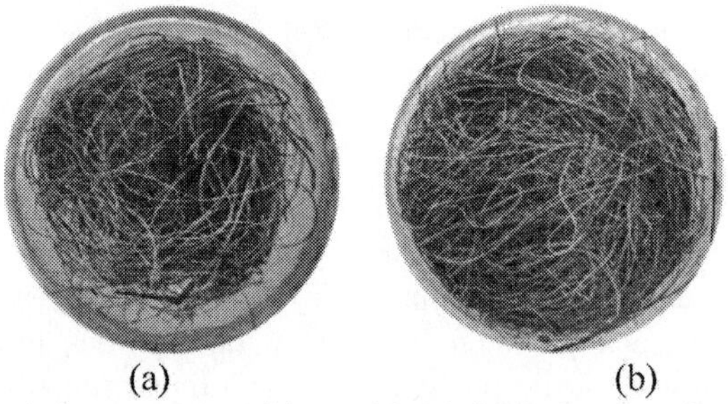

(a) (b)

Figure 1. Stack from coconut fiber with (a) 30% density, (b) 70% density

Many researchers have noted that the performance of TAR depends on the material and geometrical of stack. Random shape of material was selected because it is easy to create from waste material, low cost, and it is not possible to control all stack parameters.

Table 2 The density condition of stack

Condition of stack (%)	Mass of stack (g)	Density (kg/m^3)
30	2.58	18.79
50	4.25	30.94
70	5.91	43.03

The experiment apparatus is shown in figure 2. The apparatus consists of acoustic driven, resonator and stack. The acoustic driven is connected to amplifier and function generator. The resonator was made from acrylic and has tube length of 40 cm, diameter of 5.4 cm and thickness of 3 mm. The acrylic was used due to its strength, lightweight and transparent so that the placement of stack was easier to visualize. Stack was made from coconut fiber. The temperature was measured by thermocouple type K and collected every second by Data Acquisition (DAQ) connected to computer. Acoustic driven placed in a box was also made from acrylic. The position of stack varied from 0 until 32 cm and the frequency varied from 150, 200, 250 and 300 Hz.

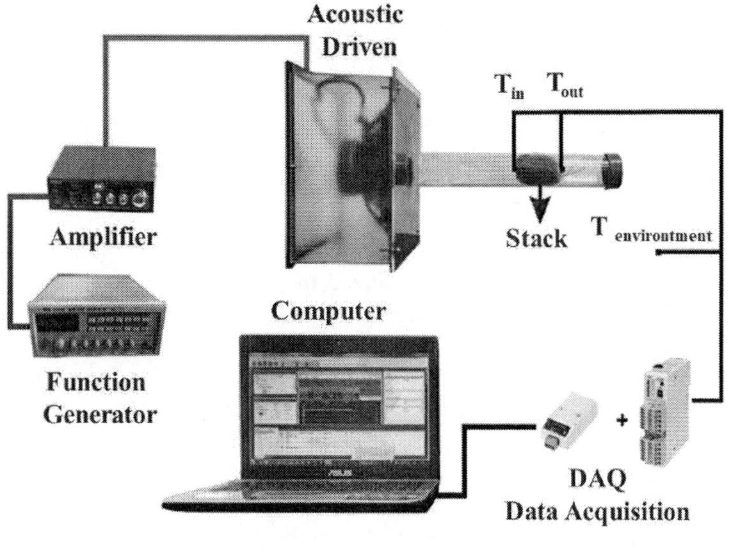

Figure 2 The apparatus for TAR system with stack from coconut fiber

2.2. *Thermo-acoustic Refrigeration Analysis*

An optimized λ/4 length parameter was used for resonator design [17], and then introduced using the equation below:

$$f_n = \frac{nv}{4\left(L + \frac{1}{2}\frac{4D}{3\pi}\right)} \tag{1}$$

Where f_n is frequency (Hz), v is velocity (m/s), L is length (m), and D is diameter (m). The thermo-acoustic effect generated in stack is an important component in TAR system. Stack length, porosity, position and acoustic frequency are all variables which influence the efficiency or the performance of the TAR. The COP is used in this experiment analysis to evaluate the performance of TAR system. The maximum of relative coefficient of performance refrigeration (COP_R) [18] is given by:

$$COP_R = \frac{COP}{COP_{carnot}} \tag{2}$$

And the COP can be calculated by:

$$COP = \frac{Q_L}{W} \tag{3}$$

The cooling load [19] can be obtained from:

$$Q_L = A_{stack}.K.\frac{\Delta T}{L_{stack}} = (1-\Omega).A_{pipe}.K.\frac{T_o - T_L}{L_{stack}} \tag{4}$$

The theoretical maximum coefficient of performance named Carnot [18] can be written as:

$$COP_{Carnot} = \frac{T_H - T_C}{T_C} \tag{5}$$

Where Q_L is cooling load (kJ/s), A is Area of stack (m^2), ΔT is temperature difference (°C), is porosity of stack (m), W is work of acoustic generation (kJ/s), T_C is temperature at the cooling side (°C), and T_H is temperature at the hot side (°C).

3. Result and discussion

3.1. *The Profile of Node (N) and Antinode (AN) Region*

Figure 3 until 5 describe the profile of Node (N) and Anti Node (AN) region made from temperature differences between cold and hot sides of stack using coconut fiber and plastics on the different frequency. The high temperature differences between cold and hot side is about 5°C occurred in coconut fiber stack with 70% density at AN region with frequency of 200 Hz and stack position of 27 cm. The lower achievement of about 1.8°C occurred in coconut fiber stack with 30% density at the same condition of frequency of 200 Hz and stack position of 27 cm. The phenomenon of Node (N) and Anti Node (AN) was noted to be occurred in all research in TAR system for some material which is used for stack (plastic, mylar, PVC) [6,7,20]. The optimum result of temperature differences for AN region in this experiment which occurred at stack position of 27 cm (coconut fiber with density 70%) was different from stainless steel wool, copper scourers, carbon foam, and mylar stack, where the AN region occurs at 1.1 m [12]. The different of AN region depended on the length and diameter of tube resonator and frequency. The experiment using stack from stainless steel wire-mesh mentioned that the different position of AN and N region occurs due to the pressure amplitude [21].

IOP Conf. Series: Earth and Environmental Science **147** (2018) 012037 doi:10.1088/1755-1315/147/1/012037

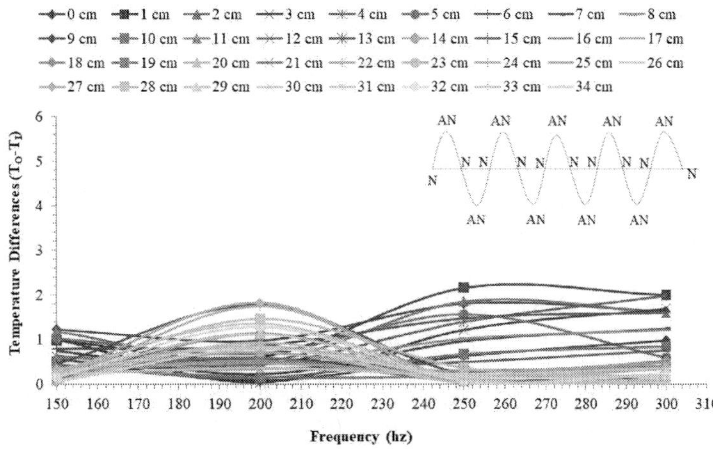

Figure 3 Profile of N and AN region of coconut fiber with density 30%

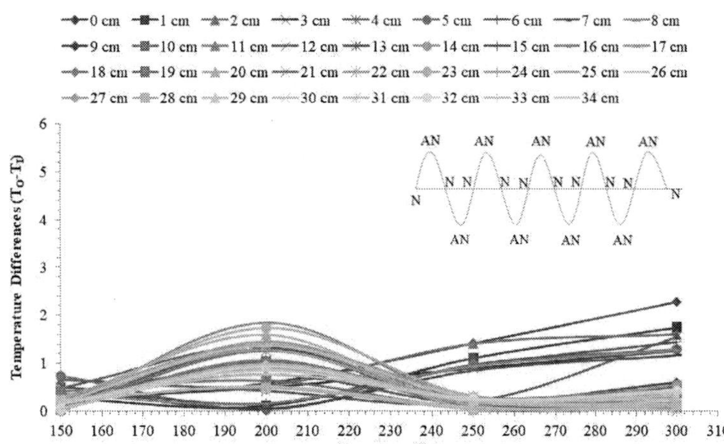

Figure 4 Profile of N and AN region of coconut fiber with density 50%

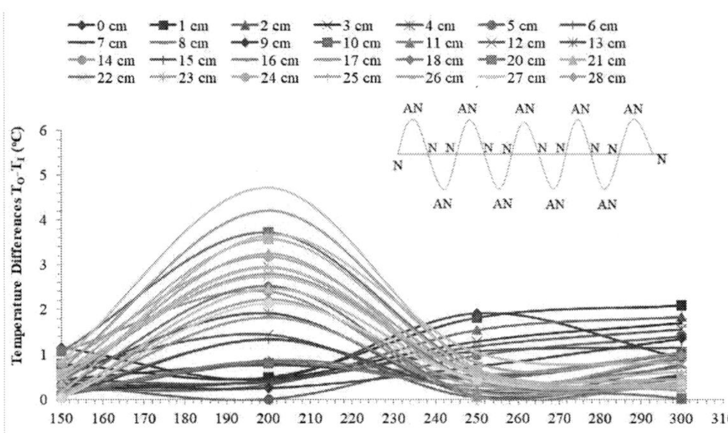

Figure 5 Profile of N and AN region of coconut fiber with density 70%

3.2. *Profile of Temperature at Node (N) and Anti Node (AN)*

Figure 6 to 8 show the outside and inside temperature position at AN and N region for coconut fiber stack with density of 30%, 50% and 70%. Profile of temperature at figure 6 for both Node (N) and Anti Node (AN) occurred at the position of 27 cm from loudspeaker with the 30% density of coconut fiber stack. The result showed that the higher acoustic effect occurred at the frequency of 200 Hz at AN region and the lower acoustic effect occurred at the frequency of 150 Hz at N region. The acoustic effect at AN region for frequency of 200 Hz and position of stack 27 cm produced outside temperature (cold side) of 25.8°C and inside temperature (cold side) of 27.6°C. Different phenomenon occurred at N region, which showed no acoustic effect because there is no differences between inside and outside temperature. The higher acoustic was achieved for coconut fiber with density of 70%. At AN region in frequency of 200 Hz and stack position of 27 cm and density of coconut fiber of 70% produced outside temperature (cold side) of 29.3°C and inside temperature (hot side) of 32.7°C. All profile of temperature showed that there was no acoustic effect at N region. The AN region produced temperature at hot side of 33°C and the cold side of 28°C. The N region showed different result; where the cold and hot side produced the same temperature of about 31.1°C until 31.6°C. There was no temperature difference in N region due to the high amplitude pressure differences at AN region [6]. Compared to the other stack materials, such as pineapple fiber, squash, and rice straw [14] the temperature difference of using coconut fiber stack is smaller, due to the different power inserted to the system [22,23].

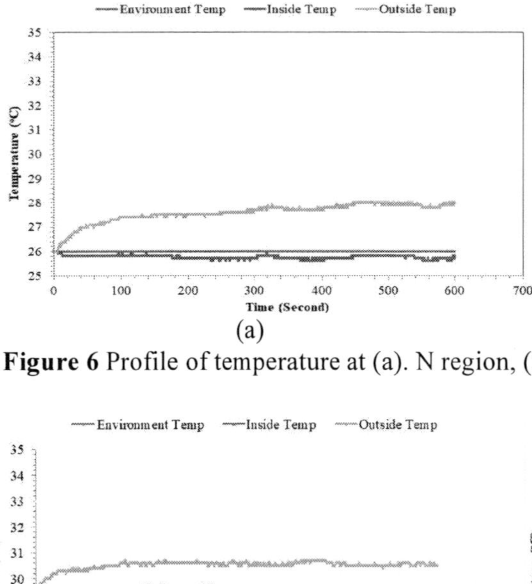

(a)

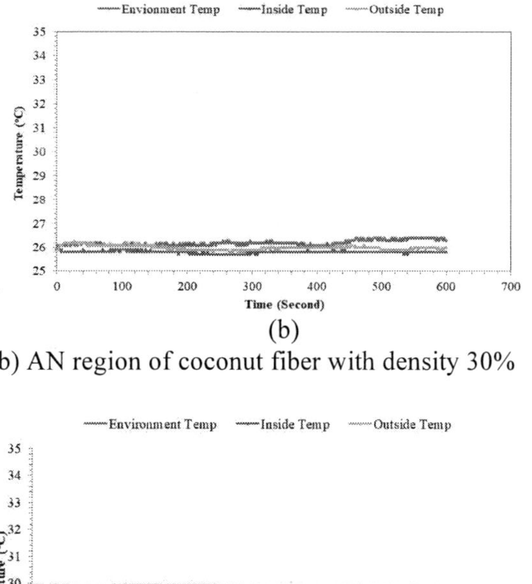

(b)

Figure 6 Profile of temperature at (a). N region, (b) AN region of coconut fiber with density 30%

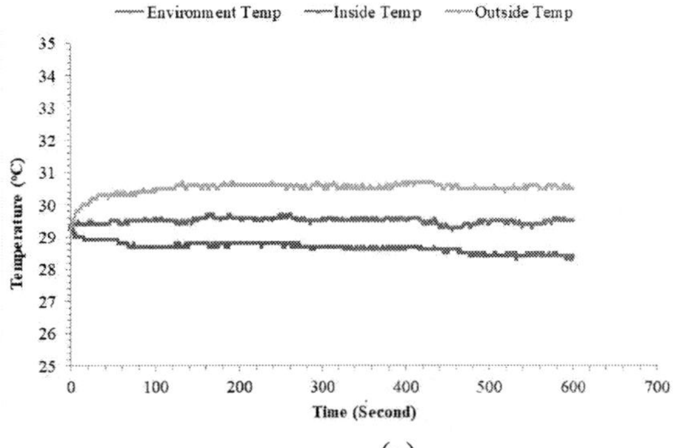

(a)

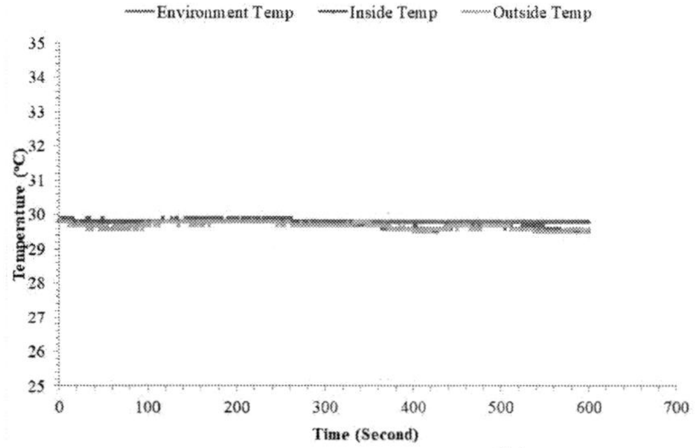

(b)

Figure 7 Profile of temperature at (a). N region, (b) AN region of coconut fiber with density 50%

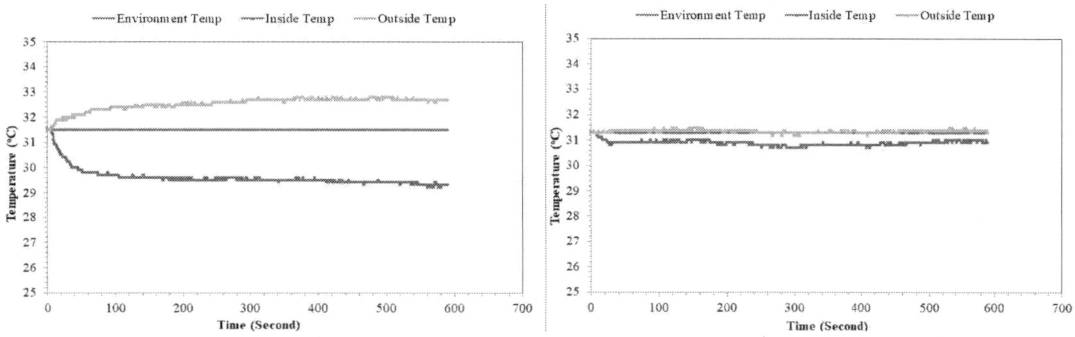

(a) (b)

Figure 8 Profile of temperature at (a). N region, (b) AN region of coconut fiber with density 70%

3.3. *Performance of TAR system*

The performance of TAR system in this experiment is evaluated by COP, COPC, and COPR. Table 3 shows the calculation result of coconut fiber performance varied from density of 30%, 50%, and 70%. The COP, COPC, COPR were calculated based on equation 1, 2, 3 and 4. The maximum COP reached using coconut fiber stack was 0.00115, whereas using plastic was 0.00882. These values are very small in comparison with conventional refrigeration technology such as vapor compression, absorption refrigeration, thermoelectric, and pulse-tube refrigeration. The COP will increase if the temperature differences between hot and cold side increased and the other researcher also mentions that the value of COP will increase linearly with the cooling load [12].

Table 3 The performance of Coconut Fiber as Stack at TAR System

Density of Coconut Fiber (%)	Frequency (Hz)	COP	COPC	COPR
30	150	0,00020	16,1875	1,24E-05
	200	0,00029	10,96	2,68E-05
	250	0,00030	9,357143	3,19E-05
	300	0,00027	9,769231	2,78E-05
50	150	0,00021	30,5	6,87E-06
	200	0,00045	12,30435	3,66E-05
	250	0,00032	16,77778	1,91E-05
	300	0,00049	11,25	4,33E-05
70	150	0,00050	15,76471	3,16E-05
	200	0,00148	5,555556	2,66E-04
	250	0,00062	10,48	5,94E-05
	300	0,00066	9,481481	6,93E-05

4. Conclusions

The stack from coconut fiber had been built and tested with varying density. Increasing the density will increase the temperature differences between hot and cold side. The Node (N) and Anti Node (AN) have an important role for temperature achievement. AN region will produce higher temperature differences. The use of coconut fiber as stack in this experiment has low COP compared to other refrigeration systems. Therefore it needs further development to use coconut fiber as stack in TAR System.

IOP Conf. Series: Earth and Environmental Science **147** (2018) 012037 doi:10.1088/1755-1315/147/1/012037

Acknowledgement
We would like to thanks the Directorate of Research, Ministry of Research, Technology and Higher Education for supporting this research by PUPT-PUD 2017 with Contract Number 1383/IT3.11/PN/2017

References

[1] Nsofor E C and Azrai A 2009 Experimental study on the performance on the thermoacoustic refrigeration system (Elsevier) *Applied Thermal Engineering* **29** 2672-79
[2] Hariharan N M, Sizashanmugam P and Kasthurirengan S 2015 Studies on performance of thermoacoustic prime mover *Experimental Heat Transfer* **28** 267-281
[3] Tertibu L K, Sun B and Kaunda M A E 2015 *Proc. Int. Conf. on Soft Computing and Software Engineering, Procedia Conference Proceeding* pp 611-618
[4] Atiqah Z N, Mohd-Ghazali N and El-Fawal M H 2016 *Renewable and Sustainable Energy Reviews* **54** 626-634
[5] Alcock A C, Tartibu L K and Jen T C 2017 *Proc. Int. Conf. on Sustainable Material Processing and Manufacturing (SMMP), Procedia Manufacturing Conference Proceeding* vol 7 pp 79-85
[6] Tijani M E H 2011 *Loudspeaker-driven thermo-acoustic refrigeration (Ph.D. thesis)* (Technische Universiteit Eindhoven)
[7] Swift G W 2001 *J. Acoust Soc Am* **113** 2379-81
[8] Zink F, Hamish W, Rosalina A and Laura S 2009 *International Journal of Thermal Science* **48** 2309–22
[9] Bai X, Jin T and Chen G B 1998 *Proc. Int. Conf. on cryogenics and refrigeration* p 522-525
[10] Qing E, Feng W, Duanyong L and Fangzhong G 2009 *Proc. Int. Conf. on Power and Energy Engineering, IEEE* pp 1-3
[11] Kamble B V, Kuzhiveli B T, Jain R, Prasad S, Kasthurirengan S and Behera U 2014 *Proc. Int. Conf on Advance in Cryogenic Engineering, AIP Conference Proceeding* vol 1573 pp 768-775.
[12] Yahya S G H, Xianoan M and Jaworski A J 2017 *International Journal of Refrigeration* **75** 52-63
[13] Putri E K N 2013 *Study the influence of heat exchanger to cold and hot side on the performance of a cooling thermoacoustic system using stack from squash* (*Luffa acutangula*) Magister thesis (Yogyakarta: University of Gadjah Mada)
[14] Hidayah Q 2015 *The Influence of stack from organic material on a thermoacoustic cooling system* Bachelor thesis (Yogyakarta: University of Gadjah Mada)
[15] Hao X H, Ju Y L, Behera U and Kasthurirengan S 2011 *Cryogenic* **42** 559-561
[16] Hariharan N M, Sizashanmugam P and Kasthurirengan S 2012 *Applied Acoustics* **73** 1052–58
[17] Manohar K, Ramlakhan D, Kochhar G and Haldar S 2006 *Journal of the Braz Soc of Mech Sci & Engineering* 2845-47
[18] Balonji S, Tartibu L K and Jen T C 2017 *Proc. Int. Conf. on Sustainable Material Processing and Manufacturing (SMMP), Procedia Manufacturing Conference Proceeding* vol 7 pp 131-137
[19] Stoll S P 2008 *Investigation of temperature gradients in a demonstration model thermoacoustic refrigerator* Ph.D thesis (University of Central Arkansas)
[20] Agustina D 2013 *The performance of thermoacoustics cooler with variation in stack geometry and drive ratio* Magister thesis (Depok: University of Indonesia)
[21] Ikhsan S, Achmadin W N, Murti P and Nohtomi M 2016 *Journal of Physics: Conference Series* **710** 1-8
[22] Putra N, Agustina D 2013 Influence of stack plate thickness and voltage input on the performance of loud speaker-driven thermoacoustic refrigeration *Journal of Physics: Conference Series* **423**(1) 1-9

Mapping The Temporal and Spatial Variability of Soil Moisture Content Using Proximal Soil Sensing

S Virgawati[1], M Mawardi[2], L Sutiarso[2], S Shibusawa[3], H Segah[4] and M Kodaira[3]

[1] Dept. of Agrotechnology, University of Pembangunan Nasional "Veteran", Yogyakarta, Indonesia
[2] Dept. of Agricultural and Biosystem Engineering, University of Gadjah Mada, Yogyakarta, Indonesia
[3] Dept. of Environmental and Agricultural Engineering, Tokyo University of Agriculture and Technology, Tokyo, Japan
[4] Dept. of Forestry, Faculty of Agriculture, University of Palangka Raya, Central Kalimantan, Indonesia

E-mail: sari_virgawati@upnyk.ac.id

Abstract. In studies related to soil optical properties, it has been proven that visual and NIR soil spectral response can predict soil moisture content (SMC) using proper data analysis techniques. SMC is one of the most important soil properties influencing most physical, chemical, and biological soil processes. The problem is how to provide reliable, fast and inexpensive information of SMC in the subsurface from numerous soil samples and repeated measurement. The use of spectroscopy technology has emerged as a rapid and low-cost tool for extensive investigation of soil properties. The objective of this research was to develop calibration models based on laboratory Vis-NIR spectroscopy to estimate the SMC at four different growth stages of the soybean crop in Yogyakarta Province. An ASD Field-spectrophotoradiometer was used to measure the reflectance of soil samples. The partial least square regression (PLSR) was performed to establish the relationship between the SMC with Vis-NIR soil reflectance spectra. The selected calibration model was used to predict the new samples of SMC. The temporal and spatial variability of SMC was performed in digital maps. The results revealed that the calibration model was excellent for SMC prediction. Vis-NIR spectroscopy was a reliable tool for the prediction of SMC.

1. Introduction

Soils vary in space and time due to the combined effects of physical, chemical, and biological processes that operate at different scales and with different intensities [1]. Some soil properties are very stable, changing slightly by the time, such as texture and soil organic matter content. The other soil properties, such as nitrate (NO_3) and moisture content can fluctuate rapidly. Soil moisture is the major factor of the soil in plant growth. The gross effects of deficient and of excessive soil moisture on plant growth are well known. With irrigation and, where necessary with drainage, the farmer has the opportunity to exercise greater control over soil moisture than he does over any of the other soil physical factors [2]. Various measurable aspects of plant growth do not respond in the same manner to increasing moisture stress. The water that comes out of the ground system through evapotranspiration, drainage, and subsurface water flow, first had to interact with the soil and partly stored in it within a few moments up to many years. It is important to measure and monitor soil moisture to improve crop yield forecasting and irrigation planning. The problem is how to provide reliable, fast and inexpensive information of soil moisture in the subsurface or root zone from numerous soil samples and repeated measurement.

Gravimetric method is currently the most commonly accepted and most reliable method for soil moisture determination and calibration of all indirect measurement methods [3]. Despite its advantages of accuracy and high reliability, the gravimetric method is time and resource consuming, destructive, and unrepeatable. The development of sensing technology coupled with advances in information and communications technology is very supportive of the use of soil sensing. Soil sensing can facilitate the measurement and monitoring of the soil's physical and biochemical attributes to better understand their dynamics and interaction with the environment while considering their large spatial heterogeneity [4].

The proximal soil sensing (PSS) techniques have been developed to obtain georeferenced data on many soil parameters at different scales and times to better understand the variability [5]. Proximal soil sensing is field-based sensors to obtain signals from the soil when the sensor's detector is in contact with or close to (within 2 m) the soil [6]. The sensors provide soil information because the signals correspond to physical measures, which can be related to soils and their properties.

In precision agriculture, techniques such as diffuse reflectance spectroscopy and geostatistics are considered increasingly for the prediction of soil characteristics and to characterize within-field spatial variation [7]. Rossel et al. [8] demonstrated the potential of diffuse reflectance spectroscopy, using the VIS, NIR, MIR and combined spectra, and showed how qualitative soil interpretations might aid with the identification and assignment of spectral bands to soil constituents. The Vis-NIR real-time soil sensor also had great potential for determining the soil properties at 10, 15, 20 cm soil depth [9]. Vasques et al. [10] investigated the potential use of Vis-NIR diffuse reflectance spectroscopy to classify soils in areas with soil, geology, and topography that varied in southeastern Brazil from three depth intervals (0-20, 40-60 and 80-100 cm). This novel approach can improve soil classification and survey in a cost-efficient manner, supporting sustainable use and management of tropical soils.

The objective of this research was to develop calibration models based on laboratory Vis-NIR spectroscopy and partial least-square regression (PLSR) analysis to estimate the soil moisture content (SMC) at four different growth stages of the soybean crop in two small farms at Yogyakarta Province. The SMC was measured using the Gravimetric method, while the soil reflectance was scanned with an ASD Field-spec 3 (range from 350 nm to 2500 nm). The PLSR with full cross-validation was performed to establish the relationship between the SMC with the pre-treated Vis-NIR soil reflectance spectra. The selected calibration model was used to predict the other new samples of SMC. The temporal and spatial variability of SMC was performed in digital maps using inverse distance weighted (IDW) interpolation method. These maps gave much information to be interpreted carefully due to many factors affect the SMC. They were useful in supporting the process of decision making in field management.

2. Material and methods

2.1. *Site description*

The research was conducted at soybean farms in two locations, i.e. Natah Village, Nglipar District, Gunung Kidul Regency or G-field (7°51'39.0"S, 110°39'19.4"E) and Jatimulyo Village, Dlingo District, Bantul Regency or B-field (7°55'22.5"S, 110°29'08.7"E), in Yogyakarta Province (figure 1). The elevation of Nglipar ranges from 200 to 210 m asl., while Dlingo elevation ranges from 190 to 200 m asl. The slope varies between 5° to 10° which Dlingo was steeper than Nglipar. The variability of soil classification in the research area was high, even occurred in the same landform. Soils in the study area were tentatively classified as *Hapludults* and *Dystrudepts* at Nglipar, while soils at Dlingo were classified as *Hapludalfs, Eutrudepts,* and *Udorthents* (table 1).

Nglipar and Dlingo had the tropical climate and classified as *Am* by Köppen and Geiger. The average annual temperatures of Nglipar and Dlingo were 25.2 °C and 25.8 °C, and the average rainfalls were 2,083 mm and 2,019 mm [11]. table 2 shows the average of monthly rainfall of the past 10 years. During the research, the monthly rainfalls from October 2016 to January 2017 were: 253, 526, 305 and 369 mm at Nglipar, and 232, 312, 420 and 411 mm at Dlingo [12].

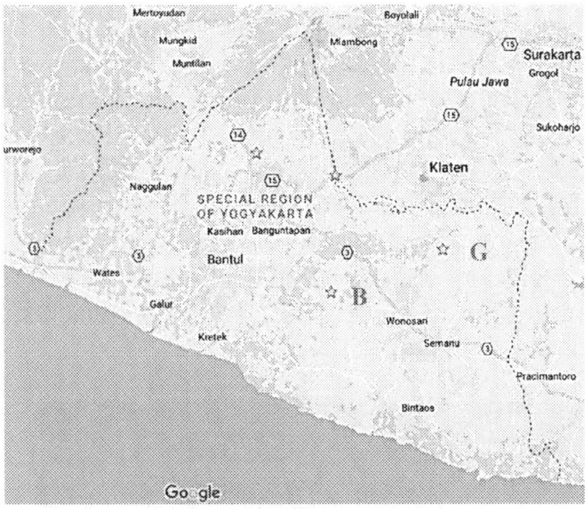

Figure 1. Location of the research area:
G-field: Nglipar, Gunung Kidul Regency
B-field: Dlingo, Bantul Regency
(Source: Modified from Google Map 2017)

Table 1. Soil class and landform of Nglipar and Dlingo

Soil Class (Great group)	Proportion (%)	Landform	Parent material	Relief (% slope)
Nglipar, Gunung Kidul: Tectonic Group				
Hapludults	50-75	Undulated tectonic plain	claystone, sandstone	undulated (8-15)
Dystrudepts	25-50			
Dlingo, Bantul: Karst Group				
Hapludalfs	50-75	Karst hill	limestone	Small hilly (15-25)
Eutrudepts	25-50			
Udorthents	10-25			

Source: Indonesian Center for Agricultural Land Resources Research and Development (BBSDLP, 2016)

Table 2. The monthly rainfall average of 2007-2016 (mm)

Month / *Location*	Jan	Feb	Mar	Apr	May	Jun	Jul	Aug	Sep	Oct	Nov	Dec
Nglipar(G)	314	334	277	190	123	79	53	38	117	80	219	496
Dlingo (B)	381	438	335	306	216	121	45	25	75	89	295	386

Source: BMKG, Sleman DIY (2017)

The activities and crop pattern in the farm at Nglipar and Dlingo were almost the same every year.

Table 3. The yearly crop pattern at Nglipar and Dlingo

Month / *Location*	Jan	Feb	Mar	Apr	May	Jun	Jul	Aug	Sep	Oct	Nov	Dec
Nglipar(G)	GN	GN	GN	SB	SB	SB	-	-	-	SB	SB	SB
Dlingo (B)	SB/C	SB/C	SB/C	VB	VB	VB	-	-	-	SB	SB	SB

GN: Ground Nut; SB: Soy Bean; C: Corn; VB: Velvet Bean

2.2. *Soil sampling*
Due to the irregular and terrace shapes of the fields, the layout of sample points was set up using the grid method combined with a transect line of 5 meter interval (figure 2). There were 30 sample points for each field marked with bamboo sticks. The soil was sampled 4 stages within one cropping season of soybean from October 2016 to January 2017, i.e., before planting (I), vegetative stage (II), generative stage (III) and after harvesting (IV). Soil at each point was taken using auger at a depth of 5-15 cm about 500 grams and stored in a labeled zip lock plastic bag. All samples were air-dried, then gently crushed to break up larger aggregates, afterward removed the visible roots, and each sample was sieved at 2 mm strainer. Eighty soil samples (10 x 2 x 4) were prepared for soil moisture analysis as the referenced parameter, and all 240 soil samples (30 x 2 x 4) were prepared for spectroscopic measurements as the predicted parameter.

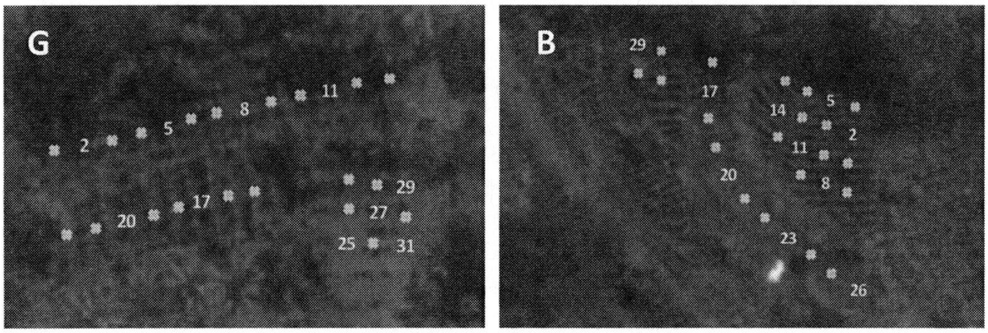

Figure 2. Field layout
left: Nglipar field (1,500 m^2), *right*: Dlingo field (1,300 m^2)
(Modified from Google Earth 2012)

2.3. *Soil moisture content analysis*
The soil moisture content was analyzed by the Soil Analytical Services Laboratory at UPN "Veteran" Yogyakarta using the Gravimetric method, i.e., calculates soil moisture by the difference between the fresh weight and dry weight of a given soil sample. A 40 g fresh weight subsample was taken from the soil sample of each plot was weighed and placed into a tin and then oven dried at 105° - 110° C for 24h. After drying the samples are reweighed.

2.4. *Laboratory Vis-NIR Spectroscopy*
The spectroscopy measurement was performed at the University of Palangka Raya, Central Kalimantan, using ASD Field-spec®3 350-2500 nm (Analytical Spectral Devices Inc., Boulder, Colorado, USA). Each soil sample was placed into a 5 cm diameter of ring sample (*Eijkelkamp*) and flattened the surface. A black aluminum ring plate (modified by TUAT Laboratory, Japan) was fitted on the top of ring sample to hold the ASD probe of the optic sensors and keep the same distance from the probe tip to the sample surface (figure 3).

Figure 3. Soil reflectance measurement
left: soils in ring sample
right: The ASD probe was inserted into a black aluminum ring plate on the sample surface

The reflectance of each sample was scanned 10 times with different positions by moving the ring sample circularly, and the results were averaged in post-processing (figure 4). Every 15 minutes the instrument was calibrated by measuring the reflectance of the white panel as a white reference. The reflectance value of each spectrum was recorded in the computer and translated from binary to ASCII using *ViewSpecPro* software.

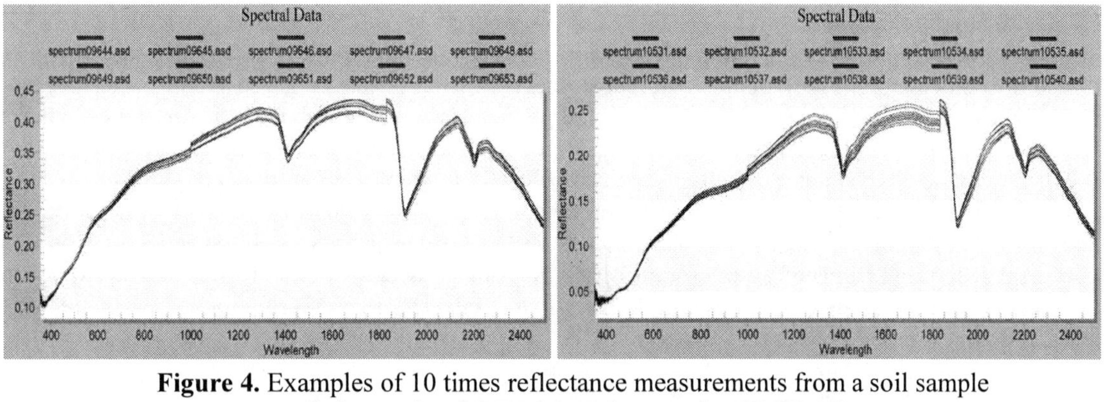

Figure 4. Examples of 10 times reflectance measurements from a soil sample (*left*: sample of G IV-25; *right*: sample of B III-15)

2.5. *Pretreatment, calibration and prediction process of spectral data*

The first step in developing the prediction models was the pretreatment of the spectral data. The process of pretreatment, calibration, and validation were performed using Unscrambler X software. The measured reflectance spectra were transformed in absorbance through the log (1/R) to reduce noise, offset effects, and to enhance the linearity between the measured absorbance and soil properties [13]. To enhance weak signals and remove noise due to diffuse reflection, the absorbance spectra were pre-treated using the second derivative Savitzky and Golay method [14]. Moreover, both edges of the spectra were removed as these parts of the spectra were unstable and rich in noise [9].

The calibration models were subsequently developed by applying the partial least-square regression (PLSR) technique coupled with full cross-validation to establish the relationship between the amount of soil moisture content (reference values) with the pre-treated Vis-NIR soil absorbance spectra from the corresponding locations [9].

Two calibration models were developed, i.e., Nglipar (G) and Dlingo (B) SMC models. The models combined the dataset of 40 referenced SMC values (10 x 4) and the treated spectra of 600-2300 nm wavelength for each SMC model. In the PLSR analysis, sample outliers were detected by checking the residual sample variance plot after the PLSR. Individual sample outliers located far from the zero line of residual variance were considered to be outliers and excluded from the analysis. Due to the small number of data set for calibration (40 samples), the number of outliers was limited to maximum 6 samples.

The selection criteria of any pretreatments were the largest coefficient of multiple determinations (R^2) and the smallest of Root Mean Square Error (RMSE). The full cross-validation ability of PLSR was given by the value of residual prediction deviation (RPD). The ability of NIRS to predict values of soil properties can be grouped into three categories based on RPD values: category A or excellent (RPD $\geq$2.0), category B or good (RPD = 1.4~2.0), and category C or unreliable (RPD <1.4) [15]. RPD was given by the ratio of standard deviation (SD) of the reference dataset to the root mean square error of full cross-validation ($RMSE_{val}$), as in equation (1) [9].

$$RPD = SD.\ RMSE_{val}^{-1} \qquad (1)$$

Each selected calibration model was used to estimate the moisture value of the new samples in associated location.

2.6. *Mapping the temporal and spatial of SMC variability*

The support of GIS is very helpful in giving visual information to support the decision making in the management of site-specific farming. In this research, the ArcGIS Ver. 10.2 software was used to perform the temporal and spatial of SMC variability maps. The inverse distance weighted (IDW) method was applied to do the spatial interpolation.

3. Results and discussion

To generate spectrum data, radiation containing all relevant frequencies in the particular range is directed to the sample. The radiation will cause individual molecular bonds to vibrate, and they will absorb light, to various degrees, with a specific energy quantum corresponding to the difference between two energy levels. As the energy quantum is directly related to frequency, the resulting absorption spectrum produces a characteristic shape that can be used for analytical purposes [16].

The soil has an easily distinguishable characteristic reflectance pattern in the visible, near-infrared and mid-infrared wavelengths [17]. Water has a strong influence on vis-NIR spectra of soils. The dominant absorption bands of water around 1400-1900 nm are characteristic of soil spectra [18]. Water also reduces the reflectance of regions in the short-wave infrared, particularly around 900, 1400, 1900 and 2200 nm [19].

3.1. *Soil properties description*

There were many factors affect the variability of SMC, for example, the texture and structure. Soil structure, texture, and depth determine the total capacity of the soil for storing available water for plant growth [2]. Clay soils hold large amounts of nutrients and water tightly, slow infiltration or high runoff means much erosion and may shrink/swell, depends upon the type of clay minerals present [20]. table 4 describes the properties of the soil used in this study. The soil properties of all 80 samples were classified as *clay* with very low organic matter content (SOM) and moderate cation exchange capacity (CEC).

Table 4. Properties of the soils at research area

Property	Clay (%)	Sand (%)	Silt (%)	pH	SOM (%)	CEC (me%)
Nglipar	64.33	22.04	13.63	07.22	0.98	16.92
Dlingo	69.09	22.67	8.24	07.09	1.10	20.06

The landform such as terraced field is also affecting the SMC. Soil moisture amount was greater in terrace channels as compared to terrace intervals [21]. Research in terrace land showed that rainfall significantly affected the temporal stability of soil water storage in the depth of 0–0.6 m and increased from the upper to lower slope [22].

Both farmers at Nglipar and Dlingo used the same variety of Soybean (Grobogan) which was ready to harvest about 80 days. They grew the soybean at the rainy season from October 2016 to January 2017, and soils were sampled before planting (October 28), at vegetative stage (November 14), at generative stage (December 1) and after harvesting (January 19). During the research, the monthly rainfalls from October 2016 to January 2017 were: 253, 526, 305 and 369 mm at Nglipar, and 232, 312, 420 and 411 mm at Dlingo.

Rainfall was the only supply for the crop water requirement in these farms. High amounts of rainfall that occur during vegetative growth are normally not beneficial unless soil water levels are extremely low before or after planting [23]. Soybean requires the most water from flowering through seed fill. By considering other factors affecting evapotranspiration, the SMC information given in figure 5 could contribute to predicting the crop water requirements based on growth stages. Therefore, the actions to maintain water needs can be decided properly before the next stage.

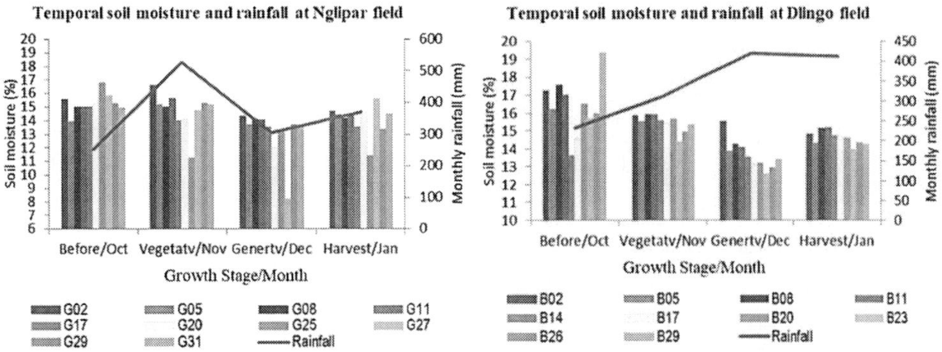

Figure 5. Soil moisture of 10 referenced samples at different stages and locations

Eighty graphs (10 x 2 x 4) from the total of 240 soil spectral measurements are shown in figure 6 and 7.

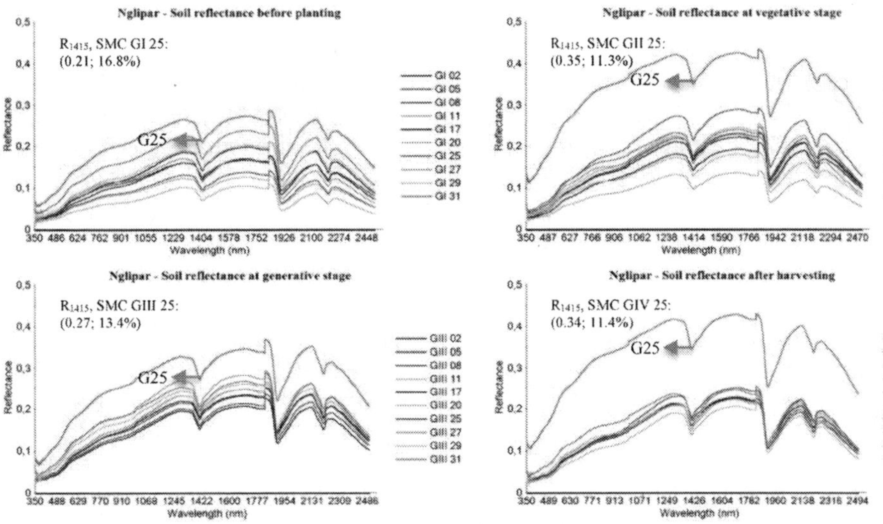

Figure 6. The soil reflectance of 10 referenced samples at different stages in Nglipar (G)

Soil reflectance can be influenced by some factors, such as soil texture, surface roughness, organic matter content, color and moisture content [3]. Field soil reflectance is reduced, particularly in the visible portion of the spectrum, when the moisture content is high [17]. Soil moisture and organic matter increase soil absorbency and result in overall lower soil reflectance [19]. To study the correlation between the soil reflectance, SMC, and growth stage. The sample G25 (figure 6) and B26 (figure 7) at about 1415 nm wavelength were picked for examples (table 5).

Table 5. The example of correlation between soil reflectance (R), soil moisture content (SMC) and growth stage at 1415 nm wavelength

Sample point	G 25 (Nglipar farm)		B 26 (Dlingo farm)	
Parameter	R_{1415}	SMC (%)	R_{1415}	SMC (%)
Stage I	0.21	16.8	0.15	15.9
Stage II	0.35	11.3	0.17	14.9
Stage III	0.27	13.4	0.19	12.9
Stage IV	0.34	11.4	0.18	14.3

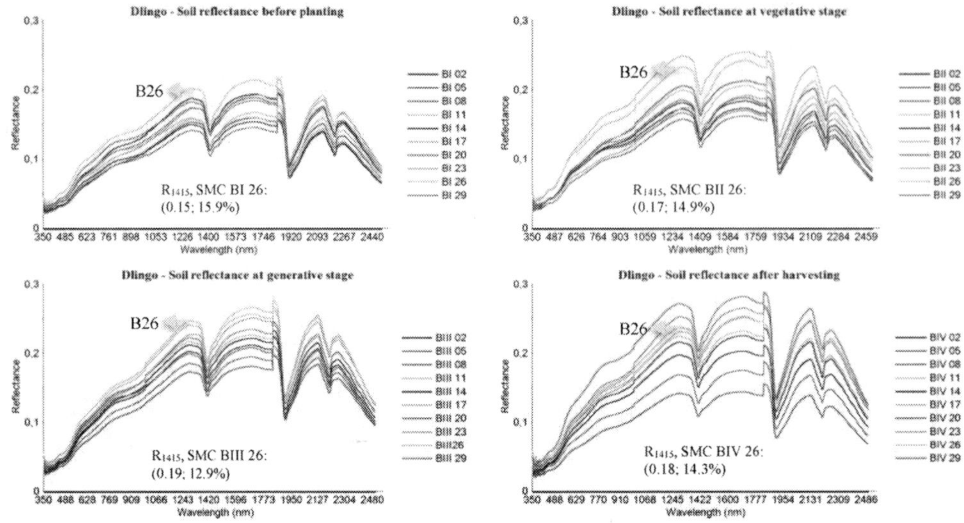

Figure 7. The soil reflectance of 10 referenced samples at different stages in Dlingo (B)

The increase of soil reflection values on samples G25 and B26 followed the decline of SMC. The effect of whitish color at G25 was strong to increase the reflectance, however high moisture might lower it. The soil color of B26 was darker than G25, so the reflectance value was lower. The higher the SMC, the wetter or darker the soil so the reflectance value was lower.

3.2. *Multivariate Statistical Analysis*

The most common calibration methods applied are based on linear regressions, namely stepwise multiple linear regression (SMLR), principal component regression (PCR), and partial least squares regression (PLSR). PCR and PLSR techniques can cope with data containing large numbers of predictor variables that are highly collinear [16]. However, PLSR is often preferred by analysts because it relates the response and predictor variables so that the model explains more of the variance in the response with fewer components, it is more interpretable, and the algorithm is computationally faster.

There were four steps of pretreatment to the spectral data before proceeding them for calibration to enhance weak signals and remove noise due to diffuse reflection, i.e.:

a. Spectroscopic transform: reflectance to absorbance
b. Derivative Savitsky-Golay transform: second derivative and 2 polynomial order
c. Remove both edges of the spectra (<600 nm and > 2300 nm)
d. Convert spectral data from 1 nm interval to 10 nm interval.

The statistical descriptions of SMC data for calibration process are shown in table 6. The SMC at Nglipar ranges from 8.25 to 16.80%, with a mean value of 14.28%, while the SMC at Dlingo ranges from 12.60 to 19.34%, with a mean value of 15.03%.

Table 6. Descriptive statistics of soil moisture content data

SMC	Sample	Mean	Max	Min	Range	SD	Var.	RMS	Skew
Nglipar	40*	14.28	16.80	8.25	8.55	1.52	2.31	14.36	-18.41
Dlingo	40*	15.03	19.34	12.61	6.73	1.37	1.88	15.09	0.76

* from 10 referenced samples x 4 stages

The summary of developing calibration models for SMC using PLSR method and the RPD category are shown in table 7.

Table 7. Summary of PLSR results for SMC calibration models

PLSR results	Nglipar (G) SMC model	Dlingo (B) SMC model
Calibration samples	40	40
Prediction samples	120	120
Optimal factors	3	5
R^2_{cal}	0.76	0.97
$RMSE_{cal}$	0.50	0.21
R^2_{val}	0.61	0.73
$RMSE_{val}$	0.66	0.61
SD	1.52	1.34
RPD (SD/ $RMSE_{val}$)	2.32	2.25
RPD Criteria	A (excellent)	A (excellent)

The Vis-NIR predicted values using PLSR for SMC are described as regression models in figure 8. The data points of the measured moistures (reference) and the predicted are indicated excellent model performance (RPD > 2). The two soils measured by the Vis-NIR reflectance sensor and by the gravimetric method are compared, and they are highly correlated ($R^2 = 0.76$ and 0.97).

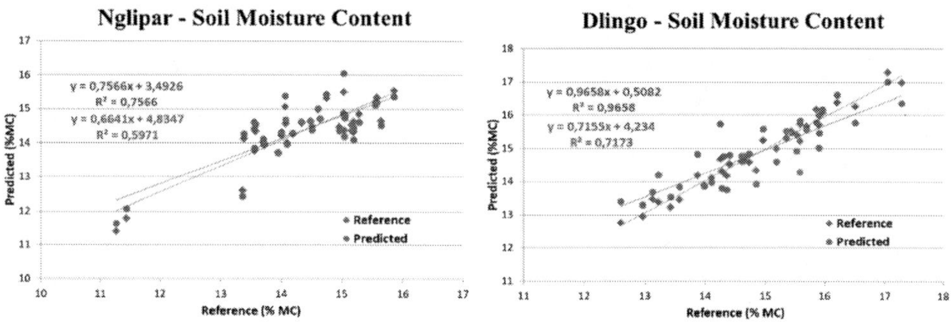

Figure 8. The regression model of soil moisture content at Nglipar and Dlingo

The better results obtained by using the PLSR method are clearly because PLSR takes advantage of the use of the entire spectral signature [24]. The regression coefficient plotted in figure 9 shows the investigated spectrum that should be considered important for the prediction of SMC. The size of the regression coefficients represents the importance of the absorption band.

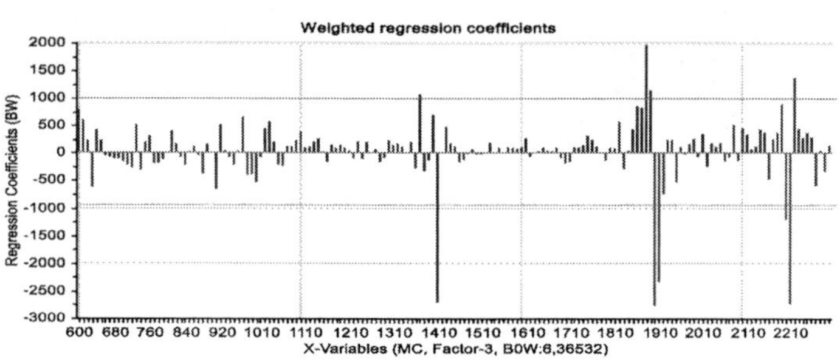

Figure 9a. The regression coefficients of SMC model for Nglipar (G)

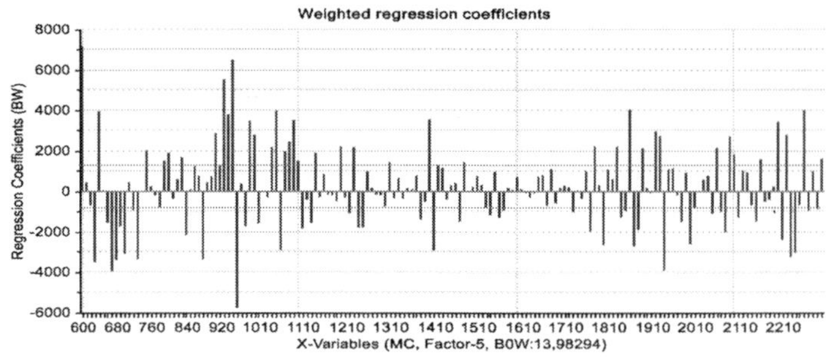

Figure 9b. The regression coefficients of SMC model for Dlingo (B)

3.3. *Temporal and spatial map of SMC variability*

The prediction value of SMC was then applied to generate the temporal and spatial map of SMC. The inverse distance weighted (IDW) method was applied to do the spatial interpolation. figure 10 and 11 show the SMC variability map of Nglipar and Dlingo for each stage. The SMC at Nglipar of all stages were dominantly about 14-16% and a small part about 12-14% at the generative stage. The SMC of G25 at vegetative stage and after harvesting were the lowest (11%). It had been noticed that soil at G25, besides the whitish color soil, it also had very shallow topsoil over the bedrocks compared to surrounding soils. These conditions might cause differences in soil properties regarding moisture storage.

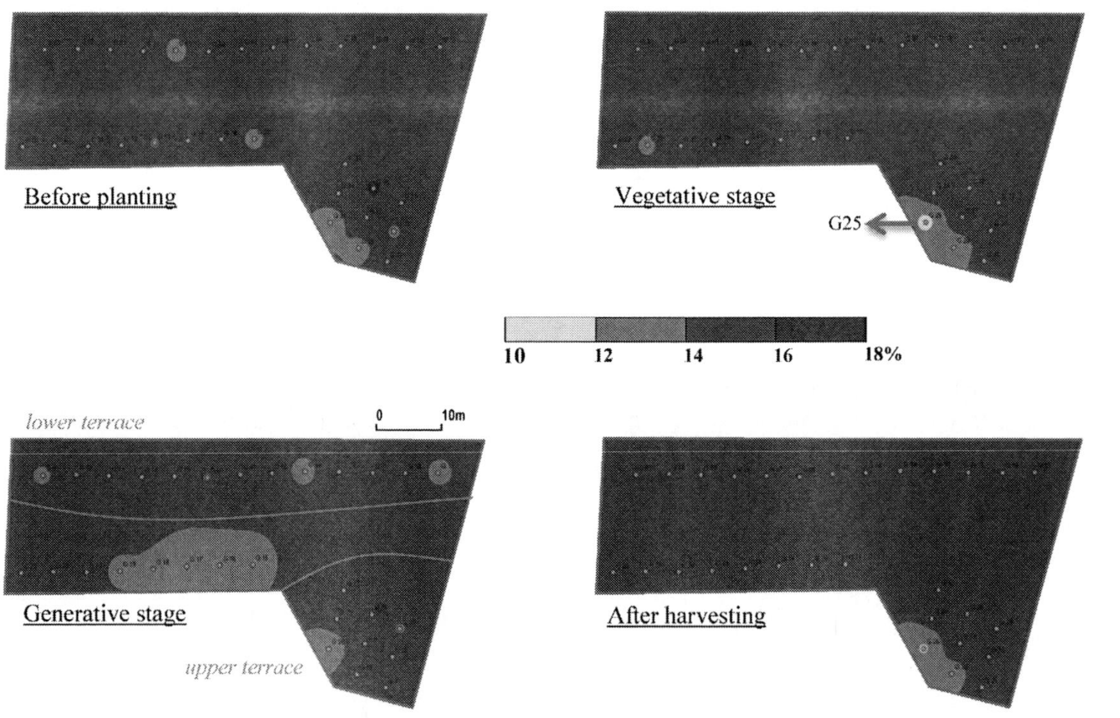

Figure 10. The temporal and spatial map of SMC variability at Nglipar (G)

The SMC at Dlingo was more varied than at Nglipar with the range from 12 to 18%. The SMC decreased gradually from before planting to the generative stage. The runoff over the terrace might be the factor that affected the decreasing of SMC, besides the increasing crop water requirement at the generative stage.

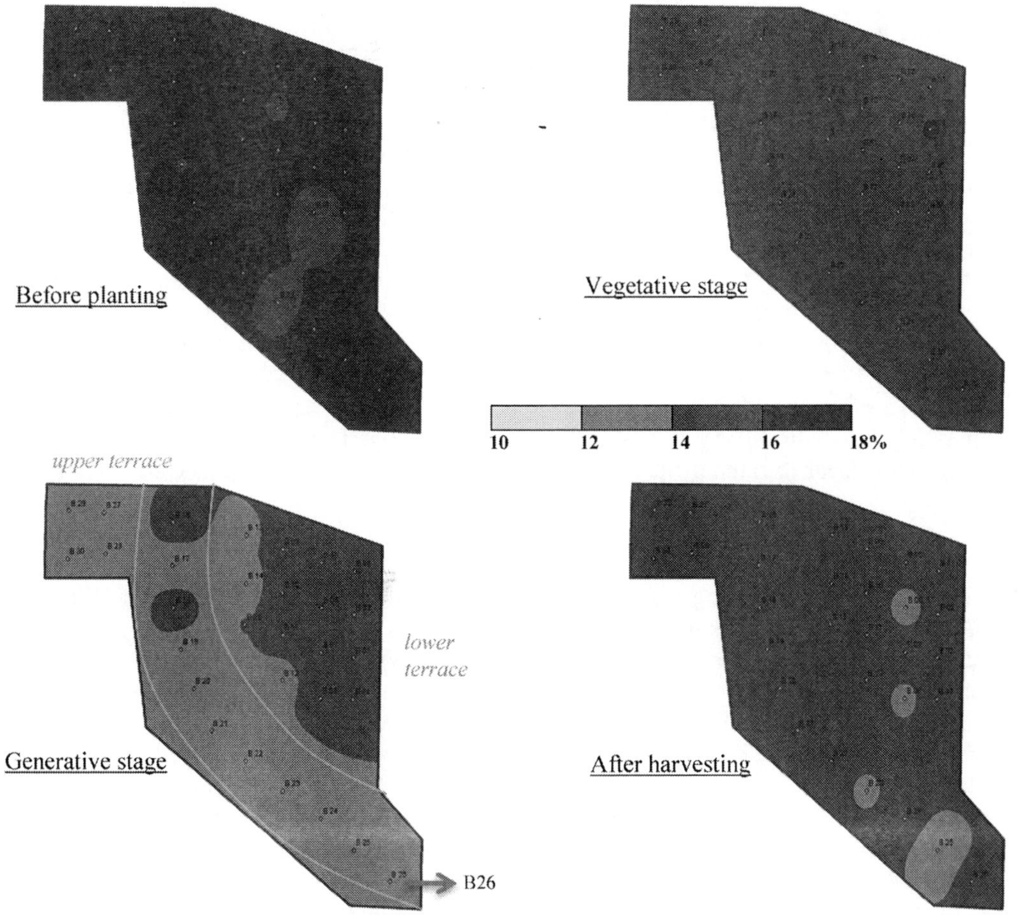

Figure 11. The temporal and spatial map of SMC variability at Dlingo

4. Conclusions

The performance of Vis-NIR reflectance spectroscopy to estimate soil moisture content (SMC) using PLSR method resulted in satisfactory level. It was proven from the RPD values that the calibration models were "excellent" to predict SMC. Dlingo prediction model had higher accuracy compared to Nglipar. In this study, soil proximal sensing using Vis-NIR spectrum was a reliable tool for the prediction of SMC of unknown soil samples. Different pretreatment process should be performed to improve the correlations between the measured soil moisture content and the spectra. The temporal and spatial maps of SMC variability gave much information to be interpreted carefully due to many factors affect the SMC. They were useful in supporting the process of decision making in field management.

References

[1] Douaik A, van Meirvenne M and Tóth T 2011 Statistical methods for the analysis of soil spatial and temporal variability *Principles, Application and Assessment in Soil Science* ed Gungor B E O (Croatia: InTech) chapter 15 pp 279–308

[2] Hagan R M 1995 Factors affecting soil moisture - plant growth relations *Report of the XIVth Int. Hortic. Cong. The Netherlands* (Holland: Veenman H and Zonen) pp 82–102

[3] Yin Z, Lei T, Yan Q, Chen Z and Dong Y 2013 A near-infrared reflectance sensor for soil surface moisture measurement *Comp. Electr. in Agric.* **99** 101–107

[4] Rossel R A V and Bouma J 2016 Soil sensing: A new paradigm for agriculture *Agric. Sys* **148** 71–74

[5] Adamchuk V I, Allred B A and Rossel R A V 2012 Proximal soil sensing: global perspective *FastTimes* (Colorado: EEGS) **17** 1 13–16
[6] Rossel R A V, Adamchuk V I, Sudduth K A, McKenzie N J and C Lobsey 2011 Proximal soil sensing: An effective approach for soil measurements in space and time *Adv. in Agron.* **113** 237–282
[7] Bilgili A V, Akbas F and van Es H M 2011 Combined use of hyperspectral VNIR reflectance spectroscopy and kriging to predict soil variables spatially *Prec. Agric.* **12** 395–420
[8] Rossel R A V, Walvoort D J J, McBratney A B, Janik L J and Skjemstad J O 2006 Visible, near infrared, mid infrared or combined diffuse reflectance spectroscopy for simultaneous assessment of various soil properties *Geoderma* **131** 59–75
[9] Aliyah B S N, Shibusawa S and Kodaira M 2015 Multiple-depth mapping of soil properties using a visible and near-infrared real-time soil sensor for a paddy field *Engin. in Agric., Envi. and Food* **8** 13–17
[10] Vasques G M, Dematte J A M, Rossel R A V, Lopez L R and Terra F S 2014 Soil classification using visible/near-infrared diffuse reflectance spectra from multiple depths *Geoderma* **223-25** 73-78
[11] Merkel A 2017 Climate data for cities worldwide website (Germany: AM Online Projects) http://en.climate-data.org/
[12] Badan Meteorologi Klimatologi dan Geofisika 2017 Monthly rainfall 2007-2017 of Dlingo and Nglipar datasheet (Yogyakarta: BMKG)
[13] Conforti M, Froio R, Matteucci G, and Buttafuoco G 2015 Visible and near-infrared spectroscopy for predicting texture in forest soil: an application in southern Italy *iForest* **8** 339–347
[14] Gholizadeh A, Amin M S M and Saberioon M M 2014 Potential of visible and near-infrared spectroscopy for prediction of paddy soil physical properties *J. of Appl. Spectroscopy* **81**(3):534–540
[15] Chang C W, Laird D A, Mausbach M J and Hurburgh Jr C R 2001 Near-infrared reflectance spectroscopy-principal components regression analysis of soil properties *Soil Sci. Soc. Am. J.* **65** 48–90
[16] Stenberg B, Rossel R A V, Mouazen A M and Wetterlind J 2010 Visible and near-infrared spectroscopy in soil science. *Advances in Agronomy* ed Sparks D L (Burlington: Academic Press) **107** 163-215 http://dP.doi.org/ 10.1016/S0065-2113(10)07005-7
[17] Todd S W and Hoffer R M 1998 Responses of spectral indices to variations in vegetation cover and soil background. *Photogrammetric Engineering & Remote Sensing* **64**(9):915-921
[18] Liu W, Baret F, Gu X, Tong Q, Zheng L and Zhang B 2002 Relating soil surface moisture to reflectance *Remote Sensing of Environment* **81** 238-246
[19] Beck R H 1975 Spectral characteristics of soils related to the interaction of soil moisture, organic carbon, and clay content *LARS Technical Reports* Paper 100 http://docs.lib. purdue.edu/larstech/100
[20] Broome S W 2016 Principle of soils science *Soil Sci. Courseware* (online lecture note) http:// broome.soil.ncsu.edu/ssc012/Lecture/topic8.htm
[21] Miller E L 1970 Effect of level terraces on soil moisture distribution and utilization by corn *Retrospective Theses and Dissertations* (Iowa State University Digital Repository) 16289
[22] Xu G, Ren Z, Li P, Li Z, Yuan S, Zhang H, Wang D and Zhang Z 2016 Temporal persistence and stability of soil water storage after rainfall on terrace land *Envi. Earth Sci.* **75** 966
[23] Kranz W L and Specht J E 2012 Irrigating soybean *NebGuide* (Univ. of Nebraska-Lincoln Extension) **G1367** http://extention.missouri.edu
[24] Curcio D, Ciraolo G, D'Asaro F and Minacapilli M 2013 Prediction of soil texture distributions using VNIR-SWIR reflectance spectroscopy *Procedia Envi. Sci.* **19** 494-503

IOP Conf. Series: Earth and Environmental Science **147** (2018) 012039 doi:10.1088/1755-1315/147/1/012039

Analysis of Groundwater Reserved in Dusun Ngantru Sekaran Village East Java

N H Pandjaitan[1], R S B Waspodo[1], T U Karunia[1] and N Mustikasari[2]

[1] Department of Civil and Environmental Engineering, Bogor Agricultural University. Indonesia

[2] Department of Soil Science and Land Resources, Bogor Agricultural University. Indonesia

E-mail: norahp@apps.ipb.ac.id.

Abstract. Limited capacity of fresh water in some areas in Indonesia made some regions had drought problem or lack of surface water. One of the solutions was increasing ground water used. This research aimed to identify aquifer and the pattern of ground water flow and also to determine potential of groundwater reserved in Dusun Ngantru. The result would be use to find the right location to be used as groundwater wells. The method used in this research was geoelectric method. This method was used to determine the condition of aquifer and rocks under the soil and to define hydrogeological condition of Dusun Ngantru.The analysis results can be used as a reference of where and what kind of groundwater runs underneath, in order to be optimally utilized. The results of hydrogeological studies and the distribution of aquifer showed that there were unconfined and semi aquifers. The direction of the groundwater flow in the study site varied greatly as the lithologic arrangement varied just as much. In the study locations there were Ledok formation, Mundu formation, and Lidah formation. Groundwater potential ware predicted of 55.33 m^3/day or 0.64 lt/s. Based on water quality standard in Indonesia, the water quality of wells were classified as first class quality.

1. Introduction

Water is an abundant natural resource on earth, but was distributed not equally according to place and time. Fresh water are found in the soils, rivers and lakes. Water sources that can be utilized to meet domestic, agricultural and industrial needs are surface water and groundwater. In addition to the economic growth, water demand are increasing, both in quantity and quality. This condition caused problem in water resources distribution. The limitation of raw water was caused, among others, by the development and changes in land use, which often does not consider the conservation of the ecosystems[9].

Aquifer is a layer, formation, or compactness of a permeable, consolidated geological unit (sand) with water-saturated conditions and has a hydraulic conductivity value (K), that can describe the soil ability to pass water. The value of hydraulic conductivity (K) is strongly influenced by soil characteristics. The aquifer store large amounts of groundwater. Confined aquifer is a water-saturated aquifer bounded on the top and bottom layers by aquiclude and the water pressure is greater than atmospheric pressure. Unconfined aquifer is a water saturated aquifer (saturated), and generally the limiting layer at the bottom is aquitard but there are no aquitard borders on the top. The spatial prediction of groundwater potential using the standard method is important for groundwater resources management[6].

Groundwater being a hidden subsurface treasures resource not easily seen with the naked eyes are very difficult to quantify potentially. Subsequently, planners and hydrological engineers have resolved to engage the efficacy of the predictive standard methods in groundwater potentiality mapping[5]. To analyse soil layers several subterranean methods of inquiry can be undertaken, including geological methods, gravity methods, magnetic methods, seismic methods, and geoelectric methods. Of these methods, the geoelectric method is a widely used method and the result is reliable. The difference between these methods lies in the methods and tools of the ground surface investigation. In seismic methods, artificial mechanical waves are used to investigate underground layer, while magnetic methods used the direction of the magnetic poles that are pressed on igneous rocks as well as the electromagnetic waves. This geoelectric estimation is intended to obtain an overview of subsoil and the presence of groundwater and minerals at a certain depth. The geoelectrical method is based on the fact that different materials will have different types of resistance when electrified. Groundwater has a lower type resistance than mineral rocks[11].

This study aimed to (1) identify aquifer in the form of position, typology, distribution, and thickness, (2) to know the pattern of groundwater flow and its distribution and (3) to determine potential of groundwater reserved and finding the right location to be used as groundwater wells .

2. Methods

This research was conducted in Dusun Ngantru, Sekaran Village, Kasiman Sub-district, Bojonengoro District (figure 1). Dusun Ngantru located in 7.2195° S and 111.7466° E. The data used in this study were primary data and secondary data. Primary data consisted of discharge measurements at monitored wells, measurement of rock type resistance and an existing hydrogeological data inventory. Secondary data consisted of geological map, hydrogeology map topography map and regulation according to water quality standard in Indonesia.

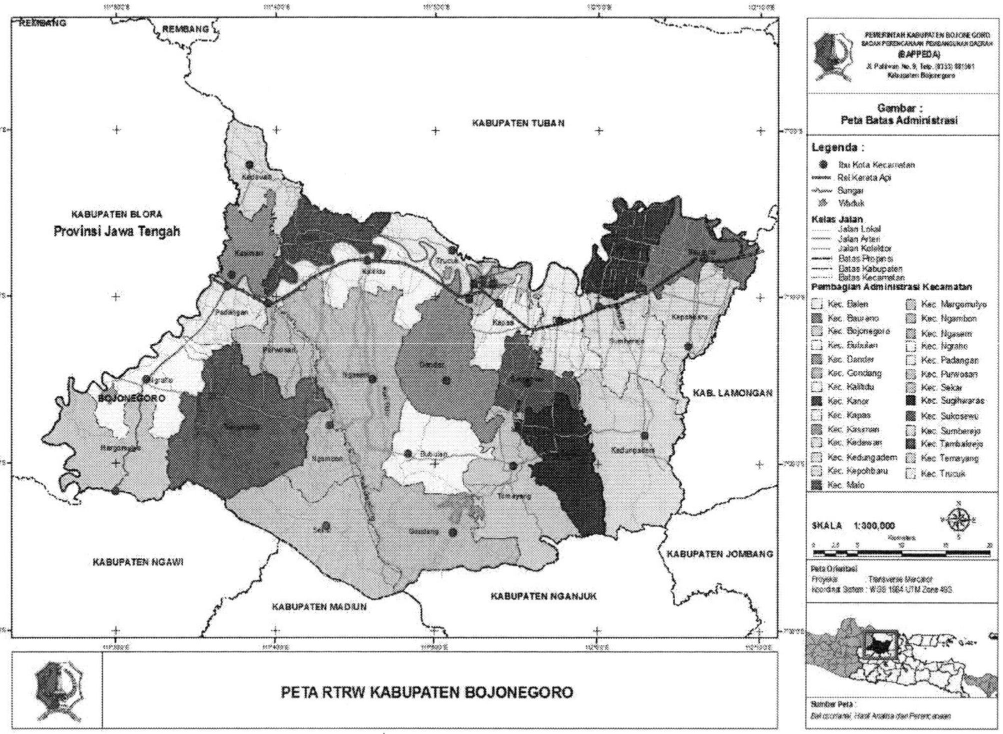

Figure 1. Bojonegoro District [1]

2.1. *Groundwater Level Measurement*

The groundwater level was measured from 5 monitored existing wells. Measurements are done manually, i.e. measuring the distance between the groundwater surface in the well monitor with the ground surface where the monitor well site was located.

2.2. *Measurement of Resistance Type*

The primary data collection required was the measurement of the type of aquifer, thickness of aquifer using geoelectric equipment. The general resistance measurement was done by injecting the current into the ground through two current electrodes (A and B), and measuring the potential difference generated at 2 potential electrodes (M and N). From the current (I) and potential (V) values data the apparent resistance value could be calculated using the Schlumberger configuration formula. Data analysis includes analysis of determination of geometry factor by using equation as follows:

$$\rho = \left[\pi \times \frac{\left(\frac{AB}{2}\right)^2 - \left(\frac{MN}{2}\right)^2}{MN} \right] \times \frac{\Delta V}{I} \qquad (1)$$

Where AM, AN, BM and BN are the electrode distances in the Wenner-Schlumberger configuration with the unit length (m).From the parameters that have been obtained, the false resistance value (ρa) could be calculated in Ωm unit. The calculated resistance value was not an actual subsurface resistance value, but was an apparent value which was a resistance of a homogeneous earth that provided the same resistance value for the same electrode arrangement. To determine the actual subsurface resistance value required inversion or forward calculation process using computer assistance (software progress version 3.0).

After resistance value was calculated, the type of soil composing layer can be determined based on the typical ranges of electrical resitivity for selected earth materials in Palacky [4]. The geoelectric data acquisition in this study used the Wenner-Schlumberger configuration with the fixed-electrode potential and current electrode to obtain variation in the direction of depth (sounding). The analysis of geoelectric data were done using Progress Version 3.0 software with respect to the resistance value of type, the distance between electrode then in-inverse modeling to get error value or RMSmin.

2.3. *Water quality test*

Water quality test aimed to determine the feasibili ty of surface water in Dusun Ngantru. Water quality test was conducted based on Regulation of the Ministry of Health number 416/MENKES/X/PER/1990 (Permenkes 416/1990) about drinking water quality[7]. There are 3 parameters used as the quality standrad in this reglation such as physic, chemical and microbiology[8].Water quality test was done for 5 samples which were took from several water sources, located from upstream to downstream in Dusun Ngantru.

3. Result And discussion

Dusun Ngantru Desa Sekaran is one of the villages in Kasiman, Bojonegoro Regency, East Java. It has a surface of 5,423,000 m^2 with 3,000 inhabitants. Almost 60% of its population works as farmer and owns cows as livestock. Local economics are far from being advanced due to the low income that is barely enough to cover daily necessity. One of the factors that contribute to the situation is the location of the village. Dusun Ngantru is rather remote with poor infrastructure to support proper transportation and enhance trading. Other factors that may indirectly influence the economics are the limited school available in the area. The village only have elementary school, whereas to earn higher education the kids have to go as far as 12 km to the nearest junior high school, with 3 km of it is only accessible by foot.

In agriculture and farming, local people make use of the reserved rainwater in sadang and kali Bayangan water springs. For cows living in communal farm, they use water from the 3 available wells.

The water supply from those wells are not sufficient for all the necessity, especially for green areas irrigation for cows food. The people of Dusun Ngantru have tried to make drilled well as a new water resource by using ground water. The drilling process resulted in producing oil instead of water due to its location being above the layers of oil-producing soil. Next to Dusun Ngantru is Desa Wonocolo, which is one of the oil-producing location in Indonesia. Apparently the oil supply seeps into Dusun Ngantru area. The circumstances have forced people in Dusun Ngantru to fully rely on rainwater and make use of the available surface water sources.

In 2017 the cow farming in Dusun Ngantru increased up to 600 cows, and need the formation of local cooperative as business platform. This trend seemed to be parallel with the increased water demand for green areas irrigation of 2-3 acres. One of the strategies that can be done to fulfill water demand is to use ground water. In order to make a drilled well, a hydrogeology analysis is needed to identify potential spots with enough soil water and does not contain oil. According to this condition, geoelectric analysis was used in this research in finding the right drilling spots. This research aimed to give the information to Dusun Ngantru villagers in locating the right drilling spots. Geoelectrical analysis and wells identification is a part of research activities with the endpoint of creating a module of pipe hydraulics design criteria in water supply system using ground water. This paper presented the analysis result of the first year of the total 2 years research.

3.1. Potential of Water Source

The water source found in Dusun Ngatru is a well (sumur) that has a depth of about 5-10 m below the soil surface and there is an inundated water source called a *sadang*. There are 2 *sadang* used daily for household needs. In addition, water from Kali Bayangan is also used by residents for daily household needs. The need for water for animal feed is still very little and the water requirement is closely related to the requirement of animal feed. Cattle feed require 50 kg for 1 tail per day, so it requires a lot of greening area. Residents of Ngantru still have problems to meet the water requirements for reforestation areas.

Dusun Ngantru also had 2 water springs : Kali Bayangan at 133 m above mean sea level and Kerincing at 120 m above mean sea level. The water of Kali Bayangan came from the village located above Ngantru so the water were used for daily household needs, but not as drinking water. On the other side the water from Kerincing spring have the potential to be utilized for drinking water, and has not been utilized yet by local people.

3.2. Geoelectric and Pump Test Measurements

Measurements were carried out in areas representing Ngantru which became the research area. The purpose of this measurement was to know the condition of the subsurface by looking at the nature of the flow of electricity in the rocks beneath the earth's surface. Resistance of rocks was very sensitive to water content, so with this geoelectric measurement the existence of groundwater could be detected. The location is about 10 km at the north of Bengawan Solo River. The height of the place is between 80-130 m above mean sea level with the slope of the steep terrain.

Tabel 1. Measurement results of water level in Ngantru

No	Well location	Elevation (m)	Water level of well (m)	Groundwater level (m)
1	SUMUR 1	109,31	8,28	101,03
2	SUMUR 2	107,42	9	98,42
5	SUMUR 3	91,41	1,3	90,11
6	SUMUR 4	90,34	1,3	89,04
7	SUMUR 5	89,61	2,6	87,01
8	SUMUR 7	108,96	1,4	107,56
9	SUMUR 8	117,52	7	110,52
10	SUMUR 9	119,11	7,4	111,71
11	SUMUR 10	113,71	4,6	109,11
12	SUMUR 11	110,85	2	108,85
13	SUMUR 12	108,67	3,8	104,87
14	SUMUR 13	100,27	1,2	99,07
15	SUMUR 14	106,72	3,1	103,62
16	SUMUR 15	103,10	2,7	100,40
17	SUMUR 16	109,46	4,6	104,86
18	SUMUR 17	110,50	7,6	102,90
19	SUMUR 18	110,77	6,3	104,47
20	SUMUR 19	105,90	2,6	103,30
21	SUMUR 20	104,56	1,6	102,96
22	SUMUR 21	92,80	5,6	87,20
23	SUMUR 22	89,81	5,6	84,21
24	SUMUR 23	87,52	2,3	85,22
25	SUMUR 24	107,74	4,3	103,44
26	SUMUR 25	102,31	0,5	101,81

3.3. Physiographic Conditions

Physiographic location of the study included in sub section of Mandala Geological Rembang consisting of Ledok Formation (TML) in North, Mundu Formation (Tpm) in the middle, and Lidah Formation (OTI) in the South. Physiography of Rembang Geology Mandala was an area with a landscape of plains to hills. The north of the study site was composed of calcarenzite, sandstone and napal. While in the middle was the arrangement of napal. As for the southern part was composed of clay, sandstone and limestone. The slope of land was between 15% - 70%.

Regionally, members of Dander of Lidah Formation ware included in the Mandala Geological Rembang. In general, the stratigraphic sequence from old to young ware Ngimbang Formation, Prupuh Formation, Tuban Formation, Tawun Formation, Bulu Formation, Wonocolo Formation and Ledok Formation, Mundu Formation and Lidah Formation. Lidah Formation had three members namely Tambakromo Members, Turi Members and Malo/Dander Members[10].

3.4. Groundwater Basin

Based on the groundwater basin map of the Directorate of Geological Environment and Mining Areas, Department of Energy and Mineral Resources Sheet VII, the study area including Rembang groundwater basin composed by: quartz sandstone and limestone. Location of 11 geoelectric measurements was showed in figure 2 and the results were showed in table 2.

Table 2. Position of Aquifer at Geoelectric Assessment Location

No	Measurement Point			Level of Aquifer (m)			Information
	Name	Coordinat	Elevation (masl)	Unconfined	Semi Confined	Confined	
1	GLB-1	07°03'51.0"S-111°40'07.1"E	144	0.5-2	-	-	
2	GLB-2	07°03'53.8"S-111°40'12.5"E	148	2.1-6.8	-	-	
3	GLB-3	07°03'55.6"S-111°40'05.7"E	129	0.5-2	21-39.1	-	
4	GLB-4	07°04'18.6"S-111°39'38.0"E	105	0.5-1.5	22.2-40	-	drilling recommendation
5	GLB-5	07°04'13.2"S-111°39'36.0"E	118	0.3-5	66-98	-	
6	GLB-6	07°04'19.0"S-111°39'33.3"E	113	1-17.3	-	-	
7	GLB-7	07°04'24.3"S-111°39'54.3"E	107	0.2-9	33.6-50	-	drilling recommendation
8	GLB-8	07°05'09.1"S-111°39'34.6"E	85	1.0-8.0	-	-	
9	GLB-9	07°05'09.4"S-111°39'31.2"E	79	0.7-3.9	6.0-14	-	
10	GLB-10	07°04'50.8"S-111°39'41.1"E	87	0.4-5.5	69-110	-	drilling recommendation
11	GLB-11	07°04'19.0"S-111°39'33.3"E	113	2.2-6.1	10-28.9	-	drilling recommendation

Based on its physiography and morphological form, the study area was included in the Mandala groundwater with groudwater flow net in the north and very few groundwater in the center and south. Dusun Ngantru area is composed of various rock types, consisting of clay and napal stones incorporated in Ledok Formation in the north, Mundu in the middle and Lidah in the south and generally very low-grade. In terms of hydrogeology, based on the type of rock and its magnitude, aquifer lithology in the study area was typology of aquifer with flow through cracks or nests and rare groundwater.

Based on geoelectric assessment location and borehole data, the study area had two types of aquifer with various depths of unconfined aquifer and semi-confined aquifer. As an example figure 3 shown the interpreted borehole data on GLB-4. Unconfined aquifer was found less than 10 m below soil surface, and a semi-confined aquifer is found between 10 to 30 m below soil surface. Unconfined aquifers were found in some places.

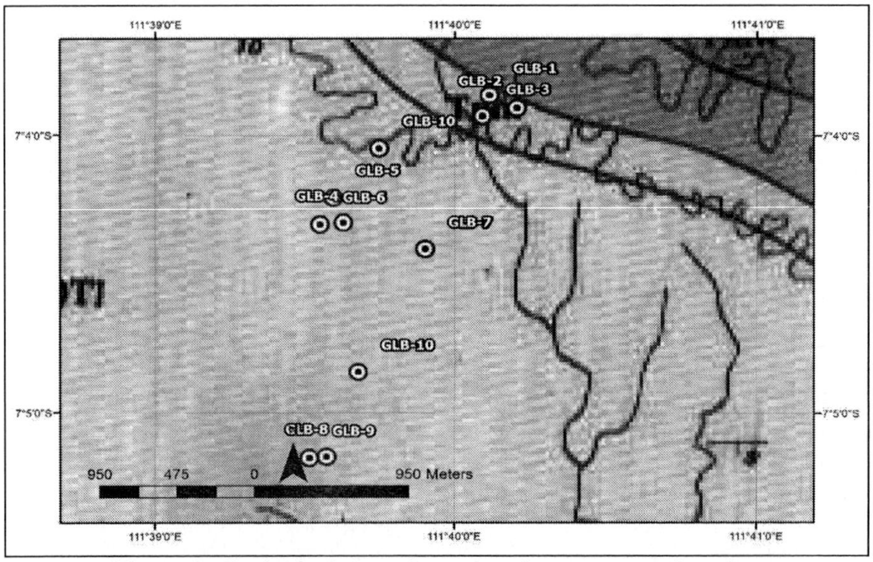

Figure 2. Geological map of geoelectric assessment location

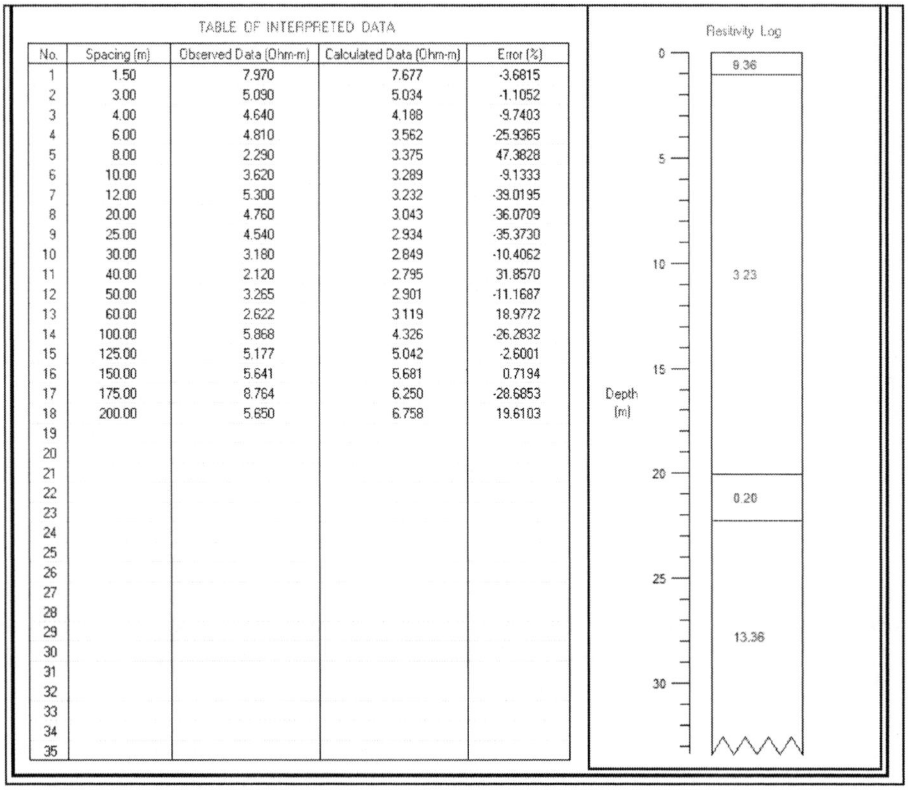
TABLE OF INTERPRETED DATA

No.	Spacing (m)	Observed Data (Ohm-m)	Calculated Data (Ohm-m)	Error (%)
1	1.50	7.970	7.677	-3.6815
2	3.00	5.090	5.034	-1.1052
3	4.00	4.640	4.188	-9.7403
4	6.00	4.810	3.562	-25.9365
5	8.00	2.290	3.375	47.3828
6	10.00	3.620	3.289	-9.1333
7	12.00	5.300	3.232	-39.0195
8	20.00	4.760	3.043	-36.0709
9	25.00	4.540	2.934	-35.3730
10	30.00	3.180	2.849	-10.4062
11	40.00	2.120	2.795	31.8570
12	50.00	3.265	2.901	-11.1687
13	60.00	2.622	3.119	18.9772
14	100.00	5.868	4.326	-26.2832
15	125.00	5.177	5.042	-2.6001
16	150.00	5.641	5.681	0.7194
17	175.00	8.764	6.250	-28.6853
18	200.00	5.650	6.758	19.6103
19				
20				
21				
22				
23				
24				
25				
26				
27				
28				
29				
30				
31				
32				
33				
34				
35				

Figure 3. Interpreted borehole data on GLB-4

3.5. Prediction of Groundwater Potential

The resistivity parameter of the weathered formation (aquiferous unit) and the delineated overburden unit thickness have been successfully evaluated for groundwater potential rating in an area[3].The parameters K and TR (transverse resistance) on the other hand, have been established as the major driver of groundwater discharge in an area and determined the variability of groundwater potential from place to place[2]. Dusun Ngantru area was investigated using 1:25000 scale of hydrogeology map. Potential of groundwater was measured by darcy equation. Based on Darcy's law, groundwater discharge is directly proportional to hydraulic gradient (i) of aquifer and hydraulic conductivity (K). Darcy equation can be noted as follows:

$$Q = -K \times A \times \frac{\delta h}{\delta L} \quad (2)$$

Where:
A = area of the aquifer cross section (m^2)
K = hydraulic conductivity (m/day)
$\delta h / \delta L$ = hydraulic gradient of aquifer
Q = groundwater discharge (m^3/day)

Dusun Ngantru had soil that contain sandstone (fine-grained) with a hydraulic conductivity value of 0.2 m / day. The area of aquifer cross section was 20750 m^2, and the potential of groundwater discharge was 55.33 m^3/day or equal to 0.64 lt/s (table 3).

Table 3. Prediction of groundwater's potential

Parameter	Volume	Unit
K	0.2	m/day
h1	93	m
h0	68	m
δh	-25	m
L	1875	m
δh/L	-0.013	
Length (l)	2075	m
Width (W)	10	m
Area (A)	20750	m^2
Discharge (Q)	55.333	m^3/day
	55333.333	l/day
	0.640	l/s

3.6. Pump Test

Test pump can only be done in shallow ground water wells. This is looking at the condition of the study sites that have no deep groundwater wells. Recovery of groundwater wells shall reach steady state after 47 minutes with very little water discharge, approximately 27 ml/s. This small water discharge recovery occurs due to the lithology of clay and napal clay and the location of the study that has the characteristics of apertured apertures and nesting. However, the condition is very diverse. So the groundwater exploration is needed in the semi-confined aquifer layer (10-30 m depth below the soil surface by adjusting the geoelectric survey results). The test pump results also indicate that the recommended pumps used for shallow wells are regular water pumps or maximum jet pump. As for deep ground water wells can use submergible pump.

3.7. Water Quality Test

Water samples were taken from 5 wells in Dusun Ngantru. These five wells were assumed can represent the water quality from upstream to downstream in Dusun Ngantru. Based on physical, chemical and microbiology parameter standard on Regulation of the Ministry of Health Republic of Indonesia number 416/MENKES/X/PER/1990 (Permenkes 416/1990), the water in Dusun Ngantru were classified in the first class water quality. The result of water quality test were shown on table 4.

Table 4. The result of water quality in one of well-point

No	Paramater	Unit	Standard *)	Analysis result	Method
Microbiology					
1	Total of Coliform	AMP100ml	0	0	MPN
2	E. Coli	AMP100ml	-	0	APHA ed. 22nd 9222 D, 2012
Chemical					
1	pH **)	-	6.5-9.0	7.02	APHA ed. 22nd 4500-H^+ B, 2012
2	Clorida (Cl^-)	mg/l	600	108	APHA ed. 22nd 4500-Cl^- B, 2012
3	Hardness as $CaCO_3$ **)	mg/l	500	81.14	APHA ed. 22nd 2340 C, 2012
4	Ferro (Fe) **)	mg/l	1.0	<0.016	APHA ed. 22nd 3111 B, 2012
5	Mangan (Mn) **)	mg/l	0.5	<0.017	APHA ed. 22nd 3111 B, 2013
6	Nitrat (NO_3-N)	mg/l	10	0.153	APHA ed. 22nd 4500-NO_3 B, 2012
7	Nitrit (NO_2-N)	mg/l	1.0	0.006	APHA ed. 22nd 4500-NO_2 B, 2013
8	Sulfat (SO_4^{2-}) **)	mg/l	400	44	APHA ed. 22nd 4500-SO_4^{2+} E, 2014
9	Organic matter	mg/l	10.00	1	APHA ed. 22nd 4500-$KMnO_4$ B, 2015
10	BOD	mg/l	-	8	APHA ed. 22nd 5210 B, 2012
11	COD **)	mg/l	-	15	APHA ed. 22nd 5220 C, 2012
Physical					
1	Color	PtCo	50	13	APHA ed. 22nd 2120 C, 2012
2	Electrical Conductivity	Umhos/cm	-	1182	APHA ed. 22nd 2520 C, 2013
3	Turbidity	NTU	25	0	APHA ed. 22nd 2130 C, 2014
4	The amount of solids dissolves **)	mg/l	1.500	864	APHA ed. 22nd 2540 C, 2015

*) standard of raw water based on Regulation of the Ministry of Health Republic of Indonesia number 416/MENKES/X/PER/1990
**) Committee of the National Accreditation (KAN)
(source: Laboratory of Departement of Agroindustrial Technology)

4. Conclusions and suggestions

4.1. Conclusions

Based on the result analysis it can be concluded that Dusun Ngantru had potential ground water source, originating from Krincing springs, that can be utilized to fulfil irrigation requirement especially for green areas for cows food. Groundwater potential were predicted of 55.33 m^3/day or 0.64 lt/s. Based on geoelectric measurements the result showed that:

- Unconfined aquifers were highly variable, whereas the semi-aquifers were only at 7 points of measurement (GLB-3, GLB-4, GLB-5, GLB-7, GLB-9, GLB-10, and GLB-11).
- Drilling Recommendation was on GLB-4, GLB-7, GLB-10, and GLB-11.

- The direction of groundwater flow in the study site varied greatly because the lithologic arrangement also varied. In one study location there were three formations, namely Ledok Formation, Mundu Formation and Lidah Formation.

4.2. Suggestions

From the results of geoelectric studies, it could be suggested that:

- Location of groundwater drilling would be decided according to the results of hydrogeological studies.
- Irrigation water distribution should use sprinkler system, because it was estimated more effective and efficient as well as because the lithology is very diverse.

Acknowledgements

The authors acknowledge the Ministry of Research, Technology and Higher Education of Indonesia for funding this research as university priority research in 2017 and 2018.

References

[1] BAPPEDA (Badan Perencanaan Pengembangan Daerah) Pemerintah Kabupaten Bojonegoro 2013 *Map of Administration* (in Indonesian) (Bojonegoro)

[2] Okogbue C O, Omonona O V 2013 Groundwater potential of the Egbe-Mopa basement area, central Nigeria *Hydrol. Sci. J.* **58**(4): 826–840

[3] Oyedele K F, Fatuna O T, Coker O J 2011 Geophysical assessment of subsurface coastal sediments and their engineering implications *Arab. J. Geosci* **4** 543– 550

[4] Palacky G J 1988 *Resistivity Characteristics of Geologic Targets, in Investiations in Geophysics vol. 3: Electromagnetic Method in Applied Geophysics-theory, vol 1, edited by M N* (Nabighian: Soc. Expl. Geophys) pp 53-129

[5] Pourghasemi H R, Beheshtirad M 2015 Assessment of a data-driven evidential belief function model and GIS for groundwater potential mapping in the Koohrang Watershed *Iran. Geocarto Int.* **30**(6), 662–685

[6] Rahmati O, Naser T, Farhad N, Saro L 2016 Spatial analysis of groundwater potential using weights-of-evidence and evidential belief function models and remote sensing *Arab. J. Geosci* **9** 79

[7] Regulation of the Ministry of Health number 416/MENKES/X/PER/1990

[8] Shofiani and Nur Endah 2003 *Recontruction of Indonesia's Drinking Water Utilities:Assessment and Stakeholder's Perspectives of Private Sector Participation in the Capital of Jakarta* (Stockholm: Royal Institute of Technology)

[9] Sosrodarsono S and Takeda K 1993 *Hydrology For Irrigation* (in Indonesian) (Jakarta: Pradnya Paramita)

[10] Srijatno 1980 *Applied Geophysics* (in Indonesian) (Bandung: Physic Department ITB)

[11] Todd D K 1995 *Groundwater Hydrology Second Edition* (Singapore: John Wiley & Sons)

Developing Automatic Water Table Control System for Reducing Greenhouse Gas Emissions from Paddy Fields

C Arif[1], Fauzan M I[1], Satyanto K S[1], Budi I S[1] and Masaru M[2]

[1] Department of Civil and Environmental Engineering, Bogor Agricultural University, Indonesia
[2] Department of Global Agricultural Sciences, The University of Tokyo, Japan

E-mail: chusnulipb@gmail.com

Abstract. Water table in rice fields play important role to mitigate greenhouse gas (GHG) emissions from paddy fields. Continuous flooding by maintenance water table 2-5 cm above soil surface is not effective and release more GHG emissions. System of Rice Intensification (SRI) as alternative rice farming apply intermittent irrigation by maintaining lower water table is proven can reduce GHG emissions reducing productivity significantly. The objectives of this study were to develop automatic water table control system for SRI application and then evaluate the performances. The control system was developed based on fuzzy logic algorithms using the mini PC of Raspberry Pi. Based on laboratory and field tests, the developed system was working well as indicated by lower MAPE (mean absolute percentage error) values. MAPE values for simulation and field tests were 16.88% and 15.80%, respectively. This system can save irrigation water up to 42.54% without reducing productivity significantly when compared to manual irrigation systems.

1. Introduction

Rice (*Oryza sativa* L) is the primary staple food in Indonesia and it covers the largest agricultural area. The total area was 8.1 million ha with rice production was 70.8 million tons in 2014. Its production decreased 0.6% in 2013 [1]. In Indonesia, paddy rice is known as cultivated plant with more water need, thus continuous flooding irrigation is applied by keeping water depth 2-5 cm above the soil surface. Irrigation consumes highest fresh water in agriculture by up to 89% following by fisheries (7%), domestic and industrial (4%) and livestock (0.2%) [2]. More water consumed for irrigation does not always have positive correlation to the increase in yield and biomass production as reported by previous studies [3, 4, 5]. On the other hand, sustainability of rice cultivation is threatened as raising water resource competition in rice production affected by regional climate change [6]. In addition, inefficiency of continuous flooding irrigation increases by more water loss from surface runoff, seepage and percolation [7].

Flooding water continuously in rice field promotes anaerobic soil condition that will contribute to negative effect to the environment by releasing greenhouse gas emissions particularly methane (CH_4) [8, 9]. More CH_4 concentration in the atmosphere contributes to global warming. CH_4 is produced by methanogens during organic matter decomposition in the soil, under an environment where the oxygen and sulfate are scarce in anaerobic soil condition [10, 11, 12]. In fact, N_2O is also emitted from paddy field with different characteristic to CH_4 emission. N_2O gas is primarily produced from the nitrification and denitrification processes in soil by microbial [13]. It is emitted more to the atmosphere when aerobic soil condition with less water availability in the fields. Both N_2O and CH_4 are two of contributors to global warming from paddy fields and their level contribution represented by global warming potential (GWP) [14].

Currently, the system of rice intensification (SRI) is known as promising rice farming with more benefits compare to that conventional farming. SRI is well-known as a set of crop management practices by changing the management of plants, soil, water, and nutrients for increasing rice productivity. Many critics were addressed to SRI [15,16,17], however, its benefits have been validated in 42 countries of Asia, Africa and Latin America including in Indonesia [18]. Here, water irrigation is supplied by applying intermittent irrigation in which the field is allowed to dry during particular time instead of keeping them continuously flooded. Intermittent irrigation is also known as a mitigation option for greenhouse gas emissions including CH_4 and N_2O gasses from paddy fields [19,20,21]

Crucial factor in SRI application is to determine optimal water management during planting season. Water management is needed to maintain water table within a proper soil moisture range [22]. Soil moisture conditions in SRI paddy field usually should be kept around saturation level to air-entry level. In addition, optimal water table also can be used as mitigation strategy to reduce CH_4 and N_2O emissions. Achieving this condition is not easy and to be the limitations in disseminating SRI in Indonesia as well as developing irrigation water control system [23]. Fuzzy based control system is promising method to control water table in paddy field since it has ability to solve complex system problems that can't be solved by conventional control system such as on-off system [24]. Fuzzy based control system adopts fuzzy logic by means specific mathematical model is not needed. Fuzzy logic works based on the rules that have been extracted from human knowledge as expert [25]. In addition, it has more accuracy compare to another control systems [26]. Therefore, the objectives of the current study were to develop fuzzy based water table control system and then to evaluate the performances.

2. Methodology

2.1. Experimental Design

Automatic water table control system was developed in the pot experiment during Feb – Aug 2017 in field laboratory of Dept Civil and Environmental Engineering in Bogor. There were four pots prepared in which only one pot was controlled by developed fuzzy-based control system (PS). The developed water table control system was used to control optimal water table that was determined by previous study [27]. Based on this study, dry regime was the best regime to mitigate greenhouse gas emissions from paddy field. This regime reduced by 34% of global warming potential compared to that flooding regime. In the dry regime, water table was kept at 0 cm water depth from the beginning to 20 days after transplanting, then the water was drained at -5 cm water depth until one week before harvesting time, and finally the water was drained until harvesting time. For comparison, there were three uncontrolled pots (P1, P2, P3).

2.2. Field Measurements

Water table was measured by calibrated E-Tape water level sensor. Meanwhile, weather parameters were measured by a Davis weather station consisting of a rain gauge, pyranometer, air temperature, humidity, air pressure, and wind speed and direction sensors with the interval measurement was 30 minutes. Plant growth parameters consisted of plant height, number of tillers/hill and number of panicles/hill that was measured every week. Collected data were then used in analyzing water balance for water control simulation process according to the following equation [28]:

$$WL_i = WL_{i-1} + P + I - ETc - DP - RO - D \quad (1)$$

where WL is water depth (mm), P is precipitation (mm), I is irrigation (mm), ETc is crop evapotranspiration (mm), DP is percolation (mm), RO is runoff (mm) and D is drainage (mm).

IOP Conf. Series: Earth and Environmental Science **147** (2018) 012040 doi:10.1088/1755-1315/147/1/012040

2.3. Fuzzy based control system

The fuzzy logic program was developed using Python code. The benefits of phyton code are large library and have provided modules to facilitate users and have a clear grammar [29]. Inference system was used based on sugeno reference with minimum implication function since it's more efficient and proper for mathematical analysis [30]. Here, we used 15 rules for decision matrix as presented in table 1.

Table 1. Decision matrix for output fuzzy system

Delta Error \ Error	Very Low	Low	Zero	High	Very High
Negative	High Irrigate	Medium Irrigate	Stop	Low Drainage	Medium Drainage
Zero	High Irrigate	Low Irrigate	Stop	Low Drainage	High Drainage
Positive	Medium Irrigate	Low Irrigate	Stop	Low Drainage	High Drainage

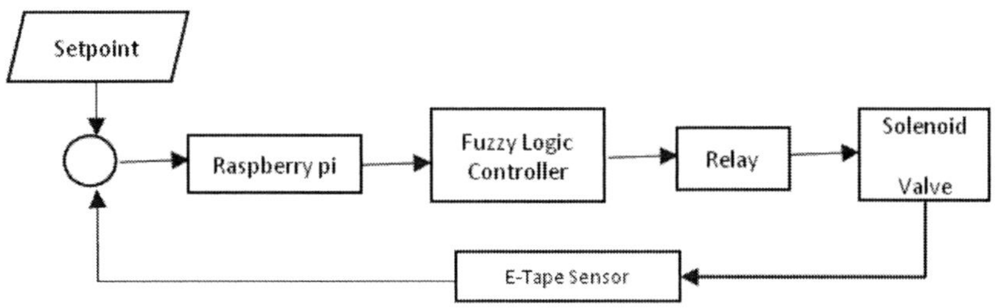

Figure 1. Schematic of fuzzy based water table control system

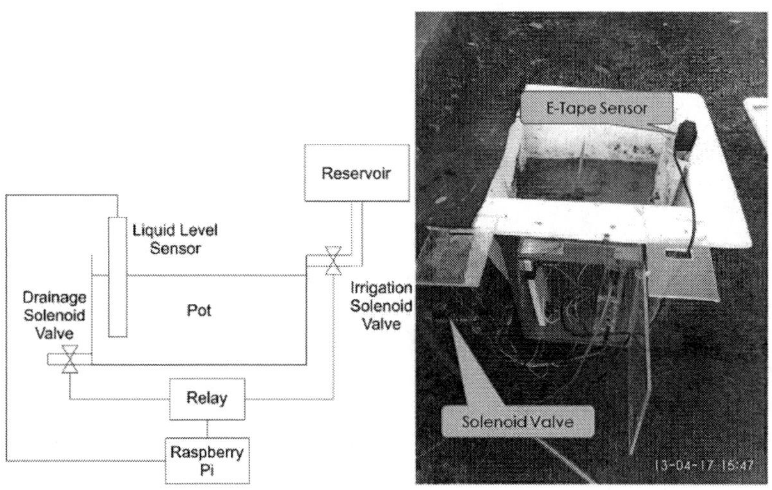

Figure 2. Layout fuzzy based water table control system in the field

For the hardware, we used Raspberry Pi microcontroller equipped with analog-digital converter (ADC). Solenoid valve was used as actuators in both irrigate and drainage system to keep expected water table (set point). Based on figure 1, actual water table will be measured by E-Tape water level sensor, then its value will be compared to the set point and we will get error and delta error. According to error and delta error values, fuzzy logic program (in fuzzy logic controller) determine degree of membership and decision matrix based on the rules. The output of fuzzy logic controller then will be

IOP Conf. Series: Earth and Environmental Science **147** (2018) 012040 doi:10.1088/1755-1315/147/1/012040

sent to the relay and give order to open/close solenoid valve in both irrigate and drainage gates. Layout fuzzy based water table control system in the field can be referred to figure 2.

The performance of system either for simulation and field test was evaluated based on the following mean absolute percentage error (MAPE) [31]:

$$MAPE = \frac{1}{n}\sum_{i=1}^{n}\left|\frac{Xa-Xi}{Xa}\right| x100\% \quad (2)$$

where n is number of data, Xi is reference data (set point) in cm and Xa is actual data (cm). The system can be accepted when MAPE value lower than 25%.

3. Results and discussion

3.1. Sensor Calibration and monitored data

The output of E-Tape water level sensor is voltage and it has negative correlation to water depth as presented in calibration graph (figure 3). It has strong correlation as indicated by R2 of 0.96 (>0.90), thus the sensor can be used to measure water depth precisely. The interval measurement of this sensor was 200-400 mV with water depth in between 0-30 cm. However, the sensor output was less sensitive when water depth less than 5 cm, thus water depth should be designed to be higher than 5 cm in the field.

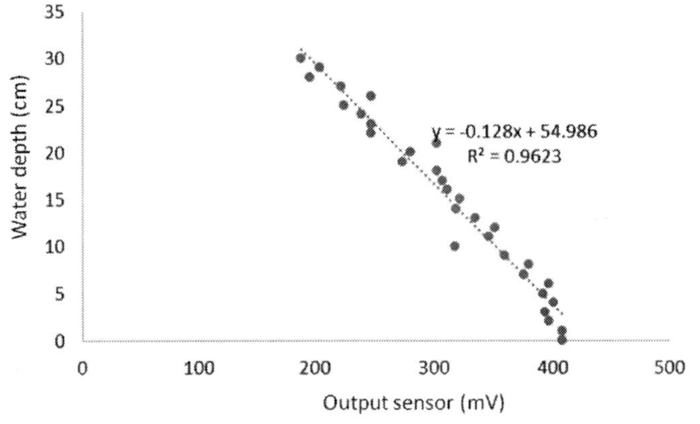

Figure 3. Calibrated water table sensor

Dynamic changes of weather parameters such as air temperature, relative humidity, solar radiation and reference evapotranspiration are shown in figure 4. In left side showed that daily maximum, average and minimum air temperatures fluctuated and their trend had a negative correlation to relative humidity. The maximum air temperature was relatively higher than 30oC and it reached maximum by 36.4oC occurred on 10 April 2017. However, the trend was declined and dropped on 10 May 2017. The same situation also happened on average and minimum air temperature that indicated declining trend. During planting season, the average air temperature was 26.6oC with minimum air temperature was 20.1oC. Meanwhile, the interval of relative humidity was 68.6% to 92.3% and the trend was also slightly declined. Maximum relative humidity occurred when maximum air temperature lower than that 30oC. On the other hand, minimum relative humidity occurred when maximum air temperature lower than 30oC. It was indicated that when air temperature reaches peak, actual vapor pressure increased, thus relative humidity decreased.

Figure 4b (right side) shows dynamic changes of solar radiation and evapotranspiration during planting season. They had a positive correlation by means solar radiation had direct effect on reference evapotranspiration. It is indicated that solar radiation is the main factor in the process of evapotranspiration both from soil and plant surfaces [32]. In addition, a sensitivity analysis of evapotranspiration model showed that solar radiation strongly governs evapotranspiration [33]. During planting season, their trend was slightly declined as well as air temperature and relative humidity. Maximum solar radiation and reference evapotranspiration were 20.14 MJ/m2/d and 5.03 mm, respectively. This event occurred when air temperature reaches peak on 10 April 2017.

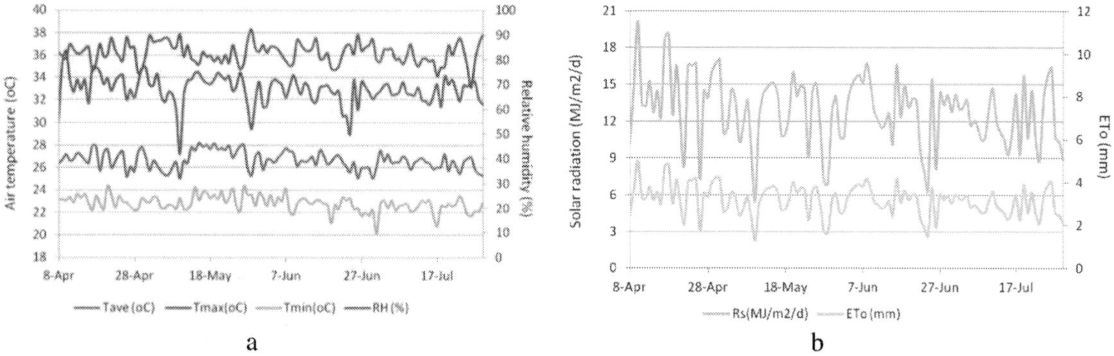

Figure 4. Monitored data: a) Air temperature and relative humidity, b) solar radiation and reference evapotranspiration

3.2. Performance of developed control system

Performance of fuzzy based water table control system in both simulation and field testing was shown in figure 5 and figure 6, respectively. The simulation was conducted during 18 minutes with initial water table was 1.91 cm. Figure 5 showed that the system can reach expected water table 1 cm in the early simulation during 18 seconds. Then, when set point was set to be 0 cm, actual water table can be reached that set point during 10 seconds. Overall, actual water table was closed to the set point indicated that the system was working well. In addition, the simulation reached MAPE by 16.88% (<25%) indicated that the system showed good performance and can be applied in the field. However, over and underestimation frequently occurred when the set point was set to be -5 cm.

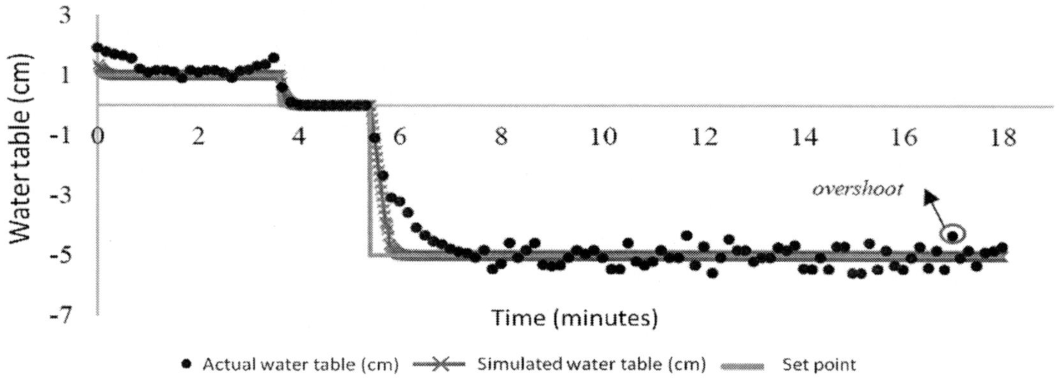

Figure 5. Simulation of the developed control system

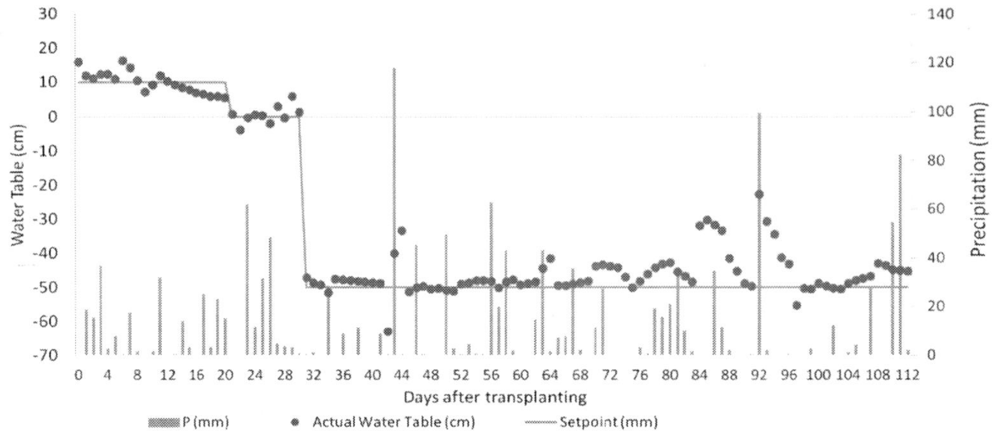

Figure 6. Performance of developed control system on field testing

The performance of the developed system was also satisfied in the field test during 112 trial days. Actual water table was closed to the set point particularly in the early planting season (0-30 days after transplanting). However, its values over fluctuated when high precipitation occurred. This event occurred several times particularly in during 40-44 days after transplanting and 84-96 days after transplanting as shown in figure 6. For set point above ground (during 0-32 days after transplanting), MAPE was 26.41%, meanwhile when set point was set below ground MAPE was 11.79%. In general, MAPE of this system was 15.80% (<25%) that indicated the system can be accepted and it was worked effectively.

3.3. Plant growth and water efficiency

Plant growths among pots were comparable and not significantly different as indicated by plant height and tiller numbers in figure 7. Plant height for controllable pot (PS) was 145.6 cm in 15 weeks after transplanting. This high was higher than that uncontrollable pot P1 and P2 and slightly lower than that pot P2. PS also produced moderate tiller numbers by 30 tillers in which it was more than those P2 and P3 and less than that P1. However, controllable pot produced heaviest biomass that indicated water was supplied efficiently for fulfilling plant water requirement, thus it evapotranspiration rate highest (table 2). This finding was supported by the previous study that has reported evapotranspiration rate under intermittent irrigation with SRI higher than that conventional rice farming with continuous flooding irrigation [34].

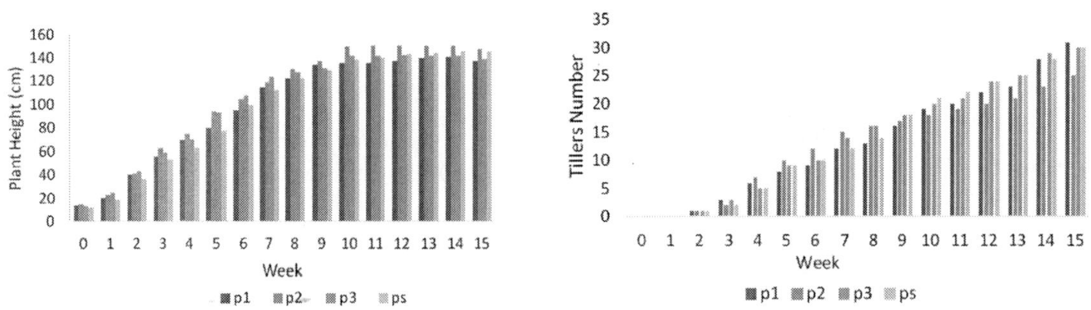

a b

Figure 7. Plant growth: a) plant height, b) tiller numbers

Table 2. Calculation of water balance and water use efficiency between pots experiment

Pot	Irrigation (mm)	Drainage (mm)	Precipitation (mm)	ETc (mm)	Percolation/Runoff (mm)
PS (controlled pot)	256.29	295.36	1386.40	312.98	1034.35
P1 (uncontrolled pot)	436.00	112.00	1386.40	136.72	1201.87
P2 (uncontrolled pot)	455.00	12.00	1386.40	136.72	1392.68
P3 (uncontrolled pot)	447.00	155.00	1386.40	208.53	1241.68
Water saving (%)	42.54%				42.57%

The developed system was more efficient in water use as presented in table 2. The system can save water by average of 42.54%. Also, the controllable water table can significantly reduce water loss due to percolation/runoff by 42.57%. Although supplied less water, crop evapotranspiration (ETc) was highest among pots that indicated the system was effective in utilization of water irrigation. Different ETc values in each plot gave different yield. For application in wider fields, it is recommended to develop proper drainage system such as subsurface drainage to keep expected water table, thus water input from precipitation can be well managed.

4. Conclusion and recommendation

The developed fuzzy based water table control system was effective. The performance was acceptable as indicated by MAPE (mean absolute percentage error) values lower than 25%. MAPE values for simulation and field tests were 16.88% and 23.06%, respectively. This system can save irrigation water up to 42.54% without reducing productivity significantly when compared to manual irrigation systems. For application in wider fields, it is recommended to develop proper drainage system such as subsurface drainage to keep expected water table, thus water input from precipitation can be well managed.

Acknowledgments

Authors wishing to acknowledge to Directorate of Higher Education, Ministry of Research, Technology and Higher Education of the Republic of Indonesia for generous financial support through grant of International Research Collaboration and Scientific Publication 2015 - 2017.

References

[1] BPS - Statistic Indonesia. 2017. Access online on 27 October 2017: https://www.bps.go.id/dynamictable/2015/09/10/895/luas-lahan-sawah-menurut-provinsi-ha-2003-2014.html

[2] Radhika S, Amirwandi, Hidayat R, Fauzi M and Hatmoko W 2012 Indonesia water demand (in Indonesian) (Bandung: Kolokium Hasil Litbang Sumber Daya Air 2012)

[3] Lin X Q, Zhu D F and Lin X J 2011 Effects of water management and organic fertilization with SRI crop practices on hybrid rice performance and rhizosphere dynamics Paddy Water *Environ* **9** 33-39

[4] Sato S, Yamaji E and Kuroda T 2011 Strategies and engineering adaptions to disseminate SRI methods in large-scale irrigation systems in Eastern Indonesia Paddy Water *Environ* **9** 79-88

[5] Li X L, Yuan W P, Xu H, Cai Z C and Yagi K 2011 Effect of timing and duration of midseason aeration on CH4 and N2O emissions from irrigated lowland rice paddies in China *Nutr Cycl Agroecosys* **91** 293-305

[6] De Silva C S, Weatherhead E K, Knox J W and Rodriguez-Diaz J A 2007 Predicting the impacts of climate change - A case study of paddy irrigation water requirements in Sri Lanka *Agr Water Manage* **93** 19-29

[7] Bouman B AM 2001 Water-efficient management strategies in rice production *International Rice Research Note* **26**(2): 17-22

[8] Cicerone R J, Delwiche C C, Typer S C and Zimmermann PR 1992 Methane emissions from

California rice paddies with varied treatments *Global Biogeochem Cy* **6** 3:233-248
[9] Setyanto P, Makarim A K, Fagi A M, Wassman R and Buendia L V 2000 Crop management affecting methane emissions from irrigated and rainfed rice in Central Java (Indonesia) *Nutr Cycl Agroecosys* **58**:85-93
[10] Epule E T, Peng C and Mafany N M 2011 Methane Emissions from Paddy Rice Fields: Strategies towards Achieving a Win-Win Sustainability Scenario between Rice Production and Methane Emission Reduction *Journal of Sustainable Development* **4**(6): 188-196
[11] Bouwman A F 1990 *Introduction In: Bouwman AF (ed) Soil and the Greenhouse Effects* (New York : John Wiley & Sons)
[12] Cicerone R J and Oremland R S 1988 Biogeochemical aspects of atmospheric methane *Global Biogeochem Cy* **2** 229-238
[13] Mosier A R, Duxbury J M, Freney J R, Heinemeyer O and Minami K 1996 Nitrous oxide emissions from agricultural fields: Assessment, measurement and mitigation *Plant Soil* **181** (1): 95-108
[14] Intergovernmental Panel on Climate Change (IPCC) 2007 In: In: Solomon S, Qin D, Manning M, Chen Z, Marquis M, Averyt K B, Tignor M, Miller HL (Eds.) Climate Change 2007: The Physical Scientific Basis, Contribution of Working Group I to the Fourth Assessment Report of the Intergovernmental Panel on Climate Change (Cambridge: Cambridge University Press)
[15] Sinclair T R and Cassman K G 2004 Agronomic UFOs *Field Crop Res* **88** (1):9-10
[16] Sheehy J E, Peng S, Dobermann A, Mitchell P L, Ferrer A, Yang J C, Zou Y B, Zhong X H and Huang J L 2004 Fantastic yields in the system of rice intensification: fact or fallacy? *Field Crop Res* **88** (1):1-8
[17] Dobermann A 2004 A critical assessment of the system of rice intensification (SRI) *Agr Syst* **79** (3):261-281
[18] Uphoff N, Kassam A, Harwood R 2011 SRI as a methodology for raising crop and water productivity: productive adaptations in rice agronomy and irrigation water management. *Paddy Water Environ* **9** :3-11
[19] Van der Hoek W, Sakthivadivel R, Renshaw M, Silver JB, Birley MH and Konradsen F 2001 Alternate wet/dry irrigation in rice cultivation: a pratical way to save water and control malaria and Japanese encephalitis? Research Report 47(Colombo: International Water Management Institute)
[20] Li X L, Yuan W P, Xu H, Cai Z C and Yagi K 2011 Effect of timing and duration of midseason aeration on CH4 and N2O emissions from irrigated lowland rice paddies in China *Nutr Cycl Agroecosys* **91**(3): 293-305
[21] Husin Y A, Murdiyarso D, Khalil M A K, Rasmussen R A, Shearer M J, Sabiham S, Sunar A and Adijuwana H 1995 Methane Flux from Indonesian Wetland Rice - the Effects of Water Management and Rice Variety *Chemosphere* **31** (4):3153-80
[22] Setiawan B I, Arif C, Rudiyanto and Askari M 2015 Optimizing Irrigation and Drainage Rates in SRI Paddy Fields *Jurnal Teknologi* **76**(15): 139-145
[23] Gani A, Kadir T S, Jatiharti A, Wardhana I P and Las I 2002 The System of Rice Intensification in Indonesia, In: Uphoff N, Fernandes E, Yuan L P, Peng J M, Rafaralahy S, Rabenandrasana J (Eds.), Assessments of the System of Rice Intensification (SRI). Cornell International Institute for Food, Agriculture and Development, Ithaca, NY, Sanya, China, pp. 58-63
[24] Rachman F N 2012 *Design and manufacture monitoring and control temperature in ncubator for baby with fuzzy logic basis* (in Indonesian). (Surabaya: Universitas Airlangga)
[25] Thiang, Resmana and Wahyudi 2001 Fuzzy logic control aplication to control universal motor speed (in Indonesian) *Jurnal Teknik Elektro* **1**(1): 33-42.
[26] Bachri S 2004 Hybrid control system PID – fuzzy logic to control DC motor speed (in Indonesian) *Jurnal Makara Teknologi* **8**(1): 25-34.
[27] Arif C, Setiawan B I, Munarso D T, Nugraha M D, Simarmata P W, Ardiansyah and Mizoguchi M 2016 Global warming potency from paddy rice system of rice intensification with variation

of ground water level (in Indonesian) *Jurnal Irigasi* **11**(2): 81-90
[28] Hillel D 1980 *Fundamentals of Soil Physics* (New York: Academic Press)
[29] Perkasa T R, Widyantara H and Susanto P 2014 Design and manufacture of motion detection using image subtraction method in single board computer (SBC) (in Indonesian) *JCONES* **3**(2):90-97
[30] Rofiq M 2013 Design internet bandwidth management with fuzzy sugeno method (in Indonesian) *Jurnal Ilmiah Teknologi dan Informasi ASIA* 7(1): 1-15
[31] Makridakis S, Wheelwright SC, McGee VE 1983 *Forecasting: Methods and Application (2nd Ed)* (New York: Wiley)
[32] Hargreaves G H and Allen R G 2003 History and Evaluation of Hargreaves Evapotranspiration Equation *Journal of Irrigation and Drainage Engineering* **129**: 53-63
[33] Bois B, Pieri P, Van Leeuwen C, Wald L, Huard F, Gaudillere JP and Saur E 2008 Using remotely sensed solar radiation data for reference evapotranspiration estimation at a daily time step *Agricultural and Forest Meteorology* **148**(4): 619-630
[34] Ardiansyah, Arif C and Wijaya K 2016 Nitrogen uptake of SRI Paddy Field compare to conventional field *Jurnal Teknologi* **78**(1-2): 45-49

Biomass Development in SRI Field Under Unmaintained Alternate Wetting-Drying Irrigation

Ardiansyah[1], Chusnul A[2], Krissandi W[1] and Asna M[1]

[1] Department of Agricultural Engineering, Jendral Soedirman University, Purwokerto,Indonesia
[2] Department of Civil and Environmental Engineering, Bogor Agricultural University,Bogor, Indonesia

E-mail: ardi.plj@gmail.com

Abstract. The aim of this research is to observe biomass development of SRI on farmers practice in three plots with different level. This research observes the farmer practice of SRI and Non-SRI during the uncertainty of irrigation water supply and its effects on paddy biomass development during growth stages and final stage of crop. A farmer group that already understand the principle of SRI, applied this method into several plots of their rented paddy field. Researcher interventions were eliminated from their action, so it is purely on farmers decision on managing their SRI plots. Three plots from both SRI and Non-SRI were chosen based on the position of the plot related their access to water. First plots had direct access to water from tertiary irrigation channel (on farm). Second plots were received water from previous upper plots and drainage water into other plots. Third plots were in the bottom position, where they received water from upper plot, and drainage water into farm drainage channel. Result shows there are similar patterns of root, straw, and leaves of biomass during crop growth. On the other hand, during generative phase, grain development shows different pattern and resulting different biomass in harvest time. Second plot, (of SRI) that has water from first plot has the average of biomass grain per plant of 54.4, higher than first plot and third plot, which are 33.8 g and 38.4. Average biomass in second plot is 74.6 g, higher than first and third plot, which are 49.9 g and 52.3 g.

1. Introduction

SRI cultivation method emphasizes on increasing rice productivity by improving the surrounding environment, be it above or below the ground [1]. Over 50 countries reported the effectiveness of SRI method to increase yield [2] as well as decrease water consumption in paddy field [3,4]. The high yield observed as influenced by the activity of microbe in aerobic condition [5–7]. Alternate wetting and drying irrigation is the key factor to provide aerobic condition.

Even though SRI practice said to be excellent in water-saving, still water availability during vegetative stage is important to maintain alternate wetting-drying irrigation pattern. The irrigation pattern should be managed well during the early stage of crop season. The actual problem farmer face in the field, that is during early stage of crop, when alternate wetting-drying irrigation applied, irrigation schedule is not well maintained due to water availability. Often, water is not available when irrigation scheduled, and conversely water is plenty during drying time.

Training on correct SRI methods has been one of interest for government instructor (*Penyuluh Pertanian Lapang*) and university's teacher (under university community service program). The training usually conducted to make farmers aware to practice SRI well according to its principles. However, after the training, farmers adopt the technology according to their wish to match their schedule, time, or local practice. Water availability during vegetative stage is important to maintain good plant structure growth and preparing for generative stage [8]. According to common practice by the farmer, continuous flooding is good practice [9,10]. However, when the water management shifting from continuous flooding to alternate wetting and drying, farmers need adjustment in their practice.

The aim of this research are:

1. Observe biomass development on SRI rice field under farmer's alternate wetting and drying irrigation practice
2. Comparing SRI to Non-SRI fields, under farmers management
3. Creating baseline for growth model calibration from water management point of view

2. Material and methods

This research observes the farmer practice of SRI during the uncertainty of irrigation water supply and its effects on paddy biomass development during growth stages and final stage of crop. A farmer group that already understand the principle of SRI, applied this method into several plots of their rented paddy field, while at the same time, they're also applied Non-SRI (the conventional practice of continuous flooding) into several plots of theirs. Researcher intervention was eliminated from their action, so it is purely on farmers decision on managing their SRI and Non-SRI plots. Three plots were chosen from both SRI and Non-SRI practice to make total six plot. The selection is based on the position of the plot related their access to water. First plots of both SRI and Non-SRI had direct access to water from tertiary irrigation channel (on farm). Second plots of both were received water from previous upper plots and drainage water into another plot. Third plots were in the bottom position, where they received water from the upper plot and drainage water into farm drainage channel.

Biomass of leaves during growth stages was estimated by using image processing technique [11,12] (figure 1). For calibration, sample image of leaves was scanned to obtained the leaves areas, while their dry biomass was measured using laboratory scale. Biomass of root was approximated using exponential growth equation model [13]. Initial value was measured from young crop seedling, while final value measured from after harvested crop samples. Model parameters were obtained by fitting parameters procedure.

Figure 1. Image processing method to obtain plant's leaf area and biomass

2.1. *Biomass Model for Plant Growth and Development*

Growth pattern for all plots was evaluated by fitting growth pattern data into biomass growth model. The biomass growth model used in this research is Shierary Rice Model [14]. This model converts radiation intercept by plant into biomass production through coefficient called radiation efficiency to biomass production (ε). Crop growth and development occur by the concept of Thermal Unit (TU) [15–17]. The phase of development needs certain amount of TU. If the amount of TU is not fulfilled, there will be no biomass improvement in the phase.
Potential biomass production (dW) modeled by equation (1):

$$dW = .Q_{intercept} \tag{1}$$

Where $\varepsilon = 1.65\ g\ MJ^{-1}$ is efficiency of radiation conversion into biomass. Intercepted radiation ($Q_{intercept}$, $MJ.m^{-2}$) is calculated using equation (2):

$$Q_{interceptin} = (1-\tau).Q_s \tag{2}$$

Where τ ($\tau = e^{-k\ LAI}$or $\tau = Q_{trans}/Q_s$) is the ratio between transmitted radiation by crop canopy to daily incoming radiation, and Q_s Q_s is daily incoming radiation ($MJ.m^{-2}$).

Biomass growth in Equation (1) the allocated into organs namely root, straw, leaves, and grains. Proportion of allocated biomass into organs was different following growth phase. Equation (3) describe the allocation mechanism:

$$dW_x = \eta_x dW - R_g - R_m \tag{3}$$

dW_x = amount of biomass gain for certain organ (x) ($g\ m^{-2}$), and η_x = biomass proportion allocated for certain organ (root, straw, leaves, or grains). Growth respiration (R_g, $g\ m^{-2}$) and maintenance respiration (R_m, $g\ m^{-2}$) [18] is taken as biomass subtracter, that calculated using Equation (4) and Equation (5):

$$R_g = k_g.\eta_x.dW. \tag{4}$$

$$R_m = k_m.W_x.Q_{10} \tag{5}$$

where k_g = growth respiration coefficient = 0.0108, k_m = maintenance respiration coefficient = 0.13,. W_x = organ biomass at computing time (g), Q_{10} [19] were formulated as $Q_{10} = (K_2/K_1)^{10/(T_2-T_1)}$ which accordingly can be stated as :

$$Q_{10} = 2^{\frac{10}{(T-20)}} \tag{6}$$

3. Result and discussion

3.1. *Irrigation Pattern*

As previously observed in the field, irrigation for SRI plots highly depended on water availability in paddy field channel. Figure 2 shows Non-SRI plots applying continuous irrigation and drainage as water always exists in the field, while SRI plots applying intermittent irrigation and drainage. Water availability were scarce in the 50th days after transplanting. It creates drying condition for both SRI and Non-SRI plots. However, in the days after transplanting below 60, water scarcity were no longer problem. Irrigation were returned to normal according to farmers practice.

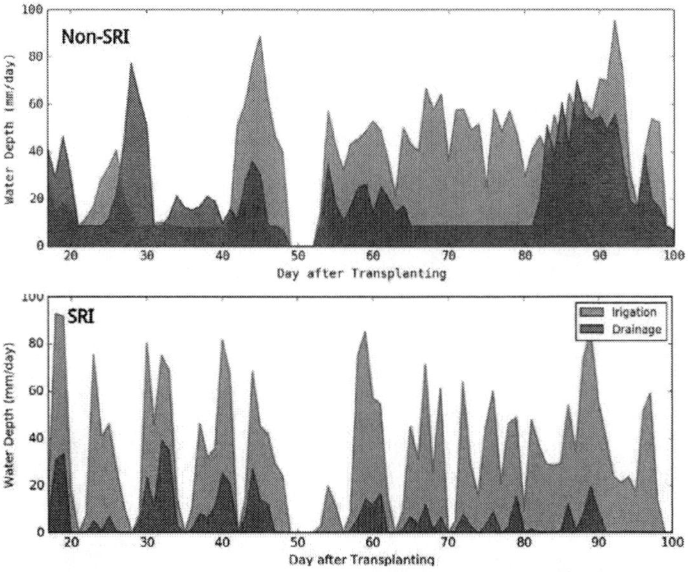

Figure 2. Irrigation and drainage pattern for SRI and Non-SRI plots

3.2. *Biomass and Growth*

Statistical analysis between Non-SRI and SRI shows there is not significant biomass difference between SRI plots and Non-SRI plots. Totally, biomass developed in Non-SRI plots and SRI plots are similar. However, looking deeper into plants organs, there is significant biomass difference between Non-SRI and SRI in Root and Straw, while leaves and Grains are statistically equal (table 1). During vegetative phase of plant development, root and straw biomass in Non-SRI plot were develop higher than in SRI plots, while leaves were developed equally for both groups. Entering generative phase, where grains biomass developing, there was no difference between Non-SRI and SRI groups.

Table 1. Level of significance difference between Non-SRI and SRI

	Root	Straw	Leaves	Grains	Total Biomass
Non-SRI to SRI Significance Level	+	+	-	-	-

+ : significant, - : not significant

Table 2 shows significance among 6 plots. Straw shows significant biomass difference for all plots, while leaves, grains, and total biomass shows no biomass difference. Average biomass for each plants organ for 6 plots was shown in Figure 3, which visually confirms the results of table 1 and table 2. Plot 2 in SRI shows highest grain biomass of all 6 plots, however, the other SRI plots (Plot 1 and Plot 3) considered similar to other Non-SRI plots.

Table 2. Level of significance difference between Non-SRI plots and SRI plots

Cropping Practice	Plot No.	Root	Straw	Leaves	Grains	Total Biomass
Non-SRI	Plot 1	-	+	-	-	-
	Plot 2	+	+	-	-	-
	Plot 3	+	+	-	-	-
SRI	Plot 1	-	+	-	-	-
	Plot 2	-	+	-	-	-
	Plot 3	+	+	-	-	-

+ : significant, - : not significant

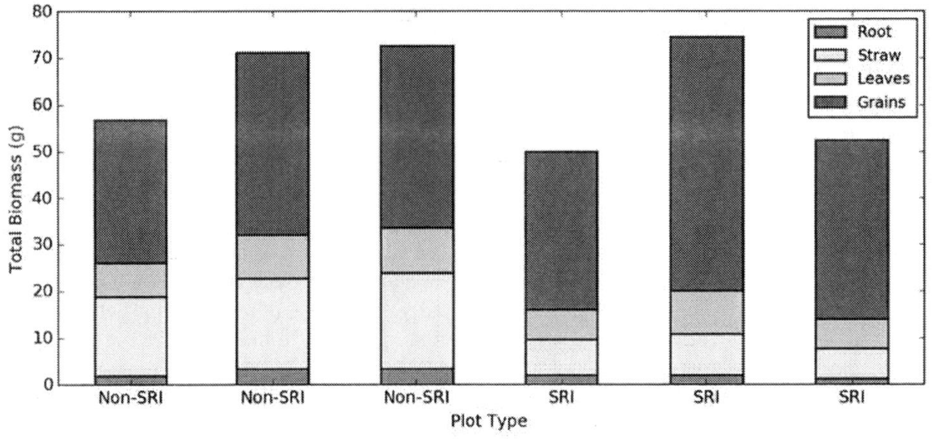

Figure 3. Total biomass of both SRI and Non-SRI plots

Result shows in figure 4 confirms the similar pattern in the growth of root, straw, and leaves of biomass during crop growth and development. On the other hand, during generative phase, grain development shows different pattern and resulting different biomass in harvest time. For example, second plot in SRI, which has water from first plot has the average of biomass grain per plant of 54.4, higher than first plot and third plot, which are 33.8 g and 38.4. Average biomass in second plot is 74.6 g, higher than first and third plot, which are 49.9 g and 52.3 g. SRI group develop less straw than all Non-SRI group (figure 3 and figure 4), however the yield (grain) consistent on the statement that "higher yields (higher grain biomass) are in line with the higher leaves biomass". From figure 3, we can see second plot in SRI group, that has higher grain biomass, develop comparatively smaller straw biomass. In spite of that, leaves biomass are similar to third plot in Non-SRI group, which make grain biomass for both quite similar.

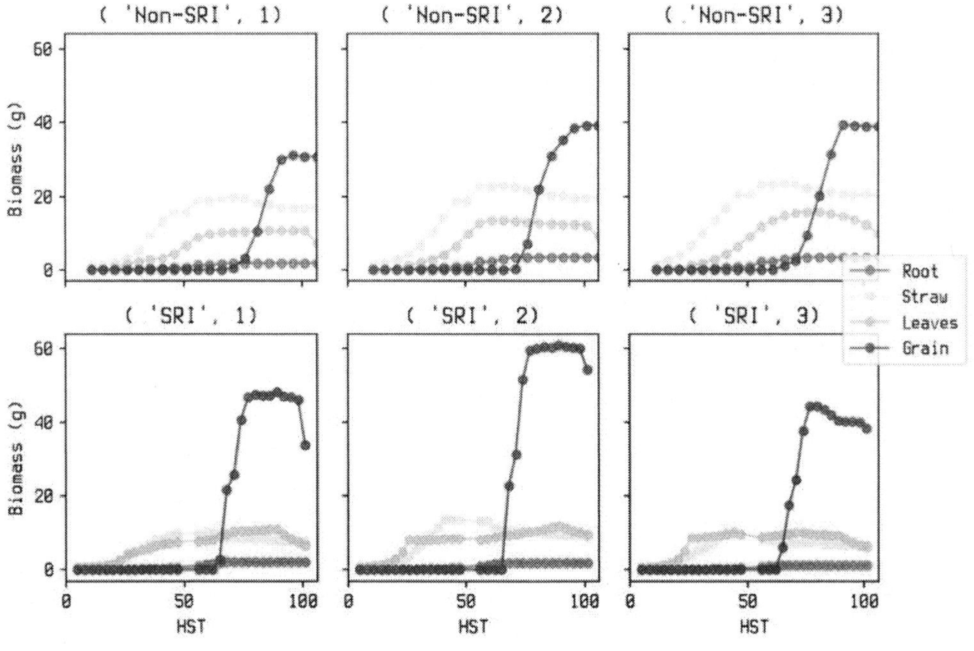

Figure 4. Biomass development of plant organs during crop growth

3.3. *Growth Model and Simulation*

Realizing that biomass depends on the availability of water, there is a need to include "water management factor" the existing growth model. Shierary Rice Model [14] separated growth phase by distinguishing its biomass allocation to plant organ. Each phase marked by a different portion of biomass allocation to root, straw, leaves, and grains. Figure 5 and figure 6 shows the result of model without calibration for water management. During vegetative stage, when farmers applying water management on rice plant, whether continuous flooding or alternate wetting and drying, the development of plant biomass is concentrated on root, straw, and mainly leaves. Entering generative stage, the biomass development shifting to grain, while root and leaves have slowed its growth. The principle of biomass growth in vegetative stage and generative stage were inline between observation and calculation. However, there is some variation among several conditions in the experimental plots that should be taken into account in the model. Hence there are some errors between the observation and the model because experimental condition (including water management) should be translated into parameters that involved. That is the challenge for the subsequent use of the growth model to predict paddy biomass on any condition.

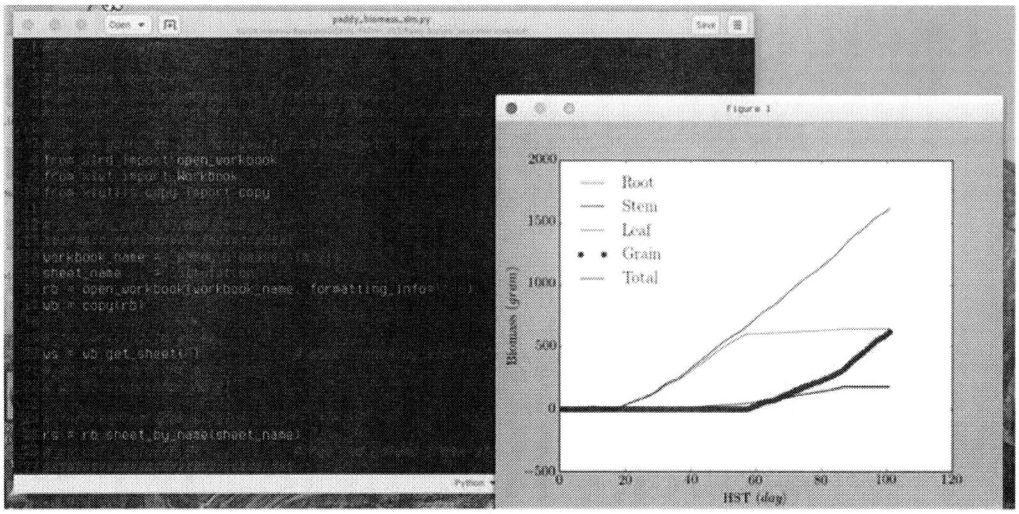

Figure 5. Running growth model without calibration for water management

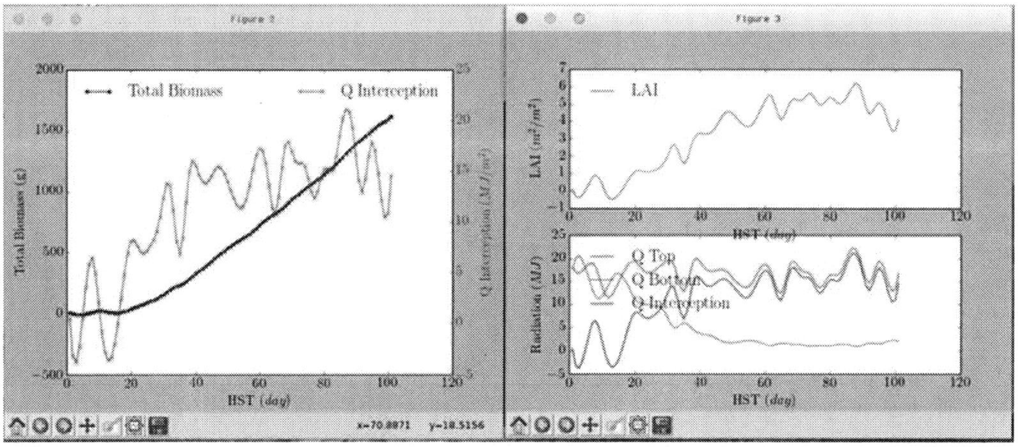

Figure 6. Total biomass growth resulting from growth model

According to plant growth model, each phase (s) of crop growth has different allocation value for plant organ development. Equations needed to describe the nature of this allocation. The portion of allocation (η) can be 0 (no biomass allocation) or 1 (all biomass allocation goes to the organ) or the value in between. For example, the model that result shows in figure 5 and figure 6, the allocation formula is as follows:

1. In plant emergence phase (beginning of the crop, s<0.25), biomass allocation to plant organs is following the equation of :

$\eta_{\text{root}} = 0.01e^{(1.53.s)}$, for root

$\eta_{straw} = 0.01e^{(6.42.s)}$, for straw

$\eta_{leaves} = 1 - \eta_{root} - \eta_{straw}$, for leaves

$\eta_{grain} = 0$, for grain

2. During vegetative stage (s between 0.25-0.5) biomass allocation to plant organs is following:

$\eta_{root} = 0.01e^{(1.24.s)}$, for root

$\eta_{straw} = 0.02e^{(3.55.s)}$, for straw

$\eta_{leaves} = 1 - \eta_{root} - \eta_{straw}$, for leaves

$\eta_{grain} = 0$, for grain

3. In generative stage (s between 0.5-0.75) biomass allocation to plant organs is following:

$\eta_{straw} = 0.08\log(s) + 0.24$, for straw

$\eta_{straw} = 0.11\log(s) + 0.12$, for leaves

$\eta_{straw} = 0.25\log(s) + 0.60$, for grain

$\eta_{root} = 1 - \eta_{straw} - \eta_{leaves} - \eta_{grain}$

4. Finally, in plant maturity stage (s>0.75), biomass allocation equations are:

$\eta_{straw} = 0$, for straw

$\eta_{root} = 0$, for root

$\eta_{leaves} = 0$, for leaves

$\eta_{grain} = 1$, for grain, where all the biomass growth allocated to grain

From the model point of view, total biomass or allocated biomass to plant organs is the circular effect of photosynthetic activity and radiation efficiency (ε) that influenced by leaf area index (figure 7). Therefore, high biomass can be resulting from good growth vegetative stage, which depend on nutrient uptake by plant root. As generally known well, soil nutrient is available (can be uptaken) by the root in the forms of ion [20–22]. Soil solution is important in nutrient uptake process, thus water availability is also important. Alternate wetting and drying irrigation is the way to shifting from water-filled porosity in the soil to partially air-filled porosity that can influence active decomposition bacteria. The calibration for "water management" should influence allocation equations in vegetative stage (s between 0.5-0.75).

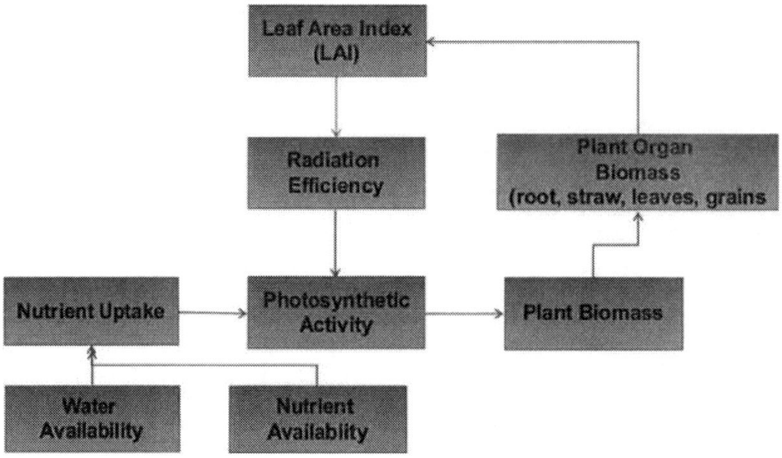

Figure 7. Plant biomass development concept

4. Conclusions

From the result, it can be concluded that unmaintained alternate wetting-drying in SRI make insignificant biomass increase in SRI. It probably causes by the soil water were not fully alternating from wet to dry. It sometimes creates extremely dried soil water in upper root zone layer that is intolerable for early stage of growth. Aeration is important in drying paddy field in vegetative stage, so that it can create different environment for soil bacteria responsible for providing nutrition. The biomass itself, is a result of complex and interdependency among soil water availability, nutrient availability, and radiation efficiency. The latter very much depends on leaf area (further Leaf Area Index), that higher leaf area will lead to higher grain biomass. Calibration for water management should be further concern in estimating biomass using growth model.

References

[1] Hidayati N, Triadiati and Anas I 2016 Photosynthesis and transpiration rates of rice cultivated under the system of rice intensification and the effects on growth and yield. *HAYATI J Biosci*

[2] Katambara Z, Kahimba FC, Mahoo HF, Mbungu WB, Mhenga F and Reuben P 2013 Adopting the system of rice intensification (SRI) in Tanzania: A review *Agric Sci* **4** 369

[3] Mati B and Nyamai M 2012 System of Rice Intensification (SRI): Growing More Rice While Saving on Water

[4] Ndiiri J A, Mati B M, Home P G, Odongo B and Uphoff N 2012 Comparison of water savings of paddy rice under system of rice intensification (SRI) growing rice in Mwea, Kenya *Int. J Curr. Res. Rev.* **4** 63–73

[5] Ardiansyah, Setiawan B I, Arif C, Saptomo S K and Mizoguchi M 2012 Soil macro nutrient (N, P, K) during growth stages under conventional and SRI (System of Rice Intensification) practices in tropical soil. *PAWEES 2012 Chall. Water Environ. Manag.*

[6] Ardiansyah A, Arif C and Wijaya K 2016 Nitrogen uptake of sri paddy field compared to conventional field *J Teknol.* **78**

[7] Case S D C, McNamara N P, Reay D S and Whitaker J 2012 The Effect of biochar addition on N2O and CO2 emissions from a sandy loam soil – the role of soil aeration *Soil Biol Biochem* **51** 125–134

[8] Virk P, Virman S S, Lopena V and Cabangon R 2014 Enhancing Water Productivity in Irrigated Rice *Proceeding 4th Int. Crop Sci. Congr.*

[9] Datta S K D 1981 Principles and practices of rice production. *Int. Rice Res. Inst.*
[10] Mambani, Datta and Redulla 1990 Soil physical behaviour and crop responses to tillage in lowland rice soils of varying clay content *Plant Soil* **126** 227–235
[11] Lati R N, Filin S and Eizenberg H 2011 Robust methods for measurement of leaf-cover area and biomass from image data *Weed Sci* **59** 276–284
[12] Tackenberg O 2007 A New method for non-destructive measurement of biomass, growth rates, vertical biomass distribution and dry matter content based on digital image analysis *Ann Bot* **99** 777–783
[13] Goudriaan J and Monteith J L 1990 A Mathematical function for crop growth based on light interception and leaf area expansion *Ann Bot* **66** 695
[14] Yuliawan T and Handoko I 2016 The Effect of temperature rise to rice crop yield in indonesia uses shierary rice model with geographical information system (GIS) feature *Procedia Environ Sci* **33** 214–220
[15] Marques B S, Silva A P P, Lima R S O, Machado E C R, Gonçalves M F and Carvalho S J P 2014 Growth and development of sourgrass based on days or thermal units *Planta Daninha* **32** 483–490
[16] Murakami T 1973 Paddy rice ripening and temperature *Jpn Agric Res Q* **7** 1
[17] Parthasarathi T, Velu G and Jeyakumar P 2013 Impact of crop heat units on growth and developmental physiology of future crop production: a review *ResearchGate* **2** 2319–95
[18] Lötscher M, Klumpp K and Schnyder H 2004 Growth and maintenance respiration for individual plants in hierarchically structured canopies of medicago sativa and helianthus annuus: the contribution of current and old assimilates *New Phytol* **164** 305–316
[19] Adu-Bredu S, Yokota T and Hagihara A 1997 Temperature effect on maintenance and growth respiration coefficients of young, field-grown hinoki cypress (chamaecyparis obtusa) *Ecol Res* **12** 357–362
[20] Roose T, Fowler A C and Darrah P R 2001 A Mathematical model of plant nutrient uptake *J Math Biol* **42** 347–360
[21] Barison J 2002 Nutrient-Use Efficiency and Nutrient Uptake in Conventional and Intensive (SRI) Rice Cultivation Systems in Madagascar
[22] Lu Y X, Li C J and Zhng F S 2005 Transpiration, potassium uptake and flow in tobacco as affected by nitrogen forms and nutrient levels *Ann Bot* **95** 991–998

IOP Conf. Series: Earth and Environmental Science **147** (2018) 012042 doi:10.1088/1755-1315/147/1/012042

Development of Rice Reprocessing to Strengthen Small Scale Rice Mills in Indramayu West Java

Y R Firdaus[1], R Hasbullah[2] and S Djohar[3]

[1] Business School, Bogor Agricultural University, Bogor
[2] Department of Mechanical and Biosystem Engineering, Bogor Agricultural University, Bogor
[3] PPM School of Management, Jakarta

E-mail: yasserfirdaus@gmail.com

Abstract. Small Rice Mill (SRM) has a very important role in rice production of strong institutional relationships to farmers and rice markets. Nevertheless, the rice produced in low quality and changing consumer preferences cause SRM to have difficulty in maintaining the role. Development of a reprocessing business - called Rice to Rice Processing Plant (R2RP) - as a separate business unit will support their role and existence. This study aimed at analyzing the feasibility of R2RP business that integrates SRM and market as an independent business unit and determines mutual partnership pattern. The study was conducted with special reference to West Java Province. The qualitative method used for non-financial aspects analysis includes raw material, market, technical-technological, management and regulation and partnership pattern. The financial aspect used the quantitative method of Net Present Value (NPV), Net Benefit Cost Ratio (Net B/C), Internal Rate of Return (IRR), Payback Period (PP) and Switching Value to check their sensitivity.The results showed R2R business is feasible for non-financially, technical-technological and financial aspects. Technology has evolved to produce various qualities (premium or medium) after the quality of raw materials (low quality or off-grade rice) using profit optimization. Value of the financial parameters was NPV of Rp 137 billion, Net B/C of 5.80, IRR of 84.27 percent and PP of 2.18 years at capacity of 19,800 tons/year with total investment of Rp 30 billion (Rp 13,500/USD). The switching value analysis showed that a decrease in product prices is sensitively influencing the financial feasibility. To strengthen cooperation that enhancing mutually beneficial relationship, R2R assists equipment investment in and buy raw material from SRM at a rational agreed price.

1. Introduction

1.1. Background of the Study
The rice-centered agricultural development program in the late 1960s caused farmers to experience a surplus of grain production. This green revolution was continuing, so rice millers were required to process grain into rice. Since then the development of rice milling industry began at the end of 1960 [1]. At that time the government focused on building a more modern SRM (small-scale rice milling) with low rice quality but better than mashed rice in accordance with current conditions of income and preference [2][3].

The number of rice mills in 2012 reached 182 thousand units were dominated by SRM by 94 percent [4]. SRM activities tended to increase post-harvest [5]. Most SRM produced at 60 percent of yield level and broken above 20 percent [1]. Patiwiri (2008) stated that the yield of milling in Indonesia was lower than Thailand (69.1 percent) and Vietnam (66.6 percent) [6]. In addition to the yield problem, PERPADI estimated the milling capacity of paddy mill to exceed the annual grain production (75.4 million dried milling paddy while the ideal rice mill was to be 45,452 units. This caused only 40 percent of the units to operate at full capacity and had to stop working in certain months or 3-4 months of production per year [7][8].

Sawit (2014) stated the tight competition of raw material did not reduce the number of SRM equally which continues to increase over the last 10 years. The increase of SRM occurred especially in SMGM (small mobile grinding mill) which had a capacity of 0.6 tons of rice/hour. The government (through BULOG) bought low-quality rice at an increased purchase price (HPP). This made the market share of low-quality rice was still large slowing modernization so did not meet the changing market demand [9]. In Wibowo (2007) study, people tended to select medium quality rice (quality requirement can be seen in table 5)[10]. Shifting demand to better quality of rice (premium) was partly due to the increased of community income [11]. Premium rice production requires modern technology such as RMU (Rice Milling Unit) or RPC (Rice Processing Complex) that could not be achieved by SRM [12]. In addition, SRM was experiencing a high loss in the milling and drying stage due to the limitations of milling technology [13].

In 2015 West Java Province has 1.9 million ha of rice fields with productivity of 6.1 tons/ha. Indramayu, Subang, and Cirebon were the largest rice production centers in West Java. The total land area of Indramayu in 2014 reached 117,792 ha followed by Subang with 84,750 ha and Cirebon of 53,368 ha [14]. Rice production increased in Indramayu in 2015 was a result of improvements in agricultural infrastructure that has been done since 2015. The improvements made were 276 agricultural infrastructures such as farm-level irrigation network (JITUT), farm-level network (JUT), village irrigation networks (Jides) and two units of seed center. This improvement program to increase rice, corn and soybean production (Pajale) in Indramayu Regency had to be achieved in 2017. Supporting rice cultivation facilities in Indramayu had to be balanced by decent postharvest handling [15][16].

1.2. Problems of the Study

One of the problems the milling faced was the high of grain water content [17]. In addition, the number of rice mills exceeded the production (table 1). This excess led to seizure and raised the price of grain. PERPADI (2014) proposed equalization of the ideal number of rice mills in Indonesia. The first alternative was 45,452 units (SRM 94 percent, PPM 4 percent and PPB 2 percent), second alternative of the number of rice mill was 42,979 units (SRM 85 percent, PPM 10 percent and PPB 5 percent) and third alternative of the number of milling 40,035 units (SRM 75 percent, PPM 15 percent and PPB 10 percent) [8]. All these alternatives still leave SRM production quality problems that did not meet market demand.

Table 1. The number of mills in Indonesia.

Description	Type of Milling			
	Small	Medium	Large	Total
Mills (unit)	171,495	8,628	2,076	182,199
Percentage (%)	94	4.7	1.1	100

Source: BPS 2012

Decreasing of number of rice mills only reduced grain competition between mills but did not solve problems of quality and post-harvest loss. It was in this perspective that organizing SRM investments through cooperation with R2RP required special attention. SRM played a role in the post-harvest loss which was still at 11-20% [18]. The highest loss occurred in drying and rice milling processes [19][20][21]. SRM uses drying with drying floors so it will be physically scattered and in quality, those were unevenly drying causing the rice to break in milling process easily [20]. The high loss was due to engine configuration which was generally a one pass or two pass (two times husking and two times polishing), without being followed by cleaning that made the grain was dirty. The configuration of mills with twice the breaking of grain skin caused a lot of rice cracked, so the percentage of rice broken was high [12].

SRM quality improvement could be performed by perfecting the current SRM engine configuration from H-P (husker and polisher) into C-H-S-P (cleaner, husker, separator, and polisher). The addition of cleaner and separator improved the quality and yield of rice by 0.9-1.9% [20]. The unit refinement by SRM had capital constrain because there were many SRM that got difficulties in purchasing grain continually [22]. Another solution was the current SRM product that was rice (poor quality) was purchased and polished by RMU into high-quality rice or known as R2RP (Rice to Rice Processing Plant). This scheme allows SRM to produce only brown rice.

SRM was expected to gradually extinct as demand for rice quality changed, the gradual competition of grain between both SRM and PPB [23]. The synergy between the two could be developed to maintain SRM as a rural enterprise [24]. The development of cooperation schemes between SRM and R2RP should be economic-based on sustainable growth [25]. Therefore, feasibility studies should be conducted as a basis for developing partnerships between SRM and R2RP.

2. Materials and methods

Averages for the costs and revenues were calculated to be considered the core of the analysis when determining the financial feasibility and other economic considerations. The total fixed, variable costs, and cash flows, as well as the total revenues, were calculated for the next 10 production cycles. Optimization of production can be solved using linear program which has two kinds of function that is objective function and constraint function. The net present value (NPV), internal rate of return (IRR), benefit-cost (B/C) ratio and Payback Period (PP) for the investigated R2RP were the financial indicators calculated in the study [25]:

1. NPV (net present value) = $\sum_{t=1}^{n} \frac{B_t - C_t}{(1+i)^t}$
2. IRR (internal rate of return) = $i_1 + \frac{NPV_1}{NPV_1 - NPV_2}(i_2 - i_1)$
3. Net B/C (net benefit-cost ratio) = $Net\frac{B}{C} = \frac{\sum_{t=1}^{n} \frac{B_t - C_t}{(1+i)^t}}{\sum_{t=1}^{n} \frac{B_t - C_t}{(1+i)^t}}$
4. PP (payback period) = $\frac{Capital\ Investment}{Net\ Benefit}$
5. Optimization of production = $Z_{max} = C_1X_1 + C_2X_2 + ... + C_nX_n$

Where: Bt = Benefit (Rp)
Ct = Cost (R)
n = Economic age (year)
i = Discount rate (%)
t = The time of the cash flow (year)

Therefore, data required to solve the above equations among others are capable of the R2RP, machinery investment cost, operational cost, working hours, working days, R2RP yield, economic age of facilities, raw material or low rice quality prices, selling price, operational cost, dollar exchange rate, tax and discount rate.

3. Results and discussion

3.1. Raw material sufficiency analysis R2RP

West Java is known to be one of the main rice producing provinces in Indonesia within which Indramayu is the biggest regency of rice producers. Therefore, Indramayu was taken as the case study for analyzing the possibility to establish cooperation or partnership between SRM and R2RP.

3.1.1. Rice Production West Java and Indramayu. Rice production in West Java in 2015 increased compared to the previous year which reached about 11 million tons. Rice production of West Java Province in 2015 was recorded at 6.9 million tons with the value of conversion of GKG to rice at level of 62.4%. If the consumption rate of rice was assumed to be 98 kg/capita /year [26] with a population of 47 million, the total requirement of West Java's rice was about 4.6 million tons of rice per year or a surplus of 33% of the demand. As a producer that always have surplus, then the potential to develop post-harvest technology and efficiency improvements (volume and quality) is very potential and needed to be optimally utilized.

Indramayu contributed to 11-12% of West Java or 33-35% from total production of the surroundings (Cirebon, Subang, Sumedang, and Majalengka later called as cluster region). This indicated that the regency has strong position in the region and regional. The production was produced from an area of 118,000 ha (figure 1). Therefore, all regencies within the region could be the sources of raw materials for the small rice reprocessing industry to produce medium, low quality and brown rice for further processing in R2RP.

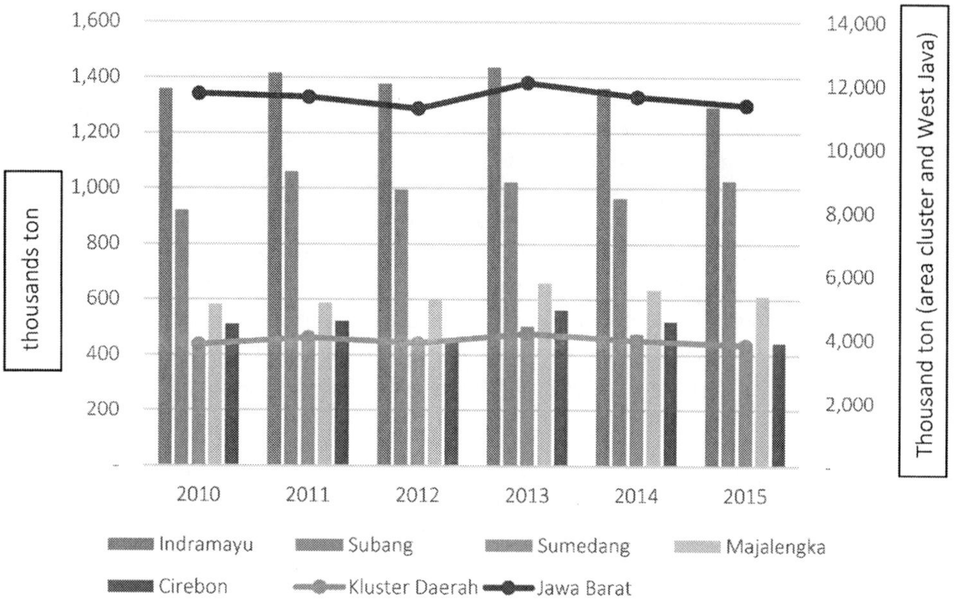

Figure 1. Paddy production of Indramayu, surrounding regencies (cluster) and West Java

The Indramayu's rice fields have not changed much over the year, which means that the conversion of rice fields to other uses was relatively small [26]. Indramayu along with Subang and Majalengka were strong producers in the region. Procurement of low quality, medium, and brown rice could come from the surrounding area which was not far apart and can even be treated as a source of raw materials in the same area. The location of Indramayu located on the Pantura (northern part of Java) Road has a strategic role to distribute the surplus of rice from regions in the cluster to national distribution channels outside the province. Therefore, all regencies within the area could be a source of raw material for R2RP. In 2014, the total area of rice fields in Indramayu was 117,792 ha which was about 35% of the cluster and 13% of West Java.

3.1.2. Sufficiency of raw materials. Indramayu was a competitive and prospective production area because, in addition to having good planting area and productivity, it serves continuous cultivation as well. Many government programs are allocated to this area because of its status as a food barn. Therefore, the availability of raw materials to meet the capacity of the R2RP was easily met (table 2).

Table 2. Calculation of R2RP raw materials necessity relative to raw materials of rice (%).

Capacity	Ton/year	Indramayu	Area cluster	West Java
Rational	19,800	2.5	0.8	0.3
Optimist	36,000	4.5	1.5	0.5
Indramayu	807,555			
Area Cluster	2,377,710			
West Java	7,096,841			

Assuming the planned R2RP capacity for rational operating conditions (11 hours per day) was 1,650 tons per month requiring 19,800 tons/year raw material (low quality, medium, and brown rice) regardless their quality. This amount was only about 2.5% of Indramayu rice production. If the purchasing area was expanded to include Subang, Sumedang, Majalengka, and Cirebon, then the raw material needs were about 0.8% of the rice production. If the assumed planned R2RP capacity to be optimistic operating conditions (20 hours per day), it is 3,000 tons per month requiring 36,000 tons/year of low-quality rice at any quality. This amount will be only about 4.5% of Indramayu rice production. If the purchasing area were expanded to cover the surrounding area (cluster), then the raw material needs were about 1.5% of the production of rice. It indicated that the establishment of R2RP is not determined by the production of raw materials of Indramayu and Area Cluster rather than depending on the management and strategy of raw material procurement.

Indramayu Regency had 1,054 mills consisting of 647 small-scale units, 381 units of medium scale and 26 units of large scale. Of all the mills there were 1,048 units operating normally with milling capacity varying from 0.8 tons - 30 tons per day and 6 mills in damaged condition. The improvement and organization of SRM could serve as the raw material base for R2RP. Millers located outside Indramayu Regency could become potential partners as raw material suppliers. The movement of raw materials of this rice is facilitated by the access of Cipali toll and the Pantura line that passed through the area of Indramayu Regency to expand the source of raw materials. R2RP should consider the diversity of rice varieties by adding the blending process as has been done by RMU.

3.2. Market and marketing

Production of R2RP was approximately 36,000 tons/year under optimistic 20 hours operating per day. This production might decrease when using rational operating hours of 11 hours per day into 19,800 tons/year (table 3). R2RP optimistic and rational capacity of the regency rice needs is about 45% based on the predicted figures of consumption or about 30% of the Susenas figure. Theoretically, marketing within the regency could be performed but not fully absorbed. Competition, both with local and foreign mills, is relatively tight. The market orientation should be out of the region due to the large production surplus.

Table 3. Projection of rice demand in Indramayu Regency.

Year	Population (lives)	Necessity (ton/year)		Optimistic R2RP Proportion (%)	
		Prediction	Susenas	Prediction	Susenas
2014	1,708,551	93,116	158,212	38.66	22.75
2015	1,718,495	88,846	154,665	40.52	23.28
2016	1,728,119	89,344	151,210	40.29	24.4
2017	1,737,796	87,411	147,539	41.18	27.33
2018	1,747,528	85,629	143,822	42.04	25.03

West Java was a province of rice producers and surplus although some districts (mainly cities) serve as consumer areas. Intra-region distribution could be easily met. Surplus ranges from 3-4.5 million tons annually. The regional orientation was still possible to fulfill the cities and deficit regions within the province (table 4).

Table 4. Calculation of supply and demand of rice in West Java.

Years	Production (ton)	Availability (ton)	Consumption (kg/capita/year)		Population (jiwa)	Rice Necessity (ton)		Surplus (ton)	
			Prediction	Susenas		Prediction	Susenas	Prediction	Susenas
2014	7,735,422	7,251,898	54.5	92.6	46,075,445	2,512,012	4,266,586	4,739,886	2,985,312
2015	7,821,399	7,366,287	53	90	46,828,323	2,484,224	4,214,549	4,882,063	3,151,738
2016	7,888,512	7,467,704	51.7	87.5	47,580,407	2,462,009	4,163,286	5,005,695	3,304,418
2017	7,966,182	7,570,316	50.3	84.9	48,334,948	2,432,404	4,103,637	5,137,912	3,466,679
2018	8,065,087	7,681,735	49	82.3	49,094,385	2,404,830	4,040,468	5,276,905	3,641,267

R2RP investments could be positioned as buyers and sellers that tended to face oligopsony and oligopoly market structures. As rice buyer with the presence of R2RP facility had a strong bargaining position because it was more efficient and far beyond the technical and economical scale, compared to farmers, middlemen, even large collectors and small grinders. As a seller, the number of rice factories was also very small compared to the number of distributors. Its market share is quite good and reaches the distributor outside the region.

The main products produced by R2RP will be medium quality 1 and premium rice. The marketing channel of R2RP is divided into two channels namely the first local marketing channel in Indramayu Regency, the second channel is marketing to meet the clusters and West Java Province and met the demand outside the Province (figure 2). The first channel was local oriented marketing starting from R2RP selling its products through retailers. The retailer then sold the rice to the last consumer. In the second channel, R2RP marketed its products in clusters up to outside the province, so its marketing is through distributors. After that the distributor market the product to the retailer and to the final consumer.

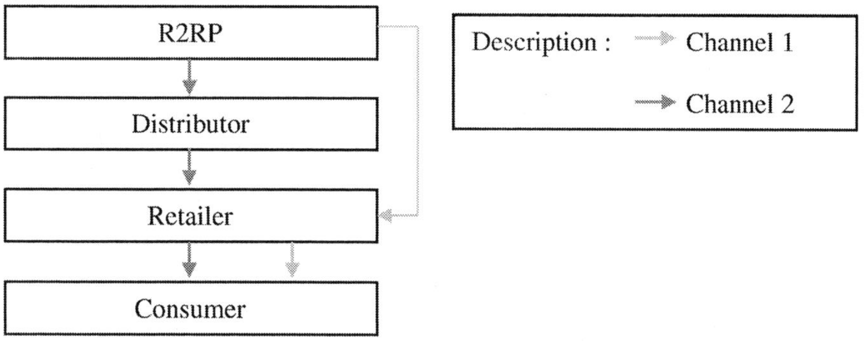

Figure 2. R2RP marketing channel

3.3. Technical and technological

3.3.1. Quality of rice, process flow, and equipment. The quality of rice in the market varied widely and the name of which varied depending on each region. This is due to the different ways of classification. Some of the most widely applied and practiced classification methods are: (i) based on rice varieties, (ii) based on their local origin, (iii) based on their processing, (iv) based on their degree of milling, and (v) based on a combination of rice varieties and its polishing. These quality requirements include both qualitative and quantitative requirements (table 5).

Table 5. Quality standard of national rice (SNI No. 01-6128-2015).

No.	Quality Components	Quality Class			
		Premium	Medium		
			1	2	3
1	Level of polishing (%, min)	100	95	90	80
2	Water content (%, max)	14	14	14	15
3	Head rice (%, min)	95	78	73	60
4	Broken (%, max)	5	20	25	35
5	Grain grouts (%, max)	0	2	2	5
6	Red grain (%, max)	0	2	3	3
7	Yellow/ damaged grains (%, max)	0	2	3	5
8	Grains of whitewash (%, max)	0	2	3	5
9	Foreign objects (%, max)	0	0.02	0.05	0.2
10	Grains (grains/100g, max)	0	1	2	3

The selection of technological level used in the establishment of R2RP is based on the characteristics of raw materials (medium quality of rice 2 and 3) and by considering the quality class to be achieved (premium rice and medium quality 1). Based on the quality and characteristics of raw rice as raw material, it could be used as a basis for selecting technology level (table 6).

Table 6. Characteristics of raw materials relatives to selection of technology level (see figure 3).

Quality parameter	Decision in selecting technology level
Level of polishing	The value of level of polishing is low namely less than 95 %, as to required polishing machine and/or shining machine.
Water content	Water content ranging from 14.0-15.0%, it indicates that the water content of rice is good enough to be processed directly without additional drying. However, it is recommended that R2RP management require a raw water content of no more than 15.0%, so it does not require dryer investment.
Head rice, whole grain and broken	Content of head rice varied between 60-78% as to improvement to higher rice quality required length grader machine as separator of rice head and broken. The percentage of head rice could be used as a basis to determine the yield of rice quality class that is going to be produced.
Grain grouts	Grains grouts ranging from 2-5% or more so rotary shifter machine to separate the grouts was required. Separation of the groats is performed before the shinning to make the machine work more efficiently. The resulting grout could be sold or processed into flour as additional income (by-product).
Grains of whitewash, yellow /damaged grains, and red grains	Random rice contained grains of whitewash, yellow/damaged grains, and red grains so color sorter is required if premium rice production is desired. However, if it is desired to produce medium quality 1 of rice only, by passing without using a color sorter could be performed.
Grains and foreign objects	Generally, rice contained grains and foreign objects so paddy separator is required to separate grain and destoner to separate the foreign objects (stone).

In designing a process flow, it is necessary to consider the condition of raw materials (quality and variety), available space and product quality to be achieved as well as efficiency. The production process run continuously in a closed system where the material flow is controlled using a bucket elevator and piped through pipes.

With this consideration, the design of the process flowchart on the ideal R2RP was as shown in figure 3. The raw materials at any quality could be processed into high-quality products including premium and medium 1. If the quality to be produced is limited to the quality of medium 1, some process stages could be eliminated like a color sorter machine. Generally raw materials of raw rice still contained unhusked rice and also foreign objects (stone). Incomplete husked rice could be overcome using a shining machine with two processes. Destoner is applied to separate stone.

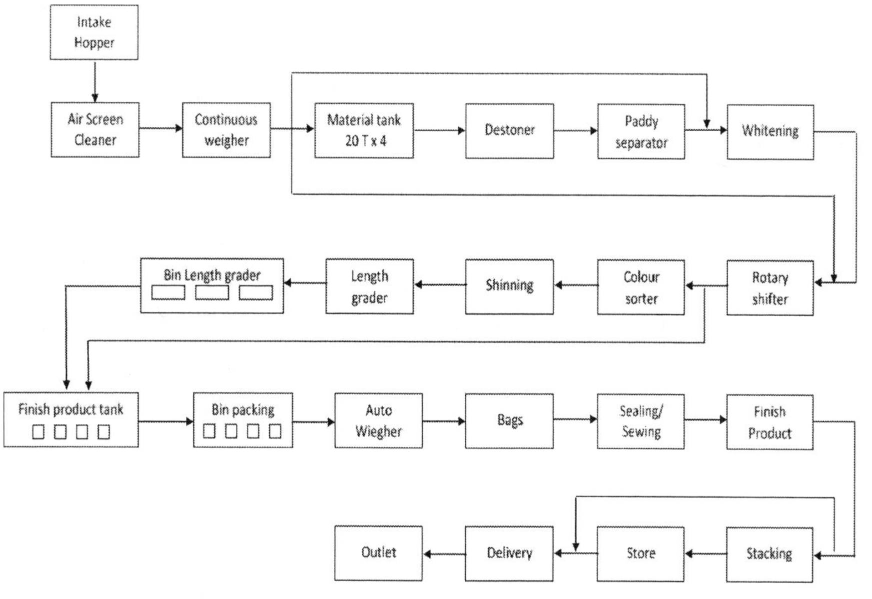

Figure 3. Design of flowchart of R2RP process

3.3.2. Production capacity. Theoretically, the production capacity must be determined based on the ability of raw material procurement (random rice), rice distribution (market absorption), and machine capacity at each stage of the process and financial feasibility. R2RP used as a reference with capacity of 6 tons/hour of rice and is expandable. Assuming the machine operated for 11 hours/day and 25 days/month then the R2RP have a capacity of about 1,600 tons/month.

3.3.3. Equipment. Overall the complete tool/machine configuration for R2RP could be categorized into 4 parts: (i) raw material handling part, (ii) milling parts, (iii) packing parts, and (iv) dust collecting parts and other supporting equipment.

3.3.4. The layout of machinery and equipment. The machine has diverse dimensions and sizes, so the layout in the area depended on the type and shape of the selected machine. Each engineering company has different technical specifications for each machine and tool at the same capacity. Based on this consideration, the area used is the optimal dimension of the machine and the tools already manufactured by the main manufacturers based on the widely used process flow.

It is important to note that the calculation of the minimum and sufficient area could be developed for the ideal area that could be rounded up to 125 x 50 m2 or equivalent to 6.250 m2 area. (0.625 Ha). The layout could be adapted to the conditions and contours of the land (figure 4).

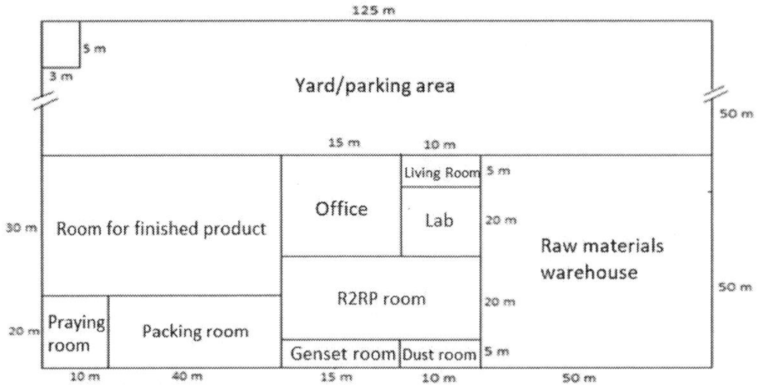

Figure 4. Design of R2RP factory layout on land (m)

3.3.5. Building needs. Machinery and equipment should be placed in a building or room to protect from weather disturbances (rain and heat). Each manufacturer had different specifications and tool dimensions. In general, the height of the R2RP tool/machine was relatively similar so that it could be placed in one building with the office and laboratory. However, to avoid contamination, the enclosure of R2RP space with offices, laboratories, raw material warehouses, packing chambers, and finished product warehouses are required. Security post and supporting buildings, if required, are made separately from factory buildings. The height of the production room followed the highest two layer equipment which was about 12 m, i.e., 25 m x 20 m x 12 m.

3.4. Need for Workforce
Most of the workforce tasks such as lifting, shifting, moving, observing, cleaning, and the like have been replaced by machines. The series of machines and devices are connected to a computer control system so they could be executed and stopped by one operator. Likewise, the loading and unloading activities could be done mechanically, so it does not require a lot of workforces. Based on these considerations, the core workforce required consist of management, operators, and technicians, supporting workforce and security (figure 5).

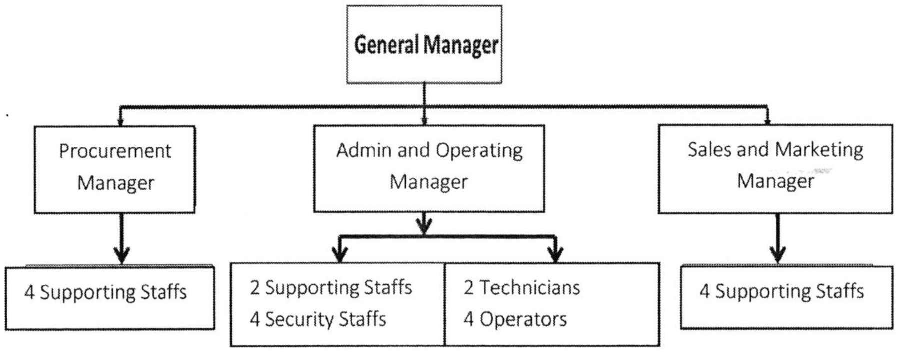

Figure 5. Organization structure of Rice to Rice Processing Plant

3.5. Optimization of rice production based on quality
Characteristics of raw material quality determined the potential of product quality that could be produced. Estimated product quality is prepared based on material balance (mass balance) corrected with shrink coefficient and damage though. The calculation of the mass balance of R2RP processing is performed by approaching the example of rice analysis from West Java, especially Pantura and the efficiency of the tool (assumption). Processing of raw rice from raw materials of Medium 2 and Medium

3 to produce Premium (X1) and Medium 1 (X2) quality products is shown in table 7. Profit coefficient was obtained from the calculation of product cost (including the price of grouts) plus the selling price minus the sum of operating costs plus raw material costs. The calculation of the profit coefficients a and b can be seen in table 8.

Table 7. The stages of random rice processing in variety of quality class.

Quality of raw material	Process stages								Quality of product	Profit coefficient
	a	b	c	d	e	f	g	h		
Medium 2	■	■	■	■	■	■	■	■	Premium	1.718
	■	■	■	■			■	■	Medium 1	1.238
Medium 3	■	■	■	■	■	■	■	■	Premium	1.534
	■	■	■	■			■	■	Medium 1	863

Description: a = destoner, b = paddy separator, c = whitening, d = rotary sifter, e = color sorter, f = shining, g = length grader, h = packaging. Price of raw materials Medium 2 = Rp. 7,300,- Medium 3 = Rp 7,150 Selling price of Premium quality = Rp. 11,000, Medium 1 quality = Rp 10,000

Table 8. Calculation of profit coefficient.

Quality of Raw Materials	Quality of Product	Operating Cost (Rp/kg)	Price of Raw Materials (Rp/kg)	Raw Materials Coefficient	Yield	Grout Coefficient
Med 2	Premium	750	7,300	1.22	0.82	0.18
	Med 1	650	7,300	1.12	0.89	0.11
Med 3	Premium	750	7,150	1.3	0.77	0.23
	Med 1	650	7,150	1.25	0.8	0.2

Table 8. Calculation of profit coefficient (continued)

Quality of Raw Materials	Quality of Product	Cost of Raw Materials (Rp/kg)	Packing Cost (Rp)	Selling price of grouts (Rp)	Product Selling price (Rp/kg)	Profit (Rp/kg)	Margin Percentage (%)
Med 2	Premium	8,902	350	720	11,000	1,718	17.20
	Med 1	8,202	350	440	10,000	1,238	13.50
Med 3	Premium	9,286	350	920	11,000	1,534	14.80
	Med 1	8,938	350	800	10,000	863	8.70

Profit estimation was obtained by taking loss coefficients, operational costs, by-products and raw material prices, selling prices and by-product prices into account. Suggestion of processing based on raw materials with profit maximation purpose function was as follows:

Purpose Function (Z) = a X1 + b X2 where a and b were the coefficients of profit (table 9).

Table 9. Profit coefficient of premium and medium 1 rice production.

Quality class of raw materials	Premium (X1)	Medium 1 (X2)
M2	1.718	1.238
M3	1.534	863

Notes:
Price of raw materials Medium 2 = Rp7,300.- Medium 3 = Rp7,150.-
Selling Price of Premium quality = Rp11,000.- Medium 1= Rp10,000.-
Price of Grouts = Rp4,000,-

Limiting Function (table 10)
(1) Capacity: m X1 + n X2 $\leq$ 66 where m and n were coefficients of raw materials.
(2) Raw Materials: the number of rice head on each quality class of raw materials

Table 10. Limiting function on optimization of premium and medium 1 rice production.

Quality class of raw materials	M2	M3
X1 $\leq$	50.7	41.7
X2 $\leq$	61.8	50.8

From the above data, the optimization calculation of each class of quality class of Medium 2 and Medium 3 became product of Premium (X1) and Medium 1 (X2) quality with production capacity of 66 ton/day and if the raw material was Medium 2 quality then it was processed into premium rice by 93 percent and Medium 1 quality rice by 7 percent. For Medium 3, it was processed into premium rice by 76 percent and Medium 1 by 24 percent (table 11).

Table 11. Rice production optimization.

Quality of	Quality of product (%)	
Raw materials	Premium	Medium 1
Medium 2	93	7
Medium 3	76	24

3.6. Financial analysis

3.6.1. Costs of investment. The R2RP reference capacity for calculation was 19,800 tons/year or about 66 tons/day equivalent to 6 tons/ hour. The investment cost consisted of purchasing machinery, factory buildings, supporting buildings and other infrastructures (table 12).

3.6.2. Fixed cost. Fixed costs included the salaries of permanent employees, office operations, administration and depreciation. All these costs were calculated in an annual period. The economic life was 10 years with depreciation using the straight-line method (table 13).

3.6.3. Variable costs. Cost of production calculated and with real influence were raw materials, packing materials and energy. The ideal R2RP raw material was rice in brown rice form. The variable cost component of R2RP was shown in table 14.

Table 12. Costs of R2RP investments

Investments	Amount (Rp)
Machines and equipment	11,606,520,000
Factory building	1,750,000,000
Supporting building	15,140,000,000
Other infrastructures	1,304,326,000
Land purchase	1,500,000,000
Land preparation	500,000,000
Total investment out of land	29,800,846,000

Table 13. Fixed cost of R2RP.

Type of cost	Total (Rp/month)	THR (Rp/year)	Total (Rp/year)
Salary of leadership and employees	55,500,000	126,500,000	1,644,500,000
Operating costs	5,000,000	-	60,000,000
Marketing	2,000,000	-	24,000,000
Equipment and machine shrinkage	91,676,250	-	1,100,115,000
Building shrinkage	14,583,333	-	175,000,000
Equipment and machine maintenance	48,360,500	-	580,326,000
Total of fixed cost	217,120,083		3,583,941,000

THR = Bonus

Table 14. Variable costs of R2RP.

Costs component	Volume (unit/year)	Price per unit (Rp)	Total (Rp/year)
Raw rice	19,800	7,150	141,570,000,000
Fuel	281,472	8,600	2,420,659,200
Oil	250	40,000	10,000,000
Packing materials	3,840,000	500	1,920,000,000
Electricity	30,000	1,467	44,010,000
Total of variable costs (Rp/year)			145,964,669,200

3.6.4. Financial assumption. The local assumptions used in the calculation were the price of raw rice and the price of the rice (premium). Raw materials used were only those in medium 2 and 3 quality for premium and medium 1 rice production:

a. The purchase price of rice was Rp7,150.- / kg with the sale of premium rice at Rp 10,000 / kg at the factory site. The market price ranged from Rp. 11,000 - Rp 12,000,
b. The yield of random rice to premium rice was 82% to anticipate the quality of diverse raw rice.
c. The by-product obtained was calculated at 18% of the unrefined rice or 3,456 tons at an average price of Rp 4,000.-

3.6.5. Total revenue. The main product was premium rice with excellent quality because it used destoner, length grader, and color sorter. In the polishing there was no foreign object so the obtained bran as well as grouts and colored rice were clean. Therefore, these by-products could also be sold at a relatively good price as shown in table 15.

Table 15. Projection of R2RP revenue.

Revenue	Total (ton/year)	Price (Rp/kg)	Revenue (Rp/year)
Premium Rice	16,236	10,000	162,360,000,000
By-product	3,564	4,000	14,256,000,000
Total			176,616,000,000

3.6.6. Profit and loss. R2RP with medium and large capacity, when it was managed and run well, might produce an adequate profit. Profit started to be earned in the second year of the project or the first year of production until the 10th year at the rate of Rp12,361,574,550.- per year. Total revenue for 10 years reached Rp123,615,745,500.-.

3.6.7. Investment Criteria. Revenue earned was five times more than the investment value at a rate of 84.27% of the investment value. This business could grow quickly if the market could be developed, so the return of capital was relatively fast which was about 2.18 years (table 16).

Table 16. The result of calculation of investment criteria of R2RP.

Investment criteria	Result of criteria		
	R2RP investment	R2RP investment + Husker	R2RP investment + Dryer
NPV (Rp)	136,669,185,414	131,699,185,414	124,699,185,414
Net B/C	5.80	4.93	4.08
IRR (%)	84.27	71.52	58.85
PP (year)	2.18	2.39	2.68

3.6.8. Sensitivity analysis (switching value). Switching value was to perform a sensitivity analysis by simulating changes to financial feasibility parameters (NPV, Net B / C, IRR, and PBP) (table 17).

Table 17. Sensitivity Analysis.

Sensitivity	Increase of Purchase Price of Random Rice	Decrease of Selling Price of Rice
Change of price (%)	14.06	12.26
Level of Price (Rp)	8,115	8,774

3.7. Partnership between SRM and R2RP

SRM was a small-scale business both technical and economical, so it had many limitations and advantages. R2RP is a relatively large scale business that could help small businesses to overcome the problems while utilizing the advantages to reduce the constraints. The advantages of SRM included having cultural attachment with the upstream (farmers) and the downstream (traditional rice market) in the value chain, more flexible to changes in production level (operational capacity). The disadvantage is the limited capital so the use of technology and market run is relatively limited. In contrast, R2RP only has a business relationship in the value chain so it is very vulnerable to change and the occurrence of price competition (table 18).

Table 18. Strengths and Weaknesses of SRM and R2RP

Strengths	
SRM	R2RP
• Having cultural relationship with the farmers • Having cultural relationship with the market (sellers) • Is spread in production center of raw materials • Flexible toward operating capacity	• The capital was relatively adequate • Advanced technology • High-quality product and could be adjusted • Production cost was efficient and competitive • Broad market reach
Weaknesses	
• Capital limitation • Narrow market reach • Relatively low product quality • Low production efficiency • Limited technology	• Relationship in value was pure business • Did not have cultural relationship with actors in value chain • Raw materials access was through free competition

The cooperative relationship between SRM and R2RP could overcome weaknesses while utilizing each other's strengths. Strengthening the power to overcome weaknesses would result in an integrated business unit as follows:

(1). SRM would utilize cultural links with farmers to obtain relatively reliable raw materials at market prices and diverse quality;

(2). SRM with flexibility on operational scale (production) could continuously reproduce to produce raw materials (rice and or rice) to support R2RP production throughout the year;

(3). R2RP would assist SRM in financing to improve the ability to produce higher quality raw materials through equipment investment (dryer and huskers);

(4). R2RP continuously bough random rice and prospective rice from SRM in any amount at an agreed level of price (including market price),

(5). R2RP developed a premium rice market and/or medium 1, so the partnership relationship grew and developed continually on the basis of business.

The relationship could be mutualistic (or referred to as Symbiotic Innovative Relationships) [27], i.e., the weakness of SRM is covered by the power of R2RP. Likewise, the weakness of R2RP is covered by the power of SRM. The synergy of this relationship would make SRM's business survived as a rural economic power based on local resources, culture and ownership (the synergy of large and small enterprises in the strategic alliance relationship according to Rothkegel [28]. R2RP grow and develop the strength of SRM, so market growth impacted rural economic growth (figure 6).

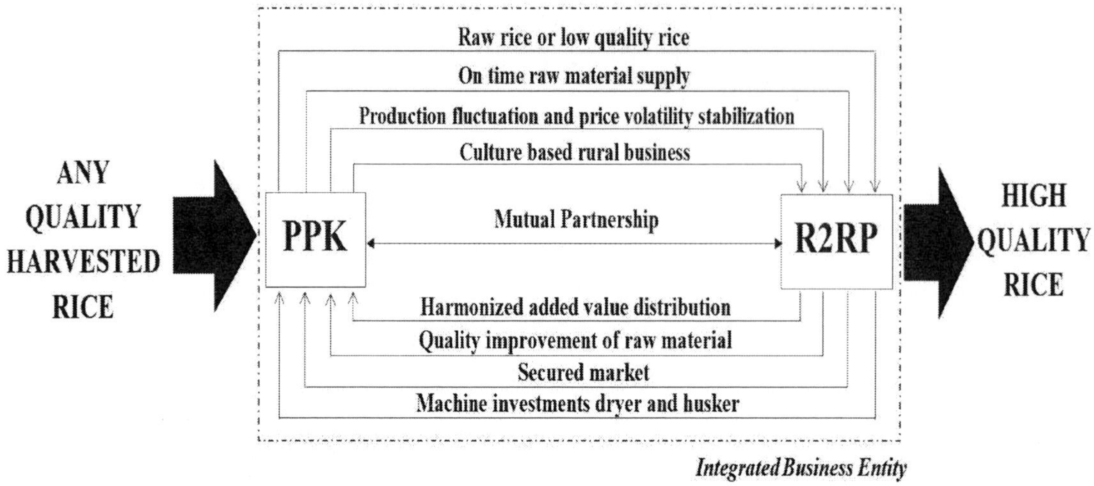

Figure 6. Partnership relationship base between SRM and R2RP

A mutually beneficial relationship (partnership) that relied on strength optimization and weakness minimization (eradication) of each side would create a strong and sustainable, new business entity (James 2014). This relationship could only be maintained if the business profits grew and adapted to a highly dynamic environmental change. Adaptability [28] should be built on upstream competition (raw material procurement) and downstream (rice market). Therefore, strengthening SRM with dryers to be able to purchase raw materials at any time at any quality and any market price through drying investments was helpful.

On the other hand, competitiveness in the rice market could only be achieved if it were able to provide quality products at a price level that compete with good quality and good delivery of goods. All of these were rooted in the power of efficient raw material procurement. Therefore, husker investment would increase the efficiency of SRM processing at adequate price and quality level. R2RP has advanced technology with high efficiency reinforced by efficient SRM; it would form an integrated business entity with high adaptability (to changes in raw material production) and high market competitiveness (quality and price).

At a capacity of 80 tons of raw materials/day, R2RP required 10 SRM with a capacity of 8 tons/day. SRM have better level of market certainty to sell the product or investment assistance (husker or dryer). If the pricing were in accordance with market price or more competitive, then this working relationship would run well and sustainable. On the contrary, if the price was not competitive and commitment was not strong then SRM might sell the product directly to the market. The determinant of the success of cooperation is whether or not an agreement to be made and whether or not the parties implemented the contents of the agreement consistently. Therefore, the performance of R2RP as an initiator of cooperation would determine the success of this cooperation pattern. So far, SRM had a weakness in the competition of raw materials and product marketing. The cooperation between R2RP and SRM is mutually beneficial where R2RP gets the supply assurance of raw materials, and SRMs get market assurance for their products.

4. Conclusions and suggestions

4.1. Conclusions

The development of R2RP through SRM strengthening investments could be done as well as feasible in non-financial, financial and partnership. Dryer and huskers investment by R2RP on SRM forms the basis of mutually reinforcing partnership cooperation. This scheme required investment cost of Rp30 billion, fixed cost Rp 3.6 billion/year, the variable cost of Rp146 billion/year. The profit earned is equivalent to IRR 84.27% and re-investment is within 2 years 2 months at the level of raw material price of Rp 7,150.- and selling price of premium rice products is Rp10,000.-

4.2. Suggestions

Further review is required to specify the partnership agreement between the parties that determine the rights and obligations stated in the purchase price of the underlying rice. The continuation of cooperation determines the economic (financial) benefits obtained by the parties.

References

[1] Sawit M H 2011 Reformation of Producer Price Policy and the Impact on Rice Competitiveness (in Indonesian) *Jurnal Pengembangan Inovasi Pertanian* **4**(1): 1-13

[2] Nataatmadja H, Kertosastro D, and Suryana A 1988 *Production development and government policy in rice production* (in Indonesian) p. 37-54 in Ismunadji *et al.* (Ed.) Padi, Buku 1. Pusat Penelitian dan Pengembangan Tanaman Pangan, Bogor.

[3] Sumardi and Thahir R 1993 *Post-harvest handling of rice* (in Indonesian) p. 915-942. *Dalam* E. Soenarjo *et al.* (Ed.) Padi, Buku 3. Pusat Penelitian dan Pengembangan Tanaman Pangan, Bogor.

[4] BPS 2012 Data Collection of Rice Milling Industry (PIPA) Katalog BPS:6106001. Oktober 2012

[5] Paman U, Bahri S, Asrol and Liana L 2016 Distribution and Use Patterns of Small-Scale Rice Mills in Kampar Regency, Riau Province, Indonesia. *International Journal on Advanced Science, Engineering and Information Technology* **6** (2) p.149

[6] Patiwiri A W 2008 *Study on Post-harvest Process Integration for Rice Commodity* (in Indonesian) PPT Direktur SDM dan Umum BULOG: Jakarta

[7] Thahir and Ridwan 2010 Revitalization of Rice Milling through Polishing Innovation Supports Rice Self-Sufficiency and Global Competition (in Indonesian) *Jurnal Pengembangan Inovasi Pertanian* **3**(3) : 171-183

[8] PERPADI 2014 AEC Polishing Rice Milling (in Indonesian) PPT disiapkan oleh B. Pasdom, yang disampaikan pada FGD di FKPR (Forum Komunikasi Profesor Riset), Bogor tgl 28 Maret 2014

[9] Sawit 2014 Analysis of Rice Milling 2012 Census Results (in Indonesian) *PANGAN* **23**(3) September 2014: 208-219

[10] Wibowo 2007 *Consumer Preferences of Rice Characteristics and Its Conformity with the Rice Quality Standard at Central Java* (in Indonesian) Apresiasi Hasil Penelitian Padi 2007 Balai Besar Penelitian Tanaman Padi

[11] Persaulian B, H Aimon, A Ali 2013 Analysis of Indonesian Population Consumption (in Indonesian) *Jurnal Kajian Ekonomi* 1(2)

[12] Badan Litbang Pertanian 2015 *Innovation of Bio-industrial Agriculture toward Food Self-Supporting and Farmer Well-Being* (in Indonesian) Laporan Tahunan 2015 Badan Penelitian dan Pengembangan Pertanian, Jakarta.

[13] Hasbullah R and T Bantacut 2006 *Rice to Rice Processing Technology* (in Indonesian) In Prosiding Lokakarya Peningkatan Dayasaing Beras Nasional Melalui Perbaikan Kualitas. Perum BULOG dan FTP IPB: Bogor

[14] Badan Pusat Statistik Provinsi Jawa Barat 2016 *Rice production in West Java 2016* Bandung (ID): Badan Pusat Statistik Provinsi Jawa Barat.

[15] Deepak K and K Prasanta 2017 Reducing Postharvest Losses during Storage of Grain Crops to Strengthen Food Security *Developing Countries Foods* **6**(1):8. Published online 2017 Jan 15

[16] Sulardjo 2014 Post-Harvest Handling of Rice (in Indonesian). *Magistra* No. 88 Th. XXVI Juni 2014 ISSN 0215-9511.

[17] Sekhon M K, Rangi P S, Kaur and Manjeet 2003 Rice processing industry in Punjab: Problem and their remedies *Indian Journal of Agricultural Economics* **58**(3):544-553.

[18] DITJEN PPHP 2008 Paddy Post-Harvest Loss in Indonesia (in Indonesian) Direktorat Jenderal Pengolahan dan Pengembangan Hasil Pertanian. Departemen Pertanian

[19] Budihartia U, R Tjahjohutomo, Harsono, R Y Gultoma, and R S Basuki 2008 Dynamic System Approach to Find Out Mechanization Model of Rice Mill to Predict Rice Production (in Indonesian) *Indonesian Journal of Agriculture* **1**(1):7-12

[20] Bantacut T 2012 Rationally Optimum Paddy Production: Chance and Challenge (in Indonesian) *Pangan* **21**(3): 281-295

[21] Alavi H R, A Htenas, R Kopicki, A W Shepherd and R Clarete 2012 *Trusting Trade and the Private Sector for Food Security in Southeast Asia* World Bank Publications; Washington, DC, USA

[22] Maryana Y E and B Raharjo 2014 Small Milling Performances in Lempuing Jaya Regency Dry Land (in Indonesian) *Prosiding Seminar Nasional Lahan Suboptimal 2014* Palembang 26-27 September 2014 ISBN : 979-587-529-9.

[23] Chung B, F M Arshad, K M Noh, S F Sidique 2016 Cost analysis of rice milling: a case study of 7 rice mills in Malaysia *Journal of Agribusiness in Developing and Emerging Economies* **6**(2):173-190.

[24] Ajala A S and A Gana 2015 Analysis of Challenges Facing Rice Processing in Nigeria *Journal of Food Processing* Volume 2015

[25] Kasmir and Jakfar 2009 *Feasibility Study of Business* Jakarta: Kencana Prenada Media Group

[26] [BPS] Badan Pusat Statistik 2015 *Average consumption per kapita of several kind of key food* Jakarta (ID): Badan Pusat Statistik.

[27] Banwo A O, J Du and U Onokala 2015 Symbiotic Innovative Relationships of Small and Medium Enterprises *Journal of Advanced Management Science* **3**(2), 128-131

[28] Rothkegel S, L Erakovic and D Shepherd 2006 Strategic Alliances between SMEs and Large Firms: An Exploration of the Dynamic Process Special Issue: Management of Interorganizational Relationships *Management Revue* **17**(1): 50-57

IOP Conf. Series: Earth and Environmental Science **147** (2018) 012042 doi:10.1088/1755-1315/147/1/012042

Appendix A. The needs of equipment/machine on technology level (Capacity 6 ton/hour)

NO.	Description	Model	Capacity Ton/hour	Power (HP)	Total
Raw Material Revenue Section					
I-1	In-Take Hopper				1
I-2	Bucket elevator + motor		10	1	1
I-3	Brown rice Tank	1 ton			1
Rice to Rice Processing Section					
R-1	Bucket elevator + motor		10	1	2
R-2	Damper Tank	+800Kg			2
R-3	Whitening machine	Stand Pole	3	120	2
R-4	Bucket elevator + motor		10	1	1
R-5	Damper Tank	+800Kg			1
R-6	Polishing machine		3	50	2
R-7	Bucket elevator + motor		10	1	1
R-8	De-stoner + motor		3	2	2
R-9	Rotary Sifter + motor		6	1	1
R-10	Bucket elevator + motor		10	1	1
R-11	Damper Tank				2
R-12	Color Sorter	384 channel	3		2
R-13	AVR				2
R-14	Air Compressor & Air Dryer		40HP	40	1
R-15	Bucket elevator + motor		10	2	2
R-16	Length grader		1	4	8
Packing Section					
P-1	Bucket elevator + motor		10	1	1
P-2	White rice Tank				2
P-3	Auto matic Scale			1	2
P-4	Thread Packing machine			1	2
P-5	Plastic Packing machine			1	2
Building and Electricity Installation Section					
O-1	Strcuture, Piping & Elbow Part				1
O-2	Air Line Part		10HP	30	2
O-3	Cyclone				2
O-4	Air Line Part		15HP	15	1
O-5	Cyclone				1
O-6	Electric wire & wire gutter				1
O-7	General electric Control box				1
O-8	Accessory & E.T.C				1
	The Need of Workforce			**273 HP**	

IOP Conf. Series: Earth and Environmental Science **147** (2018) 012043 doi:10.1088/1755-1315/147/1/012043

Problems Analysis on Increasing Rice Production Through Food Estate Program in Bulungan Regency, North Kalimantan

Setyo P [1] and Elly J [2]

[1] Department of Mechanical and Biosystem Engineering, Bogor Agricultural University, Bogor, Indonesia.
[2] Department of Agribusiness, Borneo Tarakan University, Tarakan, Indonesia

E-mail: pertiwi@ipb.ac.id

Abstract. To increase rice production in the Province of North Kalimantan, the provincial government has launched a Food Estate Program. The program is also a central government program in relation to government policies on food security. One of the food estate development areas is the Delta Kayan Food Estate of 50,000 hectares in Bulungan Regency, where about 30,000 hectares area is a tidal land with a very fertile alluvial soil type. This policy study aims to identify and analyze problems of increasing rice production through food estate development in North Kalimantan Province and formulate priority programs as recommendations for policy making in increasing rice production. The study has identified a number of problems of increasing rice production, such as land tenure, land suitability, water system, infrastructure, accessibility of production factors, institutional, and capacity of human resources. The Analytic Hierarchy Process was applied to develop priority programs, resulting in the three most important programs being water management, improving access to production factors, and improving the capacity of human resources. Action plans related to priority programs have also been identified.

1. Introduction

Food security systems in Indonesia are mainly based on rice, corn, soybean, and meat, in which rice is as its pivotal commodity. Food self-sufficiency at national and provincial levels is considered to be the precondition to food security at the household and individual levels. Food security is also one of the determinants of economic stability so that the fulfillment of food sufficiency becomes a development framework that can encourage the development of other sectors. In fact, nowadays Indonesia still imports all of these foods each year. To fulfill the national rice consumption/demand that increased considerably (around 130 kg/capita/year), the government policy is to increase rice production, to reduce rice consumption by 1.5%/year, to accelerate food diversification consumption and/or to support rice import policy [1].

Facing the dynamics and complex problems of food security, in increasing rice production the Indonesian government has made a broader perspective and comprehensive approach, namely seven pillars of agricultural revitalization, i.e. land consolidation, high seeds productivity, infrastructure, skilled labor, financial capital, institutional development, technology improvement, and downstream industry development. Thebasic purpose of agricultural revitalization is to enhance food/rice production up to surplus of 10 million tons of rice in 2015, to achieve and sustain rice self-sufficiency, to strengthen food security and nutrition security, to increase competitiveness and added value, and to improve farmer income as well as to conserve environment and climate change adaptation [2]. To that end, the government has also undertaken various breakthrough for expansion of food land, one of them through the development of food estate. To date, there are 4 programs in 4 regions that are nationally notified as food estate development areas, those are Development of 1 Million Hectares Peatland (PLG) inCentral Kalimantan, Merauke Integrated Food and Energy Estate (MIFEE) in Merauke Regency - Papua, Delta Kayan Food Estate in Bulungan Regency - North Kalimantan, and Jungkat Agri Complex in Kuburaya Regency - West Kalimantan. However, the food estate programs seem not run as expected due to various complex constraints. Even the PLG was stopped because the technology used was rejected by environmental activists, both at national and international level [3]. In other food estate locations, land conflicts, social conflicts, availability of agribusiness infrastructure, technological issues as well as political issues can potentially hinder success [4][5].

Delta Kayan Food Estate (DeKaFE) in Bulungan Regency was launched by the provincial government in 2011 as part of a central government program in relation to government policies on food security. Traditionally, Bulungan Regency is one of the food base areas in North Kalimantan Province. Quite a lot of people's foods needs are supplied from here. The main commodities of food crops are rice, soybean, and maize. The area is also well recognized for its plantation sub-sector which produces coffee, palm oil, pepper, cocoa, coconut, and rubber. The DeKaFE was planned to cover an area of about 50,000 hectares, from which most or around 30,000 hectares are fertile ground with alluvial soil types. Despite the high spirit and hope of developing the food estate, the program has not shown encouraging results after six years of its commencement.

This policy study aims to identify and analyze problems of increasing rice production through food estate development in North Kalimantan Province and formulate priority programs as recommendations for policy making in increasing rice production with special emphasis on Delta Kayan Food Estate, Bulungan Regency.

2. Research approach

The paper is based on fieldwork undertaken in the province of North Kalimantan and across Bulungan Regency during 2016. Field visits were made to Delta Kayan Food Estates in Tanjung Buka, Bulungan Regency.

The research was carried out by using qualitative research approaches to understand the underlying policy settings, the factors shaping the implementation of the policy and its outcomes, and finally the problems of increasing the rice production. Semi-structured interviews with key provincial and district officials, company officials, and university researchers were undertaken to define the existing situation. Then repeated visits were made to Bulungan Regency to interview representatives of affected groups which were selected with purposive sampling approach, and provided them with the opportunity to identify key problems and to articulate their concerns. The findings were triangulated with an analysis of key texts, including laws, scientific and technical publications, newspaper articles, and agency reports. Focus Group Discussion with key provincial and district officials, company key persons, and university researchers was carried out to consolidate the identified key problems and to formulate the recommendations for alleviating the problems, including its priority setting. Analytic Hierarch Process [6] was employed in the priority setting on alleviating the problems.

3. Results and discussion

3.1. Delta Kayan Food Estate in Bulungan Regency

Food estate is a concept in the development of food production which is done in an integrated manner covering agriculture, plantation, even farms is in a very wide land area. Food estate program is a breakthrough production mode that is attempted to be implemented in order to pursue the fulfillment

IOP Conf. Series: Earth and Environmental Science **147** (2018) 012043 doi:10.1088/1755-1315/147/1/012043

of national need of food which then expected can realize food security in Indonesia. Increasing productivity and intensity of planting and expanding production base through new land clearing is a solution offered by the food estate program. The result of food estate development can be a supply of national food security and if excess can be exported.

Delta Kayan Food Estate (DeKaFE) was launched in 2011 with the following development goals:

1) Realizing the food estate area as the center of fast-growing economic activity in the northern part of the North Kalimantan province;
2) Developing key sectors to increase added value and strengthen economic structure;
3) Mobilizing the dynamics of growth through the centers of activities that can spur the development of residential centers/villages that are in the area of its influence;
4) Fulfilling basic needs and quality of population resources, especially in remote areas that are still low; and
5) Strengthening the realization of environmentally sound spatial planning.

In Delta Kayan master plan, the food estate is defined as 'an expanse of paddy field with modern irrigation, industrial management, housing, road, and farming infrastructure that forms a well-arranged system'. The program was designed to involve government, private sectors (investors), and the community. The government will facilitate the basic infrastructure, while investors for their part would invest their capital in the intensive application of machinery, fertilizers, technology, and irrigation to 'unused' or 'uncultivated' land. They would develop infrastructure, providing development knowledge, technologies and management. It was expected that the program proceed rapidly to help the nation meet its food production targets while also providing direct and indirect benefits to regional economies.

The Regent of Bulungan Regency has allocated an area of 50,000 hectares for DeKaFE in Tanjung Palas, North Tanjung Palas, Central Tanjung Palas, and Tanjung Selor, from which 10,000 hectares was reserved for transmigration land, 8,000 hectares for food estate managed by private companies (corporate farming), and the remaining 32,000 hectares was the area that has been cultivated by the local community. The main characteristics of the land are mostly a tidal lowland with A and B type overflow, very fertile alluvial soil type, and having sufficient water supply from Kayan watershed. Figure 1 shows the orientation map of DeKaFE location.

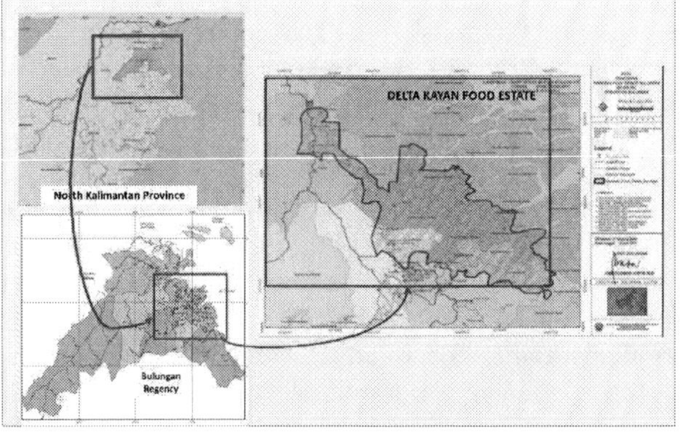

Figure 1. Orientation map of Delta Kayan Food Estate

3.2. The factors shaping the performance of food estate

Based on systems thinking perspective, the food estate can be depicted into three components: input, process, and output. The input consists of six aspects: (a) agroecology (land, water, and climate); (b) infrastructure, including transportation, irrigation system, and financial capital; (c) production factors (availability and distribution); (d) human resources (manpower), (e) technology, and (f) institutional development. These six components are the basic requirements to shape the performance of the food estate program. The process consists of two components: (a) corporate farming and (b) community farming. These components are integrated within each other. Both input and process will produce outputs in terms of (a) food (in this case rice) production; (b) distribution; and (c) people welfare. The key success factor of estate food program will be determined by harmony, synergy, and integrated

cooperation between the stakeholders involved, i.e. central government, local government, private sectors, and community. Figure 2 shows the factors shaping the performance of food estate.

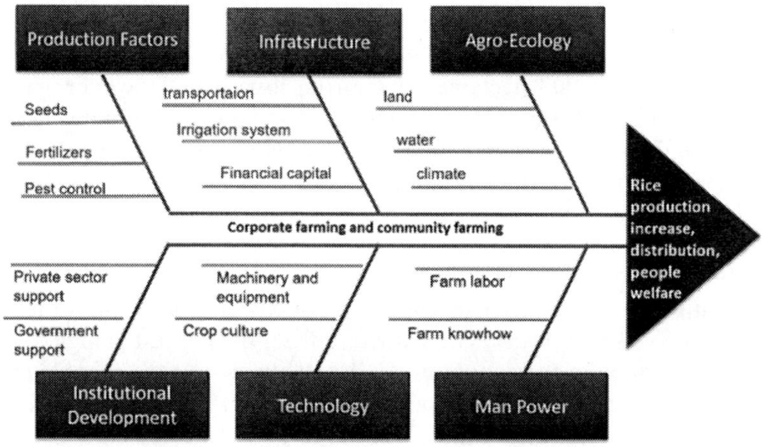

Figure 2. The factors shaping the performance of food estate

Tidal lowland development needs a proper planning, management, and utilization of land as well as proper technology application, especially water management aspect [7]. This soil has a good potential for agricultural development through proper management. The availability of sulfuric acid layer is a natural condition [8]. Macro and micro-water management is a critical determinant of successful tidal lowland management. Operation and maintenance of macro water system (covering primary, secondary and tertiary network and the water gates) should ensure the land is protected from flooding during wet season and high tide, and also from drought at dry season, while the micro-water system can reduce soil acidity and iron content which is a major constraint on tidal lowland.

There are four reasons why Delta Kayan Food Estate is considered to have a good prospect. First, it is part of central government program so that supports from central government assured. The second, Delta Kayan is a fertile area with enough water resources. By setting good irrigation infrastructure, it can be assumed that the farm must be running good. Natural potential also supports. Delta Kayan region has a fairly high rainfall that can dissolve acid content in the soil. The water of Kayan River is also proved to contain no compounds that could harm the plant. The third, the human resources are considered sufficient. The potential comes from the transmigrants. In Bulungan, 50 percent of transmigrants are locals, i.e. residents from surrounding areas who live in three or four families in one house. Other 50 percent are those coming from outside Bulungan, for example from West Java and East Java. The transmigrants from East Java already had good farming experiences, so they can transmit the technology to the local transmigrants. The fourth, the DeKaFE program has successfully attracted a private sector to invest.

3.3. Performances and problems encountered

Systematic steps have been taken by local governments to realize the intended food estate. Over the past five years, the Bulungan Regent has implemented programs and activities in the development of the DeKaFE area, at the beginning focused on Tanjung Buka, Central Tanjung Palas. Among other efforts are the implementation of transmigration for 9 settlement units (SP1-SP9), involving

10,466 inhabitants as the workforce for food estate, developing new paddy fields, and encouraging the agricultural mechanization through the provision of hand tractors to the farmers/transmigrants through farmers groups. There was also the construction of farm roads, construction of primary canals, secondary canals, and irrigation water gates, and construction of bridges between settlement units. During 2013-2016, there

were a total of 2,416 hectares of new paddy field and 35,750 m primary irrigation canals with 78 unit water gates. Yet, the DeKaFE area can only be accessed from Tanjung Selor, the capital of North Kalimantan Province, and other community activities center through river by using small boats.

Investors' interest in the food estate program has been proved by the inclusion of four investors in 2011. They processed development permits to invest in the DeKaFE through corporate farming, and the Regent has allocated 6,903 hectares for the four investors. However, as of late 2014, only one of them actually planted rice. In the implementation of corporate farming, farmers or transmigrants do not only act as workers but also act as business owners, i.e. by integrating the land they own into the land managed by the investor.

In 2015 McCarthy and Obidzinski [5] reported that despite the capital intensive and technologically advanced investment, the company operating corporate farming in DeKaFE only achieved productivity of around 2.5-3 ton per hectare. This was well below the average productivity achieved by rice farmers in Java which can achieve productivity of about 4.5-7 ton per hectare. The key problems remained was the intrusion of salt water from the coast, the high cost and difficulty of accessing inputs due to the poor infrastructure, and pest attacks. At the other hand, the neighboring transmigrants settled around the food estate also had to struggle in growing food crops. In this tidal zone, with sulfuric acid and peat soils, drainage needs to be engineered carefully to avoid salt water intrusion. In other areas, overly effective drainage had exposed the soil to the air, oxidizing the pyrites without providing the means to effectively drain the emergent sulfuric acid. All the issues mentioned above have discouraged some parties involved in DeKaFE.

In more detail fashion, the followings are a diverse set of issues that have been identified as problems faced in DeKaFE:

1) The absence of special institution that handles the development of food estate in an integrated manner; the development and problem-solving mechanisms by related agencies were less effective due to partial and sectoral approach.
2) Quite high interests of private parties to the development of DeKaFE, but there is still remaining problems related with "clean and clear status" of the land reserved for them; in several sites there were still some overlapping land tenures.
3) Efforts to increase food production are still focused on rice, but land suitability maps for rice crops are not yet available
4) In the face of salt water intrusion, acid sulfuric soils, and extensive dry periods, the high yielding rice varieties did not thrive as well as their potentials. This is considered as the main reason of low yield of rice production.
5) The absence of effective irrigation in many places. Therefore, paddy rice field is rain fed and vulnerable to production failures due to floods, drought and unreliable rainfall.
6) Lack of agricultural infrastructure, especially for the distribution of agricultural production factors to the farms as well as for distribution of agricultural products from farms to the markets.
7) Widespread pest infestations without significant measures to overcome them, especially due to limited availability of chemicals, and also due to low purchasing power of the farmers.
8) In community (household) farming, most of the farmers cannot afford fertilizers as well as chemicals for plant protection. In an in-depth interview with them, it was revealed the farmers obtained an average yield of 1.2 - 1.5 tons per hectare, even quite a lot less than that.
9) Lack of skilled labor, especially to deal with a specific management and care needs of farming

IOP Conf. Series: Earth and Environmental Science **147** (2018) 012043 doi:10.1088/1755-1315/147/1/012043

in lowland tidal area, and also to operate and carry out maintenance of agricultural machinery.

10) After the tractors for tillage sank in the swampy land, the labor required to replace them was not readily available. The estate offered wages forty percent below the market rate found in neighboring oil palm plantations, so that the estate had trouble attracting workers. Community farms even were worse due to their inability to hire farm workers except for their family members.

3.4. Alternatives problem solving

Problems that have been identified were arranged into a problem tree with reference to the factors shaping the performance of food estate, and alternative solution for each of the problem was also formulated, resulting in the hierarchy of problems and solutions demonstrated in figure 3. Priority setting of alternative solutions was made based on the urgency level of problem encountered related to DeKaFE objectives. It was determined with consolidated pairwise comparison at each level of the hierarchy made during focus group discussion, then the resulted local priorities (weights) were synthesized to obtain the overall (global) priority of each alternative solution. The result is presented in table 1.

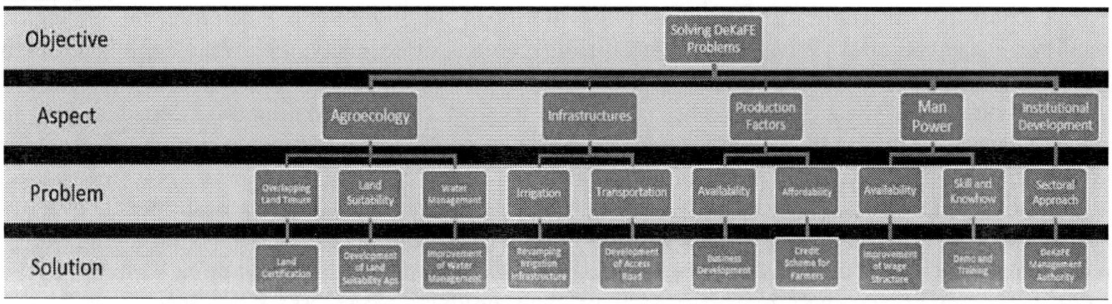

Figure 3. Hierarchy of problems and solutions identified from the DeKaFE

Table 1. Alternative solutions of DeKaFE problems and its priority setting

DeKaFE Problem Solving										
Aspect	Agroecology			Infrastructure		Production Factors		Man Power		Institutional Development
	0.409			**0.249**		**0.164**		**0.117**		**0.061**
Problem	Land Tenure	Land Suitability	Water Mngmnt	Irrigation	Trans-portation	Avail-ability	Afford-ability	Avail-ability	Skill	Sectoral Approach
	0.211	**0.109**	**0.680**	**0.500**	**0.500**	**0.600**	**0.400**	**0.350**	**0.650**	**1.000**
Solution	Certification	Suitability Mapping	Improving Water Mngmnt	Revamping Irrigation Infrastructure	Roads Development	Business Development	Credit Scheme	Wage Structure	Demo & Training	DeKaFe Management Authority
	0.086	**0.045**	**0.278**	**0.125**	**0.125**	**0.098**	**0.066**	**0.041**	**0.076**	**0.061**
Priority	4	8	1	2	2	3	6	9	5	7

From the result shown in table 1 it is discerned that the highest priority of problem solving in DeKaFE is the improvement of water management. Saragih et. al. [9] has emphasized that water management is one of the important factors for land management in the tidal land area. Rice cultivation in tidal land requires specific water management, particularly in relation to reducing high waterlogging in the wet season and anticipating the lack of water in the dry season. Macro and micro-water management is a critical determinant of successful tidal land management which will have an impact on the optimum utilization and conservation of land resources. In this case, water management is not only to reduce or to add surface water availability, but also to reduce soil acidity due to pyrites layer oxidation, to minimize salinity hazards and flooding risks, as well as to reduce toxic chemical compounds as a result of pyrite layer oxidation. To establish the above conditions, water management operation is geared toward aspects of water table retention that is always located above pyrites layer and land leaching through a controlled drainage systems.

Revamping irrigation infrastructures is also very important to be undertaken seriously. Better irrigation infrastructure will support a better water management. At the same level of priority is the provision of good access to the DeKaFE area through development of access roads from and to the neighboring community activity centers. Development of DeKaFE will deal with a lot of people and materials mobility. With good and easy access, the business on production factors may also flourish. In turn, farmers will be able to obtain the necessary production factors more easily. This is closely related to the third priority in handling the problems, i.e. the provision of production factors. Production factors such as seeds, fertilizers, and chemicals are all things that must be available on farm, including rice farming because in addition to being a technical factor in determining the yield, it will also bring social and economic effects. In addition, the access roads will also help the marketing of agricultural products that can increase farmers' income.

The next priority is resolving problems of land tenure. Overlapping land tenure would pose a risk of social conflicts that would disrupt productive activities in the food estate. Therefore, the attempt to achieve "clean and clear" status on the DeKaFE land area is of necessity, one of the approaches is by certifying the land ownership/control.

Increasing the knowledge and ability of farmers to grow crops in tidal land is needed so that the available technology can be implemented properly. This can be done by carrying out periodic training for the farmers or provision of demonstration plots.

Although the other solutions have lower priority than the five solutions described above, the difference of their weights to each other is so small so that they can be considered having equal priority. Affordability of the production factors by farmers will make them more active in doing farming, as this will affect the effectiveness and efficiency of their production process. It has been proved that unavailability and/or unaffordability of production factors has resulted to low yield that in turns will also result in low farm earnings. Increasing affordability of production factors may be achieved by provision of credit schemes that suitable with characteristics of tidal farming. Land suitability mapping will help agricultural practitioners in determining the most suitable treatments to the land. Establishment of DeKaFE management authority is not included in high priority, but it is worth to be considered. Alternatively, synergies and coordination among authorized offices/agencies need to be better realized.

Based on the above-mentioned problems, problem solutions, and priority setting of the alternative solutions, a set of action plans which are considered as recommendations for the policy makers in the province of North Kalimantan was formulated, it is presented in table 2.

Table 2. Recommended action plans for increasing rice production at Delta Kayan Food Estate

Objective	Expected Outcome	Indicator	Action Plans	Inputs	Institutions in Charge
Improvement of water management	Established water management system in lowland tidal area	Acreage of productive irrigated land	1. Development of irrigation infrastructure networks (primary, secondary and tertiary canals, water gates) 2. Development of water management seasonal operational plan at tertiary blocks	1. Rice fields 2. Irrigation infrastructures 3. Experts on water management 4. Development budget	- Office of Public Works and Spatial Planning - Office of Agriculture
Improvement of accessibility of DeKaFE area	Developed access roads with reliable transportatio n services	1. Length of access road 2. Availability of transportatio n services	1. Development of farm roads 2. Development of inter-areas access roads 3. Operation of public transportation services	1. Road networks plan 2. Land for farm road and inter-areas access road 3. Man power 4. Development budget	- Office of Public Works and Spatial Planning - Office of Transportation Affairs
Improvement of production factors supply	Accessible production factors by farms	1. Availability of production factors 2. Price of production factors	1. Development of cooperative units 2. Development of village business units based on partnership scheme with private sectors	1. Transportation mode 2. Production factors 3. Business practitioners	- Office of Agriculture - Office of Industry and Trade
Strengthening financial capital institutions	Fulfilment of capital needs for farming	Acreage of planted area	1. Incentives for agricultural sector through credit scheme for farming 2. Farm insurances	1. Expert on financial capital institution 2. Financial capital institutions 3. Financial capital	- Higher Education Institutions - Business Institutions - Office of Agriculture
Improvement of farmers and farm labors skill and knowhow	Qualified and skilled human resources of rice cultivation in lowland tidal areas	1. Number of qualified/ skilled farmers 2. Number of innovation on appropriate technology	1. Extension and training 2. Development of demonstration plots 3. Research on rice cultivation in lowland tidal areas	1. Trainers/Extens ion Workers 2. Extension /training methods 3. Crop culture technologies 4. Researchers 5. Research Funding	- Higher Education Institutions - Office of Agriculture - Research Institutions
Assurance of "clear and clean" land tenures	"Clear and clean" land tenures	1. Number of operating corporate farming 2. Percentages of certified land	1. Iventarizing and planning of land ownership 2. Exemption of land tenure 3. Land certifications	1. Allocated land 2. Land Surveyors 3. Land maps 4. Certification fund	- National Land Agency - Office of Forestry - Office of Agriculture

Objective	Expected Outcome	Indicator	Action Plans	Inputs	Institutions in Charge
Optimization of land use with appropriate rice varieties	Increased yield and production	1. Acreage of planted area with appropriate rice varieties 2. Yield per hectare 3. Total annual rice production	1. Mapping of land suitability for rice 2. Provision and development of rice varieties suitable for low land tidal areas	1. Farmland 2. Expert on land suitability mapping 3. High yield rice varieties for lowland tidal areas 4. Development budget	– Office of Agriculture – Office of Public Works and Spatial Planning
Establishment of Delta Kayan Food Estate Management Authority	Effective food estate development and management	1. Number of operating corporate farming 2. Acreage of planted area with appropriate rice varieties 3. Total annual rice production	1. Establishment of Delta Kayan Food Estate Management Authority 2. Establishment of mechanisms and working procedures within the organization	1. Government policy 2. Experts on institutional development 3. Experts on management	– Provincial Government – Regency Government

4. Conclusions

Seven pillars of agricultural revitalization to improve food security in Indonesia consist of land consolidation, high seeds productivity, infrastructure, skilled labor, financial capital, institutional development, technology improvement, and downstream industry development. It is materialized through, among others, the development of food estates. Delta Kayan Food Estate in Bulungan Regency, North Kalimantan, was launched in 2011. It covers an area of about 50,000 hectares, most of them are fertile tidal lowland with alluvial soil types. Despite fulfillment of precondition for successful food estate, the high hope, and spirit, as well as efforts that have been made, the program has not shown encouraging results after six years of its commencement. A number of problems have been identified, including the overlapping land tenures, salt water intrusion, acid sulfuric soils, ineffective irrigation infrastructure, insufficient production factors, pest infestation, and lack of access to the neighboring community activity centers. To overcome the problems, it is recommended to improve water management, improve irrigation as well as transportation infrastructures, solve overlapping land tenures through certification, and to improve provision of production factors. Knowledge and capacity of the farmers need to be improved so that available technology can be applied properly. In lack of synergy among authorized agencies, the establishment of Delta Kayan Food Estate Management Authority is also recommended. A set of action plans as recommendations for the policy makers in the province of North Kalimantan has been formulated.

References

[1] Wardis G 2014 Food Security Systems in Indonesia: Challenges and Implications to Food Security in Maluku *Occasional Paper No. 54* (Kagoshima University Research Center for the Pacific Islands)

[2] Suswono 2011 The Indonesian Food Security in the Perspective of Global Economy and National Sovereignty. Paper presented at the International Conference on "The Future of Global Food Security and Safety: Issues and Justification", IPB International Convention Center, Bogor, 27 October, 2011

[3] Mawardi I 2007 Rehabilitation and revitalization of ex-peatland development projects at Central Kalimantan (in Indonesian) *Jurnal Teknologi Lingkungan* **8**(3): 287-297

[4] Moran S 2011 INDONESIA: MIFEE: The stealthy face of conflict in West Papua. Asian

Human Rights Commission. http://www.farmlandgrab.org/post/view/20801 (assessed on September 25, 2017)

[5] McCarthy J F and K Obidzinski 2015 Responding to food security and land questions: Policy principles and policy choices in Kalimantan, Indonesia. Conference Paper No. 47. Land grabbing, conflict and agrarian-environmental transformations: perspectives from East and Southeast Asia. An international academic conference, 5-6 June 2015, Chiang Mai University

[6] Saaty T L 2008 Decision making with the analytic hierarchy process *International Journal Services Sciences* **1**(1): 83-98.

[7] Imanudin S I, M E Armanto and R H Susanto 2009 Developing Seasonal Operation for Water Table Management in Tidal Lowland Reclamations Areas at South Sumatera, Indonesia *Journal of Tropical Soils* **16**(3): 2011: 233-244.

[8] Husson O, M T Phung and M E F Van Mensvoort 2000 Soil and water indicators for optimal practices when reclaiming acid sulphate soils in the Plain of Reeds, Viet Nam *Agricultural Water Management Journal* **45**: 127-143.

[9] Saragih S I Ar-Riza and M Noer 1996 *Some Alternative Cropping Patterns Support the Optimization of Tidal Land Utilization for Food Crops. Proceedings of Seminar on Swamp and Dry Land Farming System Technology* (in Indonesian) (Banjarbaru: Balittra)

IOP Conf. Series: Earth and Environmental Science **147** (2018) 012044 doi:10.1088/1755-1315/147/1/012044

Framework for Design of Traceability System on Organic Rice Certification

P B Purwandoko[1], K B Seminar[1*], Sutrisno[1] and Sugiyanta[2]

[1] Department of Mechanical and Biosystem Engineering, Bogor Agricultural University, Bogor, Indonesia
[2] Department of Agronomy and Horticulture, Bogor Agricultural University, Bogor, Indonesia

E-mail: seminarkudangboro@gmail.com

Abstract. Nowadays, the preferences of organic products such as organic rice have been increased. It because of the people awareness of the healthy and eco-friendly food product consumption has grown. Therefore, it is very important to ensure organic quality of the product that will be produced. Certification is a series of process that holds to ensure the quality of products meets all criteria of organic standards. Currently, there is a problem that traceability information system for organic rice certification has been not available. The current system still conducts manually caused the loss of information during storage process. This paper aimed at developing a traceability framework on organic rice certification process. First, the main discussed issues are organic certification process. Second, unified modeling language (UML) is used to build the model of user requirement in order to develop traceability system for all actors in the certification process. Furthermore, the information captured model along certification process will be explained in this paper. The model shows the information flow that has to be recorded for each actor. Finally, the challenges in the implementation system will be discussed in this paper.

1. Introduction

Currently, there has been a paradigm shift in consuming food. The existences of chemical use in the process producing of food cause a healthy lifestyle awareness develop in society. A healthy lifestyle that has been grown requires the food must have characteristics to be safe for consumption, to be nutritious, and eco-friendly [1, 2]. Therefore, the production and consumer preferences of organic products continue to increase. According to Ariesusanty [3], there is so many organic products that has been developing in Indonesia, one of them is on organic rice commodity. Furthermore, in 2010 300 ha area in Indonesia was found producing the certified organic rice. The area expanded in 2013. As a result, in 2014 the organic rice products dominated the organic markets in Indonesia.

Along with the development of organic products, then the government tries to assure the quality of organic products by certifying it. The assessment in certifying of organic products not only based on the final product but also the production process which starts from cultivation practices until the final product distribution [4]. This policy is regulated by the National Standard of Indonesia (SNI) which explain that certification is a method that the organic certification organization conduct a series of continuous inspection activities, audit system, and final product inspection in order to make sure the organic product fulfill the organic quality requirements [2]. But actually, there is still an obstacle in certification of organic rice that is there has been no certification information system which is able to trace various activities in the process producing organic product. The existing system still manually to record the data, so it can cause loss of information during storage process and does not utilize to store information in big data.

The traceability development has got the attention for the last few years. Some factors have raised the interest in developing traceability system in the food sector. The existence of several issues and cases that relate to food safety has raised the consumer insecurities. Besides that, consumer needs qualified, nutritious, and non-GMO (*Genetically Modified Organism*) food materials, and also the special products like organic products have developed for the last few years [5]. Therefore, the regulation of implementing traceability system in the food producing and processing is being developed in many countries. According to European Union law, traceability system is defined as an ability to trace and follow the raw material of food, feed, and food-producing animal or substance that are used in all production process until distribution. This system enables the stakeholders or authorities to recall if the product is not safe for consumption and not according to the applicable provision.

Implementation of a traceability system is an integral part of food production chain. Primarily, used to respond the food safety cases, record production processes, fulfill the government regulations, and analyze production and logistics cost [6]. However, in its development, the developed traceability system was varied and adjusted to user necessity [7]. One of its application, traceability can be used to monitor process and product based on microbiological analysis in tuna loin production, and also can be used to monitor and prevent food loss and waste along food production [8, 9]. According to the description, the objective of this paper was to develop the basic concept of traceability information system that facilitating the certification process of organic rice. The first step taken is to identify the usage requirement of traceability information system in certification process from all actors. Furthermore, system requirement was modeled using *unified modeling language* (UML) use case diagram.

2. Organic product certification

Organic certification is one of the implementation systems and products control for organic products. All the stakeholders who play a role in a process of producing organic products including the farmers, the processors, and the organic products companies have to conduct a certification in order to maintain the integrity of organic products. The certification process is conducted by a government certification organization or a private certification organization that is acknowledged by the government. According to Indonesia Ministry of Agriculture [4], the certification process of organic products is generally shown in figure 1.

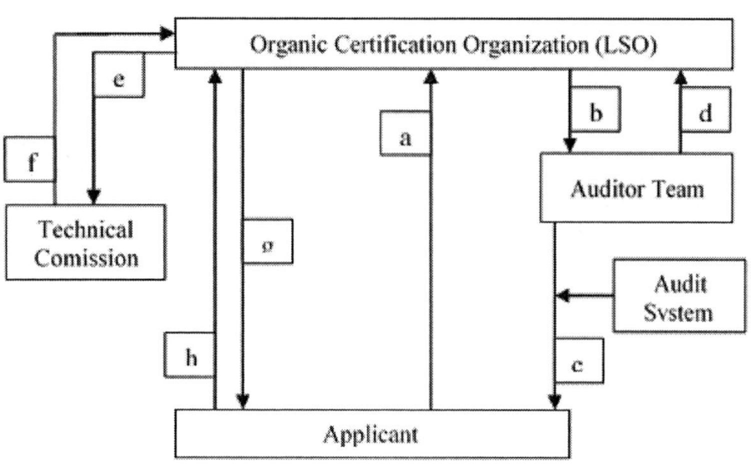

Figure 1. Organic certification process

The certification process diagram of organic production system can be explained below :

(a) The organic farm producers submit a certification proposal to the organization of organic product certification.
(b) The certification organization appoints an audit team.
(c) The audit team conducts auditing process of document adequacy.
(d) The audit team delivers the audit result to the certification organization.
(e) The certification organization delivers the audit results to the certification committee for being discussed in an internal meeting in order that some recommendations can be given.
(f) The certification committee delivers the recommendations to the certification organization.
(g) The certification organization delivers the assessment results, whether the applicant deserves the certification or not.
(h) The organization of organic product certification periodically conduct a surveillance.

3. Methodology

The basic requirement to develop an effective traceability system is to determine the information that needs to be tracked. Therefore, the information about organic rice production process was needed. The analysis was conducted in a whole of involved stakeholder. This section provides the information about the methodology of data collection on preliminary research.

- Field survey

Organic rice is one of the commodities that mostly developed in Indonesia. The field survey was conducted in January and August 2017 in Cisayong Sub District, Tasikmalaya District, the center of organic rice production in West Java. Field survey was aimed to (1) understand the structure of the production chain of organic rice, identify the involved actors and the interactions in it and to (2) map the needed information at each stage of production.

- Interview and document analysis

To collect the data was also conducted by a semi-structured interview with ICS (internal control system), the chairman of farmer group association (GAPOKTAN), and distributor of organic rice. Interviews were conducted to obtain detailed information about the production process and the other data about certification purposes. In addition, documents regarding the relevant information should be recorded.

4. Usage requirements of traceability organic certification

In the development traceability information system as a certification process, the main thing required is the ability of the system to save and communicate several important documents such as cultivation, processing, quality of the product, and the others. Therefore, it is important to define the terms of usage requirements for all actors that involved in the certification process. A system approach was used to develop a certification information system. The requirement of system usage was developed using UML (*unified modeling language*) use case diagram. According to Urva and Siregar [10], the designing of use case diagram able to give the description of involved actors, functions, and also interactions. In the use case diagram, there are 4 elements that can be identified. The explanation of each element described as follow:

- Use case: A use case describing the action shown by system that generates the measurable result for actors, mostly symbolize in ellipse form.
- Actor: Describing the person who has interaction with the system. Actor can be individual or organization and symbolize as stick figures.
- Associations: Associations describe the interaction between actors and use case, and symbolize as a solid line.
- System boundary: System boundary shows the scope of the system. In this study, a rectangular form is used as symbol for boundary system.

According to Lee and Xue [11], the designing of use case has advantages to help the complexity management since focusing on one specific usage at a time. Figure 2 shows the diagram of a use case for traceability information system in organic rice certification process. The following use case examples are defined and different actors are associated with each use case:

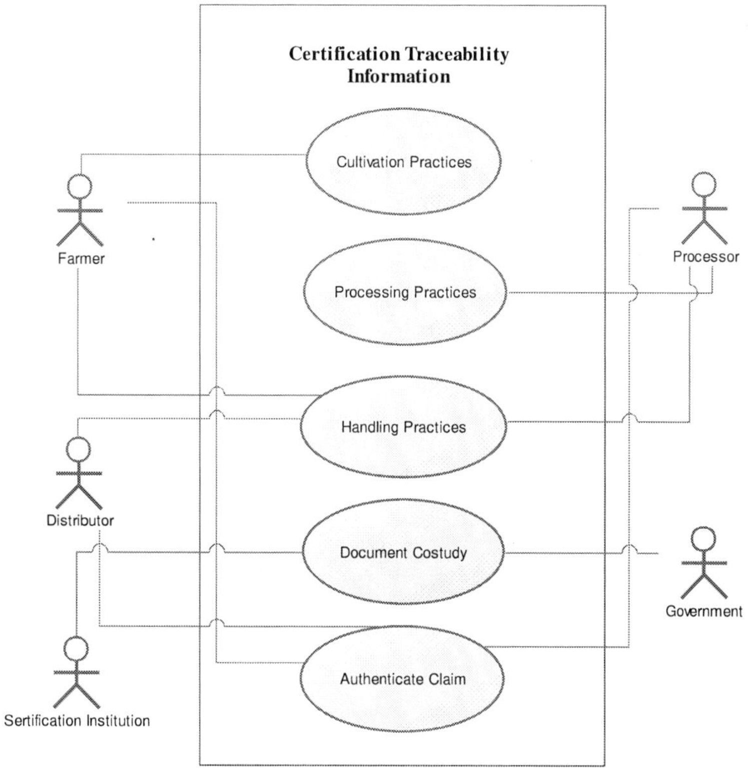

Figure 2. Traceability system in certification use case diagram

- Record of cultivation: Farmers recorded all activities during paddy cultivation. The data such as varieties used, irrigation system, plantation, fertilization process, pest and disease of plant, and etc were recorded in the system.
- Record of handling: Material handling practices such as drying, storage and etc that conducted by stakeholder, start from farmers, processor and distributor should be well recorded.
- Record of processing: Processor must record all of the processing. Several recorded parameters in processing practices are processed varieties, quantity, and quality of product.
- Authenticate claim: All actors who play a role in organic rice production able to take the data and claim their activities that conducted has been appropriate with the determined organic standard based on storage system.
- Document study: Based on the demand, the certification system should give an information holistically about the certification actors who has the role in organic rice production.

5. Information capture model

The traceability system was developed to solve the abnormal situation, create the transparency in production process and give the information to consumer and a whole of involved stakeholder including food certification institution [12]. Therefore, for developing and implementing this system, a whole information about raw material origin, production and processing, product quality, packaging material and etc is required. The developed traceability system for organic rice certification process begins from cultivation practices until whole process based on three phases shown in figure 3.

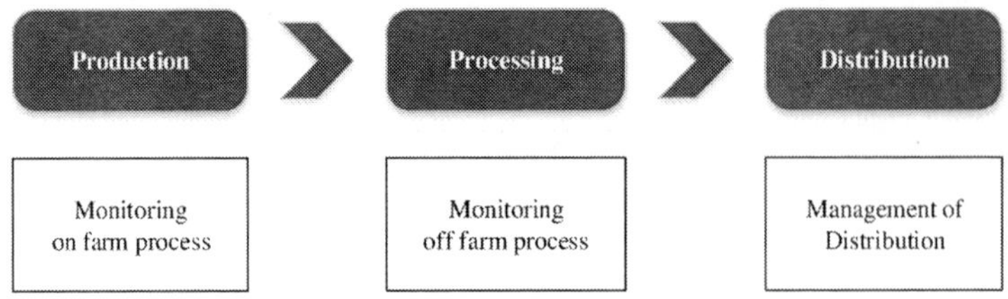

Figure 3. Organic rice traceability timeline

The development of traceability system currently varies and it is adjusted to its application purpose. In this case, traceability is used to track the internal information about the organic rice production process. The ability to trace internally can be defined as the ability to track the production information internally to ensure that the production process conforms to the organic requirement from upstream to downstream. There are three categories of information that should be recorded by each stakeholder in the organic rice production chain, those are process information, product information, and quality information. The method to capture information for each stakeholder can be different. Figure 4 shows the information model that should be captured in the organic rice processing based on three phases.

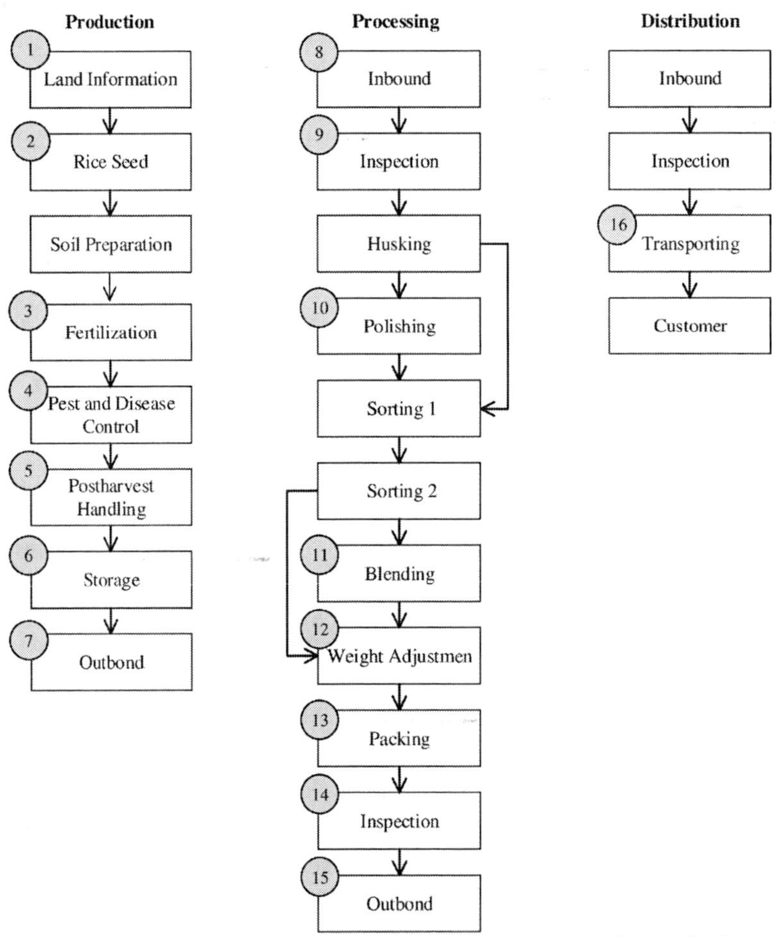

Figure 4. Information captured in process producing of organic rice

The input and output of each process are shown by the model presented above. Information captured at each stage is numbered. These numbers represent the general information that should be captured in the organic rice production chain. The general points of information captured point were identified through the preliminary study based on descriptive and qualitative analysis in a whole of stakeholders that involved in the process of producing organic rice in Tasikmalaya District. This information can be the basis for certification institution to guarantee the organic requirement of the products. Furthermore, table 1 shows the specific information that should be recorded in each point.

Table 1. Specific information capture point

Information capture point	Key Point	Information
1	Land information	Block number, area, water source, water treatment, land status
2	Seed	Variety, seed source, seed treatment, seed status, seeding date, quantity, quality, transplanting date
3	Fertilizer	Name, source, composition, method, dose, application date
4	Pest and disease control	Pest and disease species, treatment type, total amount, rate of application application date
5	Postharvest handling	Harvest date, handling status, worker, wet production, dry production
6	Storage	location, storage date
7	Outbond	Consumer ID, date of sale, quantity
8	Inbound	Supplier ID, quantity, lot number, storage location, date entry
9	Inspection	Moisture content, foreign materials, damaged materials, empty grain, called grain
10	Polishing	Milling ID, milling date, quantity, degree of polishing
11	Blending	Mixing ID, composition
12	Weight adjustment	Product weight
13	Packing	Packing material, packing supplier
14	Inspection	Moisture content, whole kernels, medium grains, small grains, foreign material, chalked grains, paddy grains, chemical residues, nutritional content
15	Outbond	Consumer ID, date of sale, quantity
16	Transporting	Distribution ID, Vehicle ID, Customer ID, Quantity

Currently the implemented traceability system in the certification process is still conducted manually by a government certification organization or a private certification organization. During the on-farm production and off-farm stage, the required information was recorded in an activity logbook. This is time-consuming and inefficient. Further, manual recording was vulnerable to loss of data during storage. In the development of this traceability system, relational databases are used to implement the system. In the system, there is only one database for the entire production phase. Users can enter all relevant data from on-farm until distribution process. The database system then connects to data entry, performed activities and other parts that recorded. Furthermore, this data can be accessed by the certification institution to ensure that all processes performed appropriately with organic requirements.

6. Challenges

Currently, the implementation of traceability system in food production is one of the basic requirements that must be applied almost throughout the country. It is because the traceability system can be a tool for risk management that allows stakeholders and policy makers to withdraw the unsafe food products. Furthermore, the traceability system is also important because it serves as a tool to document the chain and production processes, as regulatory requirements, as transparency processes for consumers, and as a tool to analyze logistics and production costs. The absence of a traceability system will lead the un-transparent of production process and cause the absence of a stakeholder's ability to identify and isolate the errors of production and the source of contamination in case of product withdrawal from the market [13]. However, in the implementation of a traceability system, several problems will be faced, such as:

(a) Human resource
The availability of qualified human resources is the success key in system implementation. Complexity of traceability system requires expert human resource to develop, to implement and to manage the system. Several stakeholders who has different goals will make the traceability become more complex. Stakeholder who plays a role may have insufficient human resource and skills to implement and manage the traceability system effectively [14, 15].

In the developing country, such as Indonesia, quality of human resource in agricultural, especially in farmer level is very low. Most farmers have low levels of formal education or do not complete basic education. Consequently, the ability to absorb the information and to adopt technology is limited. Furthermore, in the case of small and medium industries, the traceability system can lead to more workload due to the limited capacities of employed human resources [14]. Therefore, it is necessary to improve the work competence through training.

(b) Financial resource
Develop and implement the traceability system is costly and it can cause financial problems [16]. Therefore the cost of implementation and maintenance the traceability system is an important aspect to consider due to the complexity of the production process. This complexity because of several requirements, such as: (i) traceability must be capable to handle both internal and external traceability in the information exchange process, (ii) traceability must have the ability to track the food one step forward and one step backward, (iii) the built system must be effective in its use, (iv) the traceability system must be expandable to reach the new relevant information [12].

(c) Lack of awareness
The main purpose of the traceability system is to record and to document all activities in the production process from upstream to downstream. Consequently, the stakeholders assume it as an additional burden in production. Lack of knowledge and information about the benefits of traceability and the absence of compatibility with the stakeholders who play a role are the problem in the development and implementation of the system [17].

(d) Limited information
Traceability in food production processes is closely related to the uncertainty so that difficult to obtain certain information completely, accurately, timely, and easily accessible. Whereas according to European legislation on food traceability, EC 178 (2002), that the traceability system should be able to track in all stages of production from cultivation, processing to distribution. In fact, however, the obtained information sometimes incomplete, for example, the available information only focused on the origin of the material, not on the food quality and food security [18].

(e) Limited standard
Lack of adequate standards and tools in data exchange is a problem that must be faced in system implementation. The complexity of traceability occurs due to variations in data retrieval, inconsistencies of captured data types, data variations, and lack of standardization in data transmission among stakeholders [19]. Furthermore, the variations and lack of standardization of traceability

techniques in food production processes such as numerical codes, barcodes, and RFID also contributed to the complexity [20].

(f) Managing change

Maintaining the traceability information through the changes in the system is a challenge that must be faced. The study shows that changes can occur throughout the life cycle software. This change resulted in the system to update the traceability data. Therefore, stakeholders in the system need to update the data regularly. In fact, however, strong discipline to maintain the accuracy and completeness of traceability data is seldom conducted so that in practice, many organizations ignore it. Furthermore, discipline can be nurtured through the training in understanding the importance of traceability [21].

Lately, the development of a traceability system is required because various cases of food safety often occur and the transparency process needs to be realized. In the development and implementation process, some problem may be found. Several points that can reduce these problems are (i) developing the standardization, regulation, and strength law in the development and implementation of traceability system, (ii) participation and support of government as policymakers, (iii) increasing the knowledge and awareness of the community, (iv) support of all stakeholders who involved in food production, and (v) good coordination among the stakeholders and policymakers [22].

7. Conclusions

Developing a data management system is required to facilitate the traceability process. In Indonesia, these current system is still limited, where consumers can only access general information about the product. Traceability system is also needed to realize the transparency process. To achieve the traceability, all production processes must be properly recorded. Determining the term of use of the traceability system are the first step in system implementation. Moreover, to enable the effective information exchange is necessary to collect the data models. This study provided a model for information that must be captured by institution certification of organic product in three main phases i.e production, processing, and distribution. The points of information captured are identified in each phase. Furthermore, developing and implementing the traceability system for all stakeholders is quite difficult. Therefore, the strategy of system implementation is needed, one of which is by developing standardization, regulation, and force of law regarding the development and implementation of the traceability system.

Acknowledgments

The authors would is grateful to Ministry of Higher Education and Research Technology (RISTEKDIKTI) through Master Program of Education Leading to Doctoral Degree for Excellent Graduates (PMDSU) for the scholarship and their support in all possible ways.

References

[1] Idaman N, Yuliati L N and Retnaningsih R 2014 Costumer attitudes against organic rice (in Indonesia) *Jurnal Manajemen & Agribisnis* **9** 117-126

[2] Mayrowani H 2016 Organic rice development in Indonesia (in Indonesia). In: *Forum Penelitian Agro Ekonomi* pp 91-108

[3] Ariesusanty L 2014 *Organic Agriculture Statistic in Indonesia* (in Indonesia) 2014 (Bogor: Aliansi Organis Indonesia)

[4] Indonesia Ministry of Agriculture 2016 *Technical Instruction Certification Falitilation of Organic Rice* (in Indonesia) (Jakarta: Direktorat Jenderal Tanaman Pangan)

[5] Thakur M, Martens B J and Hurburgh C R 2011 Data modeling to facilitate internal traceability at a grain elevator *Computers and electronics in agriculture* **75** 327-336

[6] Thakur M and Hurburgh C R 2009 Framework for implementing traceability system in the bulk grain supply chain *Journal of Food Engineering* **95** 617-626

[7] Dwiyitno D 2009 Tracebility system implementation for fishery product (in Indonesia) *Squalen*

Bulletin of Marine and Fisheries Postharvest and Biotechnology **4** 99-104
[8] Kresna B A, Seminar K B and Marimin M 2017 Developing a Traceability System for Tuna Supply Chains *International Journal of Supply Chain Management* **6** 52-62
[9] Seminar K B 2016 Food Chain Transparency for Food Loss and Waste Surveillance *Journal of Developments in Sustainable Agriculture* **11** 17-22
[10] Urva G and Siregar H F 2017 Pemodelan UML E-Marketing Minyak Goreng *Jurteksi (Jurnal Teknologi dan Sistem Informasi)* **1** 92-101
[11] Lee J and Xue N-L 1999 Analyzing user requirements by use cases: A goal-driven approach *IEEE software* **16** 92-101
[12] Hwang Y-M, Moon J and Yoo S 2015 Developing a RFID-based food traceability system in Korea ginseng industry: Focused on the business process reengineering *International Journal of Control and Automation* **8** 397-406
[13] Narshimalu U, Potdar V and Akur A P 2015 A case study to explore influence of traceability factors on Australian food supply chain performance *Procedia Social and Behavior Science* **189** 17-32
[14] Engelseth P 2009 Food product traceability and supply network integration *Journal of Business & Industrial Marketing* **24** 421-430
[15] Zhang X, Zhang J, Liu F, Fu Z and Mu W 2010 Strengths and limitations on the operating mechanisms of traceability system in agro food, China *Food Control* **21** 825-829
[16] V. Kher S, J. Frewer L, De Jonge J, Wentholt M, Howell Davies O, B. Lucas Luijckx N and J. Cnossen H 2010 Experts' perspectives on the implementation of traceability in Europe *British Food Journal* **112** 261-74
[17] Bosona T and Gebresenbet G 2013 Food traceability as an integral part of logistics management in food and agricultural supply chain *Food control* **33** 32-48
[18] Ackerley N, Sertkaya A and Lange R 2010 Food transportation safety: characterizing risks and controls by use of expert *Food Protection Trends* **30** 212-222
[19] McMeekin T, Baranyi J, Bowman J, Dalgaard P, Kirk M, Ross T, Schmid S and Zwietering M 2006 Information systems in food safety management *International Journal of Food Microbiology* **112** 181-194
[20] Salampasis M, Tektonidis D and Kalogianni E P 2012 TraceALL: a semantic web framework for food traceability systems *Journal of Systems and Information Technology* **14** 302-317
[21] Kannenberg A and Saiedian H 2009 Why software requirements traceability remains a challenge *Defense Software Engineering* 14-19
[22] Duan Y, Miao M, Wang R, Fu Z and Xu M 2017 A framework for the successful implementation of food traceability systems in china *The Information Society* **33** 226-242

IOP Conf. Series: Earth and Environmental Science **147** (2018) 012045 doi:10.1088/1755-1315/147/1/012045

Estimation of Soil Nitrate (NO_3) Level Using Laser-Induced Breakdown Spectroscopy (LIBS)

A R Angkat[1], K B Seminar[1], M Rahmat[2] and A Sutandi[3]

[1] Department of Mechanical and Biosystem Engineering, Bogor Agricultural University, Indonesia
[2] Department of Mechanical Engineering, Ibnu Khaldun University, Indonesia
[3] Department of Soil and Land Resources, Bogor Agricultural University, Indonesia

E-mail: rarangkat80@gmail.com

Abstract. Laser-Induced Breakdown Spectroscopy (LIBS) is a method for measuring level of nitrogen (N) in the soil in the form of N-nitrate (NO_3) rapidly without going through the process of sieving and drying. The sample soil in the form of pellets subjected to laser pulses using a wavelength of 532 nm, pulse duration 5.5 ns, repetition rate of 10 Hz, and Q-switch delay of 150 μs. Emissions are captured by the spectrometer with the wavelength range of 190-1130 nm. Spectrum characterization was processed through the second derivative in order to obtain a wavelength identity that could be rapidly used to estimate the nitrate content of the soil with a determination coefficient of (R^2) 0.9254 and a coefficient of variation (CV) of 8.41%. The results of this study are very potential to be applied for rapid measurement of soil nitrate.

1. Introduction

Laser-Induced Breakdown Spectroscopy (LIBS) is a method for measuring content of nitrogen (N) in the soil in the form of N-nitrate (NO_3) rapidly without going through the process of sieving and drying, providing a fertilizer recommendation which is immediate and specific locations in order to implement precision farming.The conventional soil testing soil analysis are relatively expensive and took a long time, in addition the availability of soil testing laboratories are very limited. This led to generalization and uniformity of crop fertilizer throughout Indonesia archipelago [1].

Tools or methods that can measure the levels of nitrogen in the soil immediately and cheap is needed today in order to provide specific fertilization recommendations. According to Yoshida [2], the plants can absorb and use various forms of N such as: N-ammonium-nitrate N, N-urea and N-amino acids. N-ammonium is the form of the N most prevalent and has stable form in flooded soils (wetland),while the N-nitrate is most often found on dry land (upland).

One way to determine the nitrogen content in the soil immediately is using electromagnetic waves (UV-Vis-NIR) which can interact with atoms and chemical molecules electromagnetically. Ehsani et al. [3] determined the levels of nitrogen in the soil in the form of ammonium and nitrate compounds using NIR with a wavelength of 1800-2300 nm. Russell et al. [4] used the NIR wavelength range of 350-2500 nm to measure the uptake of nitrogen in a single season rice crop. While Hernandez et al. [5] used NIR to measure total nitrogen and total carbon in wetlands in Ohio United States. Determination of nitrogen in corn crop soil marks using Vis-NIR wave in the wavelength range of 400-2500 nm carried out by Kusumo et al. [6]. Mahmood et al. [7] measured the total of N content in soil samples (0-30 cm depth) using a wavelength of 350-2500 nm by drying in an oven at a temperature of 40 °C for 72 hours and sieved using a 2 mm sieve. Recent research conducted by Abdi et al. [8] further reaffirmed that the NIR could predict precisely the total nitrogen and nitrogen available in the soil. In another study, Xuemei et al. [9] measured the content of organic matter, nitrogen, phosphorus and potassium in the soil using Vis-NIR wave at a wavelength of 325-1075 nm by drying and sifting the ground using 1 mm sieve. Other than the dry land, research that measured levels of nitrogen in paddy fields is done by Wenjun et al. [10] using a wavelength of 350-2500 nm with a sampling soil from fields that have been dried for 10 days and sieved using a 2 mm sieve. Until now, studies to determine the composition of the nutrient content in the soil using NIRS in Indonesia is only done by Angkat [11] at a wavelength of 1000-2500 nm as well as determined a single wavelength 1506 nm to predict soil N content in the form of N total.

All methods that have been done are still in the development phase, that makes this study has not yet realized in a measurement tool that can be used in the field. This condition caused by several things such as expensive equipment as well as the high level of difficulty in the research and development of these methods. So that there is still room to determine new methods that are easier to be realized as a measuring tool in the field, one of them using the technique of Laser-Induced Breakdown Spectroscopy (LIBS).

Laser-Induced Breakdown Spectroscopy (LIBS) is a potential method to be developed to measure levels of N in the soil. LIBS is useful for determining the elemental composition of materials in solid, liquid and gas by focusing a high power laser pulse at the sample to create a spark plasma or laser to be captured by the detector. LIBS method provides an opportunity to measure the levels of nitrate (NO_3) in the form of solid or soil samples as the original form field without going through the process of drying and sieving. Although the soil samples used in several studies using LIBS methods still going through drying and sieving process, such as research conducted by Yang [12], Knadel et al [13] and Haddad [14]. In general LIBS produces a spectrum UV-Vis and NIR fraction. UV-Vis spectrum can be used to determine the characteristics of the atomic wave to the N element. This process is difficult to do since NIR is molecular and very sensitive to H_2O to verify the elements of N in the form of chemical bonds, because the chemical bonds of N in the soil are very diverse.

This study aimed to characterize the soil samples LIBS spectra of the N elements without sifting and drying, so it has some single wavelengths that interact electromagnetically with atoms or molecules of N in the soil to build a prediction model of N levels in the soil in the form of N- nitrate.

2. Materials and methods

2.1. Samples preparation

The sample for the characterization of the spectral UV-Vis-NIR soil to the element N is 10 soil samples weighing 0.5 kg that taken at a depth of 0-30 cm composite of 10 different pick-up points on dry land. The soil samples were made in the form of pellets with a diameter of 3.2 cm and 0.5 cm thick with a pressure of 4245.61 psi using a manual hydraulic press without drying and sieving the soil.

2.2. Setup of LIBS and experimental steps

The spectra soil of UV-Vis-NIR is collected through Laser-Induced Breakdown Spectroscopy (LIBS) using Nd-YAG Laser Q-Smart 850 model which resulted in a wavelength of 1064 nm, 532 nm and 266 nm. In this study, a soil sample in pellet form subjected to laser pulses using a wavelength of 532 nm, pulse duration of 5.5 ns, repetition rate of 10 Hz, the Q-switch delay of 150 μs and total energy of 15 mJ/pulse. Before subjected to the laser, the soil sample was put into a vacuum chamber with pressure of 5.75 torr. The laser will pass through the focusing lens with a focal length of 10 cm. Soil samples were subjected to laser pulses will release a plasma caused by the collapse of chemical bonds plasma chemical bonds resulting in ionization process. Plasma containing information of the elements contained in the soil sample including its concentration will be collected through the collecting lens and lens filter 532 nm which captured by the detector to measure the light intensity in the wavelength range of 190 nm - 1130 nm.

2.3. *Data processing*

2.3.1. Characterization spectrum UV-Vis-NIR against the elements of soil N. Spectrum characterization of UV-Vis-NIR soil against the N elements in the soil is done by identifying the peaks wave of UV-Vis-NIR with a specific wavelength that has electromagnetic interaction with the N in the soil based on the manual book written by Sansonetti [15]. Characterization spectrum of UV-Vis-NIR soil is done on the original spectrum, the first derivative spectrum, and the second derivative spectrum through the identification of peaks spectrum at certain wavelengths according to the manual book.

2.3.2. Determination of the soil nitrate (NO_3) level estimation model. Soil nitrate (NO_3) content estimation model is built using multiple linear regression. The intensity of a specific wavelength that emerges from the process of characterizing the spectral UV-Vis-NIR in the original spectrum, first derivative spectrum, and the second derivative spectrum will be the independent variables. Wavelength intensity value is used as an independent variable in the original spectrum, first derivative spectrum, and the second derivative spectrum is the intensity of a specific wavelength in the original spectrum based on the characterization of the spectral UV-Vis-NIR. The concentration of NO_3 in soil samples will be the dependent variable. The coefficient of determination (R^2) and coefficient of variability (CV) will be the parameters of the model built to measure the concentration of NO_3 soil. If the value of R^2 is getting closer to 1 the changes of NO_3 level in soil will cause changes in the content of NO_3 amount, otherwise, if the CV value is declining close to 0, thus the data generated will be more homogeneous.

3. Results and discussion

3.1. *Characterization of spectrum UV-Vis-NIR of the elements of soil N*

The detector captures the wave beam in the form of plasma in the range of 190-1130 nm. The spectrum is obtained in the range of UV (130-400 nm), Vis (400-750 nm) and NIR (750-1130 nm). Certain wavelengths that have strong electromagnetic interaction of the elements of N is verified based on wave peaks that appear based on the manual book written by Sansonetti [15]. Identification of the emerging wave peak produced 3 wavelengths on raw spectrum, i.e. 463.054 nm, 567.602 nm and 843.874 nm (figure 1).

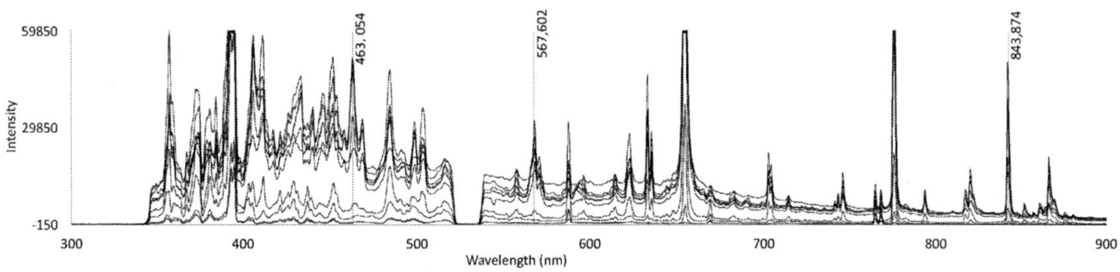

Figure 1. Raw UV-Vis-NIR spectrum.

Identification of wave peaks is also performed at the first derivative spectrum from the raw spectrum to see specific wavelengths that have interactions with atoms or molecules of N soil. First derivative spectrum can be used to remove the effects of additive baseline [16]. First derivative spectrum can reduce the effects of baseline and improve absorption peaks (figure 2). Normalization is used to reduce the variations of baseline in order to identify the wave peaks (figure 3). Normalization spectrum obtained by scaling each of spectrum in the range of 0 and 1 [17]. The identification of wave peaks in the first derivative spectrum generates 5 wavelengths i.e. 395.585 nm, 499.436 nm, 504.510 nm, 571.077 nm and 862.924 nm.

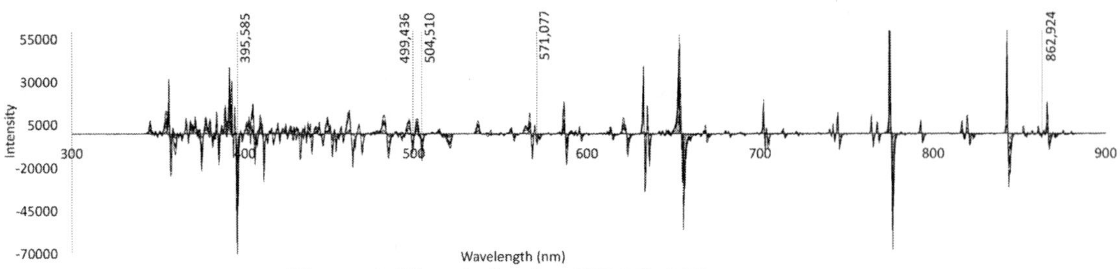

Figure 2. First derivative UV-Vis-NIR spectrum.

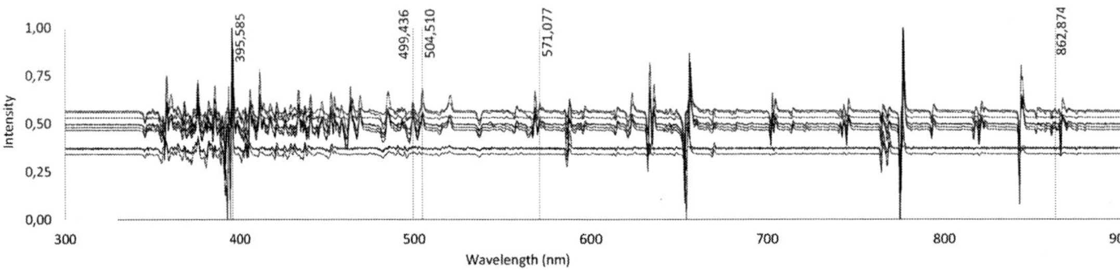

Figure 3. First derivative UV-Vis-NIR spectrum normalization.

The same principle is also applied on the second derivative spectrum to increase the absorption peaks in order to identify peak-peak waveform appears (figure 4). The second derivative to dump slope additive baseline [16]. Normalization is also performed on the second derivative spectrum to eliminate variations of baseline (figure 5). Identification of the wave peak in the second derivative produces 8 pieces wavelength, i.e. 399.500 nm, 444.703 nm, 463.054 nm, 496.398 nm, 567.480 nm, 742.364 nm, 822.314 and 843.874 nm.

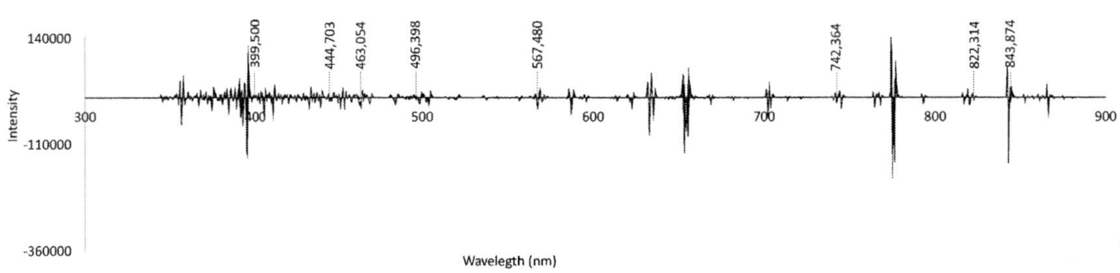

Figure 4. Second derivative UV-Vis-NIR spectrum.

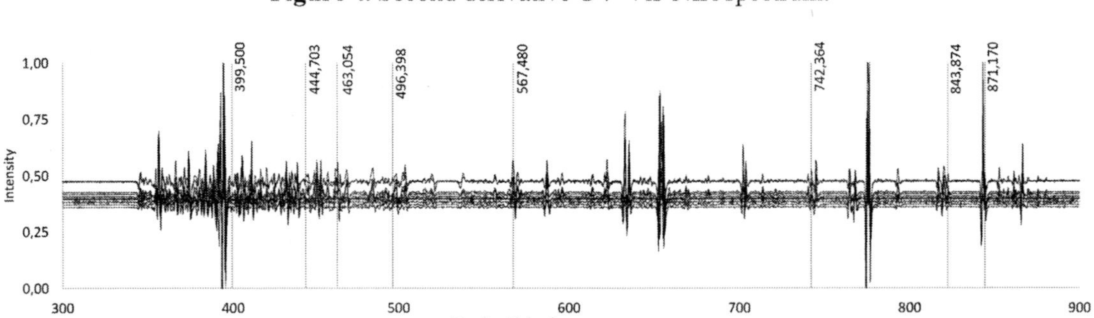

Figure 5. Second derivative UV-Vis-NIR spectrum normalization.

3.2. Nitrate (NO_3) Level Estimation Model

The NO3 level estimation model is built using multiple linear regression model with intensity value from a specific wavelength as an independent variable. NO_3 concentration values are collected through the calculation of soil laboratory as the dependent variable. Minitab software is used as a tool to process the data to build a prediction model equations levels of NO_3 soil. The best prediction model based on coefficient of determination (R^2) and coefficient of variation (CV).

In the raw spectrum uses three variable independent which is the intensity value of the wavelength 463.054 nm, 567.602 nm and 843.874 nm 10 locations soil samples to produce the following equation :

$$NO_3 = 52.84\ to\ 0.00211\lambda_{463.054} + .00557\lambda_{567.602} - 0.00115\lambda_{843.874} \quad (1)$$

While in the first derivative spectrum using 5 independent variables which is the intensity value of the wavelength 395.585 nm, 499.436 nm, 504.510 nm, 571.077 nm and 862.924 nm of 10 soil samples to produce the following equation :

$$NO_3 = 51.1 + 0.000936\lambda_{395.585} - 0.00568\lambda_{499.436} - 0.00496\lambda_{504.510} + 0.0180\lambda_{571.077} - 0.0116\lambda_{862.924} \quad (2)$$

Then second derivative spectrum uses 8 independent variables is the value of the intensity of the wavelength 399.500 nm, 444.703 nm, 463.054 nm, 496.398 nm, 567.480 nm, 742.364 nm, 822.314 and 843.874 nm of 10 samples of soil which produces the following equation :

$$NO_3 = 57.8 + 0.0119\lambda_{399.500} + 0.0238\lambda_{444.703} - 0.00358\lambda_{463.054} - 0.0152\lambda_{496.398} - 0.00087\lambda_{567.480} - 0.0332\lambda_{742.364} - 0.0487\lambda_{822.314} + .00895\lambda_{843.874} \quad (3)$$

Where λ is the wavelength identity intensity. equation 1, 2 and 3 are used to calculate the value of NO_3 estimated for comparison with NO_3 measured. The relationship between NO_3 measured, and NO_3 estimated presented in figure 6.

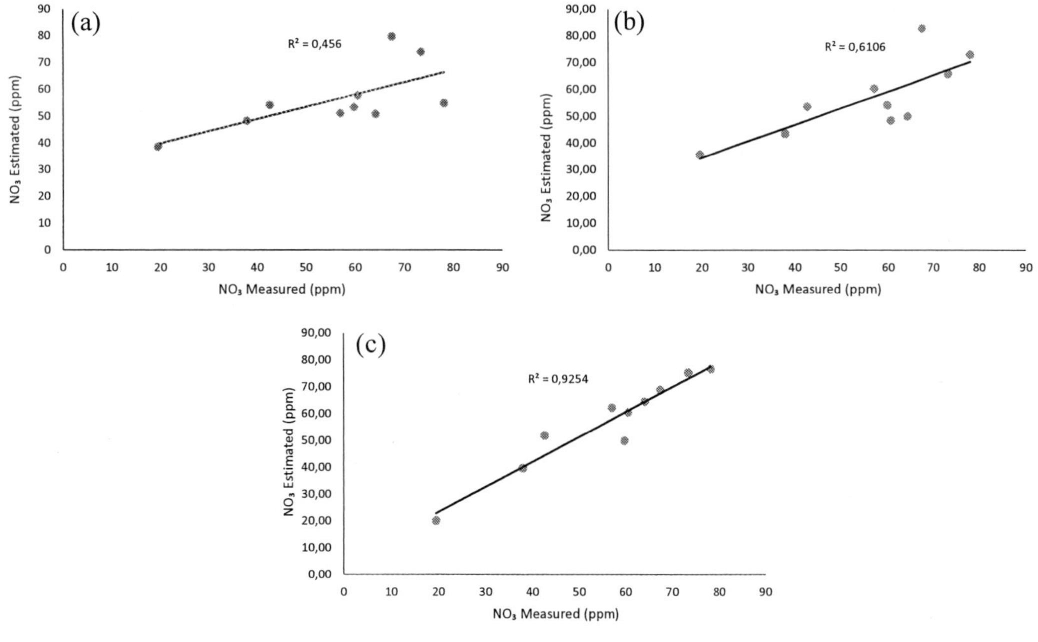

Figure 6. The relationship between the measured and the estimated NO_3 in the (a) raw spectrum, (b) first derivative spectrum and (c) second derivative spectrum.

The results show the value of R^2 is the highest in the second derivative spectrum of 0.9254 with the smallest CV value of 8.41%. So the best model to estimate the levels of NO_3 ground using equation 3 with the intensity at a wavelength of 399.500 nm, 444.703 nm, 463.054 nm, 496.398 nm, 567.480 nm, 742.364 nm, 822.314 and 843.874 nm as independent variables.

4. Conclusions

Second derivative UV-Vis-NIR spectrum has more wave peaks that consist of electromagnetic interaction with soil N elements. The level of NO3 in the soil can be estimated by value intensity of wavelength 399.500 nm, 444.703 nm, 463.054 nm, 496.398 nm, 567.480 nm, 742.364 nm, 822.314 and 843.874 nm on soil nitrate estimation equation model in the second derivative spectrum.

Acknowledgements

The author would like to thank the Indonesian Ministry of Agriculture for the Doctoral study program scholarship, and the Center for Physical Research, Indonesian Institute of Sciences for assistance in the experiment using LIBS at the Center for Physical Research.

References

[1] Subiksa I G M, Ladiyani R W and Setyorini D 2007 *Paddy Soil Test Kit* (Bogor: Soil Research Center)
[2] Yoshida S 1981 *Fundamentals of Rice Crop Science* (Los Banos: IRRI)
[3] Ehsani M R, Upadhyaya S K, Slaughter D, Shafii S and Pelletier M 1999 A NIR technique for rapid determination of soil mineral nitrogen *Precision Agriculture* **1** 217-234
[4] Russell C A, Angus J F, Batten G D, Dunn B W and William R L 2002 The potential to predict nitrogen mineralization in rice soils *Journal Plant and Soil.* **247** 243-252
[5] Hernandez M E, Pausch J and Mitsch W J 2004 Use of near infrared reflectance spectroscopy (NIR) to determine total carbon and total nitrogen in wetland soils *Near Infrared of Soil* pp 99 – 102
[6] Kusumo B H, Hedley M J, Hedley C B and Tuohy M P 2011 Measuring carbon dinamics in field soil using soil spectral relectance : prediction of maize root density, soil organic carbon and nitrogen content *Journal Plan Soil* **338** 233-245
[7] Mahmood et al 2012 Sensor data fusion to predict multiple soil properties *J. of Precision Agriculture* **13** 628-645
[8] Abdi D, Tremblay G F, Ziadi N, Belanger G and Parent L E 2012 Predicting soil phosporus related properties using near infrared reflectance spectroscopy *Soil Science Society of America Journal* **76** 2318-26
[9] Xuemei L and Jianshe L 2013 Measurement of soil properties using visible and short wave-near infrared spectroscopy and multivariate calibration. *J. of Measurement* **46** 3808-14
[10] Wenjun et al 2014 In situ measurement of some soil properties in paddy soil using visible and near-infrared spectroscopy *J. of PLoS ONE* **9**(8): e105708
[11] Angkat A R 2012 *Development of infrared sensor based data acquistion system for predicting soil nitrogen content as a reference to determine fertilizing dosage* (Bogor: Graduate School of Bogor Agricultural University)
[12] Yang N 2009 *Elemental analysis of soils using laser-induced breakdown spectroscopy (LIBS)* (Knoxville: University of Tennessee)
[13] Knadel *et al.* 2015 Comparing predictive ability of laser-induced breakdown spectroscopy to visible near-infrared spectroscopy for soil property determination *Journal of Biosystem Engineering* **156** 157-172
[14] Haddad *et al.* 2013 Artificial neural network for on-site quantitative analysis of soils using laser induced breakdown spectroscopy *Spectrochimica Acta Part B* 79–80: 51–57
[15] Sansonetti J E and Martin W C 2005 Handbook of Basic Atomic Spestroscopic Data (Gaithersburg: National Institute of Standards and Technology)
[16] Rinnan Å, Van den Berg F and Engelsen S B 2009 Review of the most common preprocessing techniques spectra. *Trends Anal Chem* **28** 1201-22
[17] Buddenbaum H and Steffens M 2012 The effects of pre-treatment on chemometric analyses of soil profiles using laboratory imaging spectroscopy *Appl Environ Soil Sci* 1-12

IOP Conf. Series: Earth and Environmental Science **147** (2018) 012046 doi:10.1088/1755-1315/147/1/012046

Vapor Measurement System of Essential Oil Based on MOS Gas Sensors Driven with Advanced Temperature Modulation Technique

A Sudarmaji[1], A Margiwiyatno[1], R Ediati[1] and A Mustofa[1]

[1] Department of Agricultural Engineering, Jenderal Soedirman University, Indonesia

E-mail: arief.unsoed@gmail.com

Abstract. The aroma/vapor of essential oils is complex compound which depends on the content of the gases and volatiles generated from essential oil. This paper describes a design of quick, simple, and low-cost static measurement system to acquire vapor profile of essential oil. The gases and volatiles are captured in a chamber by means of 9 MOS gas sensors which driven with advance temperature modulation technique. A PSoC CY8C28445-24PVXI based-interface unit is built to generate the modulation signal and acquire all sensor output into computer wirelessly via radio frequency serial communication using Digi International Inc., XBee (IEEE 802.15.4) through developed software under Visual.Net. The system was tested to measure 2 kinds of essential oil (Patchouli and Clove Oils) in 4 temperature modulations (without, 0.25 Hz, 1 Hz, and 4 Hz). A cycle measurement consists of reference and sample measurement sequentially which is set during 2 minutes in every 1 second respectively. It is found that the suitable modulation is 0,25Hz; 75%, and the results of Principle Component Analysis show that the system is able to distinguish clearly between Patchouli Oil and Clove Oil.

1. Introduction

The aroma of essential oils is complex compound; hundreds of components can be present, and most of the components which confer aroma or flavor may be present only at ppm levels [1]. The vapor of essential oils is one of important attributes to indicate the quality of essential oil which depends on the content and amount of the gases and volatiles generated from essential oil. There are 30 volatiles components were identified from patchouli oils of Java, Sumatra and Sulawesi, with 5 main components such as patchouli alcohol, alpha guanine, seychellene, alpha patchouelene and alpha bulnesene [2]. On Bangladesh clove, it was found 38 components in its leaf oil with portion of main components, i.e. eugenol (74.3%), eucalyptol (5.8%), caryophyllene (3.85%) and a-cadinol (2.43%), while from its bud, 31 volatile components were identified and the main components were eugenol (49.7%), caryophyllene (18.9%), benzene,1-ethyl-3-nitro (11.1%) and benzoic acid,3-(1-methyl ethyl) (8.9%) [3].

There are a number of causes in which the quality of essential oils can be varied, from the genetic variation and planting conditions to their harvesting, distillation processing, packing, and storage handling [4,5]. As a consumer, farmer or even supplier, it is difficult to assess quality sensuously.

Some analytical methods for qualitative and quantitative analysis of essential oils such as Gas Chromatograph and Mass Spectrometer provide accurate and selective readings. Capillary Gas chromatography-Mass spectrometry (GC-MS) proves to be an efficient and precise method for almost all combinations of components in such mixtures down to minute traces. It commonly used capillary column for the analysis of essential oils is Polyethylene Glycol (PEG). Most of the components of essential oils are identified using capillary GC with Mass Detector [4,6]. The data can be compared to an established profile or fingerprint for that particular essential oil to finally determine the purity of that oil. However, those instruments are high cost, time-consuming (mostly requires sample preparation), large size, and maintenance cost [7]. Hence, it is difficult for rapid and on-site measurements.

This paper presents a design of quick, simple, and low-cost static measurement system to acquire vapor 'fingerprint' of essential oils. Nine MOS gas sensors are employed to capture vapors of essential oils. And, three environment sensors are used to measure temperature and humidity in sensor chamber, and temperature ambient. The MOS gas sensors are operated in dynamic mode by advanced Temperature Modulation technique. A Programmable System-On-Chip (PSoC) based interface system is built to generate the modulation and acquire all sensor data to a personal computer wirelessly through a developed program under Visual Studio. The PSoC is a hybrid (mixed-signal) device, beyond the traditional microcontroller, which has all traditional microcontroller features and reconfigurable analog and digital blocks. It is configured at programming to get specific use digital and analog components, provide flexibility in choices for the benefit and optimum particular systems [8].

2. Materials and methods

2.1. Material and Sample

The materials used to build the vapor measurement system of essential oils are generally divided into 3 units: sensors, interface/acquisition, and samples of essential oils.

1. Sensor material.

   Gas and environment sensors used are shown in table 1. Gas sensors are circuited with simple voltage divider circuit and operated in dynamic mode by technique of Temperature Modulation-Specified Detection Point (TM-SDP) as shown in figure 1. Gas sensors are placed in chamber which made of 5 mm transparent acrylic and formed into square box by size (11x9x7) cm^3.

Table 1. Gas sensors and environments sensor to capture essential oils profile.

No	Type	Main Target	Sensing Range
1.	TGS2602	Odorous gases (Ammonia, Ethanol)	1-100 ppm
2.	TGS 2620	Solvent (organic) vapors	50 – 5,000 ppm
3.	TGS 2600	Air Contaminants (H_2, CO)	1-30 ppm
4.	MQ5	Natural gas, Coal gas	200-10,000 ppm
5.	MQ135	Air Quality Control	10-200 ppm
6.	MQ138	Wide volatile compound	200-10,000 ppm
7.	FIS12A	Methane	300-7,000 ppm
8.	FIS30SB	Alcohol	1-100 ppm
9.	FISAQ1	Volatile organic compound	10-10,000 ppm
10.	LM35dz	Temperature	0-100 ^{o}C
11.	HSM30G	Humidity	10-90 %
12.	KE-25	Oxygen	0-100 %

Temperature modulation alters the kinetics of the MOS gas sensor through changes in the operational temperature of device through applying modulated voltage on its heater unit. The operating modulation voltage, also consequently the operating temperature, of the sensor changes periodically either by rectangular or triangular or sine waveform [9].

Principally Temperature Modulation-SDP is similar with general temperature modulation, which besides it modulates the temperature of heater unit, it also modulates the sensing (detector) unit that associated and in same phase with temperature modulation on the heater unit. SDP means the output detection (acquiring) of MOS gas sensor is put at specified point (i.e. at middle of sensing unit modulation) [10]. The signal of TM-SDP is shown in figure 1(c).

2. Interface/Acquisition.
   Acquisition materials comprise of i) power supply unit, that provides voltage: -5V, +0.9V, +2.5V, +5V (3 pcs) and +12V, ii) PSoC CY8C28445-24PVXI, iii) Conditioning circuit for sensors, and iv) a pair of XBee (IEEE 802.15.4) with XBee USB Adapter Board 32400.
3. Samples of essential oils.
   In this preliminary work, we tested 2 kinds of essential oils (Patchouli Oil and Clove oil). These two oils are potential commodities in Banyumas regency. The samples are taken from a supplier in Banyumas regency.

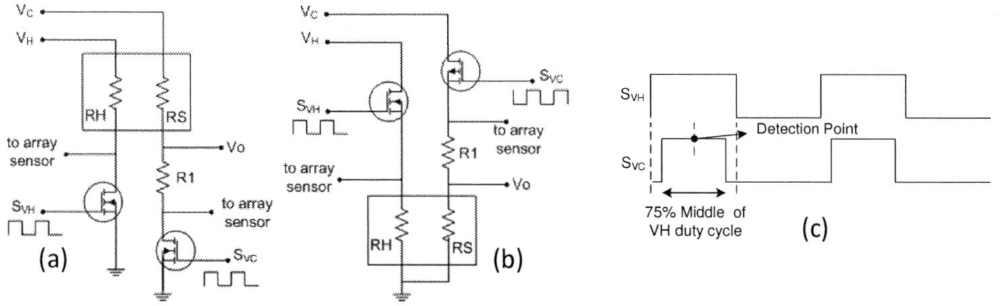

Figure 1. Sensor Circuitry of (a) TGSs, MQs and (b) FISs, driven in dynamic mode by (c) Temperature Modulation-SDP signal for capturing vapor profile of essential oils.

2.2. Methods

2.2.1. System Diagram. The measurement system of essential oils uses static headspace and static measurement (figure 2), which adopted from [11,10]. Sample Headspace (SH) of essential oils is a 15 ml glass vial with rubber cap. The vapor is delivered to sensor chamber manually using a 1 ml syringe. And as The Reference is dry air (filtered by silica gel) which pumped constantly to sensor chamber. The Mass Flow Controller (MFC) is set on 0.4 liter per minute (0.67 cm3/s).

A PSoC CY8C28445-24PVXI is employed as the center of interface unit to generate the TM-SDP signal for MOS gas sensors and acquire the outputs of all sensors to a computer wirelessly through Radio Frequency using XBee serial communication (IEEE 802.15.4) Digi International Inc. It has 24 pins GPIO, General Programmable Input Output [8]. We configured 20 GPIOs in PSoC as follow: 14 analog inputs for sensors, 4 digital outputs for TM-SDP signals, and 2 as Tx/Rx for serial communication, to accommodate the system (figure 3). We utilized PSoC Designer 5.4 to build the program into PSoC CY8C28445-24PVXI and Visual Basic.Net 2012 to develop interface software for acquisition in Personal Computer.

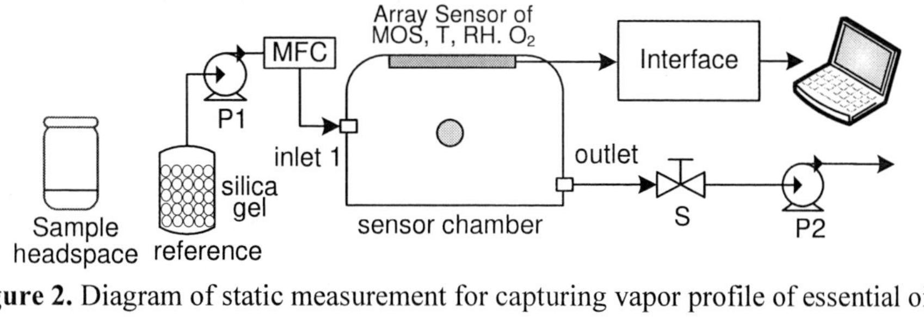

Figure 2. Diagram of static measurement for capturing vapor profile of essential oils.

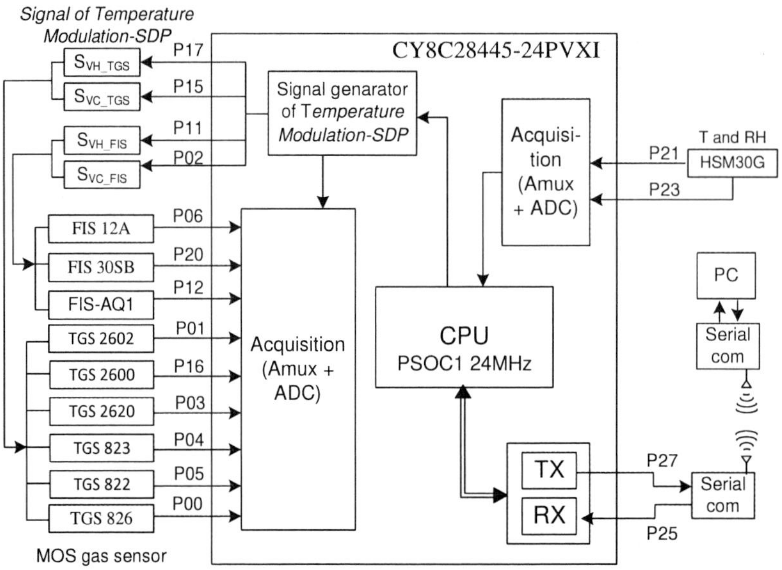

Figure 3. Diagram of PSoC CY8C28445-24PVXI based interface.

2.2.2. Measurement steps. There are 4 phases in measuring vapor of essential oils with static headspace and measurement, i.e. Idle, Reference capturing (R_O), Vapor capturing (R_V) and Purging. The Idle, Ro, and Cleaning phases are the same, i.e. when the sensor does not measure the essential oils. The flow is set from the silica gel container to the purging pump by adjusting The Switch (S), Pump-1 (P1), Pump-2 (P2) are switched on. While The R_V (measuring the vapor of essential oils) is by injecting the vapor to sensor chamber, the Switch (S), Pump 1 (P1), Pump 2 (P2) are switched off.

As sample preparation, 1 ml of essential oil solution is injected into the SH vial. And the amount of vapor of essential oils from SH vial to be injected into sensor chamber is 3 ml. One cycle measurement of essential oil sample, as shown in figure 4, consists of acquiring R_O for 1 minute and followed by R_V for 1 minute. After that the purging of sensor chamber is done in 10 minutes and back to idle phase. Each measurement of essential oil is repeated 5 times.

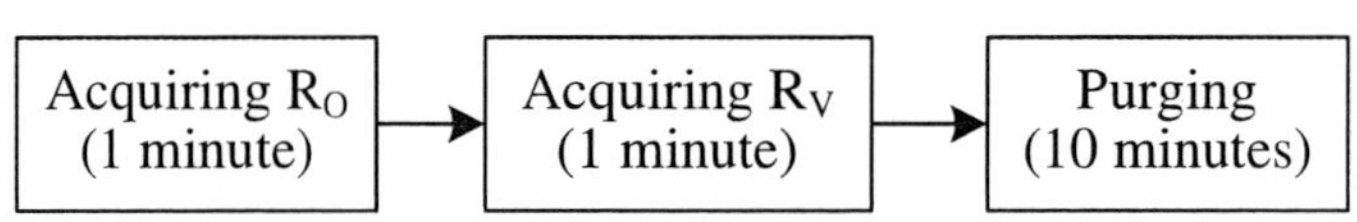

Figure 4. Steps of one cycle measurement.

Response of MOS gas sensors is presented by Sensitivity (S), where R_O is MOS resistance when measuring the reference (dry air) and R_V is MOS resistance when measuring the vapor of essential oils.

$$S = \frac{R_O}{R_V} \tag{1}$$

3. Result and discussion

3.1. The measurement system of vapor of essential oils

Figure. 4 shows the built-system, program interface, static headspace (SH) vial, and injection process of essential oils vapor. The configurations of PSoC CY8C28445-24PVXI are shown in table 2 and figure 5. We set the pair of wireless serial communication XBee (IEEE 802.15.4) into direct transmission (point to point) using XCTU software, where one is set as coordinator while another act as a router [12]. Both XBees are configured 19200 bps, 8 bit of data, no parity, 1 stop bit and none flow control which accorded with PSoC serial communication.

The TM-SDP signal and output sensors are successfully generated and acquired respectively into personal computer wirelessly using developed software under Visual Basic.Net 2012. The software (figure. 4(b)) is connected to Microsoft Excel to store and process data, such as: (a) create file, read and write data, and (b) determine automatically the average value of each sensor for each measurement mode (R_O and R_V). Our acquisition software makes automatically 3 worksheets to store 2 mode measurement and fractional data at once procedure measurement.

Table 2. PSoC CY8C28445-24PVXI configuration for interface unit

Part	Important Configured Features
Global Resource	Sysclock=5V;24MHz, V_{c1}= 2MHz, V_{c2}=153,846kHz, V_{c3}= 10kHz, Ref Mux=±Vdd/2, Analog SC On/Ref High.
ADC Incremental	Clock=V_{c1}, 14 bits, Unsigned data format, normal clock phase.
Dual ADC	Clock=V_{c1}, 13 bits, Unsigned data format, normal clock phase.
Timer8	Clock=V_{c3}, Capture=High, Terminal count and compare out= none, Period=9, Interrupt type=Terminal count Invert capture=normal.
PGA	2 PGAs connected to Dual ADC and 1 PGA to ADC Inc, Ref=Vss
UART	Clock=V_{c2}→ 19200 kbps, 8 bit, no parity, 1 stop bit and none flow control
GPIO	- 14 pins connected to Analog Mux Input, drive as High Z analog. - 4 pins as Digital Output (StdCPU), drive Strong.

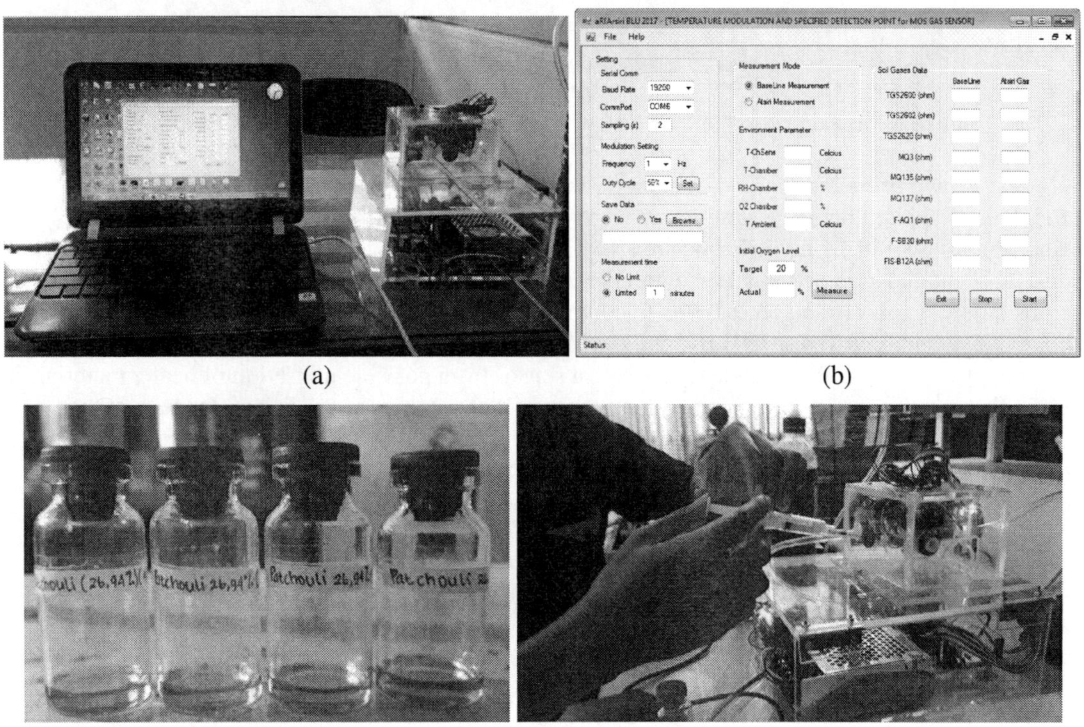

(a) (b)

(c) (d)

Figure 5. (a) Vapor measurement system , (b) acquisition program interface, (c) SH vial, and (d) vapor injection process.

IOP Conf. Series: Earth and Environmental Science **147** (2018) 012046 doi:10.1088/1755-1315/147/1/012046

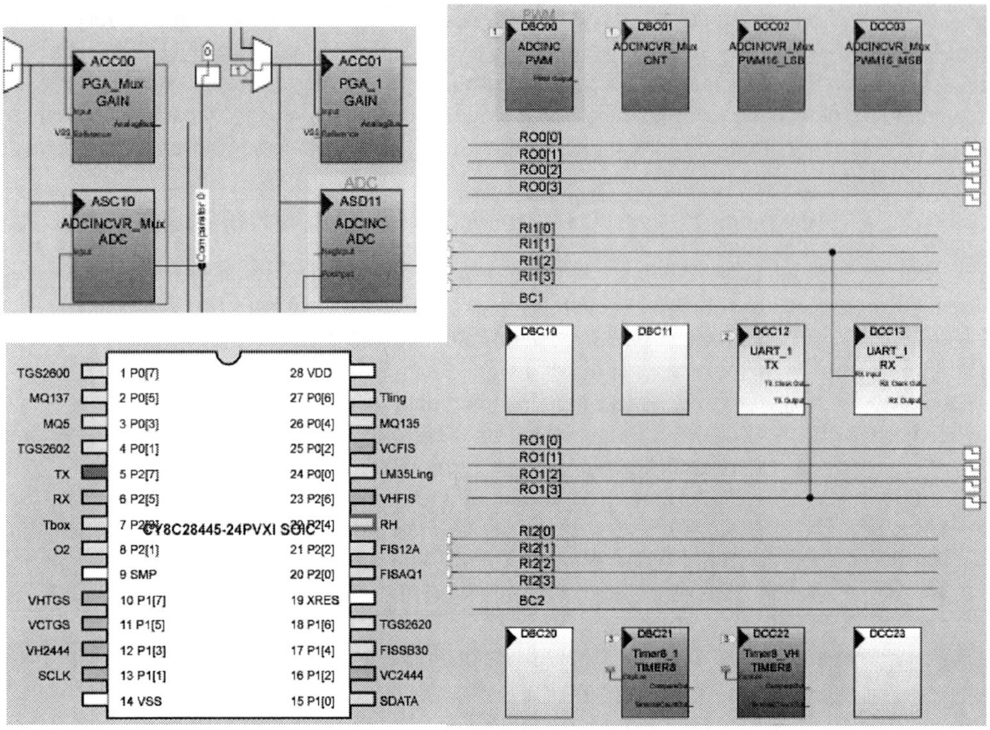

Figure 6. (a) Configuration of PSoC CY8C28445-24PVXI using PoSC designer 5.4.

3.2. The Temperature Modulation-Specified Detection Point (TM-SDP)

Modulations applied on MOS gas sensor were checked with oscilloscope Tektronix TDS-2024B. For instance as shown in figure 7, for desired TM-SDP 0.25Hz;75%, the measured frequency of V_H was 0.2510 Hz and the high of V_C is laid in middle 75% of high of V_H. The acquiring of all MOS (in array) begins at middle of VH and takes 0.08s to complete it. The high of V_H of TGS and FIS were measured about 4.98 V and 0.95 V respectively, and the V_C of both TGS and FIS were 4.98 V.

By applying TM-SDP, it may lead to prevent sensor from possible migration of heater materials into the sensing material which could causes long-term drift of sensing material's resistance to higher values. It means that a pulsed-V_C giving less force to drive migration than a constant V_C, rendering negligible possibility of migration, particularly when operated under high humidity and temperature [10].

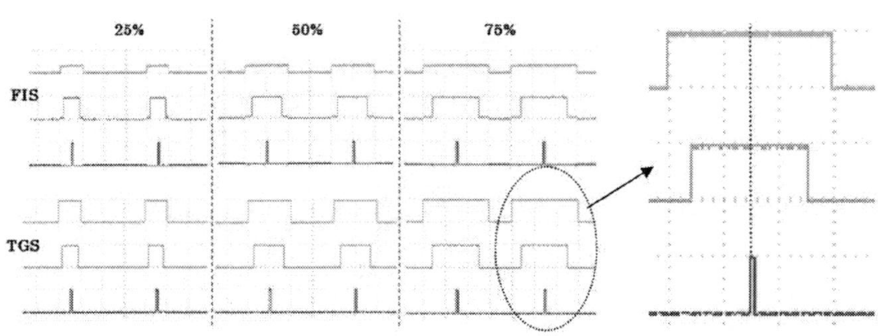

Figure 7. Captured signal of TM-SDP on MOS gas sensors of 0.25 Hz with duty cycle 25%, 50% and 75%, where: VH (orange)= 2V/div; VC (blue) =5V/div ; SDP (purple) =5V/div ; Time-Div= 1s.

3.3. Response of MOS gas sensors under various TM-SDP for Patchouli Oil and Clove Oil.

We tested the vapor measurement system to Clove Oil and Patchouli Oil for various modulation of TM-SDP, namely without modulation, frequency modulation (0.25 Hz, 1 Hz, and 4 Hz), and duty cycle modulation (25%, 50%, and 75%). During measurement, the condition of temperature environment was at 18.6 °C to 26.4 °C. While the oxygen level was at to 21.3 % to 25.8% (average 23.8%). The oxygen concentration in the chamber was conditioned manually by flowing oxygen gas inside the chamber when its level is below 21%. Typically the MOS gas sensors (such as TGSs) require the presence of minimum 21% (ambient) oxygen in their operating environment in order to work and properly [13].

The presence of minimum required ambient oxygen is essential to the sensor's operation which means oxygen plays an important role to reducing gases and its concentration effected to detection of combustible which mediated by reaction with adsorbed oxygen on the sensor surface [14,15]. The behavior of steady-state resistance of MOS gas sensor is greatly influenced by the concentration of ambient oxygen and the reduced oxygen pressure will lead the decrement of the resistance of sensor [13].

Firstly, we investigated the responses of each MOS at a particular frequency of TM-SDP (for instance 0.25 Hz) with different duty cycle. We found that at TM-SDP of frequency 0.25 Hz, among duty cycle 25%, 50%, and 75%, the duty cycle 75% provided most distinguished response to Clove Oil and Patchouli Oil. This result also shown by the TM-SDP of frequency 1 Hz and 4 Hz (figure 8). Secondly, by applying the 75% duty cycle we investigated the different frequency (0.25 Hz, 1 Hz, or 4 Hz). The results show that frequency 0.25 Hz provide most distinguished response (figure 9). Hence it reveals that the suitable TM-SDP which lead most selective response is the frequency 0.25 Hz; duty cycle 75%.

From the figure 8 and 9, it also shows that some MOS gas sensor has of each are a cross-sensitivity response. This is considered as a limitation of MOS gas sensors that potentially caused from temperature variation which changes the baseline of the sensor signal shifts. And the most significant, we found that under TM-SDP 0.25H; 75% the TGS2802 and MQ5 were the two most sensitive and selective to response and strongly discriminate between Clove Oil and Patchouli Oil. While FIS12A was very less sensitive to discriminate Clove Oil and Patchouli Oil. Intrinsically, the FIS12A are designed to sense the methane although it also may sense other gases. In other hand, it can be used only TGS2802 and MQ5 to distinguish between Clove Oil and Patchouli Oil.

3.4. Performance of array MOS gas sensor to distinguish Patchouli Oil and Clove Oil.

We utilized the Principal Component Analysis (PCA) to evaluate the selectivity performance of the system that uses array 9 of MOS gas sensors in identifying the Clove Oil and Patchouli Oil on each modulation. PCA is commonly used as feature extraction part to test distinguish (selectivity) performance and a powerful linear classification technique that is usually employed in correlation with cluster analysis and visualization the difference in similarities or differences among the treatments. The large dimension of interrelated variables is reduced into few important Principal Components (PCs).
We use the function of PCA and Graph Plot in Matlab 7.12.0 (R2011a). The input of PCA is the output (S value) of nine MOS gas sensor that operated under TM-SDP 0.25Hz; duty cycle 75%. And the PCA results show that collectively the two PCs hold 81.2% of data. However, by using those first two PCs, the system has been able to discriminate clearly as shown in figure 10. The first two or three uncorrelated components hold most significant of variation present in all variables and widely used in various application [16].

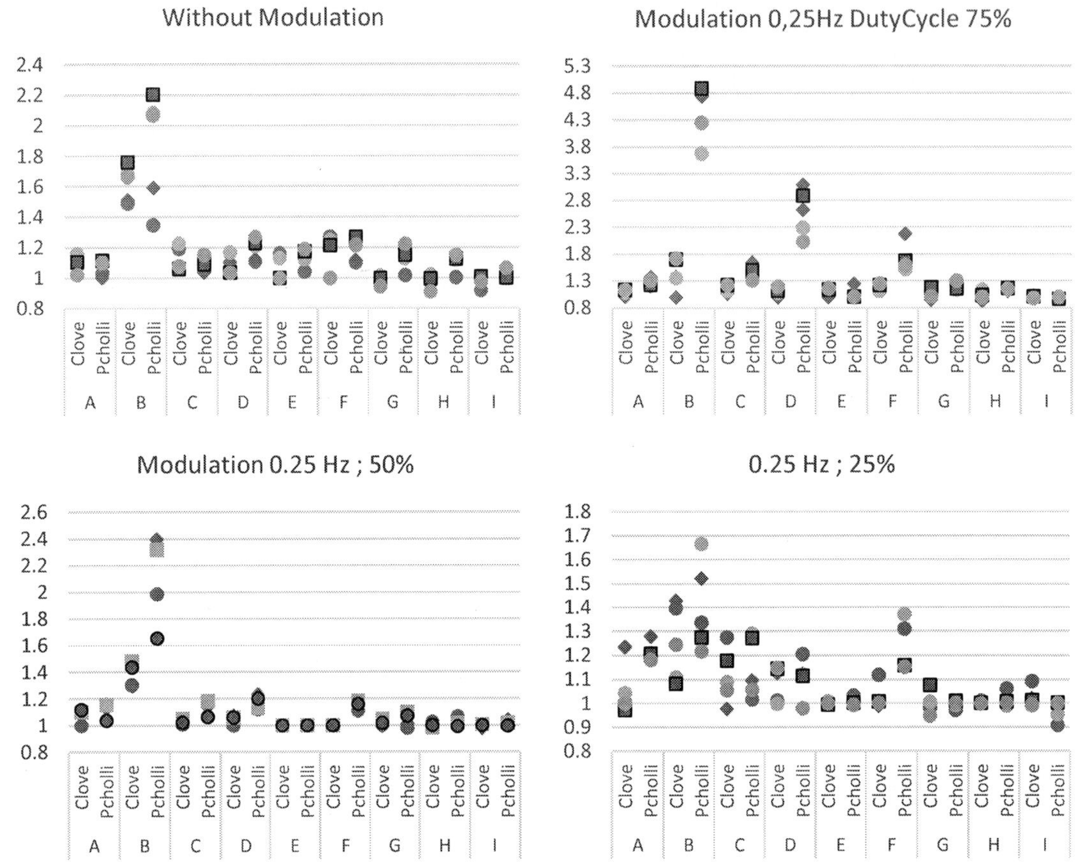

Figure 8. Responses of individual MOS gas sensor (S) to vapor of Clove Oil and Patchouli Oil under no modulation vs TM-SDP 0.25Hz duty cycle 25%, 50%, and 75% where: A=TGS2600 ; B=TGS2602 ; C=TGS2620 ; D=MQ5 ; E=MQ135 ; F=MQ138 ; G=FISAQ1 ; H=FISSB30 ; I=FIS12A.

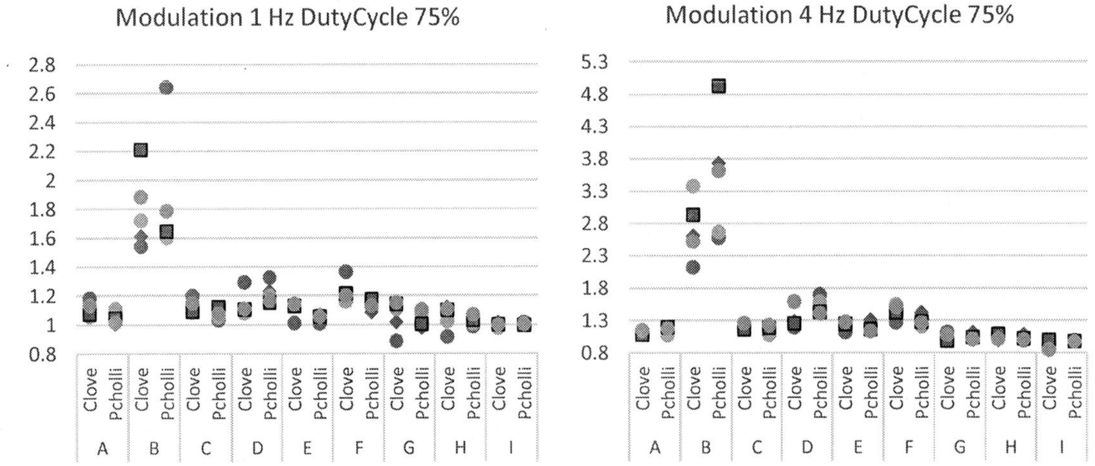

Figure 9. Responses of individual MOS gas sensor (S) to vapor of Clove Oil and Patchouli Oil under TM-SDP Duty Cycle 75% with different Frequency, where: A=TGS2600 ; B=TGS2602 ; C=TGS2620 ; D=MQ5 ; E=MQ135 ; F=MQ138 ; G=FISAQ1 ; H=FISSB30 ; I=FIS12A.

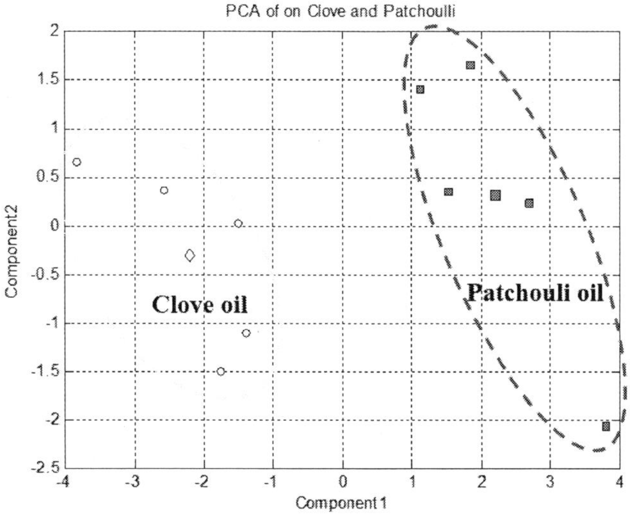

Figure 10. PCA plot of array gas sensors operated under TM-SDP 0.25Hz; 75% in discriminating between Clove Oil and Patchouli Oil.

4. Conclusions

In this paper, we present a preliminary work of potential application of MOS gas sensors which operates in a dynamic mode using technique of Temperature Modulation-Specified Detection Point (TM-SDP) to measure vapor of essential oils. Nine MOS gas sensors which designed to sense some volatile compounds were employed. We tested various TM-SDP both frequency and duty cycle to capture the volatile fingerprint of Clove Oil and Patchouli Oil. We found that the suitable TM-SDP which lead most distinctly response to Clove Oil and Patchouli Oil is the frequency 0.25 Hz with duty cycle 75%. PCA result shows that the system is able to distinguish clearly between Patchouli Oil and Clove Oil.

Moreover, it is also found that the TGS2602 and MQ5 are the two most sensitive and selective sensor to distinguish between Patchouli Oil and Clove Oil. Hence it is possible to use only them to distinguish those two essential oils when they operate on dynamic mode under TM-SDP 0.25 Hz; 75%.

Acknowledgements

This research is supported by Public Service Agency (BLU) of Jenderal Soedirman University. Authors also would like thank to Indonesian Directorate General of Higher Education (DIKTI) and additional acknowledgement to PT. Indesso Aroma Purwokerto.

References

[1] Németh-Zámboriné É 2016 Natural Variability of Essential Oil Components *Handbook of Essential Oils: Science, Technology, and Applications: Second Edition* ed K H C Bas,er and G Buchbauer (Boca Raton, FL: CRC Press Taylor & Francis Group) pp 87–125

[2] Riyadi R 2012 *Profil Senyawa Volatil pada Berbagai Jenis Minyak Atsiri asal Indonesia* (Bogor Agricultural University)

[3] Bhuiyan N I, Begum J, Nandi N C and Akter F 2010 Constituents of the essential oil from leaves and buds of clove (Syzigium caryophyllatum (L .) Alston) *African J. Plant Sci.* **4** 451–4

[4] Dhole V R, Sitharaman B and Kaur I 2012 *Characterization and Identification of Essential Oil Components by GC-MS* (Maharashtra)

[5] Franz C and Novak J 2016 Sources of Essential Oils *Handbook of Essential Oils: Science, Technology, and Applications: Second Edition* ed K H C Bas,er and G Buchbauer (Boca Raton, FL: CRC Press Taylor & Francis Group) pp 43–78

[6] Zellner B d'Acampora, Dugo P, Dugo G and Mondello L 2016 Analysis of Essential Oils *Handbook of Essential Oils: Science, Technology, and Applications: Second Edition* ed K H

C Bas,er and G Buchbauer (Boca Raton, FL: CRC Press Taylor & Francis Group) p 195=228
[7] Lee D and Lee D 2001 Environmental Gas Sensors *IEEE Sens. J.* **1** 214–24
[8] Cypress 2010 *Programmable System-on-Chip (PSoC): CY8C28243/ CY8C28403/ CY8C28413/ CY8C28433/ CY8C28445/ CY8C28452/ CY8C28513/ CY8C28533/ CY8C28545/ CY8C28623/ CY8C28643/ CY8C28645* (San Jose, CA)
[9] Huang X, Meng F, Pi Z, Xu W and Liu J 2004 Gas sensing behavior of a single tin dioxide sensor under dynamic temperature modulation *Sensors Actuators, B Chem.* **99** 444–50
[10] Sudarmaji A and Kitagawa A 2015 Sensors & Transducers Temperature Modulation with Specified Detection Point on Metal Oxide Semiconductor Gas Sensors for E-Nose Application *Sensors & Transducers* **186** 93–103
[11] Sudarmaji A and Kitagawa A 2016 Application of Temperature Modulation-SDP on MOS Gas Sensors: Capturing Soil Gaseous Profile for Discrimination of Soil under Different Nutrient Addition *J. Sensors* **2016** 1–11
[12] Digi 2008 *X-CTU Configuration and Test Utility Software: User's Guide*
[13] Figaro Engineering Inc. 2005 *General Information For TGS Sensors: Technical Information on Usage of TGS Sensors for Toxic and Explosive Gas Leak Detectors*
[14] Clifford P K and Tuma D T 1982 Characteristics of semiconductor gas sensors II. transient response to temperature change *Sensors and Actuators* **3** 255–81
[15] Clifford P K and Tuma D T 1982 Characteristics of semiconductor gas sensors I. Steady state gas response *Sensors and Actuators* **3** 233–54
[16] Haddi Z, Bougrini M, Tahri K, Braham Y, Souiri M, Bari N El, Maaref A, Othmane A, Jaffrezic-Renault N and Bouchikhi B 2014 A hybrid system based on an electronic nose coupled with an electronic tongue for the characterization of moroccan waters *Sensors & Transducers* **27** 190–7

IOP Conf. Series: Earth and Environmental Science **147** (2018) 012047 doi:10.1088/1755-1315/147/1/012047

Development of Weeds Density Evaluation System Based on RGB Sensor

M Solahudin[1], Slamet W[1] and Wahyu W[1]

[1] Department of Mechanical and Biosystem Engineering, Faculty of Agricultural Engineering and Technology, Bogor Agricultural University, Indonesia

E-mail: msoul9@yahoo.com

Abstract. Weeds are plant competitors which potentially reduce the yields due to competition for sunlight, water and soil nutrients. Recently, for chemical-based weed control, site-specific weed management that accommodates spatial and temporal diversity of weeds attack in determining the appropriate dose of herbicide based on Variable Rate Technology (VRT) is preferable than traditional approach with single dose herbicide application. In such application, determination of the level of weed density is an important task. Several methods have been studied to evaluate the density of weed attack. The objective of this study is to develop a system that is able to evaluate weed density based on RGB (Red, Green, and Blue) sensors. RGB sensor was used to acquire the RGB values of the surface of the field. An artificial neural network (ANN) model was then used for determining the weed density. In this study the ANN model was trained with 280 training data (70%), 60 validation data (15%), and 60 testing data (15%). Based on the field test, using the proposed method the weed density could be evaluated with an accuracy of 83.75%.

1. Introduction

Weeds are hazardous plants which presence is a competitor for main plants in getting sunlight, water, and soil nutrients. The level of competition depends on rainfall, soil conditions, weed density, weed growth, and age of cultivation plants as weeds begin to compete [1]. One of the weed control efforts is to use herbicide which is a chemical substance to suppress weed growth and can even turn it off [2]. Weed control can be done by chemical (using herbicides) and non-chemicals. Herbicides are chemicals that can kill herbs or weeds. Chemical weed control is biological control, ecology, competition and emphasis in the presence of other plants.

Based on the timing of the administration of the herbicide may be given by: 1) Pre-cultivation, prior to tillage, weeds on the ground are given herbicides to facilitate processing, 2) Pre-planting, after tillage and before planting is given herbicides to inhibit weed growth and facilitate planting, 3) Pre-emergence (pre-emergence), after planting, herbicides are given before the plants and weeds appear, and 4) Post-emergence, herbicides are given after the plant or weed appears [2].

Environmentally sound weed controls no longer apply a single dose method in herbicide spraying activities. Determination of herbicide dose applied to the soil should be adjusted to the spatial and temporal diversity of weeds. Based on this method the effectiveness and efficiency of herbicide use on weed control can be improved so that weed control and land damage can be minimized which result in increased production output. Such a practice is well known as site-specific weed control which is widely adopted [3].

The current problem is that the evaluation of weed density is still done manually so that the precise map base herbicide treatment is difficult to describe. This study aims to develop a red, green, blue (RGB) sensor based to evaluate weed density in precision farming system concepts. The application of precision farming on weed control requires a device capable of identifying the location and degree of weed density in the field. The results of identification can determine the dosage at each location of herbicide spraying. Based on this concept of precision agriculture in weed control can be realized.

2. Materials and methods

2.1. Weed Density Classification

Solahudin [3] in his dissertation classifies that the weed density as a result of the filtering of RGB values is divided into four groups by multilevel method. The staggered group breaks will group the green mean value greater than half the maximum green average into the "Solid" or "Class 4" groups. While the next classes are with the value of the limit of half of the lower limit value of the upper class. Determination of the density class by means of the stratum as seen in table 1 leads the clustering method to a higher level of density, since the higher the weed density class will have a wider range of green width values.

Table 1. Weed density grade classification [3]

Class	Average of Green Value		Definition
	Lower Bound	Upper Bound	
1	0.00	38.22	Clear
2	38.22	76.45	Sparse
3	76.45	114.67	Medium
4	114.67	255.00	Solid

These values can be seen in table 1. Values 1 through 4 are given in the RGB value as a result of weed density classification based on the mean green value of the RGB color values. Value 1 for clean land condition from weed, value 2 for sparse weed, 3 for medium weed, and value 4 for solid weed where 255 is the highest value of green color component. The method used in this research is the prototype development method described in figure 1 and described as follows:

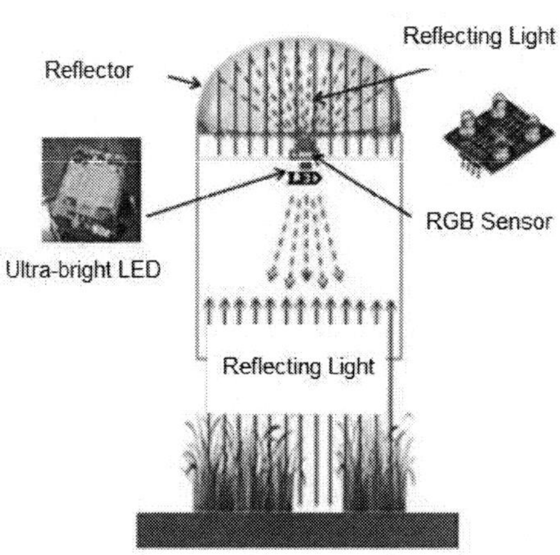

Figure 1. Illustration of acquisition of RGB value of weeds

2.1.1. RGB Reader Unit. The reading of the RGB value of weeds is done using the TCS3200 color sensor. The sensor consists of 64 sensor arrays that will convert the received reflected light into 4 outputs namely R value, G value, B value and colorless readout value. This sensor is relatively affordable and easy to use. It's just that this sensor has a shortage of effective distance and a relatively limited reading area. Therefore, the design of the sensor to the effective distance and reading area increases in accordance with the needs of research.

2.1.2. RGB Data Processing Unit. RGB value data processing is done by using open source Arduino Mega platform which has microcontroller unit ATMega 2560 as the data processing of sensor output. The output data from the sensor in the form of RGB value is processed by interpolation method and artificial neural network to classify the level of weed density in the field.

2.1.3. Acquisition of RGB Value. Acquisition of RGB value is a process of retrieving RGB values of objects read by sensors. This process involves a reader unit and a data processing unit operated against a weed object. Figure 1 shows an illustration of the working schema configuration of the RGB retrieval. The sensor device is positioned perpendicular to the color object and as the distance between the edges of the house and the color, object is used three buffers with h cm height. This height is the distance between the sensor house and the object taken. This spacing is done to simulate the position of the sensor with the weed object in the real application in the field. The results of sensor readings in the form of R, G, and B values of each color sample are processed by microcontroller (Arduino) and then sent to computer for further processing for processing by interpolation method and artificial neural network method.

2.2. Artificial Neural Network

In the sensor calibration process the value of RGB sensor readings are used as input ANN which is then used to predict the output of the RGB value. Among these inputs and outputs is placed a hidden layer with six nodes. In the evaluation of weed density, the RGB value of the sensor readings is used as an ANN input to determine the weed density level as output. Among these inputs and outputs is placed a hidden layer with ten nodes. Another important thing related to the ANN created is the activation function on the hidden layer and output layer. In this case, in the hidden layer the activation function used is the sigmoid function which can be expressed as equation 1 while in the output part the function is used linear.

The results of the evaluation of weed density are arranged into 10x50 cm rectangular area where the reading area consists of five parallel horizontal sensor readings which are then averaged as in figure 2.

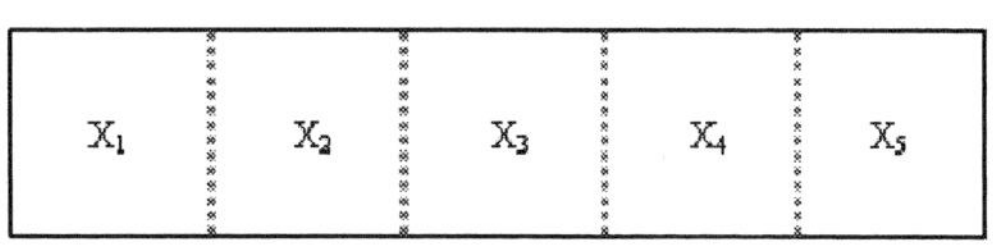

Figure 2. Merging sensor readings

The merger is made to match the spayer unit which has a sprayer width of 50 cm. The result of system reading then averaged using equation:

$$Average\ Reading\ Value = \frac{1}{n}\sum_{1=1}^{n} X_i \tag{1}$$

As figure 3 the results of the system evaluation are interpreted into colors and numbers indicating 4 levels of weed density on open land

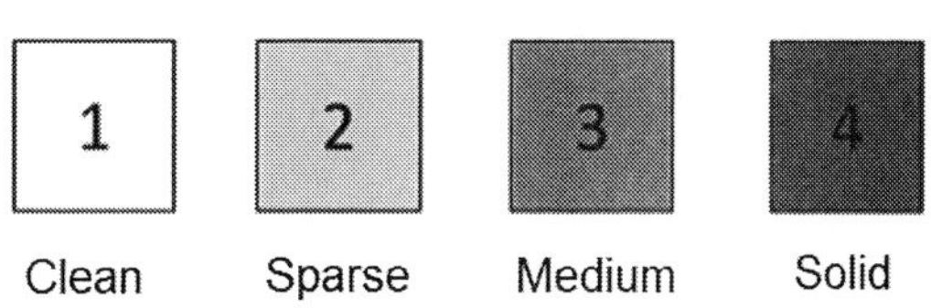

Figure 3. Interpreation of evaluation system results

2.3. Analysis and Testing

To find out the results of system design that has been made, then will be tested by taking data from the system and by analyzing it to obtain the expected results. The tests to be performed are:

2.3.1. Testing the optimum sensor distance. This test is intended to find out how the optimum distance between sensors with objects that have the best output. As a treatment differentiator, two types of type A and type B sensors have distance of sensor with object of 10 cm and 13 cm. The distance is chosen by reason of that distance which allows to be applied to real weed control in the field. In this test, black and white paper is used in black and white color where the color has the lowest RGB value (0, 0, 0) and white color has the highest RGB value (255, 255, 255). Based on the two color samples can be known maximum reading range of each type that is directly proportional to the accuracy of the sensor in discriminating the color.

2.3.2. Sensor calibration. Sensor calibration is done to compare the result of reading RGB sensor value with R, G, and B value of actual object. The color sample used is a combination of several R, G, and B values. For each color component selected 5 levels of 0, 65, 130, 195 and 255. In total obtained 125 color combinations. Testing and calibration are intended to ensure that the developed system can extract the color components of the observed object precisely before it is later used to detect weeds in the field. From 125 data then grouped into dataset for training, validation and test with each consist of 87 data (70%), 19 data (15%), and 19 data (15%) on processing with ANN method. Further analysis of the results obtained to see the improvement of the performance of each calibration method in improving the accuracy of sensor readings.

In this case, the analysis is done in two ways: 1) visually by comparing the color resulting from the combination of obtained RGB value and the reference color, and 2) quantitatively based on the delta-E (ΔE) value that represents the Euclidean distance between the obtained color and the reference color. To calculate the ΔE, first the obtained RGB value must be converted to L^*, a^*, b^* color space where each variable represents lightness, color opponents green–red and color opponents blue–yellow respectively. The ΔE value then can be calculated using the following equation:

$$\Delta E = \sqrt{(L_1^* - L_2^*) + (\, a_1^* - a_2^*\,) + (b_1^* - b_2^*)} \qquad (2)$$

The smaller the delta-E value shows the closer / similar a color to another color (reference). For convenience, the mean and standard deviation values of delta-E are calculated in groups according to the reference to the dataset distribution used in the training, validation and ANN tests.

2.3.3. Real condition test. The test was conducted with a sample of field weed object that had been prepared in the form of grid measuring 10 cm x 10 cm of 1 m x 4 m. The test is done in the morning around 09.00-10.00 WIB which is the actual spraying time of weeds in the field where the wind speed is relatively low and the intensity of sunlight is not too high so that the sprayed herbicide does not drift by the wind or evaporate into the air.

The normalization process is done each time the acquisition process of RGB value (per grid). This is intended for adaptive sensors to weather changes (sunny cloudy) so the sensor works better under various conditions. The results of the evaluation of weed density of the sensors were then validated with the evaluation results of the weed density estimation application developed and used in [4] with some modification according to the research needs. The testing stages of the field weed density evaluation field are shown in figure 4.

Figure 4. Image processing for weed density evaluation

3. Result and discussion

The sensors used in this study were the TCS3200 color sensor, which was then redesigned to improve the accuracy and optimal range of sensor readings. The design of the sensor is done with the aim of extending the reading range by increasing the light intensity received by the sensor up to a certain point. Light source of the original sensor comes from four Light Emitting Diode (LED) is usually substituted using 10 Watt ultrabright LED. As shown in figure 5 the sensor is positioned upright perpendicular to the center of the dome and the LED lamp is arranged perpendicular to the bottom to the weed object in line with the sensor's home-making design described in the next section. The result of the design of the sensor is the optimal distance of the sensor to the original object ± 2 cm is increased 5-fold to ± 10 cm and the accuracy of the sensor in differentiating the color increased significantly along with the increase of RGB sensor value reading range.

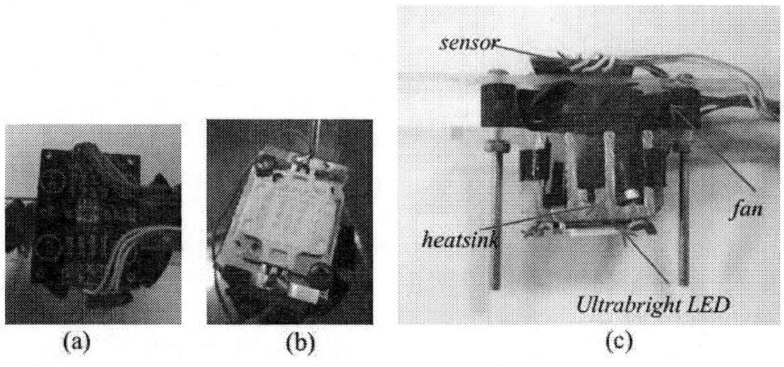

Figure 5. Sensor design, (a) RGB sensor module (b) Ultrabright LED (c) sensor device design result

Table 2. Corelation between Height of sensor and RGB Value

Sensor Height	Color Object	R	G	B
Tipe A (10 cm)	White	87	71	100
	Black	27	20	24
	Range	60	51	76
Tipe B (13 cm)	White	70	56	77
	Black	29	20	25
	Range	41	36	52

From table 2 it can be seen that the largest difference of white and black color readings on RGB values occurs in type A (10 cm) sensors. This is because the reflectance of the reflected light received by the sensor on type A is larger than the light reflectant in type B. In this case the larger the RGB

IOP Conf. Series: Earth and Environmental Science **147** (2018) 012047 doi:10.1088/1755-1315/147/1/012047

value the resolution of the sensor in differentiating one color with the other color will be better. Based on these considerations, type A sensor (10 cm) is selected as the optimum distance between the sensor with the object. Calibration is performed to compare the readings of RGB sensor values with actual object RGB values. In this research, 2 calibration methods are tested with interpolation and ANN to improve the accuracy of RGB value reading with ANN structure of RGB value of sensor readings as input and RGB value of predicted result as output. Among the inputs and outputs are placed one hidden layer with six nodes and from 125 data 70% data is used as training data, 15% as validation data, and 15% data as test. The results of the tests are then visualized as shown in figure 6 showing the color comparisons made based on the original RGB value of sensor readings, interpolated processing, processing with artificial neural network and reference value.

Visually as shown in figure 6 it is clear that sensor readings are relatively different from references (actual colors). The use of interpolation method can improve the reading result of this just the color is darker than the reference. The best results are shown by the results of the processing with ANN.

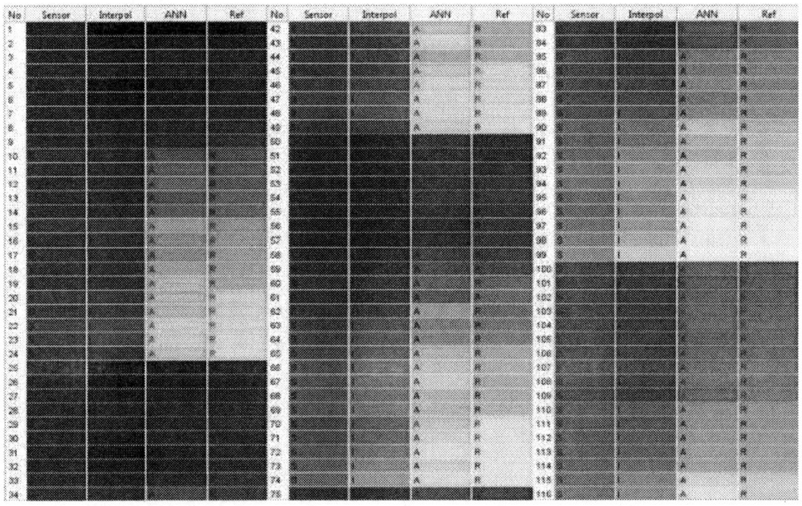

Figure 6. Color comparison result

Visually the resulting color is quite close to the reference color. Further this can be confirmed quantitatively on the basis of the E-delta-E value that represents the Euclidean distance of one color with another color referenced by the color component L *, a *, b * as shown in table 3. Based on the mean and standard deviation values of the delta- E shows that interpolation can reduce the distance between the colors produced by reference but the ANN gives better results. These results indicate that the developed system (sensors and calibrations) can read the color components of the observed object quite well.

Table 3. Mean and standard deviation values of delta-E

Data Set	Mean of delta-E		
	Sensor	Interpolation	ANN
Training	58.4 ± 24.5	45.8 ± 21.7	15.2 ± 10.1
Validation	60.1 ± 22.3	45.7 ± 19.1	13.2 ± 9.1
Test	57.1 ± 30.1	45.3 ± 25.0	13.9 ± 7.0

Evaluation test of Weed Density is intended to determine the ability of the system to evaluate the real weed density level in the field. Sudden weather changes in the evaluation process (sunny to cloudy or otherwise) potentially interfere with sensor performance. Therefore the normalization process is done on each grid during the acquisition process of RGB value so that the RGB value obtained is not affected by the solar intensity change in the field so that the sensor has adaptive ability on the weather change.

The result of RGB value from sensor reading is then processed by interpolation method and ANN with RGB value of sensor reading as input and weed density level as output and between input and output placed one hidden layer with ten nodes. From all of data taken 70% data composed as data training, 15% data as validation data, and 15% data as test data. The validation process is performed between the value of the weed density of the sensor evaluation result and the weed density value of the evaluation of weed density equation estimation application as a reference (figure 7).

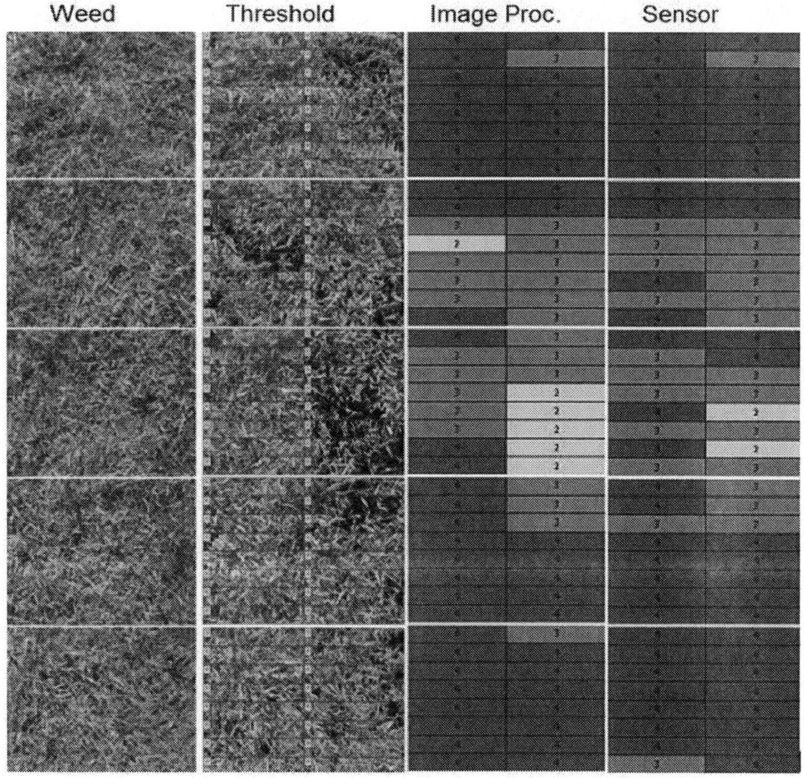

Figure 7. Modification of weed density estimation application

As shown in figure 7 on weed density evaluation testing, the field trials were able to evaluate weed density with an accuracy of 83.75% against density evaluation by weed density estimation applications. The errors that occur in the estimation of weed density levels occur only on one level only (eg weed level 2 density is read as 3) so the system can still be said to work pretty well. Based on this it can be concluded that the system is able to do a good weed density evaluation in the field.

4. Conclusions

Based on the sensor calibration obtained the result that the interpolation method and the ANN method can improve the accuracy of sensor readings in the process of color identification and the ANN method gives better results. These results can be confirmed visually and quantitatively based on the delta-E value so that the developed system is able to read the color components of the observed object well enough and sufficient to be used to detect weed density levels. An algorithm has been developed to evaluate the density of weeds using the ANN method with the input of weed RGB value and the output of weed density level which has an accuracy of 83.75% in field trial.

References

[1] Gallandt E R and J Weiner 2015 *Crop–Weed Competition* eLS 1–9

[2] Moenandir J 2010 *Weed Science* (in Indonesian) (Malang: Universitas Brawijaya Press)
[3] Christensen S H T Søgaard P Kudsk M Nørremark, I Lund, E S Nadimi and R Jørgensen 2009 Site-specific weed control technologies *Weed Research* **49** 233–241.
[4] Solahudin M 2013 *Development of Weed Control Method in Precision Farming Based on Multi Agent Computing* (in Indonesian) (Bogor: Graduate School of Bogor Agricultural University)

IOP Conf. Series: Earth and Environmental Science **147** (2018) 012048 doi:10.1088/1755-1315/147/1/012048

IoT based Growth Monitoring System of Guava (*Psidium guajava* L.) Fruits

Slamet W[1], M Irham N[1] and Sutan M S A[1]

[1] Department of Mechanical and Biosystem Engineering, Faculty of Agricultural Engineering and Technology, Bogor Agricultural University, Indonesia

E-mail: slamet_ae39@apps.ipb.ac.id

Abstract. Growth monitoring of plant is important especially to evaluate the influence of environment or growing condition on its productivity. One way to monitor the plant growth is by measuring the radial growth (i.e., the change of circumference) of certain part of plant such as trunk, branch, and fruit. In this study we develop an internet of things (IoT) based monitoring system of radial growth of plant using a low-cost optoelectronic sensor. The system was applied to monitor radial growth of guava fruits (*Psidium guajava* L.). The principle of the developed sensor is based on the optoelectronic sensor which detects alternating white and black narrow bar printed on reflective tapes. Reflective tape was installed encircling the fruit. The movement of reflective tapes will follow the radial growth of the fruit so that the infrared sensor on the optoelectronic would response reflective tapes movement. This device is designed to measure object continuously and long-term monitor with minimum maintenance. The data collected by the sensors are then sent to the server and also can be monitored in real-time. Based on field test, at current stage, the developed sensor could measure the radial growth of the fruits with a maximum error 2 mm. In term of data transfer, the success rate of the developed system was 97.54%. The result indicated that the developed system can be used as an effective tool for growth monitoring of plant.

1. Introduction

Monitoring the growth of crop radials is essential to observe the effects of environmental conditions or agronomic practices/treatment on the productivity of fruit crops. One way to monitor the plant growth is by measuring the radial growth (i.e. the change of circumference) of certain parts of plant such as trunk, branch, and fruit. There are various commercial measuring tools that have been developed to measure the growth of plant part radials such as dendrometer [1-2]. It offers advantages such as fine resolution and equipped with temperature compensation features which result in high measurement accuracy. However, it also has some limitations in the application such as expensive, large size, and the sensors must be supported by additional frame to avoid the possibility of the sensor overload the tree itself. Another limitation is that it is difficult to move the sensor from one tree to another because it has to be re-calibrated first. Therefore, the development of alternative sensors that can overcome these weaknesses until now is still an interesting topic. One of them is an optoelectronic based measuring tool as reported in [3]. This tool offers a practical and cost-effective radial growth measurement alternative.

This research tries to integrate such an alternative radial growth sensor and Internet of Things (IoT) technology. The use of IoT enables monitoring can be done in more flexible manner both in the local area and in the remote places by using internet. IoT has received a lot of attention and has been widely applied in various areas including agriculture [4]. Through this integration, it is expected that growth data can be monitored continuously even in real-time and data can be accumulated in the database so that it can be better managed and utilized. As an example of application cases, the proposed monitoring system is used to monitor the growth of guava fruits (*Psidium guajava* L.) var. crystal. Recently, guava fruits var. crystals has become popular. Through the development of this monitoring system, it is expected that farmers can study the characteristics of guava growth and its relationship with environmental factors (i.e. temperature, air humidity, light intensity, and rainfall) and also agronomic practices, so they can develop an appropriate cultivation techniques that can give an optimal results both qualitatively and quantitatively.

2. Materials and methods

The developed monitoring system including two main components: optoelectronic based radial growth sensor and data handling (i.e. transmission and display) system. Reflective Optical Sensor (Vishay CNY 70), and reflective tape with printed alternating white and black color is used as growth sensor. Reflective tapes are deliberately designed with narrow patterns with 0.5 mm width, so once the tape fully moves (i.e., due to the growth of the fruit) from the black pattern to the white one and vice versa, it means that the circumference increase by 0.5 mm. The growth sensor, as shown in figure 1, was developed with some modification from the previous work [1]. The main modification made here is the change of data reading method. In the previous study, data readings were performed periodically over a period. In the previous work, the readings were done at intervals of 5 min and the obtained values of the counters were saved at every full hour onto its EEPROM chip. In this study the sensor readings are done every color change on the reflective tape from black to white and vice versa. In addition for an efficient use of energy, the sensor works in two modes of operation that is normal and sleep mode. By default, the sensor will work in sleep mode but when there is a change in the reflective tape color reading then this change will trigger the sensor to run normal mode, i.e. transmitting data for a few seconds (± 8 seconds) and then return to sleep mode.

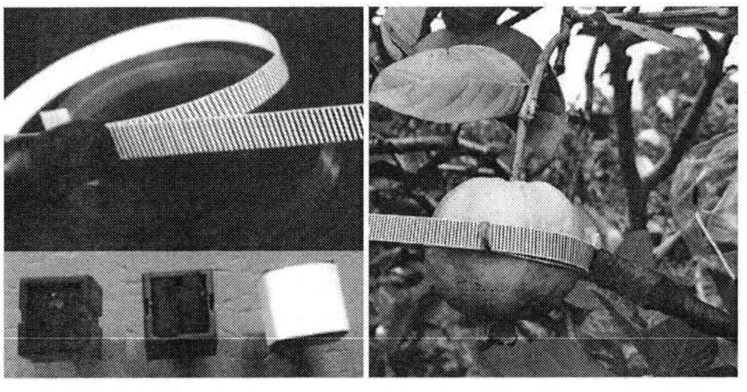

Figure 1. Optoelectronic based radial growth sensor: a) reflective tape; b) optoelectronic sensor; and c) field application for growth monitoring of guava fruit.

Before conducting the test in the field, testing is done to read the change of analog signal through the sensor. The reading is done by manually moving the reflective tapes every 1 minute. The test is performed to determine the threshold value. It is one of the critical task that will affect the overall system including measurement accuracy and operation mode (normal/sleep). The data of sensor readings is shown in figure 2. From the data, it is suggested that value of 0.7 – 0.8 Volt can be used as a proper threshold value.

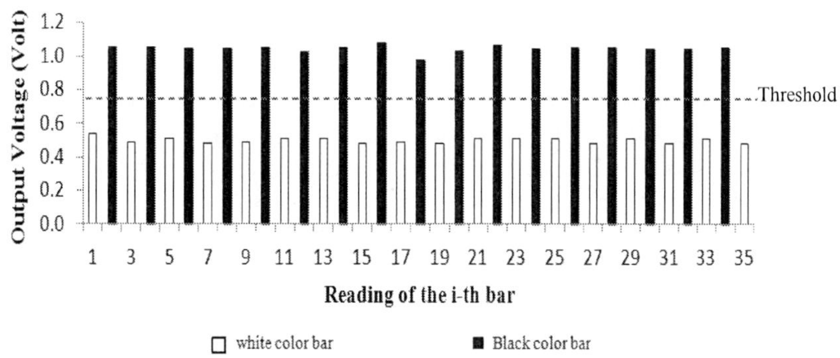

Figure 2. Determination of threshold value between black and white bars.

At each measurement point (sensor node) there is an electronic circuit consisting of growth sensor, microcontroller (Arduino nano), and RF Communication module. Meanwhile, on the coordinator, there is an electronic circuit consisting of microcontroller (Arduino Uno), RF communication module and GPRS Shield SIM900. In this case, XBee S2 modules with effective range up to 120 m for the line-of-sight outdoor condition are used. As an energy source, each circuit is equipped with 3.7 V, 5800mAh Li-Ion Battery 18650. The sensor node and the coordinator communicate using the RF module, and then all of the compiled data will be sent by the coordinator to the server and will be then processed and displayed on the display unit. In this case, ThingSpeak is used. It is an Internet of Things (IoT) platform that has been widely used to develop IoT applications especially to collect and store sensor data in the cloud. Data acquired by sensors can be sent to ThingSpeak from wide-range of available hardware such as Arduino, Raspberry Pi, etc. The data is then can be further visualized and analyzed. The data can be easily accessed from PC or mobile devices (smartphone and tablet) as long as it has an internet connection. The network architecture is shown in figure 3 below.

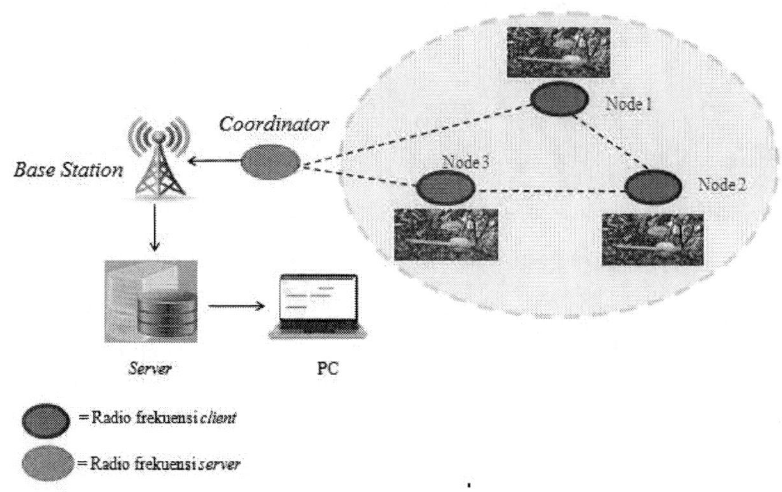

Figure 3. Schematic of network architecture.

The prototype of the developed system was tested in the field for monitoring the growth of guava fruits. The test was carried out for 30 days at Agribusiness Development Station (ADS) Bogor Agricultural University in Cikarawang, Bogor, Indonesia. There are several parameters evaluated including accuracy of the sensor, power consumption and success rate of data transfer.

IOP Conf. Series: Earth and Environmental Science **147** (2018) 012048 doi:10.1088/1755-1315/147/1/012048

3. Results and discussion

The developed prototype of growth monitoring system is shown in figure 4. The system is then installed for field test as shown in figure 5. Reflective tapes associated with optoelectronic sensors are installed around the fruit so that when the fruits grow reflective tapes simultaneously move to follow the fruit growth. Then the cable and all circuits are wrapped to avoid the rain. The sensor will read reflective tape displacement through black-and-white bars printed on reflective tapes. At the time of the bars moving from black to white or from white to black, the RF client is tasked to send the readout data to the RF server. The RF server then receives the data and sends it to the ThingSpeak network host via a GPRS signal.

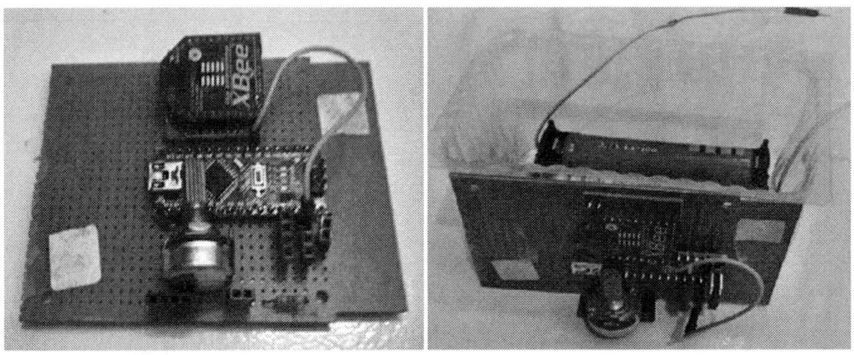

Figure 4. A prototype of the sensor connected to RF module.

Figure 5. Installation of the sensor for field test.

Before mounting the sensor on the guava fruit, the initial circumference (C_0) is measured and used as an input in the program run by the microcontroller. The results of circumference measurement using the developed system are shown in table 1. As shown in table 1, the measurement result using the developed sensor (C_t) shows a good agreement with the manual measurement (C_a) with an average error of 1 mm. The errors that occur are due to the growth of guava which is not perfectly round, causing a gap and affect the sensor reading. The error occurs, however, is relatively low and is still acceptable.

Time series data of guava fruits growth as displayed in the ThingSpeak platform are shown in figure 6. From the graph, there are differences in chart trends observed from different sensor nodes. This is due to the guava fruits measured are not uniform regarding age and the initial circumference. Each sensor (node 1, node 2, and node 3) was installed on fruits with age of 1, 1.2, and 1.3 months after flowering with an initial circumference (C_0) of 170 mm, 189.5 mm, and 160.5 mm respectively. In the measurement conducted for 30 days, the data indicated that the average radial growth rate of guava fruit is around 1 mm/day with a maximum error of 2 mm. When approaching the harvesting time , the radial growth rate slows down by an average of 0.5 mm/day.

IOP Conf. Series: Earth and Environmental Science **147** (2018) 012048 doi:10.1088/1755-1315/147/1/012048

Table 1. Accuracy of circumference measurement using the developed system

	C_0 (mm)	Ct (mm)	C_a (mm)	Error
Node1	170.5	196.0	197.0	1 mm
Node2	189.5	198.5	198.5	0 mm
Node3	160.5	192.0	194.0	2 mm
		Average		1 mm

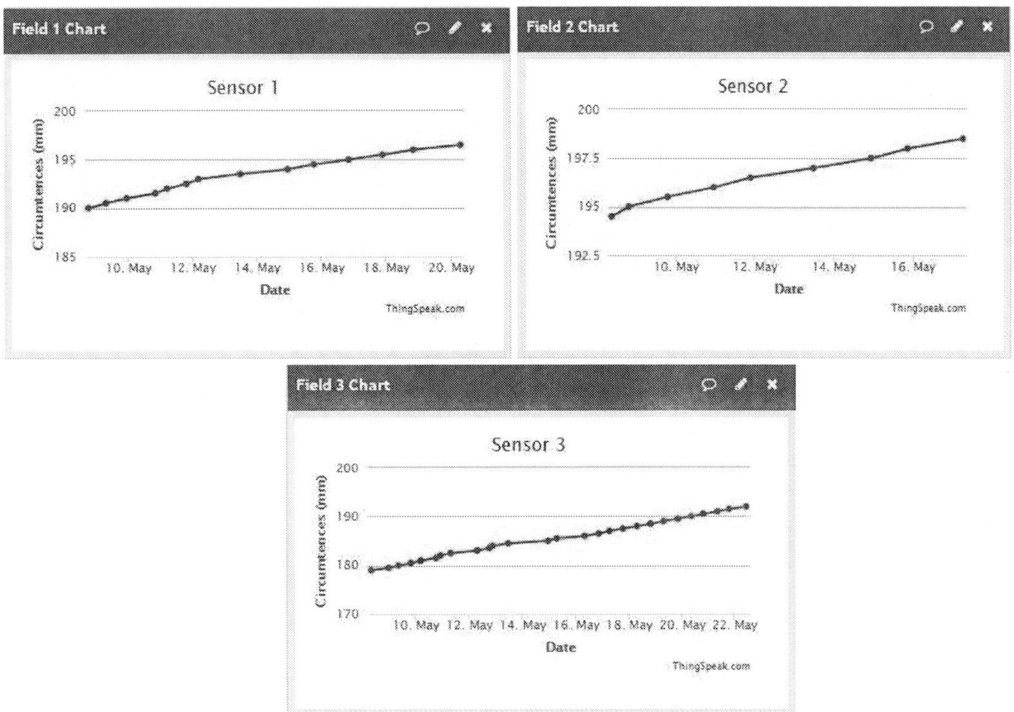

Figure 6. Display of the growth data of guava fruit on ThingSpeak platform.

In the implementation, the sensor node must be active all the time. Therefore the power supply (i.e., battery) must be well planned to avoid replacement in the middle of the measurement. It is because the replacement of the power supply may cause a disturbance to the data recording and the measurement will repeat from the initial value as input in the program run on the microcontroller. Therefore, an accurate calculation of power consumption on the node sensors is important. Table 2 below shows daily power consumption data on sensor node and server. In this study on each sensor node used 3 Li-Ion battery 18650 mounted in series with 3.75 Volt and 5800 mAh capacity for each battery. In total, the energy capacity of the battery is 65.25 Wh. Therefore based on the average daily energy consumption of each sensor node which is 1.127 Wh/day (table 2). Theoretically, the average battery life can be expected to reach of 58 days (almost 2 months).

Table 2. Power consumption of sensor node (client) and server.

	Mode	Power (W)	Daily Energy Consumption (Wh)
Node 1	*sleep*	0.046	1.111
	Normal	0.351	0.001
Node 2	*sleep*	0.048	1.149
	Normal	0.350	0.001
Node 3	*sleep*	0.047	1.120
	Normal	0.350	0.001
Server	Normal	0.470	11.281

Another important measure needs to be evaluated is the performance of data transmission. Success rate of data transmission can be observed from the total of data successfully transferred from the sensor node to the server/network host. The success rate can be seen in table 3 below. Based on table3, the system being built has a success rate with an average of 97.54%. This is due to certain condition when the server fails to receive the data. The data failed because the server failed to distinguish client data when logged in simultaneously. However, failing data has no significant effect on the measurement because the server will continue to receive data from the client and re-read it so that the graph displayed on the ThingSpeak network host will return to normal.

Table 3. Success rate of data transfer.

	Transferred Data	Received Data	Failed/Un-received data	Success rate (%)
Node 1	53	52	1	97.61
Node 2	55	55	0	100.00
Node 3	50	48	2	95.00
		Average		97.54

However, for a longer period of monitoring, it is possible that the reading error will accumulate and increase significantly. Therefore further investigation on the error is necessary to be done. Another important thing to be considered is the possibility of error on data transfer due to unstable network connection which is not uncommon in rural area. This issue has become an interesting challenge in the implementation of IoT-based technology in agriculture. An alternative solution to overcome this issue has been reported in several studies such as framework proposed in [5].

4. Conclusions

IoT based growth monitoring system using optoelectronic sensor has been successfully developed and used for monitoring growth of guava fruits. The developed system has been tested in the field for 30 days and showed a good performance, indicated by high measurement accuracy (1 mm error), 97.54% success rate of data transfer, and long battery life (almost 2 months). The developed system can be used as an alternative for monitoring the (radial) growth of plant.

References

[1] Drew D M, G M Downes 2009 The use of precision dendrometers in research on daily stem size and wood property variation: a review *Dendrochronologia* **27** 159-172

[2] Sheil D 2013 Growth assessment in tropical trees: large daily diameter fluctuations and their concealment by dendrometer bands *Canadian Journal of Forest Research* **33**(10): 2027-2035

[3] Thalheimer M 2016 A new optoelectronic sensor for monitoring fruit or stem radial growth *Computers and Electronic in Agriculture* **123** 149-153

[4] Lehmann R, R Reiche, G Schiefer 2012 Future internet and the agri-food sector: State-of-the-art in literature and research *Computers and Electronics in Agriculture* **89** 158-174

[5] Nugroho A P, T Okayasu, T Hoshi, E Inoue, Y Hirai, M Mitsuoka and L Sutiarso 2016 Development of a remote environmental monitoring and control framework for tropical horticulture and verification of its validity under unstable network connection in rural area *Computers and Electronic in Agriculture* **124** 325-339